Sensor Networks and Configuration

Nitaigour P. Mahalik (Ed.)

Sensor Networks and Configuration

Fundamentals, Standards, Platforms, and Applications

Editor

Nitaigour P. Mahalik
Gwangju Institute of Science and Technology
Department of Mechatronic
Oryong-dong 1, 500-712
Gwangju, Republic of South Korea
nmahalik@yahoo.co.uk

Library of Congress Control Number: 2006933058

ISBN-10 3-540-37364-0 Springer Berlin Heidelberg New York
ISBN-13 978-3-540-37364-3 Springer Berlin Heidelberg New York

Springer is a part of Springer Science+Business Media
springeronline.com

Typesetting: by the authors
Cover design: eStudio Calamar, Girona, Spain

Printed on acid-free paper SPIN: 11580782 VA62/3100/Integra

Dedicated to my

Teachers and family members

Preface

Advances in networking principles may indeed influence many kinds of monitoring and control systems in the most dramatic way. Sensor network and configuration (SNC) falls under the category of modern networking systems. A Wireless Sensor Network (WSN), a sophisticated, compact, and advanced networking method has emerged that caters to the need for real-world applications. Methodology and design of WSNs represent a broad research topic with applications in many sectors such as industry, home, computing, agriculture, environment, and so on, based on the adoption of fundamental principles, specifications characterisations, modeling, simulations, and state-of-the-art technology. Technological research in this field is now expanding; its design phases appear to be highly complex and involve interdisciplinary approaches.

The main objective of this book is to provide information on concepts, principles, characteristics, applications, latest technological developments, and comparisons with regard to sensor networks and configuration. This book incorporates research, development, tutorials, and case studies. Academic and industrial research and developments in networked monitoring and control (e.g., intelligent home, pet management, etc.) are being carried out at many different institutions around the world. The technological trends in this domain (e.g., design, integration, communication schemes, development methodology, current application scenarios, pros and cons, etc.) need to be extensively disseminated so that the sensor network revolution can spread to serve society in a bigger way. In particular, the book is intended to focus on describing the implicit concept of advanced networking, personal area networking, and mobile area networking, as well as application-oriented design tips and hints, as much as the techniques and methodology. This book will enable readers to understand the underlying technology, philosophy, concepts, ideas, and principles, with regard to broader areas of sensor networks. Aspects of sensor network in terms of basics, standardization, design process, practice, techniques, platforms, and experimental results have been presented in a proper order. Fundamental methods, initiatives, significant research results, as well as references for further study have also been provided. Relative merits and demerits are described at the appropriate places so that novices as well as advanced practitioners can use the evaluation to guide their choices. All the contributions have been reviewed, edited, processed and placed in appropriate order to maintain consistency so that irrespective of whether the reader is an advanced practitioner or a newcomer he or she can get most out of it. Since this book covers many aspects of SNC the importance of this order is considered significant. The roadmap of the book is as follows.

Chapter 1 is a general introduction. Chapter 2 presents the backbone of WSNs, the IEEE 802.15.4 protocol. The requirements for service-oriented sensor webs are presented in Chapter 3. Cross-layer design principles are described in Chapter 4. Grid computing has evolved as a standards-based approach for coordinated resource sharing. There are several issues and challenges in the design of sensor grids. Chapter 5 has been dedicated to the sensor grid architecture for distributed events classification. Chapters 6, 7, and 8 deal with topology controls, routing protocols, and energy aware routing fundamentals, respectively. Chapter 9 discusses the aspects of probabilistic queries and quality assurances. A statistical approach-based resilient aggregation is studied in Chapter 10. The communication performance study is presented in Chapter 11. A sensor network consists of a large number of *nodes* connected through a multi-hop wireless network. Data management is an issue discussed in Chapter 12. Localisation and location estimation are also two important design considerations. Chapters 13 and 14 introduce these. It has been variously proposed that the future of the monitoring and control will be based on sensor networks. A comprehensive description of an application driven design, ZigBee WSN and their applications, MANET versus WSN, etc. can be found in Chapters 15-17. There has been recent confusion on sensor network and industrial Distributed Control Systems (DCS). In fact, sensor networks and DCS are complementary to each other. As such, two chapters have been dedicated to introduce industrial sensor and actuator networks; (the *fieldus*) and the DCS simulation scenario. The book also contains three chapters regarding applications of WSNs. The application domains are pet management systems, agriculture monitoring, and intelligent CCTV. The last supplemental chapter reviews the modulation techniques and topology, an essential topic for novice researchers and readers.

The success story of this book 'Sensor Network and Configuration' is due to the direct and indirect involvement of many researchers, technocrats, academicians, developers, integrators, designers, and last but not the least the well-wishers. Therefore, the editor and hence the publisher acknowledge the potential authors and companies whose papers, reports, articles, notes, study materials and websites have been referred to in this book. Further, many of the authors of the respective chapters gracefully acknowledge their funding agencies, without which their research could not have been completed. Every precaution has been taken to credit their work in terms of citing their names and copyright notices under the figures and texts incorporated in this book: but in case of error we would like to receive feedback so that the same can be incorporated in the next phase of printing. In particular, persons such as Abhijit Suprem, Suprava Mishra, M Tubaishat, S Madria, Debasis Saha, M Reichardt, RS Raji, R Murugesan, KK Tan, KZ Tang, P Raja and G Noubir, Jianliang Xu, Chagwen Xie, PR Moore, Jun-Sheng Pu, Saleem Bhatti, AS Hu, SD Servetto, WW Manges, P Meheta, PH Michael and the following agencies, institutes, companies and journals are acknowledged: UiSec&Sens EU Project (contract number 026820) and the Hungarian Scientific Research Fund (contract number T046664), Hungarian Ministry of Education (BÖ 2003/70), HSN Lab., Italian Ministry for University and Research (MIUR) through the PATTERN project, Research Grants Council of the Hong Kong SAR: China (Project No. HKBU 2115/05E), Kumoh National Institute of Technology: South Korea, w3.antd.nist.gov, ZigBee Alliances, Echelon Corporation, Geospatial Solutions, Fuji Press, Maxim/Dallas, and http://www.cs.ucl.ac.uk.

Nitaigour Premchand Mahalik

Contents

Authors

Anis Koubaa
IPP-HURRAY Research Group, Loria Trio, 615, Rue du Jardin Botanique, 54602, France

Arunava Saha Dalal
Nvidia Graphics Private Limited, Bangalore, India

Barış Fidan
National ICT Australia, School of Electrical and Information Engineering, The University of Sydney, Australia

Bonnie S. Heck
School of Electrical and Computer Engineering, Georgia Institute of Technology, GA 30332-0250, USA

Brian C. Lovel
Intelligent Real-Time Imaging and Sensing Group, EMI, School of Information Technology and Electrical Engineering, The University of Queensland, Brisbane, 4072, Australia

Brian DO Anderson
National ICT Australia, School of Electrical and Information Engineering, The University of Sydney, Australia

Carla-Fabiana Chiasserini
Department of Electronics, Politecnico di Torino, Italy

Changwen Xie
Wicks and Wilson Limited, Morse Road, England, United Kingdom

Chen-Khong Tham
Department of Electrical and Computer Engineering, National University of Singapore, Singapore

Claudio Casetti
Department of Electronics, Politecnico di Torino, Italy

Dario Rossi
Department of Electronics, olitecnico di Torino, Italy

Dario Pompili
Broadband and Wireless Networking Laboratory, School of Electrical and Computer Engineering, Georgia Institute of Technology, Atlanta

Debashis Saha
MIS and Computer Science Group, Indian Institute of Management (IIM), Joka, D. H. Road, Kolkata, India

Dong-Sung Kim
Networked System Laboratory, School of Electronic Engineering, Kumoh National Institute of Technology, Gumi, Republic of South Korea

Eduardo Tovar
IPP-HURRAY Research Group, Polytechnic Institute of Porto Rua Dr. Antonio Bernardino de Almeida, 431, 4200-072 Porto, Portugal

Edward Chan
Department of Computer Science, City University of Hong Kong, Hong Kong, Peoples Republic of China

Elmoustapha Ould-Ahmed-Vall
School of Electrical and Computer Engineering, Georgia Institute of Technology, GA 30332-0250, USA

George F. Riley
School of Electrical and Computer Engineering, Georgia Institute of Technology, GA 30332-0250, USA

Guoqiang Mao
National ICT Australia, School of Electrical and Information Engineering, The University of Sydney, Australia

H. G. Goh
Multimedia University, Jalan Multimedia, 63100 Cyberjaya, Malaysia

Haibo Hu
Hong Kong University of Science and Technology, Hong Kong, Peoples Republic of China

Hong Tat Ewe
Multimedia University, Jalan Multimedia, 63100 Cyberjaya, Malaysia

Ibrahim Korpeoglu
Department of Computer Engineering, Bilkent University, Ankara, Turkey

István Vajda
Budapest University of Technology and Economics, Laboratory of Cryptography and Systems Security

J. A. Garcia-Macias
CICESE, Código Postal 22860, Ensenada, B.C. Mexico

Javier Gomez
Wireless Networking Laboratory, Electrical Engineering Department, National University of Mexico, Ciudad University, D. F. Mexico

Jianliang Xu
Hong Kong Baptist University, Kowloon Tong, Hong Kong, Peoples Republic of China

Kam-Yiu Lam
Department of Computer Science, City University of Hong Kong, Hong Kong, Peoples Republic of China

Khusro Saleem
National ICT Australia, Department of Electrical and Electronic Engineering, University of Melbourne, Australia

Kiseon Kim
Department of Information and Communications, Gwangju Institute of Science and technology, Republic of South Korea

Levente Buttyán
Laboratory of Cryptography and Systems Security, Department of Telecommunications, Budapest University of Technology and Economics, Hungary

M. L. Sim
Multimedia University, Jalan Multimedia, 63100 Cyberjaya, Malaysia

Mark Halpern
National ICT Australia, Department of Electrical and Electronic Engineering, University of Melbourne, Australia

Mário Alves
IPP-HURRAY Research Group, Polytechnic Institute of Porto Rua Dr. Antonio Bernardino de Almeida, 431, 4200-072 Porto, Portugal

Mehmet Can Vuran
Broadband and Wireless Networking Laboratory, School of Electrical and Computer Engineering, Georgia Institute of Technology, Atlanta, USA

Meng-Shiuan Pan
Department of Computer Science, National Chiao Tung University, Hsin-Chu, 30010, Taiwan, Peoples Republic of China

Muhammad Umar
Faculty of Computer Science and Engineering, Ghulam Ishaq Khan Institute of Engineering Sciences and Technology, Topi, Swabi, Pakistan

N. P. Mahalik
Department of Mechatronics, Gwangju Institute of Science and Technology, 1 Oryong dong, Buk gu, Gwangju, 500 712, Republic of South Korea

Péter Schaffer
Budapest University of Technology and Economics, Laboratory of Cryptography and Systems Security

Rajkumar Buyya
Grid Computing and Distributed Systems Laboratory, Department of Computer Science and Software Engineering, The University of Melbourne, Australia

Reynold Cheng
Department of Computing, Hong Kong Polytechnic University, Hong Kong, Peoples Republic of China

S. H. Gary Chan
Department of Computer Science and Engineering, The Hong Kong University of Science and Technology, Hong Kong, Peoples Republic of China

Sajid Qamar
Faculty of Computer Science & Engineering, Ghulam Ishaq Khan Institute of Engineering Sciences and Technology, Topi, Swabi, Pakistan

Shaokang Chen
National Information and Communications Technology Australia (NICTA)

Soo Young Shin
School of Electrical and Computer Engineering, Seoul National University, Republic of South Korea

Stan Skafidas
National ICT Australia, Department of Electrical and Electronic Engineering, University of Melbourne, Australia

Ting Shan
Intelligent Real-Time Imaging and Sensing Group, EMI, School of ITEE, The University of Queensland, Australia 4072

Tommaso Melodia
Broadband and Wireless Networking Laboratory, School of Electrical and Computer Engineering, Georgia Institute of Technology, Atlanta

Usman Adeel
Faculty of Computer Science & Engineering, Ghulam Ishaq Khan Institute of Engineering Sciences and Technology, Topi, Swabi, Pakistan

Victor Cheung
Department of Computer Science and Engineering, The Hong Kong University of Science and Technology, Hong Kong, Peoples Republic of China

Wanzhi Qiu
National ICT Australia, Department of Electrical and Electronic Engineering, University of Melbourne, Australia

Xingchen Chu
Grid Computing and Distributed Systems Laboratory, Dept. of Computer Science and Software Engineering, The University of Melbourne, Australia

Xueyan Tang
School of Computer Engineering, Nanyang Technological University, Singapore

Yu-Chee Tseng
Department of Computer Science, National Chiao-Tung University, Hsin-Chu, 300, Taiwan, Peoples Republic of China

1 Introduction

N. P. Mahalik

Department of Mechatronics, Gwangju Institute of Science and Technology, 1 Oryong dong, Buk gu, Gwangju, 500 712, Republic of South Korea

1.1 Introduction and Background

A sensor network is a network of many smart devices, called nodes, which are spatially distributed in order to perform an application-oriented global task. The primary component of the *network* is the sensor, essential for monitoring real-world physical conditions or variables such as temperature, humidity, presence (absence), sound, intensity, vibration, pressure, motion, and pollutants, among others, at different locations. Each smart device within the netwok is small and inexpensive, so that it can be manufactured and deployed in large quantities. The important design and implementation requirements of a typical sensor network are energy efficiency, memory capacity, computational speed and bandwidth. The smart device has a microcontroller, a radio tranasmeter, and an energy source. Sometimes a central computer is integrated onto the network in order to manage the entire networked system. Regardless of achieving its global task, a sensor network essentially perfoms three basic functions: sensing, communicating and computation by using the three fundamental components: hardware, software and algorithms, respectively.

Conventionally, a sensor network is considered a wireless network, however, some sensor networks are wired or hybrid types. For example, fieldbus-type systems (Mahalik 2003) are usually wired networks (optical fibers, co-axial cables, etc. are used). A wireless sensor network (WSN), as its name implies, consists of a number of microcontroller-integrated smart devices. Each node is equipped with a variety of sensors, such as acoustic, seismic, infrared, still/motion video-camera, and so on. The node has some degree of intelligence for signal processing and management of network data.

The basic goals of a WSN are to: (i) determine the value of physical variables at a given location, (ii) detect the occurrence of events of interest, and estimate parameters of the detected event or events, (iii) classify a detected object, and (iv) track an object (w3.antd.nist.gov). Thus, the important requirements of a WSN are: (i) use of a large number of sensors, (ii) attachment of stationary sensors, (iii) low energy consumption, (iv) self-organisation capability, (v) collaborative signal processing, and (vi) querying ability.

Fig. 1.1 shows the general architecture of a sensor network. As can be seen in the figure, the three important layers are the services-layer, data-layer, and physical-layer. The layers provide routing protocol, data dissemination, and aggregation. The physical-layer containing the node defines itself as either a sink node, children node, cluster head, or parent node. Parent nodes are integrated to more than two cluster heads. Messages are modeled in the data-link layer. Broadcasting of a query is carried out by the use of sink nodes. The broadcasting can be either to the sensor network or to a designated region depending on the way the query is being used. In response to a change in the physical parameter the sensor nodes, which are close to the sensed object, broadcast this information to their neighbouring sensor nodes. In effect, a cluster head will receive this transmission. The role of a cluster head is to process and aggregate this data and broadcast it to the sink node(s) through the neighboring nodes. This is due to the fact that cluster head receives many data packets from its children (Tubaishat M. and Madria).

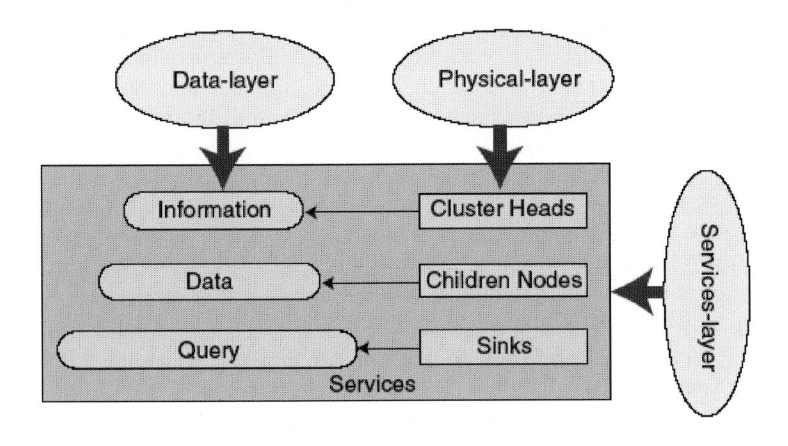

Fig. 1.1. Wireless sensor architecture (Source: Adapted from Tubaishat and Madria 2006 Courtesy: IEEE)

Some of the important application domains of WSNs are listed below.

- Military sensor networks
- Networks for detecting chemical, biological, radiological, nuclear, and explosive material
- Environmental monitoring networks
- Traffic sensor networks
- Surveillance applications
- Parking systems in shopping malls and big markets

A more comprehensive list of the research and development areas of WSNs is given below.

- Adaptive beacon placement

- Adaptive protocols (Kulik *et al.* 1999) for information dissemination
- Adaptive self-configuring topologies
- Address-free architecture for dynamic networks
- Addressing mechanisms
- Application specific protocol architectures
- Data gathering using energy delay metric
- Data-centric storage
- Directed diffusion-scalable communication paradigm
- Distributed micro-sensor systems
- Dynamic fine-grained localisation
- Dynamic power management
- Energy complexity, energy-efficient MAC (Media Access Control)
- GPS-less outdoor localisation for very small devices
- Habitat monitoring
- Impact of network density on data aggregation
- Instrumentation
- Location mechanisms
- Low-, ultra-low power systems, low power systems-on-a-chip
- Low-level naming
- Mobile networking for *Smart Dust*
- Modeling and simulation
- Negotiation-based protocols for disseminating information
- Network interfaces for hand-held devices
- Network layers
- Physical layer driven protocol
- PicoRadio
- Positioning system, convex position estimation
- Prediction-based monitoring
- Probabilistic approach to predict energy consumption
- Protocols for self-organisation
- Random, ephemeral transaction identifiers
- Recursive data dissemination
- Residual energy scans for monitoring
- Routing, Rumor routing algorithm
- Scalable computation, scalable coordination
- Self-configuring localisation systems
- Service differentiation
- System architecture directions
- Time synchronisation
- Topology discovery algorithm
- Transmission control for media access
- Upper bounds on the lifetime of WSNs

The main objective of this book is to provide information on concepts, principles, characteristics, applications, latest technological developments, and comparisons with regard to WSNs and their configuration. It incorporates current research, development, and case studies. Academic and industrial research and developments in WSNs are being carried out at many different institutions around the world. The technological trends in this domain (e.g., design and development methodology, current application scenarios, pros and cons, etc.) need to be extensively disseminated so that the revolution can spread to serve society. This book is intended to focus on describing the implicit concepts, multi-engineering principles of networking systems, state-of-the-art tools, design tips and hints, rather than solely focusing on techniques and methodology. In particular, the scope of the book, which will be briefly presented in the sequel, is as follows.

- Modulation techniques
- IEEE 802.15.4 and ZigBee protocols
- Cross layer design
- Sensor-grid computing architecture
- Topology control
- Routing protocols and energy aware routing
- Quality assurance
- Aggregation
- Communication performance
- Sensor data management
- Localisation and estimation
- SensorWeb
- Distributed controls
- Applications: pet-dog management, agricultural monitoring, and CCTV

1.2 IEEE 802.15.4

WSNs have been attracting increasing interest for developing a new generation of embedded systems. Communication paradigms in WSNs differ from those associated to traditional wireless networks, triggering the need for new communication protocols. In this context, the recently standardised IEEE 802.15.4 protocol presents some potentially interesting features, such as power-efficiency, timeliness guarantees, and scalability. When addressing WSN applications with timing requirements (soft/hard) some inherent paradoxes emerge, such as power-efficiency versus timeliness, and scalability versus communication latencies, triggering the need of engineering solutions for the efficient deployment of IEEE 802.15.4 in WSNs. This book includes the most relevant characteristics of the IEEE 802.15.4 protocol for WSNs and presents important research challenges for time-sensitive applications. Timeliness analysis of IEEE 802.15.4 will unveil relevant directions for resolving the paradoxes mentioned above. IEEE 802.15.4 is a standard defined by the IEEE (Institute of Electrical and Electronics Engineer) for low-rate, wireless personal area networks (PAN). It defines both physical and medium access layers. The physical layer and media access layers are called PHY and MAC, respectively, and the former defines a low-power spread spectrum radio operating at 2.4 GHz with a fundamental bit rate of 250 kbs. There is also an alternate PHY specification, which works around 915 MHz and 868 MHz for operating at lower data rates. The MAC specification deals with the multiple radio operation. It supports several architectures such as tree topologies, star topology, and mesh topologies as shown in the Fig. 1.2.

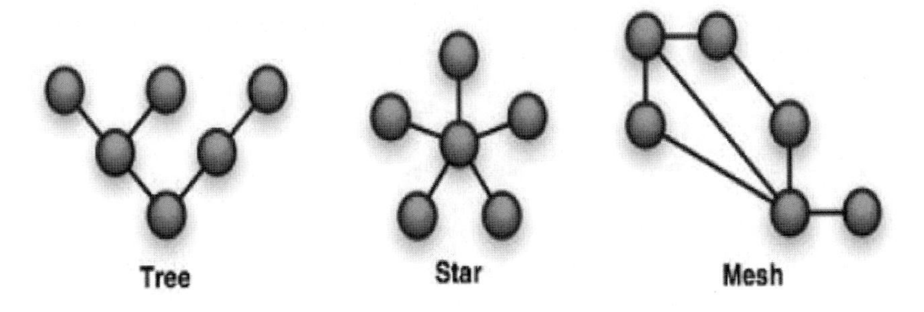

Fig. 1.2. Some of the typical IEEE 802.15.4 topologies (Source: http://en.wikipedia.org)

1.3 ZigBee

The ZigBee alliance, incorporated in August 2002, is an association of companies working together to enable reliable, cost-effective, low-power, wirelessly networked, monitoring and control products based on an open global standard. Many research institutes and industrial companies have developed sensor platforms based on ZigBee and IEEE 802.15.4 solutions. The ZigBee alliance includes leading semiconductor manufacturers, technology providers, OEMs, and end-users worldwide. Membership in the alliance is approaching 200 companies in 24 countries. The objective of this organisation is to provide the upper layers of the protocol stack, essentially from the network to the application layer, including application profiles. It also provides interoperability compliance testing, marketing of the standard, and advanced engineering for the evolution of the standard (http://www.caba.org/standard/zigbee.html).

ZigBee technology is suited for sensor and control applications, which do not require high data rates, but must have low power, low costs and ease of use in the platform, such as remote controls and home automation applications. One can note that ZigBee builds upon the 802.15.4 standard to define application profiles (Fig. 1.3) that can be shared among different manufacturers. The origin of ZigBee comes from the fact that simply defining a PHY and a MAC does not guarantee that different field devices will be able to talk to each other. As already pointed out, ZigBee starts with the 802.15.4 standard, and then defines application profiles that can allow devices manufactured by different companies to talk to one another. ZigBee supports both star and mesh topologies. One chapter is dedicated to provide detail information on ZigBee. The chapter also discusses some applications such as medical care and fire emergency system along with some prototyping systems.

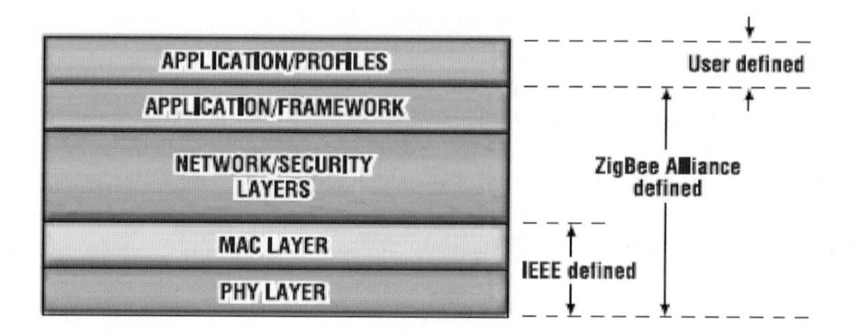

Fig. 1.3. ZigBee defined layers/specification (Courtesy: ZigBee Alliances)

1.4 Cross-layer Design

Continued study on cross-layer protocols, protocol improvements, and design methodologies for WSNs is important. A cross-layer design paradigm is necessary to achieve the desired performance of the transmission protocol stack in terms of meeting QoS (Quality of Service) (Macias 2003) requirements. Cross-layer protocol engineering is an emerging research area addressing relevant issues that support their needs. The development of concepts and technologies are critical. The protocols that focus on cross-layer design techniques are to be reviewed and classified. Methodologies for the design of cross-layer solution for sensor networks as resource allocation problems in the framework of non-linear optimisation are discussed in this book. Open research issues in the development of cross-layer design methodologies are also discussed and possible research directions are indicated. The shortcomings of design techniques such as lack of modularity, decreased robustness, difficulty in system enhancements, and risk of instability are discussed, and precautionary guidelines are presented. The research considerations are:

- Architectures and methodologies
- Exploitation of physical layer information for medium access control
- Management and congestion control
- Cross-layer adaptation for energy minimisation
- End-to-end QoS
- Simulation tools
- Scalability issues
- Signaling and interaction

1.5 Sensor-Grid Computing

Real-time information about phenomena in the physical world can be processed, modeled, correlated, and mined to permit 'on-the-fly' decisions and actions in a large-scale environment. Examples include surveillance for homeland security, business process optimisation, environment monitoring with prediction and early warning of natural disasters, and threat management (e.g. missile detection, tracking and interception). In one chapter, the sensor-grid computing concept, the SensorGrid architecture is described and the works on information fusion, event detection and classification, and distributed autonomous decision-making on SensorGrid, are presented as examples of what can be achieved. Integrating sensor networks and grid computing (Estrin *et al.* 2002) based on SensorGrid architecture is like giving 'eyes' and 'ears' to the computational grid.

1.6 Routing Protocol

Routing in WSN presents unique challenges as compared to traditional techniques in other conventional wired and wireless networks. This book dedicates a chapter, to outlining many design issues, and presents a number of state-of-the-art schemes that have been proposed during recent years. The approaches have been classified into flat, hierarchical and geographic routing based on the network structure for which they are proposed. The authors of this chapter have briefly discussed the schemes outlining the routing mechanism, both advantages and drawbacks. Although most of them are quite successful in fulfilling the requirements, there are still many challenges to overcome before WSN technology becomes mature.

1.7 Energy Efficiency

From on-going research work it has been deemed that there is a pressing demand to design new routing protocols for WSNs, since the features and applications of WSNs are substantially different from similar networks like wireless ad-hoc networks. Energy is a much more scarce resource in WSNs, and therefore all routing protocols designed for WSNs should consider the energy consumption as the primary objective to optimise. There are various energy efficient routing protocols proposed for WSNs. It is equally imperative to outline the differences between WSNs and wireless ad-hoc networks as far as routing is concerned. New approaches such as the cross layer design in the design of protocols for WSNs are also similarly important. The effect of link layer wireless technology on the design of routing protocols is considered in practical realisations of WSNs.

Energy-efficient data management for sensor networks is a challenging problem. Three topics, data storage, exact query processing, and approximate query processing, are to be the matter of discussion. The foremost design principle of an energy-efficient sensor network is to combine pull- and push-based data dissemination schemes so as to strike a balance between the query arrival rate (and query precision if it is an approximate query) and data capture rate. In other words, the sensor networks should reduce the cost of sampling, storage and transmission of unnecessary data, i.e., those data that do not affect the query result, as much as possible. The various techniques discussed in that chapter will inspire the reader in developing their own applications.

1.8 Topology Control

Energy saving by TC is identified to be one of the methods (Deb *et al.* 2002). Analysis of sensor node lifetime shows a strong dependence on battery lifetime. In a multi-hop ad-hoc WSN, each node plays the dual role of data originator and data router. The malfunctioning of a few nodes can cause significant data loss due to topological changes, and might require rerouting of packets and reorganisation of the network. Hence, power conservation and power management take on additional importance. For these reasons, we need power-aware protocols and algorithms for WSNs. In this book we have presented an overview of the topology control (TC) problem in WSNs for both stationary and mobile types, and discuss various solutions proposed to prolong the lifetime of WSNs by minimising energy requirements. We have also classified various TC protocols, and analysed their approaches. To this extent, a summary is provided in Table 1.1. As can be seen, most practical TC problems are NP-hard or NP-complete. Hence, only approximate solutions exist for some specific problems. There still remain many open problems for which no good heuristic exists. It is expected that research would be carried on to improve upon existing heuristics, and to determine good approximate algorithms for the harder open problems.

1.9 Quality Assurance

In a large network, sensors are often employed to continuously monitor changing entities like locations of moving objects and temperature. These sensor readings are usually used to answer user-specified queries. Due to continuous changes in these values in conjunction with measurement errors, and limited resources (e.g., network bandwidth and battery power), the system may not be able to record the actual values of the entities.

Queries that use these values may produce incorrect answers. However, if the degree of uncertainty between the actual value and the database value is limited, one can place more confidence in the answers to the queries. Answers to the query can be augmented with probabilistic guarantees. This book has a chapter, which deals with the quality assurance aspect of the sensor network. In particular, it emphasises how the quality of queries can be guaranteed for uncertain sensor data.

We discuss a classification scheme of queries based upon the nature of the result set, and how probabilistic answers can be computed for each class. This suggests the identification of different probability-based and entropy-based metrics for quantifying the quality of these queries.

Table 1.1: Comparison of TC protocols (Contributed by Debasis Saha)

Topology Control Algorithm	Graph Model	Methodology
MECN and SMECN	Minimum power topology di-graph, two-dimensional space.	Uses localised search and applies distributed Bellman-Ford algorithm.
Distributed Cone Based (CBTC)	Two-dimensional plane. Undirected graph	Builds a connected graph, followed by redundant edge removal.
LINT, LILT, CONNECT, Bicon-Augment	Undirected and directed	One algorithm is similar to minimum spanning tree and another to depth first search. LINT uses local while LILT global information for mobile networks.
Loyd et al.	Undirected and directed graphs with preferably monotone graph property.	Estimates total number of candidate optimal power values. For each value, constructs a directed graph followed by sorting and searching.
Krumke et al.	Undirected	Uses (MCTDC) (minimum cost tree with a diameter constraint); a combination of sort & search; and an algorithm for Steiner Trees.
SMET protocols	Undirected	One heuristic is based on a minimum edge cost-spanning tree and the second one is an incremental power heuristic.
MobileGrid	N.A.	Estimates CI for each node, checks for optimal CI and adjusts transmit power based on this information
COMPOW	N.A.	Maintains routing tables at all power levels, and then selects the optimum power level based on the routing information.
K-NEIGH	Undirected	Computes a symmetric sub-graph of the k nearest neighbours graph by ordering the list of k nearest neighbours
CLTC	Undirected	Forms clusters and selects cluster heads followed by intra-cluster and inter-cluster TC
Dist. NTC	Planar undirected graph with maximum triangulation	Determines topology using Delaunay Triangulation. Then it adjusts each node degree through neighbor negotiation so that each node has a similar number of neighbours

Two major classes of algorithms for efficiently pulling data from relevant sensors in order to improve the quality of the executing queries are provided. The first type of algorithm selectively probes sensors in order to reduce the effect of sampling uncertainty. The second method exploits the low-cost nature of sensors and selects an appropriate number of redundant sensors in order to alleviate the problem of noisy data. Both methods are aware of the limited resource of network bandwidth, and only pull data from sensors when necessary.

1.10 Aggregation

In typical sensor network applications, the sensors are left unattended for long periods of time. In addition, due to cost reasons, sensor nodes are not usually tamper resistant. Consequently, sensors can be easily captured and compromised by an adversary (e.g., bogus measurements). Once compromised, a sensor can send messages to other nodes and to the base station, but those messages may contain arbitrary data. A similar effect can be achieved by manipulating the physical environment of uncompromised sensors so that they measure false values. Bogus data introduced by the adversary may considerably distort the output of the aggregation function at the base station, and may lead to wrong decisions. The goal of resilient aggregation is to perform the aggregation correctly despite the possibility of the above mentioned attacks. A chapter in the book gives an overview of the state-of-the-art in resilient aggregation in sensor networks, and briefly summarises the relevant techniques in the field of mathematical statistics. An approach for resilient aggregation is subsequently introduced and discussed in more detail. This approach is based on RANSAC (RAndom SAmple Consensus). The authors have also presented some initial simulation results showing that a RANSAC-based approach can tolerate a high percentage of compromised nodes.

1.11 Localisation

Sensor network localisation is an important area that has recently attracted significant research interest. In a typical WSN many sensors work together to monitor conditions at different locations. Knowing the senor location where data is sensed is important not only to interpret the data, but also to efficiently route data between nodes. Localisation of a node refers to the problem of identifying its spatial co-ordinates in some co-ordinate system. Many methods have been proposed for localisation. The most commonly used method is some variation of multilateration, which is an extension of trilateration. Trilateration techniques use beacons, which act as reference points for a node, to calculate its own position. Sufficient numbers of beacons are necessary as otherwise nodes, which are not aware of their positions, will not be able to calculate their positions. At the same time, too many beacons become expensive and cause self-interference. Thus, it is necessary to have optimal beacon placement. Most WSN applications require accurately measuring the locations of thousands of sensors. The three major categories of measurement techniques are:

- Received signal strength measurements
- Angle of arrival measurements
- Propagation time based measurements

The localisation algorithms built on these measurement techniques are subsequently discussed.

In a WSN, which measurement technique and algorithm to use for location estimation will depend on the specific application. There are a number of factors that may affect the decision. These include cost, energy, localisation accuracy, efficiency and scalability of the algorithm. Some algorithms like the category of connectivity-based localisation algorithms, although not able to give an accurate location estimate, can be attractive for applications requiring an approximate location estimate only due to their simplicity. The fundamental theory underpinning distance-based sensor networks from the graph theory point of view is also presented. Two fundamental problems in the area are whether a specific sensor network with a given set of sensors and inter-sensor distance measurements is uniquely localisable, and the computational complexity of the localisation algorithm. Some important results in this area are introduced. Finally, some current researches on distance-based sensor network localisation are presented. The overview of principles to estimate nodal locations is highlighted. The way the location estimation is performed, techniques can also be classified as distance-based, angle-based, pattern-based and connectivity-based. The book illustrates their basic mechanisms through examples and provides a comparison among them.

1.12 Sensor Web

The Sensor Web is an emerging trend (Fig. 1.4), which makes various types of web-resident sensors, instruments, image devices, and repositories of sensor data discoverable, accessible, and controllable via the World Wide Web (WWW). Sensor Web enables spatio-temporal understanding of an environment through coordinated efforts between (i) multiple numbers and types of sensing platforms, (ii) orbital and terrestrial, and (iii) both fixed and mobile. A lot of effort has been invested in order to overcome the obstacles associated with connecting and sharing the heterogeneous sensor resources. One chapter in this book emphasises the Sensor Web Enablement (SWE) standard defined by the OpenGIS Consortium (OGC), which is composed of a set of specifications, including SensorML, observation and measurement, sensor collection service, sensor planning service and web notification service. It also presents reusable, scalable (Intanagonwiwat *et al.* 2000), extensible, and interoperable service oriented sensor web architecture that:

- Conforms to the SWE standard
- Integrates Sensor Web with grid computing
- Provides middleware support for Sensor Webs

In addition, the chapter describes the experiments and an evaluation of the core services within the architecture.

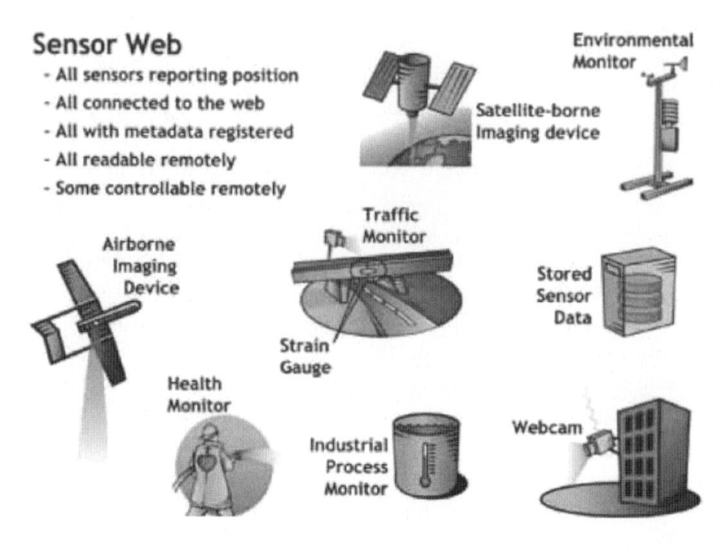

Fig. 1.4. A glance at Sensor Web (Reichardt 2003; Copyright: Geospatial Solutions)

1.13 MANET versus WSNs

MANET has been expanded as a mobile ad-hoc network, a network that has self-configuration capability in terms of configuring networks of mobile routers and hosts (stations) interfaced with wireless links. The integrated system forms an arbitrary topology. In this network, the stations can move around and change the network topology. The wireless topology of the network can change rapidly and unpredictably. It is presumed that there exists a difference between MANET and the emerging WSNs. One chapter is dedicated to review and analyse the similarities as well as the differences, between MANET and WSNs. The main focus of discussion is on how they are (or not) alike. From certain points of view a WSN is seen as a special type of MANET; the protocols, algorithms, and design issues of the former cannot be applied to the latter.

Although, wireless ad-hoc networks have been studied for nearly three decades, there are still opportunities for research. Some of these opportunities may derive from the fact that most previous researches are around military applications and that the basic assumptions for these applications may not hold for common non-military applications. Some argue that it has just barely reached the end of its exploratory stage, where the foundations have been laid out, in practice it is a fact that WSNs have been used for applications with limited scope, or just as prototypes to provide proof of concepts. The visions of ubiquitous networks with millions of nodes have yet to be realised (Gomez 1999).

1.14 Distributed Control

In recent years, industrial automation and control systems (Fig. 1.5) have been preferred to implement a Distributed Control System (DCS) instead of centralised systems, because of their advantage of greater flexibility over the whole operating range. Other benefits are low implementation costs, easy maintenance, configurability, scalability and modularity. Conventional centralised control is characterised by a central processing unit that communicates with the field devices (sensors, actuators, switches, valves, drives, etc.) with separate parallel point-to-point links. On the other hand, DCS interconnects devices with a single serial link. Since I/O points within the control systems are distributed and since the number of auxiliary components and equipments progressively increase, DCS architecture is seen to be appropriate. Each I/O point can be defined in terms of smart devices. Other operational benefits of adopting DCS schemes can be summarised. These include, sharing of the processing load for avoiding the bottle neck of a single centralised controller, replacement of complex point-to-point wiring harnesses with control networks to reduce weight and assembly costs, freedom to vary the number and type of control nodes on a particular application in order to easily modify its functionality, ability to individually configure and test segments (components) of the target application before they are combined, build and test each intelligent subunits separately, and provisions for interfacing data exchanges between the runtime control system and other factory/business systems (e.g., management information, remote monitoring and control, etc.). DCS can be leveled into five layers of automation services. The bottom layer is called the component level, which includes the physical components such as intelligent devices (PC, industrial PC, PLC (Programmable Logic Controller), microprocessor, micro-controller etc.), and non-intelligent devices (sensors, actuators, switches, A/D, D/A, port, transceivers, communication media, etc.). The interface layer is similar to a MAC sublayer of the link layer protocol. The third layer, called the process layer, includes application layer features. Since control systems do not transfer raw data through a physical media, an application layer has to exist. The top application layer defines the variables responsible for transferring data from one place to other when they are logically connected. This layer also generates object code from the source code by the use of resources and methods such as compilers and OLE/DDE, respectively. The final management layer manages the control design. There is a set of generic functions within the management layers, which are accountable for providing services for all aspects of control management. Typical management functions are installation, configuration, setting, resetting, monitoring, and operator interfacing and testing, among others.

1.14.1 DCS Realising Platform

Fieldbus is a technology for the realisation of DCS. This technology is being developed for most industrial automation and control applications. The leading fieldbuses with their characteristics can be seen in Mahalik (Mahalik 2003). The technology includes a protocol which can trans-receive digital plant data through multiple media and channels. Recently, many fieldbuses are available in the technology market place and their application areas vary. For industrial control applications, the type of bus used is described as a sensor-bus or device-bus, and usually deals with complex devices such as motors, drives, valves, etc., that provide process variable information. In summary, a fieldbus provides a good basis for distributed real-time control systems. It has been successfully tried and tested in many applications, including a wide range of general industrial applications (food processing, conveyor system automation, packaging plant automation, etc.), ship automation, FMS, monitoring and control in hazardous area, flight simulator, building automation, SCADA systems (automotive power distribution applications), deterministic applications (robotics, air craft, space), rehabilitation, and so on. Two chapters are included to describe several different architectures for implementing distributed control systems. The first chapter deals with design and implementation aspects of DCS. The second chapter analyses the architecture in terms of simulating DCS. The simulation of DCS is possible through an extension of GTSNetS. This simulation tool can be used to study the combined effects of network parameters and control strategies. One of the main features of GTSNetS is its scalability to network size. In fact, it can be used to simulate networks of several hundred thousand nodes. Another important feature of GTSNetS is the fact that it is easily extensible by users to implement additional architectural designs and models. A simple example simulation is presented to illustrate the use of simulator.

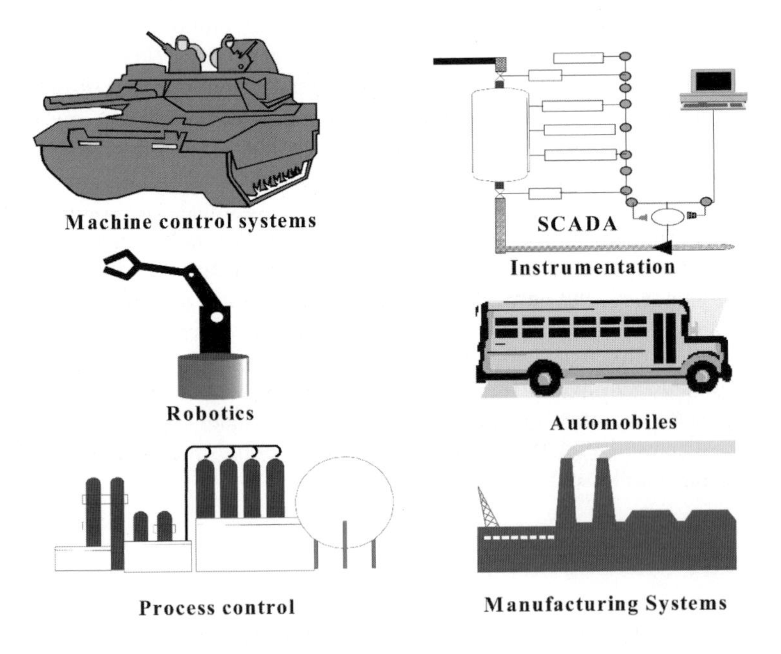

Fig. 1.5. Typical automation and control environments

1.15 Some Typical Applications

There can be thousands of applications for WSNs. This book deals with the fundamental aspects of sensor networks. To justify the applicability of sensor network technology, we have included some real-field case studies. They are:

- Pet-dog management system
- Agriculture monitoring
- Intelligent Closed-Circuit Television (CCTV) network

1.15.1 Pet-dog Management

This chapter presents a pet-dog management system based on wireless sensor networks. The intelligent wireless management system is composed of a central control system, an auto-feeder, a mini guidance robot, and wireless sensing devices. The system uses three sensed data types, including luminance, temperature, and sounds from a pet-dog or surrounded environment, respectively. The presented design method provides an efficient way to control and monitor the pet dog using WSNs. The implemented system can be used as a design framework of portable devices for other pet management systems.

1.15.2 Agriculture Monitoring

In the past few years, new trends have emerged in the agricultural sector in terms of precision agriculture. Precision agriculture concentrates on providing the means for observing, assessing, and controlling agricultural practices (Baggio 2006). It covers a wide range of agricultural concerns from daily herd management through horticulture to field crop production. It concerns as well pre- and post-production aspects of agricultural enterprises. Various field monitoring boards (Fig. 1.6) and systems as the digitalised infrastructures are very useful for ubiquitous and global sensing, computing, and networking in agriculture. Field monitoring is difficult to perform because installation sites mostly lack infrastructure for providing power or networking capabilities and precision instruments must endure exposure to heat, moisture, rain, and dust. In order to counter the adverse conditions, researchers propose a long-term remote field monitoring system consisting of a sensor network equipped with a Web server for measurement, and wireless LAN for data transmission (Fukasu and Hirafuji 2005).

1.15. 3 CCTV

Intelligent Closed-Circuit Television (CCTV) has attracted worldwide attention and government interest since such systems are being used to such great effect to track the movements of intruders or attackers. Here we explore some of the technical issues to be solved for building the natural descendant of today's integrated metropolitan CCTV systems in order to become a planetary-wide sensor network incorporating both fixed, mobile, and nomadic video sensors. In particular, the authors have present recent research that provides reliable person-location and recognition services on a planetary sensor network.

Fig. 1.6. A standard Field Server (Source: Fukasu and Hirafuji 2005; Courtesy: Fuji Press)

1.16 MICA Mote

MICA mote is a commercially available sensor network entity that has been recently used in academic and development research units. It has essential features of a mote; hence the name MICA mote. The MICA mote platform was developed at UC Berkeley in collaboration with Intel Inc. The platform includes an 8-bit Atmel ATMEGA 128L microcontroller, 132K of memory, a 512K nonvolatile flash memory, and a 40 Kbps radio operating at 433 MHz or 916 MHz. The mote is powered by commonly available standard AA batteries. The motes come in two form factors (physical size); (i) Rectangular, measuring 5.7 x 3.18 x.64 cms and (ii) Circular, measuring 2.5 x .64 cms. MICA Mote consists of accessories, and is offered at affordable price on the market. It has transmitting-receiving range of several hundred feet. The transmission rate is about 40 thousand bits per second. At resting state it consumes less than one microamp, while at the time of receiving data, the consumption is 10 milliamps. On the other hand, during transmission the consumption is 25 milliamps. The hardware components together create a MICA mote. A system designer writes software to control the mote and allow it to perform in a particular way depending upon the application requirements. The application software that is designed on the top of the so-called operating system called

TinyOS. TinyOS also deals with the power management systems. The applications are written in NesC, a high-level language for programming structured component-based applications. The NesC language is predominantly intended for embedded systems. NesC has a C-like syntax. It also supports the concurrency model and mechanisms for structuring, naming, and making static and dynamic linking to all the software components in order to form a robust network embedded system. Another platform developed by Japanese researcher (Saruwatari et. al. 2004) the PAVENET, also supports to configure various sensor network based systems. PAVENET includes hardware and software. The U^3 serves as the hardware module and U^3 SDK as the software development kit for U^3 (Fig. 1.7).

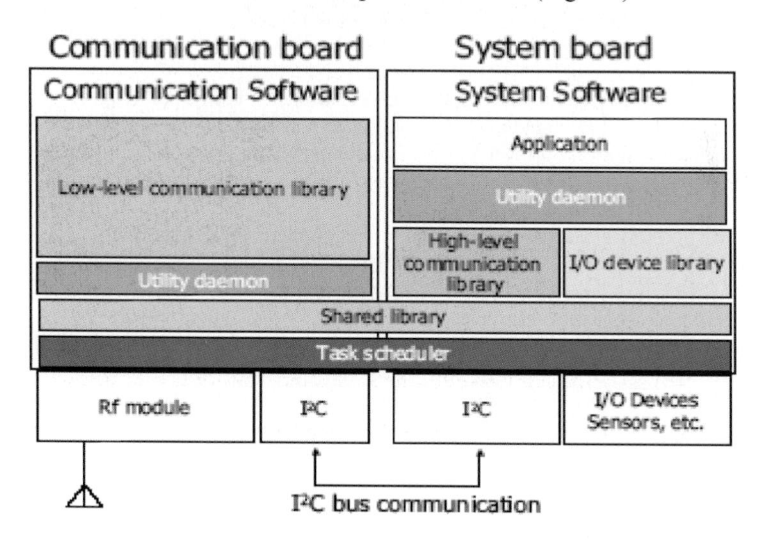

Fig. 1.7. U^3 SDK

1.17 References

Baggio A (2006) Wireless sensor networks in precision agriculture, technical report, Delft University of Technology, The Netherlands Available at http://www.sics.se/realwsn05/papers/baggio05wireless.pdf

Deb B, Bhatnagar S and Nath B (2002) A topology discovery algorithm for sensor networks with applications to network management, IEEE CAS workshop, Sept. 2002

Estrin D, Culler D, Pister K and Sukhatme G (2002) Connecting the physical world with pervasive networks, IEEE Pervasive Computing, vol.1, no.1

Fukatsu T and Hirafiji M (2005) Field Monitoring Using Sensor-Nodes with a Web Server, Journal of Robotics, and Mechatronics, vol.17, no.2, pp.164-172, Fuji press, Japan

Gomez J, Campbell AT, Naghshineh M, and Bisdikian C (1999) Power-aware routing in wireless packet networks, Proc. 6th IEEE Workshop on Mobile Multimedia Communications (MOMUC99), Nov.

Hill J, et al. (2000) System architecture directions for networked sensors. In Proc. of ASPLOS'2000, Cambridge, MA, USA, Nov. pp. 93–104.

http://en.wikipedia.org
http://www.geospatial-online.com/geospatialsolutions/article/articleDetail.jsp?id=52681
http://www.caba.org/standard/zigbee.html
Intanagonwiwat C, Ramesh G, Estrin D (2000) Directed diffusion: A scalable and robust communication paradigm for sensor networks, Proc. of the 6th annual Int. Conf. on Mobile Computing and Networks
Krumke S, Liu R, Loyd EL, Marathe M, Ramanathan R and Ravi SS (2003) Topology control problems under symmetric and asymmetric power thresholds, Proc. 2nd Int. Conf. on ad-hoc networks and wireless, LNCS, vol. 2865 Oct. pp187-198
Kulik J, Heinzelman WR and Balakrishnan H (1999) Adaptive protocols for information dissemination in WSN, Proc. 5th ACM/IEEE MobiCom Conf., Seattle
Loyd E, Liu R, Marathe M, Ramanathan R and Ravi S (2002) Algorithmic aspects of Topology Control problems for ad-hoc networks, Proc. of ACM Mobihoc pp123–134
Macias JAG, Rousseau F, Sabbatel GB, Toumi L and Duda A (2003) Quality of service and mobility for the wireless internet, Wireless Networks, vol. 9, pp341-352
Mahalik NP (2003) (ed) Fieldbus Technology; Industrial network standard for real-time distributed control, Springer Verlag, June
Mainwaring A, et al., "Wireless sensor networks for habitat monitoring. In Proc. of WSNA'2002, Atlanta
Reichardt M (2003) The sensor Web's point of beginning, Geospatial Solutions, Apr 1
Saruwatari S, Kasmima T, Kawahara Y, Minami M, Morokawa H and Aoyama T (2004) PAVENET: A hardware and software framework for WSN. Trans. of the Society of Instrument and Control Engineers, 3:1, pp 11
Tubaishat M and Madria S (2006) Sensor networks: an overview, IEEE Explore, March
w3.antd.nist.gov
ZigBee alliances (2004) The ZigBee buzz is growing: New low-power wireless standard opens powerful possibilities, A Penton Publication (A Supplement to Electronic Design), Jan.12, http://www.elecdesign.com/Files/29/7186/7186_01.pdf

2 Time Sensitive IEEE 802.15.4 Protocol

Anis Koubaa, Mário Alves and Eduardo Tovar

IPP-HURRAY Research Group, Polytechnic Institute of Porto Rua Dr. Antonio Bernardino de Almeida, 431, 4200-072 Porto, Portugal

2.1 Contexts and Motivation

Wireless Sensor Networks (WSNs) have revolutionised the design of emerging embedded systems and triggered a new set of potential applications. A WSN is typically composed of a large set of nodes scattered in a controlled environment and interacting with the physical world. This set aims at the collection of specified data needed for the monitoring/control of a predefined area/region. The delivery of sensory data for process and analysis, usually to a control station (also referred as sink), is based on the collaborative work of the WSN nodes in a multi-hop fashion (Fig. 2.1).

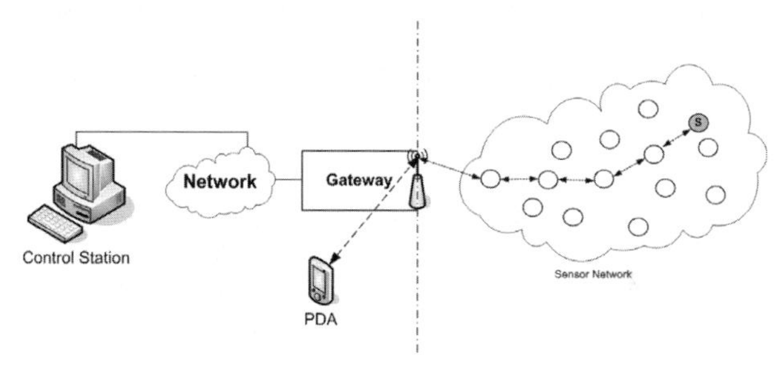

Fig. 2.1. Typical topology of a WSN

Hence, a WSN node is intended to act as:

- A *data source* with some basic capabilities, namely sensing (eventually other I/O), processing (and memory) and wireless communications, and that produces sensory data by interacting with the physical environment and collecting relevant physical parameters.
- A *data router* that transmits data from one neighbour sensor node to another, towards the control station, which processes and analyses the data collected from the different sensors/nodes in the network.

This particular form of distributed computing raises many challenges in terms of real-time communication and coordination due to the large number of constraints that must be simultaneously satisfied, including limited power, CPU speed, storage capacity, and bandwidth. These constraints trigger the need for new paradigms in terms of node/sensor design and network communication/coordination mechanisms. The design of wireless sensor networks is mainly concerned with power-efficiency issues, due to the severe limitation in terms of energy consumption (Aykildiz *et al.* 2002; Stankovic *et al.* 2003). However, the design complexity is even more significant when applications have, in addition, real-time and/or scalability requirements (Stankovic *et al.* 2003). Several research initiatives, aiming at providing different design solutions for WSNs protocols, have recently emerged (Lu *et al.* 2002; Bandyopadhyay and Coyle 2003; He *et al.* 2003; Ye *et al.* 2004; Bacco *et al.* 2004). However, we believe that the use of standard technologies pushed forward by commercial manufacturers can speed-up a wider utilisation of WSNs. In this context, the IEEE 802.15.4 protocol (IEEE 802.15.4 Standard 2003), recently adopted as a communication standard for Low-Rate Wireless Local Area Networks (LR-WPANs), shows up itself as a potential candidate for such a deployment. This protocol provides enough flexibility for fitting different requirements of WSN applications by adequately tuning its parameters, even though it was not specifically designed for WSNs. In fact, low-rate, low-power consumption and low-cost wireless networking are the key features of the IEEE 802.15.4 protocol, which typically fit the requirements of WSNs. Moreover, the ZigBee specification (ZigBee Alliance 2005) relies on the IEEE 802.15.4 Physical and Data Link Layers, building up the Network and Application Layer, thus defining a full protocol stack for LR-WPANs. More specifically, the IEEE 802.15.4 Medium Access Control (MAC) protocol has the ability to provide very low duty cycles (from 100% to 0.1 %), which is particularly interesting for WSN applications, where energy consumption and network lifetime are main concerns. Additionally, the IEEE 802.15.4 protocol may also provide timeliness guarantees by using the Guaranteed-Time Slot (GTS) mechanism, which is quite attractive for time-sensitive WSNs. In fact, when operating in beacon-enabled mode, i.e. beacon frames are transmitted periodically by a central node called the *PAN Coordinator* for synchronising the network, the IEEE 802.15.4 protocol allows the allocation/deallocation of GTSs in a superframe for nodes that require real-time guarantees. Hence, the GTS mechanism provides a minimum service guarantee for the corresponding nodes, thus enabling the prediction of the worst-case performance for each node's application. In this chapter, we describe the most important features of the IEEE 802.15.4 protocol that are relevant for WSNs, and we discuss the ability of this protocol to fulfill the different requirements of WSNs and to resolve inherent paradoxes involving power-efficiency, timeliness guarantees and scalability issues.

2.2 Overview of the IEEE 802.15.4 Protocol

The IEEE 802.15.4 protocol (IEEE 802.15.4 Standard 2003) specifies the MAC sub-layer and the Physical Layer for LR-WPAN (hereafter denoted as PAN). The IEEE 802.15.4 protocol is very much associated with the ZigBee protocol (ZigBee

Alliance 2005), which specifies the protocol layers above IEEE 802.15.4 to provide a full protocol stack for low-cost, low-power, low data rate wireless communications. This layered architecture is depicted in Fig. 2.2

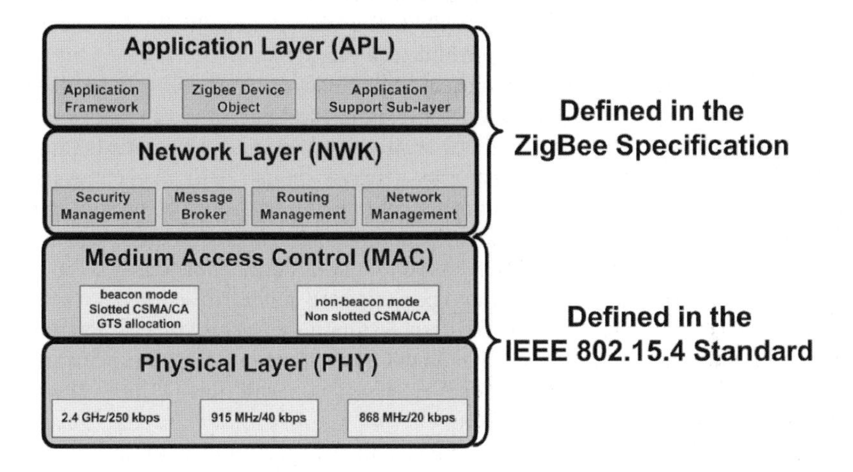

Fig. 2.2. IEEE820.15.4/ZigBee protocol stack architecture

This protocol stack results from the joint effort within the ZibBee Alliance, an organisation with over 200 member companies that has been working in conjunction with IEEE (task group 15.4) in order to achieve a full protocol specification as well as to foster its use worldwide.

2.3 Network Components and Topologies

The IEEE 802.15.4 protocol basically defines three types of nodes.

- **PAN Coordinator.** It is the principal controller (Master) of the network, which identifies its PAN, and to which other nodes may be associated. It also provides **global** synchronisation services to other nodes in the network through the transmission of beacon frames containing the identification of the PAN and other relevant information.
- **Coordinator**. It has the same functionalities as the PAN Coordinator with the exception that it does not create its PAN. A Coordinator is associated to a PAN Coordinator and provides **local** synchronisation services to nodes in its range by means of beacon frame transmissions containing the identification of the PAN defined by the PAN Coordinator to which it is associated, and other relevant information.
- **Simple (Slave) node**. It is a node that does not have any coordination functionalities. It is associated as a slave to the PAN Coordinator (or to a Coordinator) for being synchronised with the other nodes in the network.

In the IEEE 802.15.4 Standard, the first two types of nodes are referred to as *Full Function Devices* (FFD), which means that they implement all the functionalities of the IEEE 802.15.4 protocol for ensuring synchronisation and network management. The third type is referred to as *Reduced Function Device* (RFD) meaning that the node is operating with a minimal implementation of the IEEE 802.15.4 protocol. Note that a simple node can also be an FFD.

Note that a PAN must include at least one FFD acting as a PAN Coordinator that provides global synchronisation services to the network and manages other potential Coordinators and slave nodes in its range. Once a PAN is started, it must be maintained by its PAN Coordinator by generating and sending beacons frames, managing association and dissociation of other nodes to the PAN, providing synchronisation services, allowing GTS allocation and management, etc. More details on starting and maintaining PANs are available in (Koubâa *et al.* 2005a).

Two basic network topologies have been defined in the IEEE 802.15.4 specification: the *star* topology and the *peer-to-peer* topology. A third type of topology – the *cluster-tree* topology, can be considered as a particular case of a peer-to-peer topology.

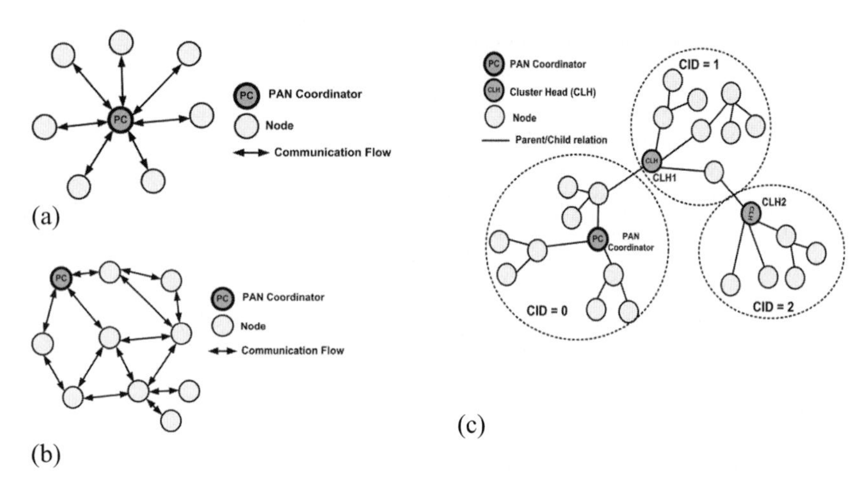

Fig. 2.3. Network topologies in IEEE 802.15.4a) Star (b) Peer-to-Peer (c) Cluster-tree

In the star topology (Fig. 2.2(a)), a unique node operates as a PAN Coordinator. The communication paradigm in the star topology is centralised; that is, each node joining the network and willing to communicate with the other nodes must send its data to the PAN Coordinator, which will then dispatch it to the destination nodes. Due to the power-consuming tasks of the PAN Coordinator in the star topology, the IEEE 802.15.4 standard recommends that the PAN Coordinator should be mains-powered, while other nodes are more likely to be battery-powered. The star topology may not be adequate for traditional wireless sensor networks for two reasons.

First, a sensor node selected as a PAN Coordinator will get its battery resources rapidly ruined. A potential bypass to this problem is to have a dynamic PAN Coordinator based on remained battery supplies in sensor nodes, such as made in LEACH protocol (Heinzelman *et al.* 2000). However, this solution seems to be quite complex, since the dynamic election of a PAN Coordinator among a large number of sensor nodes is not always efficient. Second, the coverage of an IEEE 802.15.4 cluster is very limited while addressing a large-scale WSN, leading to a scalability problem. Nevertheless, the star topology may be promising in case of cluster-based architecture, as it will be shown later.

The peer-to-peer topology (Fig. 2.3(b)) also includes a PAN Coordinator that identifies the entire network. However, the communication paradigm in this topology is decentralised, where each node can directly communicate with any other node within its radio range. This mesh topology enables enhanced networking flexibility, but it induces an additional complexity for providing end-to-end connectivity between all nodes in the network. Basically, the peer-to-peer topology operates in an ad-hoc fashion and allows multiple hops to route data from any node to any other node. However, these functions must be defined at the Network Layer and therefore are not considered in the IEEE 802.15.4 specification.

Wireless Sensor Networks are one of the potential applications that may take advantage from such a topology. In contrast with the star topology, the peer-to-peer topology may be more power-efficient and the battery resource usage is fairer, since the communication process does not rely on one particular node (the PAN Coordinator).

In the cluster-tree topology (Fig. 2.3(c) topology, one coordinator is nominated as the PAN Coordinator, which identifies the entire network. However, any node may act as a Coordinator and provide synchronisation services to other nodes or other Coordinators. The nomination of new Coordinators is the role of the PAN Coordinator.

The standard does not define how to build a cluster-tree network. It only indicates that this is possible, and may be initiated by higher layers. The cluster forming is performed as follows. The PAN Coordinator forms the first cluster by establishing itself as *Cluster Head* (CLH) with a *Cluster Identifier* (CID) equal to zero. It then chooses an unused *PAN Identifier* (PAN ID) and broadcasts beacons to neighbouring nodes.

Nodes that are in the range of this CLH may request to be associated to the network through the CLH. In case of acceptance, the CLH adds the requesting node as a child node in its neighbour list, and the newly joined node adds the CLH as its parent in its neighbour list and begins transmitting periodic beacons. Other nodes can then join the network at the latter joined node. If for some reason the requesting node cannot join the network at the cluster head, it will search for another parent node.

For a large-scale sensor network, it is possible to form a mesh out of multiple neighbouring clusters. In such a situation, the PAN Coordinator can promote a node to become the CLH of a new cluster adjacent to the first one. Other nodes gradually connect and form a multi-cluster network structure (Fig. 2.3(c). The Network Layer defined in the ZigBee specification uses the primitives provided by the IEEE 802.15.4 MAC sub-layer and proposes a cluster-tree protocol for either a single cluster network or a potentially larger cluster-tree network.

2.4 Physical Layer (PHY)

The IEEE 802.15.4 offers three operational frequency bands: 2.4 GHz, 915 MHz and 868 MHz. There is a single channel between 868 and 868.6 MHz, 10 channels between 902 and 928 MHz, and 16 channels between 2.4 and 2.4835 GHz (see Fig. 2.4).

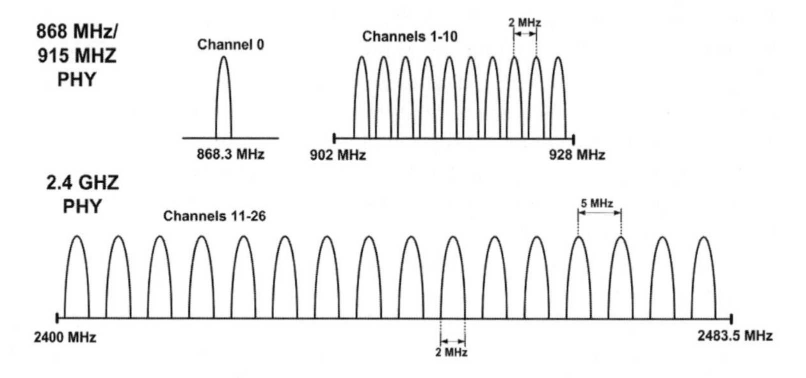

Fig. 2.4. Operating frequency bands

The data rates are 250 kbps at 2.4 GHz, 40 kbps at 915 MHZ and 20 kbps at 868 MHz. Lower frequencies are more suitable for longer transmission ranges due to lower propagation losses. However, the advantage of high data rate transmission is the provision of higher throughput, lower latency or lower duty cycles. All of these frequency bands are based on the *Direct Sequence Spread Spectrum* (DSSS) spreading technique. The features of each frequency band (modulation, chip rate, bit rate) are summarised in Table 1. Note that one '*symbol*' is equivalent to four '*bits*'.

Table 2.1. Frequency Bands and Data Rates

Freq. band (MHz)	Spreading Parameters		Data Parameters		
	Chip rate (kchip/s)	Modulation	Bit rate (kbps)	Symbol rate (ksymbol/s)	Symbols
868	300	BPSK	20	20	Binary
915	600	BPSK	40	40	Binary
2400	2000	O-QPSK	250	62.5	16-ary

In addition, the PHY of the IEEE 802.15.4 is in charge of the following tasks.

- **Activation and deactivation of the radio transceiver**. The radio transceiver may operate in one of three states: *transmitting*, *receiving* or *sleeping*. Upon request of the MAC sub-layer, the radio is turned ON or OFF. The standard recommends that the *turnaround time* from transmitting to receiving states and vice versa should be no more than 12 symbol periods.

- **Receiver Energy Detection (ED).** It is an estimation of the received signal power within the bandwidth of an IEEE 802.15.4 channel. This task does not involve any signal identification or decoding on the channel. The standard recommends that the energy detection duration should be equal to 8 symbol periods. This measurement is typically used to determine if the channel is busy or idle in the Clear Channel Assessment (CCA) procedure or by the Channel Selection algorithm of the Network Layer.
- **Link Quality Indication (LQI).** The LQI measurement characterises the Strength/Quality of a received signal on a link. LQI can be implemented using the receiver ED technique, a signal to noise estimation or a combination of both techniques. The LQI result may be used by the higher layers (Network and Application layers), but this procedure is not specified in the standard.
- **Clear Channel Assessment (CCA).** The CCA operation is responsible for reporting the medium activity state: busy or idle. The CCA is performed in three operational modes:
  - *Energy Detection mode.* The CCA reports a busy medium if the received energy is above a given threshold, referred to as *ED threshold.*
  - *Carrier Sense mode.* The CCA reports a busy medium only if it detects a signal with the modulation and the spreading characteristics of IEEE 802.15.4 and which may be higher or lower than the ED threshold.
  - *Carrier Sense with Energy Detection mode.* This is a combination of the aforementioned techniques. The CCA reports that the medium is busy only if it detects a signal with the modulation and the spreading characteristics of IEEE 802.15.4 and with received energy above ED threshold.
- **Channel Frequency Selection.** The IEEE 802.15.4 defines 27 different wireless channels. A network can choose to operate within a given channel set. Hence, the Physical Layer should be able to tune its transceiver into a specific channel upon the reception of a request from a Higher Layer.

There are already commercially available wireless sensor motes that are compliant with the IEEE 802.15.4 physical layer. For instance, the CC2420 transceiver (CC2420 datasheet 2004) from Chipcon Company provides the implementation of the IEEE 802.15.4 physical layer, operating at 2.4 GHz with a data rate of 250 kbps. This transceiver is widely used by many sensor network products such as MICAz from Crossbow Tech. (MICAz datasheet 2004).

The deployment of IEEE 802.15.4 WPANs in the presence IEEE 802.11b WLANs (IEEE 802.11 Specification 1999) triggers some inherent problems since they both operate in the 2.4 GHz frequency band. Coexistence between both technologies has become an important issue after the proposal of the IEEE 802.15.4 standard and has been subject of recent research works. In (Howitt and Gutierrez 2003), the authors analysed the impact of an IEEE 802.15.4 network composed of several clusters on an IEEE 802.11b station communicating with a WLAN access point. An expression of the probability of an IEEE 802.11b packet collision due to the interference with IEEE 802.15.4 has been proposed. The authors conclude that the IEEE 802.15.4 network has little to no impact on the performance of IEEE 802.11b, unless the IEEE 802.11b station is very close to an IEEE 802.15.4 cluster with high activity level. A later work in (Shin *et al.* 2005) analysed the packet error rate of IEEE 802.15.4 WPANs under the interference of IEEE 802.11b WLAN and proposed some coexistence criteria for both standards. The results of this work show that the interference caused by the IEEE 802.11b WLAN does not affect the performance of an IEEE 802.15.4 WPAN if the distance between the IEEE 802.15.4 nodes and the IEEE 802.11b WLAN is larger than 8 m. Moreover, if the frequency offset is larger than 7 MHz, the interference of IEEE 802.11b has negligible effect on the performance of the IEEE 802.15.4. Another experimental work by Crossbow Tech. (Crossbow Tech. 2005) considered a set of three experiments using the MICAz motes, which implement the physical layer of the IEEE 802.15.4 and the Stargate single board computer compliant with IEEE 802.11b. The first experiment run with no WiFi interference and the other experiments run under IEEE 802.11 interference with two different power levels (standard level and 23 dBm). The packet delivery rate was analysed. The experiment shows that high power transmissions of IEEE 802.11b packet reduce the packet delivery rate up to 80% for 23 dBm Wifi cards. In general, these results state that the coexistence of both IEEE 802.15.4 and IEEE 802.11b networks is generally possible with an acceptable performance, when nodes are not in a close proximity of each other and channels are adequately selected to prevent overlapping.

2.5 Medium Access Control Sub-Layer

The MAC protocol supports two operational modes that may be selected by the PAN Coordinator (Fig. 2.5).

- The *non beacon-enabled mode*, in which MAC is simply ruled by non-slotted CSMA/CA.
- The *beacon-enabled mode*, in which beacons are periodically sent by the PAN Coordinator to synchronise nodes that are associated with it, and to identify the PAN. A beacon frame delimits the beginning of a *superframe* defining a time interval during which frames are exchanged between different nodes in the PAN. Medium access is basically ruled by slotted CSMA/CA. However, the beacon-enabled mode also enables the allocation of some time slots in the superframe, called Guaranteed Time Slots (GTSs) for nodes requiring guaranteed services.

Due to its importance in the context of this chapter, we address next the main characteristics of the beacon-enabled operational mode. We also describe the slotted and non-slotted versions of CSMA/CA. The MAC sub-layer of the IEEE 802.15.4 protocol has many similarities with the one of the IEEE 802.11 protocol (IEEE 802.11 Specification 1999), such as the use of CSMA/CA (*Carrier Sense Multiple Access / Contention Avoidance*) and the support of contention-free and contention-based periods. However, the specification of the IEEE 802.15.4 MAC sub-layer is adapted to the requirements of LR-WPANs as, for instance, eliminating the RTS/CTS mechanism in IEEE 802.11, mainly due to the following reasons: (1) data frames in LR-WPANs are usually as small as RTS/CTS frames, and thus the collision probability would be the same with or without the RTS/CTS mechanism; (2) the exchange of RTS/CTS is energy consuming which is not adequate for LR-WPANs; (3) Sensor networks applications basically rely on broadcast transmissions, which do not use the RTS/CTS mechanism.

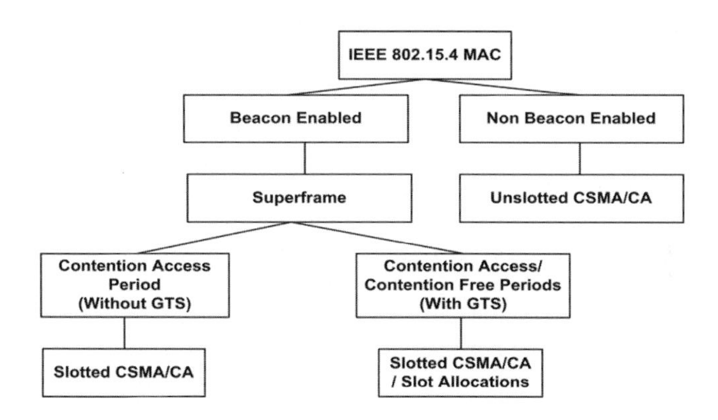

Fig. 2.5. IEEE 802.15.4 MAC operational modes

In beacon-enabled mode, beacon frames are periodically sent by the PAN Coordinator to identify its PAN and synchronise nodes that are associated with it. The *Beacon Interval* (BI) defines the time between two consecutive beacon frames. It includes an active period and, optionally, an inactive period (Fig. 2.6). The active period, called *superframe*, is divided into 16 equal time slots, during which frame transmissions are allowed. During the inactive period (if it exists), all nodes may enter into a sleep mode, thus saving energy.

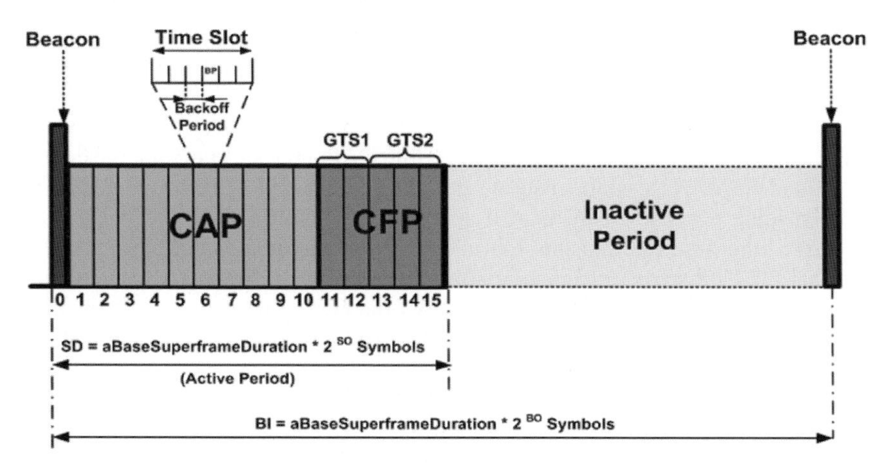

Fig. 2.6. Beacon interval and superframe concepts

The Beacon Interval and the *Superframe Duration* (*SD*) are determined by two parameters, the *Beacon Order* (*BO*) and the *Superframe Order* (*SO*), respectively. The Beacon Interval is defined as follows:

$$BI = aBaseSuperframeDuration \cdot 2^{BO} \ for \ 0 \leq BO \leq 14 \quad (2.1)$$

The Superframe Duration, which corresponds to the active period, is defined as follows:

$$SD = aBaseSuperframeDuration \cdot 2^{SO} \ for \ 0 \leq SO \leq BO \leq 14 \quad (2.2)$$

In Eqs.(1) and (2), *aBaseSuperframeDuration* denotes the minimum duration of the superframe, corresponding to $SO = 0$. This duration is fixed to 960 symbols (IEEE 802.15.4 Standard 2003) (where a symbol corresponds to four bits) corresponding to 15.36 ms, assuming 250 kbps in the 2.4 GHz frequency band. In this case, each time slot has a duration of 15.36/16 = 0.96 ms. In this chapter, we will mainly consider the features of the 2.4 GHz frequency range supported by available COTS sensor network products, such as the MICAz motes of Crossbow Tech. (MICAz datasheet 2004). By default, nodes compete for medium access using slotted CSMA/CA during the *Contention Access Period* (CAP). A node computes its backoff delay based on a random number of backoff periods, and performs two CCAs before transmitting. The IEEE 802.15.4 protocol also offers the possibility of defining a *Contention-Free Period* (CFP) within the superframe (Fig. 2.6). The CFP, being optional, is activated upon request from a node to the PAN Coordinator for allocating *Guaranteed Time Slots* (GTS) depending on the node's requirements.

The information on Beacon Interval and the Superframe Duration is embedded in each beacon frame sent by the PAN Coordinator (or any other Coordinator in the PAN) in the *superframe specification* field. Therefore, each node receiving the beacon frame must correctly decode the information on the superframe structure, and synchronise itself with PAN Coordinator and consequently with the other nodes. Observe that the CAP starts immediately after the beacon frame, and ends before the beginning of the CFP (if it exists). Otherwise, the CAP ends at the end of the active part of the superframe. During the CAP, nodes can communicate using slotted CSMA/CA while ensuring that their transactions (data frame + inter-frame spacing + acknowledgement if any exists) would finish before the end of the CAP; otherwise the transmission is deferred to the next superframe. Finally, note that the CFP starts immediately after the end of the CAP and must complete before the start of the next beacon frame. All the GTSs that may be allocated by the PAN Coordinator are located in the CFP and must occupy contiguous slots. The CFP may therefore grow or shrink depending on the total length of all GTSs. The transmissions in the CFP are contention-free and therefore do not use the CSMA/CA mechanism. Additionally, a frame may only be transmitted if the transaction (data frame + inter-frame spacing + acknowledgement if any exists) would finish before the end of the corresponding GTS.

2.6 The CSMA/CA Mechanisms

The IEEE 802.15.4 defines two versions of the CSMA/CA mechanism: (i) *slotted CSMA/CA* algorithm – used in the beacon-enabled mode and (ii) *non-slotted CSMA/CA* algorithm – used in the non beacon-enabled mode. In both cases, the CSMA/CA algorithm uses a basic time unit called *Backoff Period* (BP), which is equal to *aUnitBackoffPeriod* = 20 Symbols (0.32 ms). In slotted CSMA/CA, each operation (channel access, backoff count, CCA) can only occur at the boundary of a BP. Additionally, the BP boundaries must be aligned with the superframe time slot boundaries (Fig. 2.6). However, in non-slotted CSMA/CA the backoff periods of one node are completely independent of the backoff periods of any other node in a PAN. The slotted/non-slotted CSMA/CA backoff algorithms mainly depend on the following three variables.

- The *Backoff Exponent* (*BE*) enables the computation of the backoff delay, which is the time before performing the CCAs. The backoff delay is a random variable between 0 and (2^{BE}-1).
- The *Contention Window* (*CW*) represents the number of backoff periods during which the channel must be sensed idle before accessing the channel. The *CW* variable is only used with the slotted CSMA/CA version. The IEEE 802.15.4 standard set the default initialisation value *CW* = 2 (corresponding to two CCAs). In each backoff period, channel sensing is performed during the 8 first symbols of the BP.
- The *Number of Backoffs* (*NB*) represents the number of times the CSMA/CA algorithm was required to backoff while attempting to access the channel. It is initialised to zero before each new transmission attempt.

Observe that the definition of *CW* in IEEE 802.15.4 is different from its definition in IEEE 802.11 (IEEE 802.11 Specification 1999). In the latter, *CW* has a similar meaning to the time interval [0, 2^{BE}-1]. Fig. 2.7 presents the flowchart of the slotted and non-slotted CSMA/CA algorithms. The slotted CSMA/CA can be summarised in five steps as follows.

Step 1* - *initialisation of NB, CW and BE. The number of backoffs and the contention window are initialised (*NB* = 0 and *CW* = 2). The backoff exponent is also initialised to *BE* = 2 or *BE* = min (2, *macMinBE*) depending on the value of the *Battery Life Extension* MAC attribute. *macMinBE* is a constant defined in the standard. After the initialisation, the algorithm locates the boundary of the next backoff period.

Step 2* - *random waiting delay for collision avoidance. The algorithm starts counting down a random number of BPs uniformly generated within [0, 2^{BE}-1]. The count down must start at the boundary of a BP. To disable the collision avoidance procedure at the first iteration, *BE* must be set to 0, and thus the waiting delay is null and the algorithm goes directly to *Step 3.*

Step 3* - *Clear Channel Assessment (CCA). When the timer expires, the algorithm then performs one CCA operation at the BP boundary to assess channel activity. If the channel is *busy*, the algorithm goes to *Step 4*, otherwise, i.e. the channel is idle, the algorithm goes to *Step 5*.

Step 4 - busy channel. If the channel is assessed to be *busy*, *CW* is re-initialised to 2, *NB* and *BE* are incremented. *BE* must not exceed *aMaxBE* (default value equal to 5). Incrementing *BE* increases the probability of having greater backoff delays. If the maximum number of backoffs (*NB* = *macMaxCSMABackoffs* = 5) is reached, the algorithm reports a failure to the higher layer, otherwise, it goes back to (*Step 2*) and the backoff operation is restarted.

Step 5 - idle channel. If the channel is assessed to be *idle*, *CW* is decremented. The CCA is repeated if $CW \neq 0$ (Step 3). This ensures performing two CCA operations to prevent potential collisions of acknowledgement frames. If the channel is again sensed as idle ($CW = 0$), the node attempts to transmit. Nevertheless, collisions may still occur if two or more nodes are transmitting at the same time.

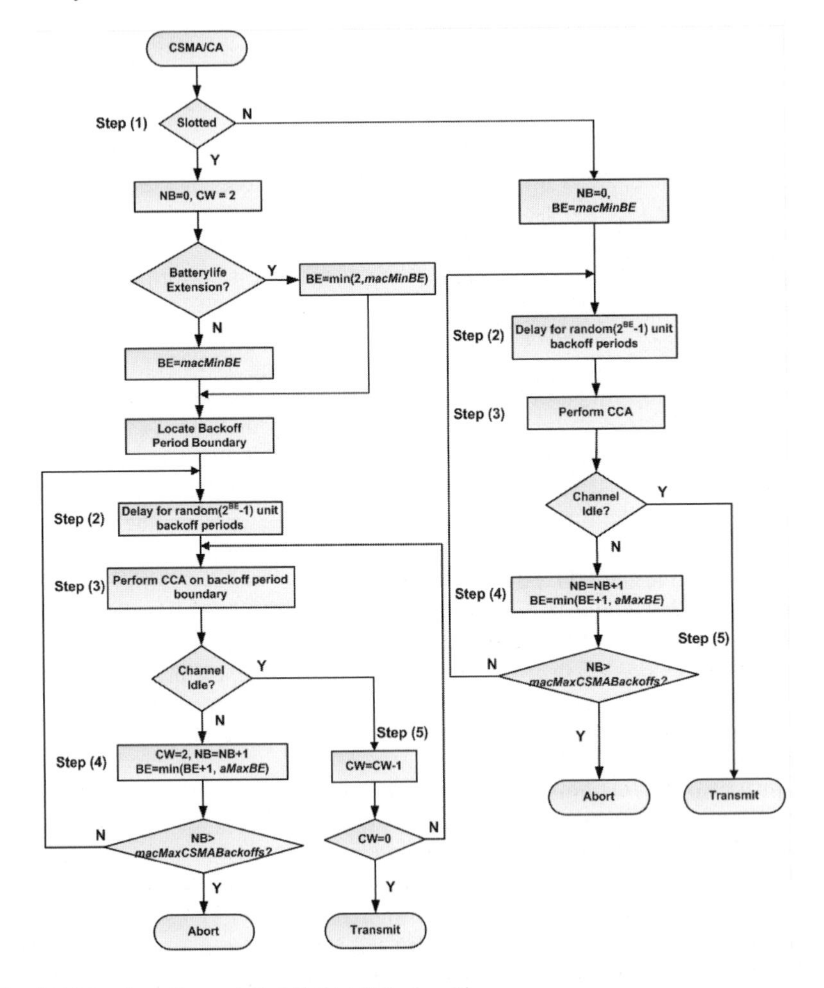

Fig. 2.7. The slotted/non-slotted CSMA/CA algorithms

The non-slotted algorithm is similar with a few exceptions as follows.

Step 1. The *CW* variable is not used, since the non-slotted has no need to iterate the CCA procedure after detecting an idle channel. Hence, in Step 3, if the channel is assessed to be idle, the MAC protocol immediately starts the transmission of the current frame. Second, the non-slotted CSMA/CA does not support *macBattLife-Ext* mode and, hence, *BE* is always initialised to the *macMinBE* value.

Steps 2, 3 and 4. It is similar to the slotted CSMA/CA version. The only difference is that the CCA starts immediately after the expiration of the random backoff delay generated in Step 2.

Step 5. The MAC sub-layer starts immediately transmitting its current frame just after a channel is assessed to be *idle* by the CCA procedure.

2.7 GTS Allocation and Management

The IEEE 802.15.4 protocol offers the possibility of having a Contention-Free Period (CFP) within the superframe. The CFP, being optional, is activated upon request from a node to the PAN Coordinator for allocating a certain number of time slots. Hence, a node that wants to allocate time slots in the CFP for an exclusive use sends its request to the PAN Coordinator, which decides whether to accept this request or not, based on the available resources. Fig. 2.8 shows the GTS characteristics field format sent within a GTS allocation request command frame by a requesting node to the PAN Coordinator.

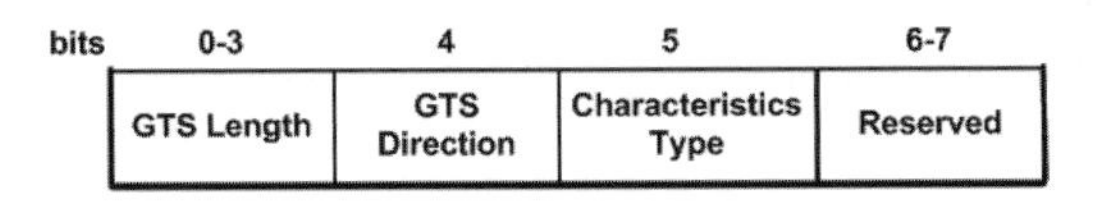

Fig. 2.8. GTS characteristics field format in IEEE 802.15.4

The node explicitly expresses the number of time slots that it wants to allocate in the *GTS Length* field. Note that the GTS length can be up to 15 time slots. The *GTS Direction* field specifies if the GTS is in receive-only mode (value = 1), i.e. data is transmitted from the PAN Coordinator to the requesting node, or in transmit-only mode (value = 0), i.e. data is transmitted from the requesting node to the PAN Coordinator. The *Characteristics Type* field refers to a GTS allocation if it is set to one or a GTS deallocation if it is set to zero.

Upon receiving this request, the PAN Coordinator checks whether there are sufficient time slots available in the superframe for this request. If the number of available time slots is smaller than the number requested, the GTS allocation request is rejected, otherwise it is accepted. The PAN coordinator must ensure that the CAP length remains always greater than *aMinCAPLength* equal to 7.04 ms according to the standard. In the former case, the corresponding node may still send its data frames during the CAP, but with no guarantee. If the GTS allocation request is accepted, the admitted node must keep track of beacon frames for checking which time slots have been allocated for that GTS in the current superframe. This information is located in the GTS descriptor field (Fig. 2.9), which is embedded in each beacon frame. A beacon frame cannot have more than seven GTS descriptors, thus inherently limiting the number of GTSs to seven.

bits	0-15	16-19	20-23
	Node Address	GTS Starting Slot	GTS Length

Fig. 2.9. GTS Descriptor Field Format in IEEE 802.15.4

The PAN Coordinator must also update the *final CAP slot* subfield of the *superframe specification* field of the beacon frame (IEEE 802.15.4 Standard 2003), which indicates the final superframe slot used by the CAP. The explicit GTS allocation adopted by the standard has the advantage of being simple. However, it may be not efficient enough in terms of bandwidth utilisation for flows with low arrival rates, which is typically the case in wireless sensor networks, since the guaranteed bandwidth of a GTS can be much higher than the arrival rates.

2.8 Time-sensitive Applications

Basically, peer-to-peer communications using flat routing protocols or broadcast transmissions is the commonly used paradigm in wireless sensor networks due to its data centric nature (Akyildiz *et al.* 2002, Stankovic *et al.* 2003). In that context, the IEEE 802.15.4 protocol can potentially fit such a paradigm, as it supports peer-to-peer topologies. On the other hand, most wireless sensor network applications are large-scale, which trigger the need to consider scalability issues in IEEE 802.15.4.

For such kind of networks, the non-beacon enabled mode seems to be more adapted to the scalability requirement than the beacon-enabled mode. In fact, when disabling the beacon-enabled mode, it is easier to construct a peer-to-peer network than when periodic beacons are sent, since in the former case all nodes are independent from the PAN Coordinator, and the communication is completely decentralised. In addition, the non-slotted CSMA/CA version in the non beacon-enabled mode will enable a better flexibility for a large-scale IEEE 802.15.4-compliant peer-to-peer network, since it does require any synchronisation, contrarily to the case where slotted CSMA/CA is used.

However, when addressing WSN applications with timing requirements, the peer-to-peer paradigm becomes controversial. In fact, with the inherently limited characteristics of wireless sensor nodes, timeliness guarantees (as well as reliability) are far from being granted. One of the problems is the use of the (non-slotted) CSMA/CA mechanism, which does not provide any guarantee for a time-bounded delivery. To illustrate the scalability/latency paradox in peer-to-peer WSNs, let us consider a typical scenario of data delivery in WSNs (Fig. 2.10).

Observe that the sensory data originated from sensor node *S* towards the data sink *D* is routed in a multi-hop ad-hoc fashion. Hence, the end-to-end delay experienced by such a data packet is basically a function of the number of hops, the distance, data rate, resources, medium access and routing protocols. On the other hand, in large-scale WSNs, the number of intermediate hops and the distance from source to data sink may be significantly high. Moreover, due to the low data rate and the need for low-power consumption, processing and transmission delays are far from being negligible, when compared to those in other types of communication networks. Adding to these facts the unpredictable behaviour of the communication protocols (e.g. the MAC sub-layer typically deploys contention-based MAC protocols, dynamic topological changes due to node failures), the end-to-end delay in large-scale WSNs is inevitably high. Even with the provision of very efficient communication protocols, the timing performance is still limited due to the stringent resource constraints and to the unreliability of individual sensor nodes.

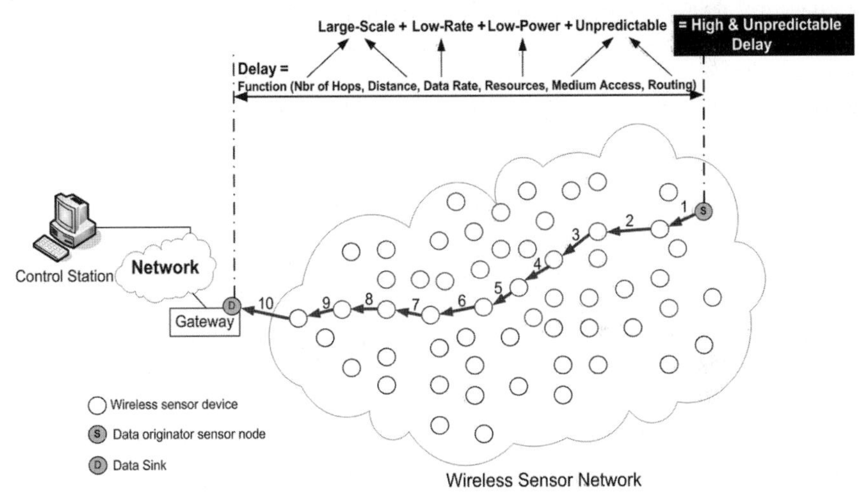

Fig. 2.10. Timing performance limitations in peer-to-peer wireless sensor networks

On the other hand, in beacon-enabled mode, the IEEE 802.15.4 enables the provision of real-time guarantees by using its GTS mechanism. Unfortunately, the use of GTSs is limited to direct communications between the PAN Coordinator and the nodes within its radio coverage, since GTS allocations are only granted by the PAN Coordinator. As a consequence, to benefit from the GTS mechanism, a node must be within the transmission range of the PAN Coordinator, which limits the configuration of the WPAN to a star topology. However, the star topology is not scalable and therefore is not adequate for large-scale wireless sensor networks, since only a small subset of WSN nodes would be in the range of their PAN Coordinator. The protocol also allows a cluster-tree topology to increase the network coverage, but still the GTSs cannot be used outside the range of the PAN Coordinator. In what follows, we present some guidelines to tackle scalability versus timing performance paradox.

2.8.1 Two-tiered Cluster-based Architecture

The previously referred limitation has been tackled by means of a two-tiered architectural approach, where an upper tier Wireless Local Area Network (WLAN) serves as a backbone to a large-scale WSN (Koubâa and Alves 2005b; Koubâa *et al.* 2005c). In (Koubâa and Alves 2005b), the authors have presented the concept of the two-tiered architecture with some important design goals. However, no communication protocol has been investigated for both tiers. Then, (Koubâa *et al.* 2005c) proposed ART-WiSe, a two-tiered cluster-based architecture for real-time communications in large-scale WSNs based on the IEEE 802.15.4 protocol. The ART-WiSe architecture provides a solution for supporting real-time communications in large-scale WSNs, while coping with the limited network coverage of IEEE 802.15.4 in beacon-enabled mode. This is achieved by organising a "traditional" WSN in two interoperable networks arrayed in a hierarchical manner (Fig. 2.11): tier-1 and tier-2.

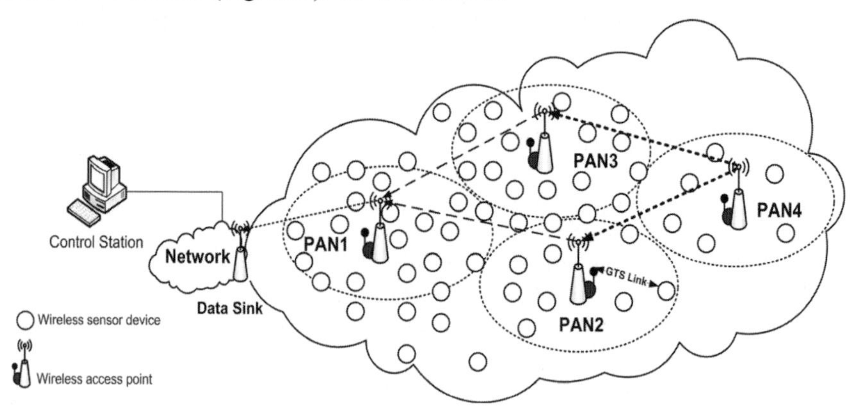

Fig. 2.11. ART-WiSe architectural approach

The Tier-1 network is a basic IEEE 802.15.4-compliant WSN interacting with the physical environment to collect sensory data. This network is characterised by a low data rate (250kbps), a short transmission range (10-70 m) and severe energy constraints, resulting in limited communication capabilities. The WSN is partitioned into several clusters, where each cluster is an independent WPAN managed by a node (access point) of the Tier-2 network. For time-sensitive applications, real-time communications are achieved with the IEEE 802.15.4 beacon-enabled mode, through the GTS mechanism inside one IEEE 802.15.4 cluster of tier-1. This mechanism provides applications with predictable minimum service guarantees, enabling to predict the worst-case real-time performance of the network (Koubâa *et al.* 2006a).

The Tier-2 network (or the *overlay network*) is a wireless network acting as a backbone for the underlying WSN. It is composed of a set of special nodes called *access points*, which act as an interface between the WSN and the overlay network. The Tier-2 network is dedicated to relaying sensory data from the sensor network to the data sink. Each node (access point) in the Tier-2 network acts as a unique PAN Coordinator of the PAN cluster that it manages. Even though the architecture is independent from the federating communication protocols, the use of IEEE 802.11b in the overlay network is envisaged since it is a mature, widely used and cost-effective solution with significant bandwidth (11 Mbps up to 54 Mbps) and long transmission ranges (>100 m). Although the basic IEEE 802.11 does not provide any Quality of Service (QoS) guarantees, it has been shown that it performs well under lightly loaded networks (Bharghavan 1998, Zheng *et al.* 2003). In ART-WiSe, it is expected that the overlay network will not be subject to high traffic load, since the difference between data rates in the WSN (250 kbps) and in the overlay wireless network (> 11 Mbps) is quite high. However, it is still possible to use the IEEE 802.11e extension (IEEE 802.11e Standard 2001) that provides additional QoS guarantees to the IEEE 802.11 protocol. Another important option for the overlay wireless network is the on-going extension of IEEE 802.15.4 made by the TG4a working group and which envisages a new physical layer based on Ultra Wide Band Impulse Radio enabling low-power consumption and high data rate.

2.8.2 Tackling the Power/Timing Efficiency Paradox

With the emergence of new WSN applications under reliability and timing constraints, the provision of real-time guarantees may be more crucial than saving energy during critical situations. The IEEE 802.15.4 protocol presents the advantage to fit different requirements of potential applications by adequately setting its parameters. Real-time guarantees can be achieved by using the GTS mechanism in beacon-enabled mode. The allocation of a GTS by a node provides it with a minimum service guarantee, enabling the prediction of the worst-case timing performance of the network. On the other hand, power-efficiency can be achieved by operating at low duty cycles (down to 0.1%). However, power-efficiency and timeliness guarantees are often two antagonistic requirements in wireless sensor networks.

This issue has been addressed in (Koubâa *et al.* 2006b). We have analysed and proposed a methodology for setting the relevant parameters of IEEE 802.15.4-compliant WSNs taking into account an optimal trade-off between power-efficiency and delay bound guarantees. To tackle this challenge, we have proposed an accurate model of the service curve provided by a GTS allocation as a function of the IEEE 802.15.4 parameters, using Network Calculus formalism. We then evaluated the delay bound guaranteed by a GTS allocation and expressed it as a function of the duty cycle. Based on the relation between the delay requirement and the duty cycle, we proposed a power-efficient superframe selection method that simultaneously reduces power consumption and enables meeting the delay requirements of real-time flows allocating GTSs. In what follows, we present the most relevant results presented in (Koubâa *et al.* 2006b) for showing a potential solution of the power/timing efficiency paradox. We consider an IEEE 802.15.4 cluster with a unique PAN Coordinator, and a set of nodes within its radio coverage. The network operates in beacon-enabled mode, thus the PAN Coordinator periodically sends beacon frames. The Beacon Interval (*BI*) and the Superframe Duration (*SD*) are defined by Eq. (1) and Eq. (2), respectively. Let C be the total data rate of the channel. In our case, the data rate is fixed to $C = 250$ kbps.

Each sensor node in the range of the PAN Coordinator runs an application that generates a data flow. We consider that each data flow has a cumulative arrival function $R(t)$ upper bounded by the linear arrival curve $\alpha(t) = b + r.t$, with b denoting the maximum burst size, and r denoting the average arrival rate. We assume that each flow has a delay requirement D.

The main challenge is the following. *Given a set of data flows within an IEEE 802.15.4 cluster, where each data flow has a delay requirement D, what is the most efficient network setting (BO and SO pair) that satisfies the delay requirement of each data flow, allocating one time slot GTS, and minimises the energy consumption?*

Hence, the purpose of our analysis is to derive an expression for the delay bound as a function of the duty cycle to evaluate the trade-off between energy consumption and delay guarantee.

2.8.3 Delay Bound Analysis of a GTS Allocation

With the Network Calculus theory, the delay bound can be computed for a given flow constrained by an arrival curve $\alpha(t) = b + r.t$, if the service curve granted for this flow is known. The *service curve* defines the minimum amount of transmitted data at a given time. Obviously, the allocation of a GTS provides a minimum guaranteed bandwidth R with a maximum latency T, which is typically known as the rate-latency service curve, denoted as $\beta_{R,T}(t) = R.(t-T)^+$, with $(x)^+=\max(0,x)$.

The first part of the work consists in determining the average bandwidth guaranteed by a GTS allocation and its maximum latency, which define the service curve of the GTS allocation. The guaranteed service of a GTS allocation is depicted in Fig. 2.12. It can be observed in Fig. 2.12(a) that only a part of an allocated time slot can be used for data transmission. The remaining part of the time slot will be idle (due to Inter-Frame Spacing (IFS)) or used by a potential acknowledgement frame. We denote by T_{data} the maximum duration used for data frame transmissions inside a GTS. T_{idle} is the sum of idle times spent inside a GTS due to protocol overheads (IFS and/or Ack frames).

As a result, a node allocating a time slot of length Ts will be able to transmit data at a rate C during T_{data} for each Beacon Interval (*BI*). Hence, the maximum service latency is then equal to $T = BI - Ts$. This behaviour is captured by the stair service curve $\beta_{C,T}^{stair}(t)$ presented in Fig. 2.12(b). In (Koubâa *et al.* 2006b), we have proved that:

$$\beta_{C,T}^{stair}(t) = \sum_{k} \beta_{C,T}^{k}(t) \qquad \forall t \tag{2.3}$$

$$\text{with } \beta^k_{C,T}(t) = \begin{cases} (k-1)\cdot C\cdot T_{data} + C\left(t-(k\cdot BI - Ts)\right)^+ \\ \quad \forall t,\ (k-1)\cdot BI \le t \le k\cdot BI - T_{idle} \\ 0 \qquad \text{Otherwise} \end{cases}$$

In addition, a simplified approximation $\beta_{R,T}(t)$ of the stair service curve has been proposed, which has the form of a simple rate-latency service curve expressed as (see Fig. 2.12(b):

$$\beta_{R,T}(t) = R\cdot(t-T)^+ \\ \text{where, } R = \frac{T_{data}}{BI}C \text{ and } T = (BI - TS) \tag{2.4}$$

Using the formulations of these service curves corresponding to a GTS allocation, the delay bound is simply the maximum horizontal distance between the arrival curve $\alpha(t) = b + r.t$ and the service curve received by the flow.

Hence, if considering the approximated service curve $\beta_{R,T}(t)$, the delay bound experienced by a data flow with an arrival curve $\alpha(t) = b + r.t$, which has allocated one time slot GTS, is computed as follows:

$$D_{\max} = \frac{b}{\frac{T_{data}}{BI}\cdot C} + (BI - Ts) \tag{2.5}$$

A more accurate delay bound can be obtained using the real service curve $\beta^{stair}_{C,T}(t)$ as depicted in Fig. 2.12(b). We show that for a burst size b such that $k\cdot C\cdot T_{data} < b \le (k+1)\cdot C\cdot T_{data}$, the delay bound of a data flow with an arrival curve $\alpha(t) = b + r.t$, which has allocated one time slot GTS is:

$$D^{stair}_{\max} = \frac{b}{C} + (k+1)\cdot BI - Ts - k\cdot T_{data} \text{ if } k\cdot C\cdot T_{data} < b \le (k+1)\cdot C\cdot T_{data} \tag{2.6}$$

2.8.4 Duty Cycle Evaluation as a Function of the Delay Bound

At start-up, the PAN Coordinator of a given WPAN cluster must choose a superframe structure. The choice of the superframe structure affects the timing performance as well as the energy consumption in the cluster. Saving energy requires superframe structures with low duty cycles, whereas improving the timing performance requires higher duty cycles. A trade-off must be achieved by choosing the lowest duty cycle that still satisfies the timing constraints of the data flow. Hence, we investigate the following question.

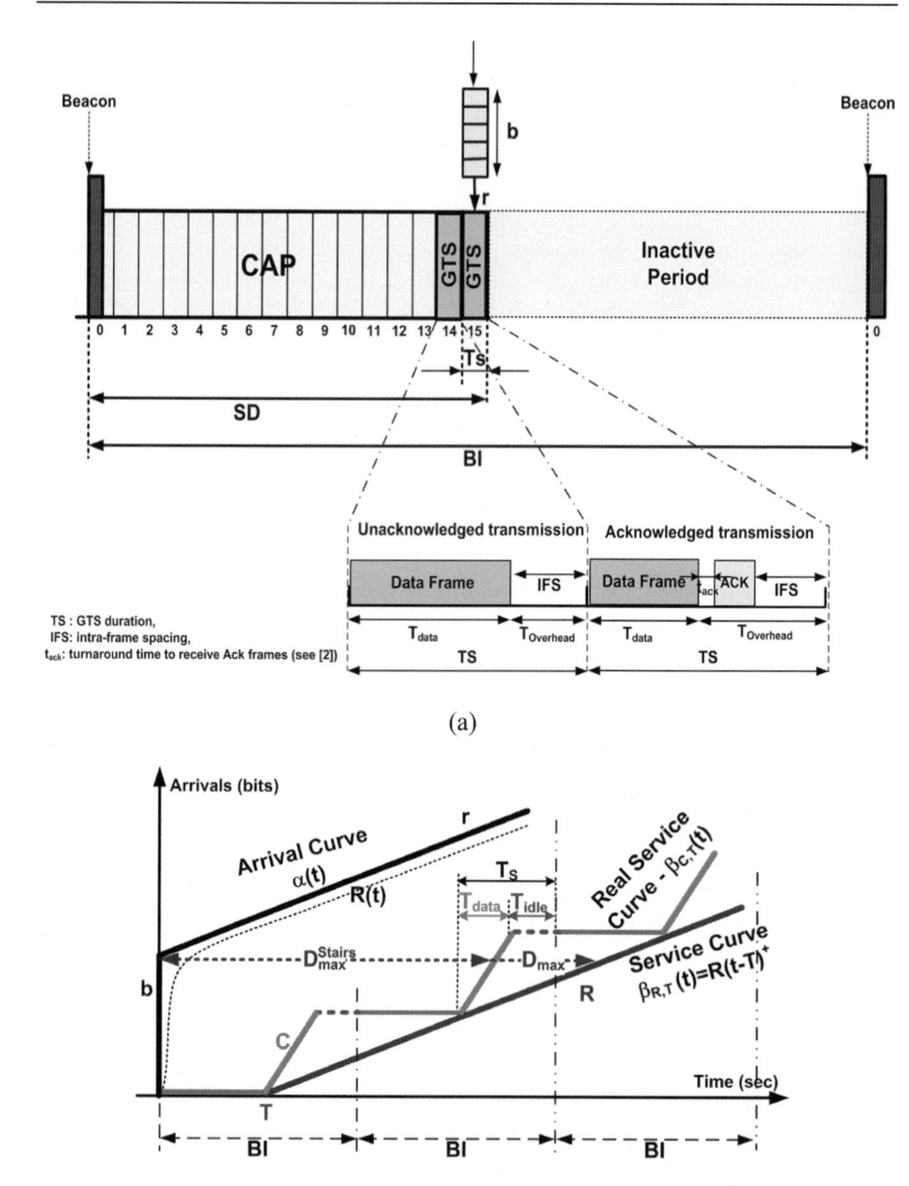

Fig. 2.12. The GTS allocation guaranteed service a) The GTS service time and transmission modes b) Service curve of a GTS allocation and the corresponding delay bound

Given a set of data flows within a PAN cluster, where each data flow has a maximum burst size b and a per-hop delay requirement D, what is the most efficient superframe structure, i.e. the combination of SO and BO, that satisfies the delay requirement of each data flow when it allocates one time slot GTS and minimises the energy consumption?

To evaluate the trade-off between energy consumption and delay guarantee, we derive the expression of the duty cycle as a function of the delay bound. For a given Superframe Order *SO* and Beacon Order *BO*, the duty cycle is defined as:

$$DC = \frac{SD}{BI} = 2^{SO-BO} \tag{2.7}$$

Where, *BI* is the Beacon Interval (Eq. (1)) and *SD* is the Superframe Duration (Eq. (2)). As a result, based on Eqs. (1), (2) and (5), we show that the duty cycle can be expressed as a function of the delay requirement as follows:

$$DC = \frac{SD}{D + \lambda \cdot SD} \cdot \left(\frac{b}{T_{data} \cdot C} + 1 \right) \quad \text{where, } \lambda = 1/16 \tag{2.8}$$

According to Eq. (7), the minimum valid value of the duty cycle is then:

$$DC = 2^{IO} \quad \text{where, } IO = \left\lfloor \log_2 \left(\frac{SD}{D - \lambda \cdot SD} \cdot \left(\frac{b}{T_{data} \cdot C} + 1 \right) \right) + 1 \right\rfloor \tag{2.9}$$

Eqs. (8) and (9) reflect the relation between energy consumption and the delay guarantee. From Eq. (9), it is possible to determine the lowest duty cycle that satisfies the delay requirement *D* of the flows.

2.9 GTS Performance Evaluation

Based on the aforementioned analysis, it is possible to evaluate the performance of a GTS allocation and capture the energy/delay trade-off. Fig. 2.13 presents the guaranteed bandwidth (*R* expressed in Eq. (4)) for different superframe orders with 100 % duty cycle (*SO* = *BO*) with unacknowledged transmissions. This figure is important to understand the effective bandwidth guaranteed while considering the impact of IFS. Note that if IFS = 0, we would have a constant guaranteed bandwidth, independently of *SO* values, equal to *Ts.C*/*BI* = 15.625 kbps for 100% duty cycle. We observe that the guaranteed bandwidth is slightly above 13 kbps, except for low superframe orders. The guaranteed bandwidth for *SO* = 0 and *SO* = 1 is relatively low compared to the others, due to a more important impact of IFS, since the time slot durations are too small for sending higher amounts of data. Our problem is to evaluate the impact of the delay bound on the duty cycle for a given superframe order *SO* and a given burst *b*. Fig. 2.14 shows the variation of the duty cycle as a function of the delay bound for different values of *SO*. The burst size is equal to 200 bits.

Observe in Fig. 2.14 that decreasing the delay requirement does not automatically increase the duty cycle. For instance, delay values in the range of [600,

1000] ms have the same 6.25% duty cycle for $SO = 0$. This fact is due to the slotted behaviour of the superframe structure defined in Eqs. (1) and (2). Hence, in some cases, relaxing the delay requirement will not automatically lead to a lower duty cycle for some IEEE 802.15.4 superframes. It is also observed from Fig. 2.14 that the number of possible superframe structure configurations (alternatives for *BO* and *SO*) increases with the delay. Hence, for low delay requirements, only the lower superframe orders (*for low burst size*) can meet these delay bounds, if it is possible, due to the increased latency for large *SO* values (that also leads to larger *BO* values).

Another interesting problem is to determine the adequate superframe order reducing the duty cycle and still meeting a delay bound *D* for a given burst *b*. Fig. 2.15 shows the variation of the duty cycle as a function of the superframe order for different values of the burst size. The delay bound requirement is assumed to be 3 seconds. Observe that for relatively low burst sizes (0.1 kbits, 1 kbits) the minimum duty cycle required to satisfy the delay bound increases with the superframe order. For a burst size equal to 10 kbits, there is no advantage of using low superframe orders for $SO \in \{0, 1, 2, 3, 4\}$. The duty cycle remains the same, since lower superframe orders have lower latencies, whereas higher superframe orders provide higher guaranteed bandwidths.

However, for a burst size $b = 35$ kbps, only superframe orders $SO \in \{2, 3, 4\}$ can satisfy the delay bound of 3 s, with a full duty cycle. This is because the guaranteed bandwidth has the most significant impact on the delay bound.

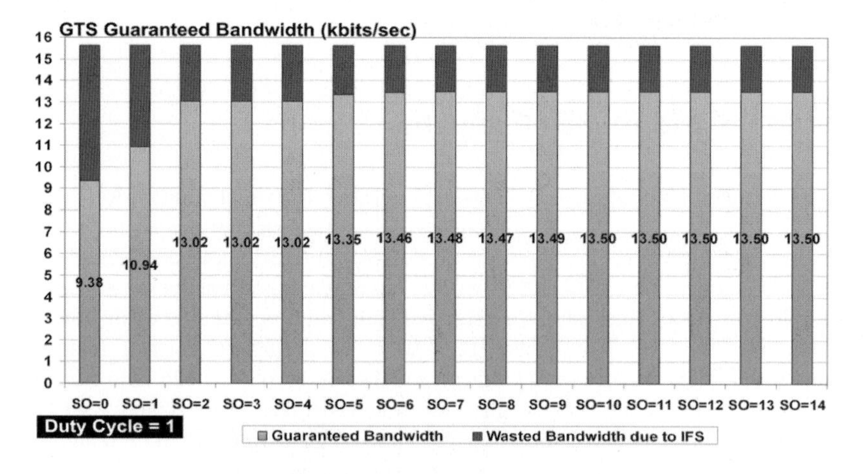

Fig. 2.13. Guaranteed bandwidth (kbps) per one GTS allocation

2.9.1 Performance Limits of the Slotted CSMA/CA Mechanism

The slotted mechanism has attracted more attention from the research community as compared to the non-slotted version. This is essentially due to two reasons. First, the beacon-enabled mode has more interesting features as compared to the non beacon-enabled mode, such as providing synchronisation services using beacon transmissions, and optionally a Contention Free Period (CFP) using the GTS mechanism. Second, in contrast to the non-slotted version, the slotted CSMA/CA mechanism has particular characteristics that are different from other well-known CSMA/CA schemes (e.g. DCF in IEEE 802.11) due to its slotted nature and its distinctive backoff algorithm using two *Clear Channel Assessments* (CCA).

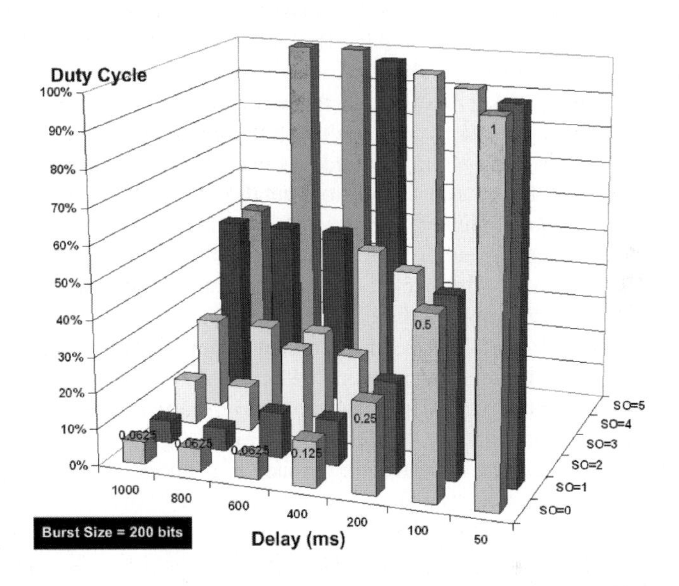

Fig. 2.14. Duty cycle versus delay bound

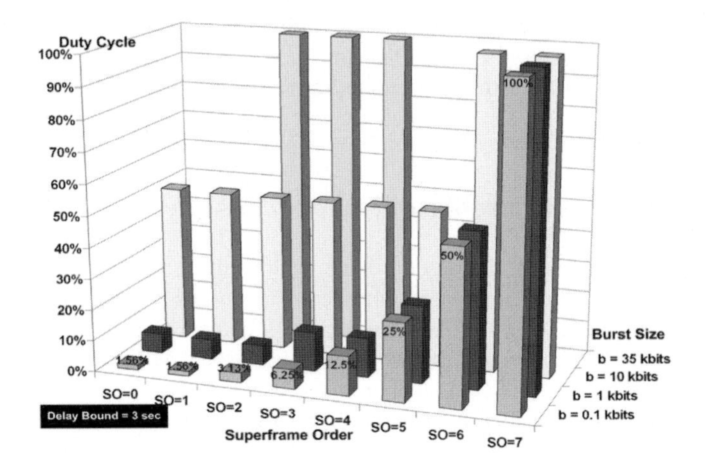

Fig. 2.15. Duty cycle versus superframe order

The performance of the slotted CSMA/CA mechanism in IEEE 802.15.4 was recently evaluated using discrete time Markov chain models (Mišic and Mišic 2005a, Mišic *et al.* 2005b, Park *et al.* 2005, Pollin *et al.* 2005). Those works provided analytic models of the slotted CSMA/CA mechanism in both saturation and non saturation modes, and also provided steady state solutions. These analytical models are interesting for capturing the behaviour of the protocol in terms of throughput and access delays. Other research works focused on improved schemes for slotted CSMA/CA to achieve higher efficiency in terms of power consumption and (soft) delay guarantees. In (Kim *et al.* 2005a), the authors proposed a new scheme for slotted CSMA/CA to enable fast delivery of high priority packets in emergency situations, using a priority toning strategy. The idea is: nodes that have high priority information to be transmitted must send a tone signal just before the beacon transmission. If the tone signal is detected by the PAN Coordinator, an emergency notification is conveyed in the beacon frame, which alerts other nodes with no urgent messages to defer their transmissions by some amount of time, in order to privilege high priority packet transmissions at the beginning of the CAP. This solution seems to improve the responsiveness of high priority packets in IEEE 802.15.4 slotted CSMA/CA, but requires a non-negligible change to the IEEE 802.15.4 MAC protocol to support the priority toning strategy. In another work (Kim and Choi 2005b), the authors proposed a priority-based scheme of slotted CSMA/CA also for reducing the delivery delay of high priority packets. High priority packets are allowed to perform only one CCA operation instead of two. A frame tailoring strategy was proposed to avoid collisions between data frames and acknowledgment frames when only one CCA is performed before the transmission of high priority data frames. In addition, they combine the priority toning scheme proposed in (Kim *et al.* 2005a) with frame tailoring to define their improved slotted CSMA/CA scheme. Also here, a non-negligible modification of the standard is needed to implement this medium access technique.

One of the interesting challenges regarding the improvement of slotted CSMA/CA is to enhance this mechanism by adequately tuning its parameters without imposing changes to the protocol, to ensure backward compatibility with the standard.

2.9.2 Performance Evaluation of Slotted CSMA/CA

In (Koubâa *et al.* 2006c), the authors have evaluated the performance of slotted CSMA/CA using simulations, as a complementary work to the aforementioned analytical approaches. The simulation work itself, using a fine model, provides an added value to the theoretical work in (Mišic *et al.* 2005a, Mišic *et al.* 2005b, Park *et al.* 2005, Pollin *et al.* 2005). It also presents results without doing restrictive assumptions and taking into account some realistic features of the physical layer (propagation delays, fading, noise effect, etc.). In (Koubâa *et al.* 2006c), it has been considered a typical wireless sensor network in a (100 m x 100 m) surface with one PAN Coordinator and 100 identical nodes (randomly spread) generating Poisson distributed arrivals, with the same mean arrival rate. Note that the Poisson distribution is typically adopted by most simulation and analytical studies on CSMA/CA. The frame size is equal to 404 bits corresponding to 300 bits of data payload and 104 bits of the MAC header according to the standard. The PAN Coordinator periodically generates beacon frames according to the *BO* and *SO* parameters. Unless it is mentioned differently, *BO* and SO are both equal to 3. Throughout the analysis, it is always assumed that *SO* = *BO* (100% duty cycle). Hereafter, when it is mentioned that the superframe order changes, it means that the beacon order is also changed and satisfies the equality *BO* = *SO*.

In WSNs, data dissemination is typically based on the diffusion of sensory data to all neighbours using broadcast transmissions. Therefore, this study the authors consider unacknowledged transmissions, since broadcast transmissions do not use acknowledgements. In order to focus on the performance analysis of the slotted CSMA/CA algorithm, it is assumed that the network is fully connected, i.e. all nodes hear each other (no hidden-node problem). In this simulation study, two performance metrics was considered: (1) The Network Throughput (*S*) is the fraction of traffic correctly received by the network analyser (a device in promiscuous mode hearing all the traffic in the network) normalised to the overall capacity of the network (250 kbps). (2) The Average delay (*D*) is the average delay experienced by a data frame from the start of its generation by the application layer to the end of its reception by the analyser. Both metrics were evaluated as a function of the offered load *G*, defined as the global load generated by all node's application layers. The offered load *G* depends on the inter-arrival times of the flows, which are exponentially distributed (Poisson arrivals). First, Figs. 2.16 and 2.17 present the impact of *BO* and *SO* values on the network throughput and the average delay, respectively. Observe that, as expected, low *SO* values produce lower network throughput. This is basically due to two factors. First, the overhead of the beacon frame is more significant for lower *SO* values, since beacons are more frequent. Second, CCA deference is also more frequent in case of lower *SO* values, leading to more collisions at the start of each superframe.

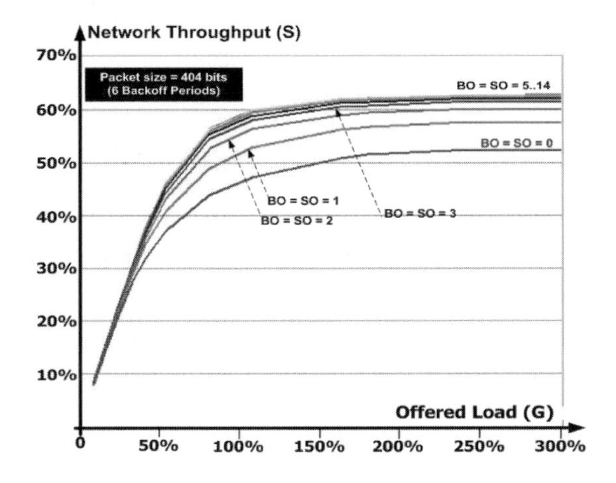

Fig. 2.16. The throughput as a function of the offered load for different (BO, SO) values

Note that the increase in the superframe order, from *SO* equal to 5 until 14, has a reduced impact on the network throughput. In fact, for high *SO* values (≥ 5), the probability of deference is quite low, which reduces the amount of collisions due to simultaneous CCA deference in multiple nodes, and thus leads to higher network throughputs. Fig. 2.17 shows that the average delays significantly increase with *SO* for a given offered load *G* higher than 50 %, as explained next. In fact, for low *SO* values, the high probability of CCA deference results in having more frequent collisions of data frames at the beginning of a new superframe. Hence, the backoff delays will not increase too much due to this frequent collision in case of low superframe orders. However, for high superframe orders the backoff algorithm will be less exposed to this problem, and then nodes will go into additional and higher backoff delays, since the backoff exponent is increased each time the channel is sensed as busy.

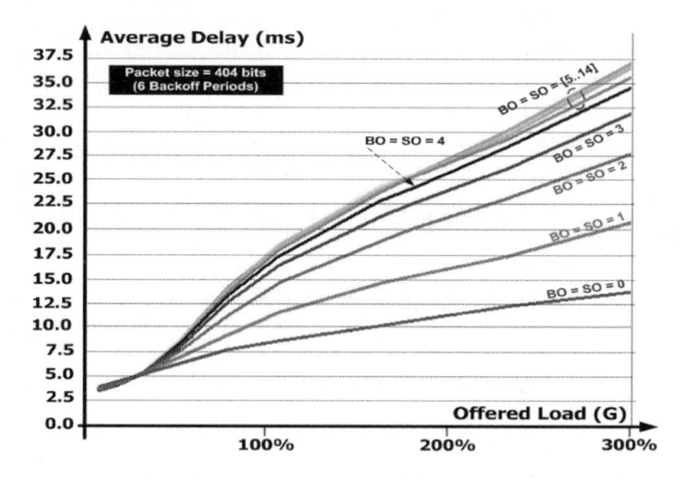

Fig. 2.17. The average delay as a function of the offered load for different (BO, SO) values

Second, Figs. 2.18 and 2.19 present the impact of the initialisation value of the backoff exponent *macMinBE* on the network throughput and on the average delay, respectively. The throughput would be improved with higher *macMinBE* since the backoff interval would be larger. However, this is not the case in this example. This result is due to the backoff algorithm behaviour of slotted CSMA/CA. In fact, for a given *macMinBE*, the interval from which the backoff delay is randomly generated at the first iteration is $[0, 2^{macMinBE} -1]$. Independently from *macMinBE*, the lower limit of the backoff delay interval is always 0 while the upper limit will be incremented each time the channel is sensed busy. Since the number of nodes is high (100 nodes), the probability that a medium is busy is also high, which leads to increasing *BE* for improved collision avoidance in the next iterations. *BE* cannot exceed *aMaxBE* = 5 and this value is reached by the competing nodes at most after 5 transmissions from other nodes. The backoff interval will tend to [0, 31] in all remaining nodes waiting for medium access and, as a result, the backoff delay distribution will not depend too much on the initialisation value of *macMinBE*.

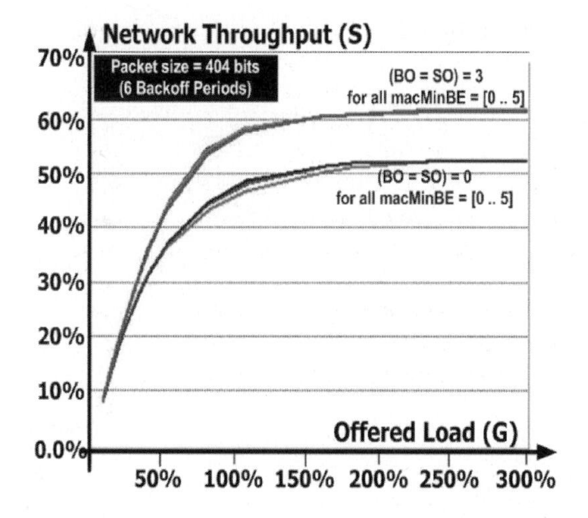

Fig. 2.18. The throughput as a function of the offered load for different *macMinBE* values

On the other hand, observe that the average delay increases with *macMinBE* for a given offered load (Fig. 2.19). Lower *macMinBE* values provide lower average delays with the same network throughputs. This is because the average backoff delays are higher for large [0, 2^{BE} -1] intervals. Observe that for low offered loads ($G < 50$ %), the variance of the average delays for different *macMinBE* is not significant (around 10 ms from *macMinBE* from 0 to 5). However, for high offered loads $G \geq 50$ %, the impact of *macMinBE* is significantly more visible. For instance, for $G = 300\%$, the average delay is higher than 110 ms for *macMinBE* = 5, whereas it is does not exceed 8 ms in case of *macMinBE* = 0.

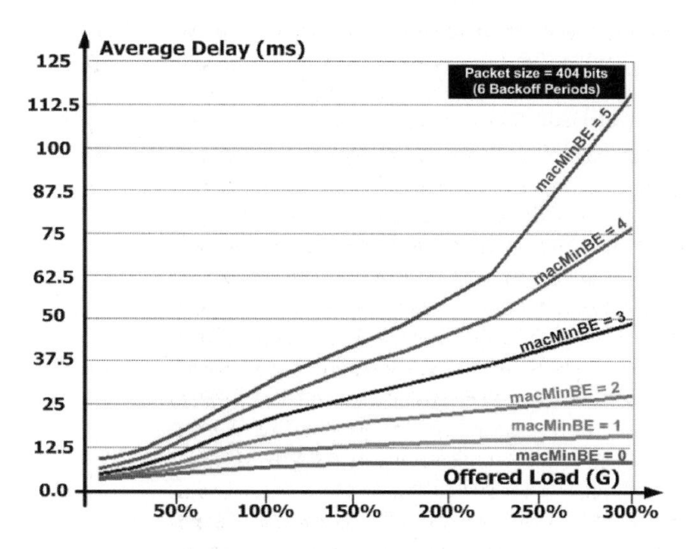

Fig. 2.19. The average delay as a function of the offered load for different *macMinBE*

2.10 Research Trends

Since its proposal in 2003, the IEEE 802.15.4 protocol has been attracting more and more research work envisaging its deployment in wireless sensor networks. It is expected that many commercial manufacturers of wireless sensor technologies will shift towards this standard solution due to its low-cost and improved performance. However, there are still some challenges that must be addressed to tune this COTS technology for WSN applications. In this section, we present some hints on potential research trends and future challenges for the use of IEEE 802.15.4 in wireless sensor networks.

One of the challenges is the deployment of the beacon-enabled mode in a multi-hop network with several Coordinators. The main problem is to find adequate synchronisation schemes for different Coordinators in the same range, to avoid beacon collisions. In such kind of networks, beacon frames of a given Coordinator may collide with beacon frames of other Coordinators (which lead to permanent collisions) or data/control frames (which lead to occasional collisions). Loosing beacon frames will lead to synchronisation failures of associated nodes in the network. This problem was analysed in (Ha *et al.* 2005), where the authors derived the probability of beacon frame collisions with other beacons, and the probability of beacon frame collisions with data/control frames. The analytical results and experiments show that, when the starting time of the first beacon frame generated by each Coordinator is uniformly distributed, multi-hop beacon enabled networks are feasible for *BO* values higher than 1, and for evenly distributed Coordinators. In such conditions, they show that synchronisation failures may be less that 1%. An interesting extension of this work is to propose improved synchronisation schemes for several Coordinators operating with beacon orders equal to 0 and 1, since these modes provides better performances in terms of delay guarantees, as it has been shown. In addition, it is important to propose a deterministic synchronisation scheme that ensures that no collisions will disturb beacon frame transmissions to reach 0 % of synchronisation failure.

Another open issue is to resolve the hidden-node problem in IEEE 802.15.4, since its MAC protocol does not use any RTS/CTS mechanism. This problem may be serious in large-scale wireless sensor networks, which typically use broadcast transmissions to disseminate data. In such conditions, the hidden-node problem would have a negative impact on the network throughput. This issue was addressed in (Hwang *et al.* 2005), in which the authors proposed a grouping strategy to solve the IEEE 802.15.4 hidden-node problem without using the RTS/CTS mechanism. This strategy groups nodes according to their hidden-node relationship such that all nodes in each group are not hidden to each other. Then, this technique allocates to each group guaranteed time slots, in which slotted CSMA/CA is used by different nodes in the group to access the channel. The PAN Coordinator is in charge of detecting the hidden-node situation and performing grouping procedure if necessary.

Concerning time-sensitive sensor networks, the improvement of the GTS mechanism is still an open issue. In fact, the protocol only supports explicit GTS allocations, i.e. a node allocates a number of time slots in each superframe for exclusive use. The limitation of the explicit GTS allocation is that GTS resources may quickly disappear, since a maximum of seven GTSs can be allocated in each superframe, preventing other nodes to benefit from guaranteed service. Moreover, the GTSs may be only partially used, resulting in a wasted bandwidth. To overcome this limitation, one possible solution is to share the same GTS between multiple nodes, instead of being exclusively dedicated to one node, if a certain schedule that satisfies the requirements of all requesting nodes exists. Sharing a GTS by several nodes means that the time slots of this GTS are dynamically allocated to different nodes in each superframe, according to a given schedule.

Another important issue regarding the deployment of IEEE 802.15.4 is the adequacy of the ZigBee routing layer for wireless network applications. In fact, the IEEE 802.15.4 protocol is intended to operate on the bottom of the ZigBee network/application stack. It is therefore quite important to analyse the suitability of ZigBee solutions, namely in terms of routing protocols and network services, with the requirements of wireless sensor networks, which are typically data centric contrarily to traditional wireless networks. One of the problems is that the full ZigBee stack seems to be heavy to be implemented in tiny sensor devices. It is also interesting to propose localisation mechanisms based on the IEEE 802.15.4 MAC services, since most WSN applications rely on location-awareness, which is not provided in the IEEE 802.15.4/ZigBee stack. In conclusion, we believe that the IEEE 802.15.4 protocol is a promising enabling technology for low-cost, low-rate and low-power consumption WSNs due to its flexibility to fulfill different requirements of various application patterns, when adequately tuning its parameters.

2.11 References

Akyildiz IF, Su W, Sankarasubramaniam Y, and Cayirci E (2002) Wireless sensor networks: A survey. Journal of Computer Networks, vol. 38, pp. 393-422.

Bacco GD, Melodia T, and Cuomo F (2004) A MAC protocol for delay-bounded applications in wireless sensor networks. In Proc. of the Third Annual Mediterranean Ad Hoc Networking Workshop (MedHoc'04), June 27-30.

Bandyopadhyay S, and Coyle EJ (2003) An energy efficient hierarchical clustering algorithm for wireless sensor networks. In Proc. of the Twenty-Second Annual Joint Conference of the IEEE Computer and Communications Societies (INFOCOM 2003), vol. 3, 30 March - 3 April

Bharghavan V (1998) Performance evaluation of algorithms for wireless medium access. In Proc. of the Int. Computer Performance and Dependability Symp. (IPDS'98), Durham, North Carolina (USA), Sep. 1998

CC2420 datasheet (2004) Chipcon Company, http://www.chipcon.com

Crossbow Tech. (2005) Avoiding RF interference between WiFi and ZigBee, Crossbow Technical Report.

Ha J, Kwon WH, Kim JJ, Kim YH, and Shin YH (2005) Feasibility analysis and implementation of the IEEE 802.15.4 multi-hop beacon-enabled network. In Proc. of the 15th Joint Conf. on Communications & Info, Jun. 2005.

He T, Stankovic JA, Lu C, Abdelzaher T (2003) SPEED: A stateless protocol for real-time communication in sensor networks. In Proc. of the 23rd IEEE Int. Conf. on Distributed Computing Systems (ICDCS'03): 46-55.

Heinzelman W, Chandrakasan A, and Balakrishnan H (2000) Energy-efficient communication protocols for wireless microsensor networks. In Proc. of the Hawaii Int. Conf. Systems Sciences, vol. 2, pp. 10, Jan. 2000.

Howitt I, and Gutierrez JA (2003) IEEE 802.15.4 low rate-wireless personal area network coexistence issues. IEEE Wireless Communications and Networking Conf. (WCNC'03), vol. 3, pp. 1481- 1486, 16-20 Mar. 2003.

Hwang L, Sheu ST, Shih YY, and Cheng YC (2005) Grouping strategy for solving hidden terminal problem in IEEE 802.15.4 LR-WPAN. In Proc. of the 1st Int. Conf. on Wireless Internet (WICON'05), pp. 26-32.

IEEE 802.11 Specification (1999) IEEE Standards for Information Technology, Telecommunications and Information Exchange between Systems - Local and Metropolitan Area Network - Specific Requirements - Part 11: Wireless LAN Medium Access Control (MAC) and Physical Layer (PHY) Specifications.

IEEE 802.11e Standard (2001) Draft Supplement to STANDARD FOR Telecommunications and Information Exchange Between Systems - LAN/MAN Specific Requirements - Part 11: Wireless Medium Access Control (MAC) and Physical Layer (PHY) specifications: Medium Access Control (MAC) Enhancements for Quality of Service (QoS), IEEE 802.11e/D2.0, Nov. 2001.

IEEE 802.15.4 Standard (2003) Part 15.4: Wireless medium access control (MAC) and physical layer (PHY) specifications for Low-Rate Wireless Personal Area Networks (LR-WPANs), IEEE Standard for Information Technology, IEEE-SA Standards Board, 2003.

Kim T, Lee D, Ahn J, and Choi S (2005a) Priority toning strategy for fast emergency notification in IEEE 802.15.4 LR-WPAN, In Proc. JCCI'2005, April 2005.

Kim T, and Choi S (2005b) Priority-based delay mitigation for event-monitoring IEEE 802.15.4 LR-WPANs, In IEEE Communications Letters, Nov. 2005.

Koubâa A, Alves M, and Tovar E (2005a) IEEE 802.15.4 for wireless sensor networks: A technical overview. IPP-HURRAY Technical Report, HURRAY-TR-050702, July 2005.

Koubâa A, and Alves M (2005b) A two-tiered architecture for real-time communications in large-scale wireless sensor networks: research challenges. In Proc. of 17th Euromicro Conference on Real-Time Systems (ECRTS'05), WiP Session, Palma de Mallorca (Spain), 5-7 July.

Koubâa A, Alves M, and Tovar E (2005c) A real-time sensor network architecture with power-efficient resource reservation. IPP-HURRAY! Technical Report, HURRAY-TR-050702, July 2005.

Koubâa A, Alves M, and Tovar E (2006a) GTS allocation analysis in IEEE 802.15.4 for real-time wireless sensor networks. In Proc. of the 14th Int. Workshop on Parallel and Distributed Real-Time Systems (WPDRTS 2006), invited paper in the special track on Wireless Sensor Networks, Apr. 2006.

Koubâa A, Alves M, and Tovar E (2006b) Energy/Delay trade-off of the GTS allocation mechanism in IEEE 802.15.4 for wireless sensor networks. IPP-HURRAY Technical Report, HURRAY-TR-060103, (Submitted), Jan. 2006

Koubâa A, Alves M, Tovar E, and Song YQ (2006c) On the performance limits of slotted CSMA/CA in IEEE 802.15.4 for broadcast transmissions in wireless sensor networks. Hurray Technical Report, (submitted)

Lu C, Blum BM, Abdelzaher T, Stankovic J, and He T (2002) RAP: A real-time communication architecture for large-scale wireless sensor networks. In IEEE Real-Time and Embedded Technology and Applications (RTAS 2002), Sept. 2002

MICAz datasheet (2004), http://www.xbow.com

Mišic J, and Mišic B (2005a) Access delay and throughput for uplink transmissions in IEEE 802.15.4 PAN. Elsevier Computer Communications Journal, vol. 28, No.10, pp. 1152-1166, Jun. 2005

Mišic J, Shafi S, and Mišic B (2005b) The impact of MAC parameters on the performance of 802.15.4 PAN. Elsevier Ad hoc Networks Journal, vol. 3, N. 5, pp.509–528, 2005.

Park TR, Kim T, Choi J, Choi S, and Kwon W (2005) Throughput and energy consumption analysis of IEEE 802.15.4 slotted CSMA/CA, IEEE Electronics Letters, vol. 41, issue 18, pp. 1017-1019, Sept. 2005

Pollin S, et al. (2005) Performance analysis of slotted IEEE 802.15.4 medium access layer. Technical Report, DAWN Project, Sept. http://www.soe.ucsc.edu/research/ccrg/DAWN/papers/ZigBee_MACvPV.pdf

Shin SY, Choi S, Park HS, and Kwon WH (2005) Packet error-rate analysis of IEEE 802.15.4 under IEEE 802.11b interference. In Proc. of Conf. on Wired/Wireless Internet Communications (WWIC'2005), Springer LNCS, vol. 3510, Xanthi (Greece), 11-13 May, 2005

Stankovic J, Abdelzaher T, Lu C, Sha L, and Hou J (2003) Real-Time communication and coordination in embedded sensor networks. Proc. of IEEE, vol. 91, No. 7, pp.1002-1022.

Ye W, Heidemann J, and Estrin D, (2004) Medium access control with coordinated adaptive sleeping for wireless sensor networks. IEEE/ACM Transactions on Networking, vol. 12 , Iss. 3, pp. 493 - 506 Apr. 2004.

Zheng L, Dadej A, and Gordon S (2003) Fairness of IEEE 802.11 distributed coordination function for multimedia applications. In Proc. of the 7th Int. Conf. on DSP and Communications and 2nd Workshop on the Internet, Telecommunications and Signal Processing, Coolangatta (Australia), DSPCS 2003 & WITSP 2003, pp. 404-409

ZigBee Alliance (2005), http://www.caba.org/standard/zigbee.html. ZigBee specification

3 Service Oriented Sensor Web

Xingchen Chu and Rajkumar Buyya

Grid Computing and Distributed Systems Laboratory, Dept. of Computer Science and Software Engineering, The University of Melbourne, Australia

3.1 Introduction

Due to the rapid development of sensor technology, current sensor nodes are much more sophisticated in terms of CPU (Central Processing Unit), memory, and wireless transceiver. Sensor networks are long running computing systems that consist of a collection of sensing nodes working together to collect information about, for instance, light, temperature, images, and other relevant data according to specific applications. Wireless sensor networks (WSNs) have been attracting a lot of attention from both academic and industrial communities around the world. The ability of the sensor networks to collect information accurately and reliably enables building both real-time detection and early warning systems. In addition, it allows rapid coordinate responses to threats such as bushfires, tsunamis, earthquakes, and other crisis situations.

However, the heterogeneous features of sensors and sensor networks turn the efficient collection and analysis of the information generated by various sensor nodes, into a rather challenging task. The main reasons for that are the lack of both uniform operations and a standard representation for sensor data that can be used by diverse sensor applications. There exists no means to achieving resource reallocation and resource sharing among applications as the deployment and usage of the resources has been tightly coupled with the specific location, sensor application, and devices used.

The Service Oriented Architecture (SOA) provides an approach to describe, discover, and invoke services from heterogeneous platforms using XML and SOAP standards. The term 'service' not only represents a software system but also refers to hardware and any devices that can be used by human beings. A service may be an online ticket booking system, a legacy database application, a laser printer, a single sensor, or even an entire network infrastructure. Bringing the idea of SOA to sensors and sensor networks is a very important step forward to presenting the sensors as reusable resources which can be discoverable, accessible and where applicable, controllable via the World Wide Web. Furthermore, it is also possible to link distributed resources located across different organisations, countries, or regions thus creating the illusion of a sensor-grid, which enables the essential strengths, and characteristics of a computational grid.

Fig. 3.1 demonstrates an abstract vision of the Sensor Web, which is the combination of SOA, grid computing and sensor networks. Various sensors and sensor nodes form a web view and are treated as available services to all the users who access the Web. Sensor Web brings the heterogeneous sensors into an integrated and uniform platform supporting dynamic discovery and access. A sample scenario would be the client (may be the researchers or other software, model and workflow system), who wants to utilise the information collected by the deployed sensors on the target application, such as weather forecast, tsunami or pollution detection. The client may query the entire sensor web and get the response either from real-time sensors that have been registered in the web or existing data from a remote database. The clients are not aware of where the real sensors are and what operations they may have, although they are required to set parameters for their plan and invoke the service (similar to when people perform a search on Google, filling in the search field and clicking the search button). The primary goal of the Sensor Web is to offer reliable and accessible services to the end-users. In other words, it provides the middleware infrastructure and the programming environment for creating, accessing, and utilizing sensor services through the Web.

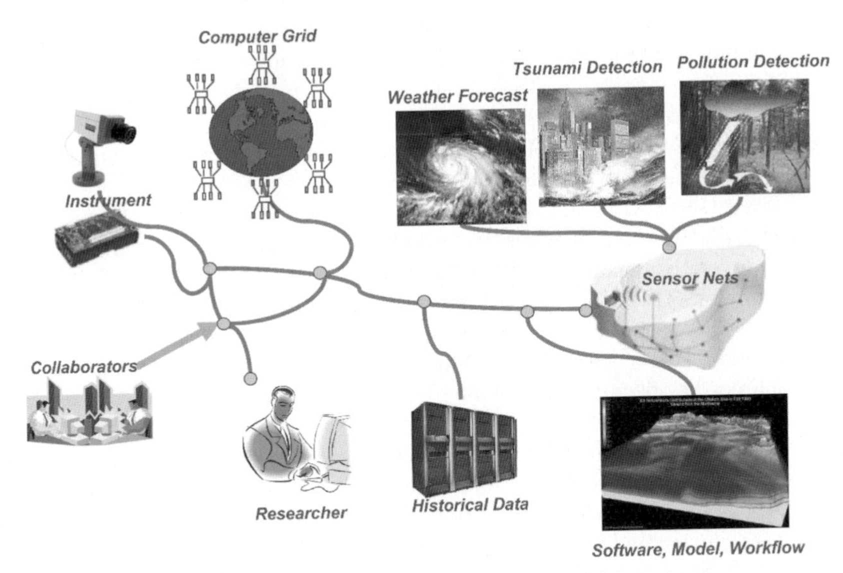

Fig. 3.1. Vision of the Sensor Web

The remainder of this chapter is organised as follows. Related work on sensor middleware support, sensor-grid, and sensor web is described in Section 2. Section 3 details the emerging standard of the Sensor Web: Sensor Web Enablement. Section 4 describes OSWA, a service oriented sensor web architecture, and the design and implementation of its core services. Evaluation of applying the middleware to a simple temperature monitoring sensor application is discussed in Section 5. This chapter concludes with the summary and the future work.

3.2 Related Work

A lot of effort has been invested in building middleware support for making the development of sensor applications simpler and faster. Impala (Liu and Martonosi, 2003) designed for the ZebraNet project, considers the application itself while adopting mobile code techniques to upgrade functions on remote sensors. The key to the energy efficiency provided by Impala is making sensor node applications as modular as possible, thus imposing small updates that require little transmission energy. MiLAN (Heinzelman et al., 2004) is an architecture that extends the network protocol stack and allows network specific plug-ins to convert MiLAN commands into protocol-specific commands. Bonnet et al., 2000 implemented Cougar, a query-based database interface that uses a SQL-like language to gather information from wireless sensor networks. However, most of these efforts concentrate on creating protocols and are designed to ensure the efficient use of wireless sensor networks. In contrast to these middleware, Mires (Soutoo et al., 2004) is a message-oriented middleware that is placed on top of the operating system, encapsulates its interfaces and provides higher-level services to the Node Application. The main component of Mires is a publish/subscribe service that intermediates communication between middleware services, which might be used as the foundation of Sensor Web middleware.

Besides middleware support for the sensor applications, integrating sensor networks with grid computing into a sensor grid is also quite important. Tham and Buyya (Tham and Buyya 2005) outlined a vision of sensor-grid computing and described some early work in sensor grid computing by giving examples of a possible implementation of distributed information fusion and distributed autonomous decision-making algorithms. Discussion about the research challenges needed to be overcome before the vision becomes a reality have also been presented. A data-collection-network approach to address many of the technical problems of integrating resource-constrained wireless sensors into traditional grid applications have been suggested by Gaynor et al. 2004. This approach is in the form of a network infrastructure, called Hourglass that can provide a grid API to a heterogeneous group of sensors. Those, in turn, provide fine-grained access to sensor data with OSGA standards. Another sensor grid integration methodology introduced by Ghanem et al., 2004 utilised the grid services to encompass high throughput sensors, and in effect make each sensor a grid service. The service can be published in a registry by using standard methods and then made available to other users.

Nickerson et al., 2005 described a Sensor Web Language (SWL) for mesh architecture, which provides a more robust environment to deploy, maintain and operate sensor networks. As they stated, greater flexibility, more reliable operation and machinery to better support self-diagnosis and inference with sensor data has been achieved with the mesh architecture support in SWL. At the GeoICT Lab of York University, an open geospatial information infrastructure for Sensor Web, named GeoSWIFT, has been presented, which is built on the OpenGIS standards. According to Tao et al., 2004, XML messaging technology has been developed, serving as a gateway that integrates and fuses observations from heterogeneous spatial enabled sensors. The IrisNet architecture at Intel Research, introduced by Gibbons et al., 2003, is a software infrastructure that supports the central tasks common to collect, filter and combine sensor feeds and perform distributed queries. There are two-tiers of IrisNet architecture including sensing agents that provide a generic data acquisition interface to access sensors, and organizing agents that are the nodes implementing the distributed database. Finally, the most important effort that has been made to Sensor Web is the Sensor Web Enablement (SWE) introduced by Reichardt, 2005. SWE consists of a set of standard services to build a unique and revolutionary framework for discovering and interacting with web-connected sensors and for assembling and utilizing sensor networks on the Web. The following section of this chapter discusses SWE standards in more in detail.

3.3 Standard: OCG Sensor Web Enablement

Many sensor network applications have been successfully developed and deployed around the world. Some concrete examples include:

- Great Duck Island Application: as Mainwaring et al. 2002 stated, 32 motes are placed in the areas of interest, and they are grouped into sensor patches to transmit sensor data to a gateway, which is responsible for forwarding the data from the sensor patch to a remote base station. The base station then provides data logging and replicates the data every 15 minutes to a Postgress database in Berkeley over a satellite link.
- Cane-toad Monitoring Application: two prototypes of wireless sensor networks have been set up, which can recognise vocalisations of at maximum 9 frog species in Northern Australia. Besides monitoring the frogs, the researchers also plan to monitor breeding populations of endangered birds, according to Hu et al. 2003.
- Soil Moisture Monitoring Application: Cardell-Oliver et al. 2004 presents a prototype sensor network for soil moisture monitoring that has been deployed in Pinjar, located in north of Perth, WA. The data is gathered by their reactive sensor network in Pinjar, and sent back to a database in real time using a SOAP Web Services.

However, none of these applications address the ability for interoperability, which means that users cannot easily integrate the information into their own applications (the Soil moisture monitoring application utilises the Web Services only for remote database operations). Moreover, the lack of semantics for the sensors that they have used makes it impossible to build a uniform Web registry to discover and access those sensors. In addition, the internal information is tightly coupled with the specific application rather than making use of standard data representations, which may restrict the ability of mining and analyzing the useful information.

Imagine hundreds of in-site or remote weather sensors providing real-time measurements of current wind and temperature conditions for multiple metropolitan regions. A weather forecast application may request and present the information directly to end-users or other data acquisition components. A collection of Web-based services may be involved in order to maintain a registry of available sensors and their features. Also consider that the same Web technology standard for describing the sensors, outputs, platforms, locations, and control parameters is in use beyond the boundaries of regions or countries. This enables the interoperability necessary for cross-organisation activities, and it provides a big opportunity in the market for customers to get a better, faster and cheaper service. This drives the Open Geospatial Consortium (OGC) to develop the geospatial standards that will make the "open sensor web" vision a reality (http://www.geoplace.com/ uploads/FeatureArticle/0412ee.asp). As the Sensor Web Enablement (SWE) becomes the de facto standard regarding Sensor Web development, understanding SWE is crucial for both researchers and developers. In general, SWE is the standard developed by OGC that encompasses specifications for interfaces, protocols and encodings that enable discovering, accessing, obtaining sensor data as well as sensor-processing services. The following are the primary specifications for SWE:

1. Sensor Model Language (SensorML): Information model and XML encodings that describe either single sensor or sensor platform in regard to discover, query and control sensors.
2. Observation and Measurement (O&M): Information model and XML encodings for observations and measurement.
3. Sensor Collection Service (SCS): Service to fetch observations, which conforms to the Observations and Measurement information model, from a single sensor or a collection of sensors. It is also used to describe the sensors and sensor platforms by utilizing SensorML.
4. Sensor Planning Service (SPS): Service to help users build feasible sensor collection plan and to schedule requests for sensors and sensor platforms.
5. Web Notification Service (WNS): Service to manage client session and notify the client about the outcome of her requested service using various communication protocols.

As Reichardt 2005 stated, the purpose of SWE is to make all types of web-resident sensors, instruments and imaging devices, as well as repositories of sensor data, discoverable, accessible and, where applicable, controllable via the World Wide Web. In other words, the goal is to enable the creation of Web-based sensor networks. Fig. 3.2 demonstrates a typical collaboration between services and data encodings of SWE.

3.3.1 SensorML

Web-enabled sensors provide the technology to achieve rapid access to various kinds of information from the environment. Presenting sensor information in standard formats enables integration, analysis and creation of various data "views" that are more meaningful to the end user and to the computing system which processes this information. Moreover, a uniform encoding benefits the integration of heterogeneous sensors and sensor platforms as it provides an integrated and standard view to the client. The Sensor Model Language (SensorML) is a new XML encoding scheme that may make it possible for clients to remotely discover, access, and use real-time data obtained directly from various Web-resident sensors. SensorML describes the geometric, dynamic, and observational features of sensors of all kinds.

SensorML is a key component for enabling autonomous and intelligent sensor webs, and provides the information needed for discovery of sensors, including sensor's capabilities, geo-location and taskability rather than a detailed description of the sensor hardware. Moreover, both in-site and remote sensors, on either static or dynamic platforms are supported by SensorML. Fig. 3.3 depicts the basic structure of SensorML. The information provided in SensorML includes the sensor name, type, and identification (identifiedAs); time, validity, or classification constraints of the description (documentConstrainedBy); a reference to the platform (attachedTo); the coordinate reference system definition (hasCRS); the location of the sensor (locatedUsing); the response semantics for geolocating samples (measures); the sensor operator and tasking services (operatedBy); metadata and history of the sensor (describedBy); and metadata and history of the document itself (documentedBy).

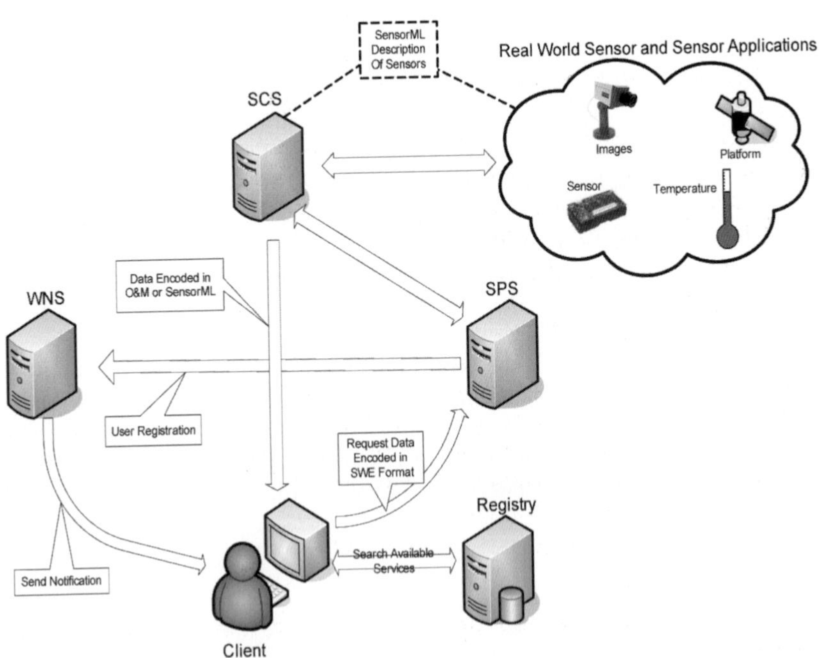

Fig. 3.2. A typical collaboration within Sensor Web enablement framework

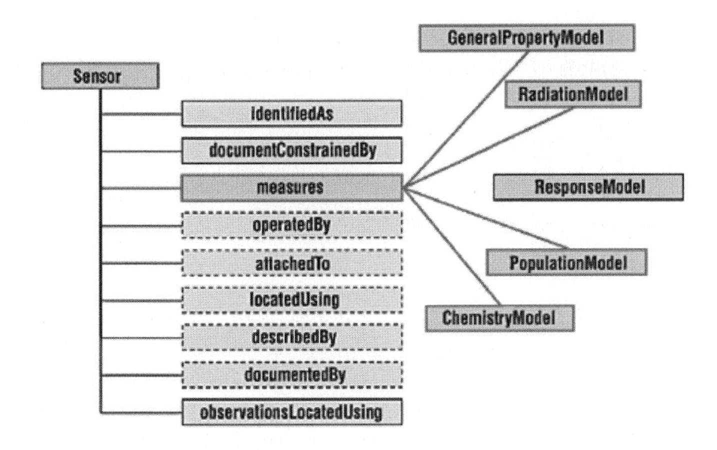

Fig. 3.3. High-level structure of SensorML (Reichardt, 2005)

Besides the importance of SensorML in SWE framework, SensorML itself is an independent standard rather than part of the SWE framework, which means that it can be used outside the scope of SWE. Other benefits of adopting SensorML include (i) enabling long-term archive of sensor data to be reprocessed and refined in the future and (ii) allowing the software system to process, analyse and perform a visual fusion of multiple sensors[1].

3.3.2 Observation and Measurement

Besides collaborating SensorML, which contains information about sensors and sensor platforms, SWE utilises Observation and Measurement (O&M)[2]. O&M is another standard information model and XML encoding that is important for Sensor Web to find universal applicability with web-based sensor networks. The O&M model is required specifically for the Sensor Collection Service and related components of OGC Sensor Web Enablement, which aims at defining terms used for measurements, and relationships between them.

As Fig. 3.4 indicates, the basic information provided by Observation includes the time of the event (timeStamp); the value of a procedure such as instrument or simulation (using); the identification of phenomenon being sampled (observable); the association of other features that are related to the Observation (relatedFeature); the common related features that have fixed role such as Station or Specimen (target); the quality indicators associated with the Observation (quality); the result of the Observation (resultOf); location information (location) and the meta-

[1] http://member.opengis.org/tc/archive/arch03/03-0005r2.pdf, Sensor Model Language IPR, OGC 03-005.

[2] http://portal.opengeospatial.org/files/index.php?artifact_id=1324, Observation & Measurement, OGC 03-022r3.

data description (metadataProperty). Moreover, the observation data can be either single or compound values that may contain a collection or an array of observations.

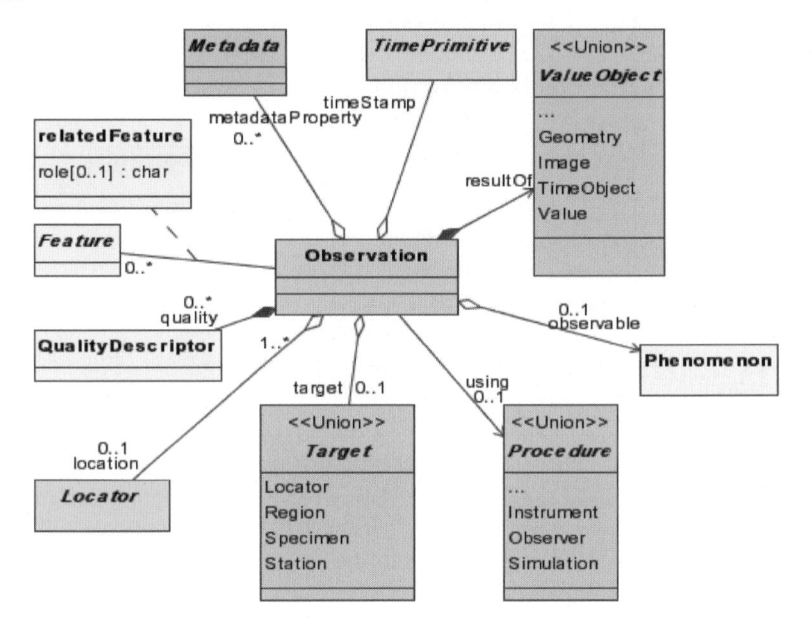

Fig. 3.4. Basic Observation structure[3]

3.3.3 SWE Services

SWE not only utilises the information model and encoding like SensorML and Observation and Measurements (O&M), but also defines several standard services that can be used to collaborate with sensor networks in order to obtain data. Currently, the SWE contains three service specifications including Sensor Collection Service (SCS), Sensor Planning Service (SPS) and Web Notification Service (WNS). As the SWE is still evolving, new services will be developed to satisfy emerging requirements of Sensor Web development. A new service called Sensor Alert Service has recently been introduced, which specifies how alert or "alarm" conditions are defined, detected and made available to interested users. Also, a new TransducerML[3] has also been defined, which is an XML based specification that describes how to capture and "time tag" sensor data. However, as these two specifications are still quite new, this chapter will only discuss the three well-known specifications in details.

One of the most important services is the Sensor Collection Service (SCS) which is used to fetch observations from a sensor or a constellation of sensors. It

[3] http://www.iriscorp.org/tml.html, Transducer Markup Language

plays a role of intermediary agent between a client and an observation repository or near real-time sensor channel. The getObservation method of SCS accepts queries from the client within certain range of time as input parameters and responses with XML data conformed to Observation & Measurement information model. The describeSensor and describePlatform methods are used to fetch the sensor's information based on SensorML. Each client that intends to invoke the Sensor Collection Service must strictly follow the SCS specification.

The Sensor Planning Service (SPS) is intended to provide a standard interface to handle asset management (AM) that identifies, uses and manages available information sources (sensors, sensor platforms) in order to meet information collection (client's collection request) goals. SPS plays a crucial role as a coordinator which is responsible for evaluating the feasibility of the collection request from the client and, if valid, submitting the request by querying the SCS about the Observation data. The DescribeCollectionRequest operation of SPS presents the information needed for the client's collection request. The GetFeasibility method of SPS accepts requests from the clients and makes a 'yes' or 'no' decision according to specified rules regarding to the feasibility of the collection. Clients can invoke the SubmitRequest method to actually schedule their requests and submit to the SCS once the GetFeasibility method responses with 'yes'. SPS also defines UpdateRequest, CancelRequest and GetStatus methods to manage and monitor the collection request made by users.

In general, the synchronous messaging mechanism is powerful enough to handle collections of in-situ sensors. However, observations that require evaluating collection feasibility or intermediate user notifications are not suitable for synchronous operations. The Web Notification Service (WNS) is an asynchronous messaging service, which is designed to fulfill the needs of supporting these complicated scenarios. Sending and receiving notifications are the major responsibilities of the WNS, which can utilise various communication protocols including HTTP POST, email, SMS, instant message, phone call, letter or fax. Besides, WNS also takes charge of user the management functionality that is used to register user and trace the user session over the entire process. Operations including doNotification, doCommunication and doReply are defined to conduct both one-way and two-way communication between users and services whereas registerUser handles user management, which allows registering users to receive further notifications.

3.4 Open Sensor Web

Open Sensor Web Architecture (OSWA) is an OGC Sensor Web Enablement standard compliant software infrastructure for providing service oriented based access to and management of sensors created by NICTA/Melbourne University. OSWA is a complete standards compliant platform for integration of sensor networks and emerging distributed computing platform such as SOA and Grid Computing. The integration has brought several benefits to the community. First, the heavy load of information processing can be moved from sensor networks to the backend distributed systems such as Grids. The separation is either saving a lot of energy and power of sensor networks just concentrating on sensing and sending information or accelerating the process for processing and fusing the huge amount of information by utilizing distributed systems. Moreover, individual sensor networks can be linked together as services, which can be registered, discovered and accessed by different clients using a uniform protocol. Moreover, as Tham and Buyya, 2005 stated, Grid-based sensor applications are capable of providing advanced services for smart-sensing by developing scenario-specific operators at runtime.

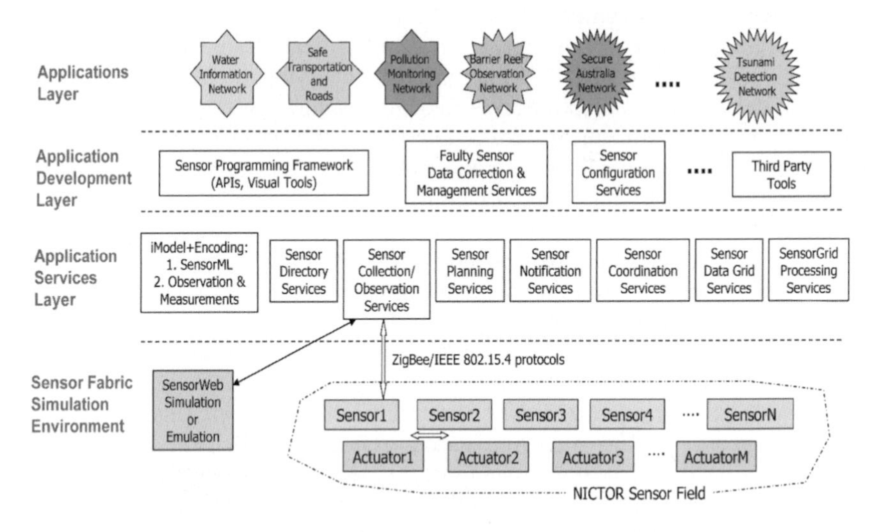

Fig. 3.5. High-level view of open sensor web architecture

The various components defined in OSWA are showed in Fig. 3.5. Four layers have been defined, namely Fabric, Services, Development and Application layers. Fundamental services are provided by low-level components whereas higher-level components provide tools for creating applications and management of the lifecycle of data captured through sensor networks. The OSWA based platform provides a number of sensor services such as:

- Sensor notification, collection and observation;
- Data collection, aggregation and archive;
- Sensor coordination and data processing;
- Faulty sensor data correction and management, and;
- Sensor configuration and directory service

Besides the core services derived from SWE, such as SCS, SPS and WNS, there are several other important services in the service layer. Sensor Directory Service provides the capability of storing and searching services and resources.

The SCS enables the interaction between groups of sensors, which monitor different kinds of events. The Sensor Data Grid Service provides and maintains the replications of large amount of sensor data collected from diverse sensor applications. The SensorGrid Processing Service collects the sensor data and processes them utilizing grid facilities. The development layer focuses on providing useful tools in order to ease and accelerate the development of sensor applications. The OSWA mainly focuses on providing an interactive development environment, an open and standards-compliant Sensor Web services middleware and a coordination language to support the development of various sensor applications.

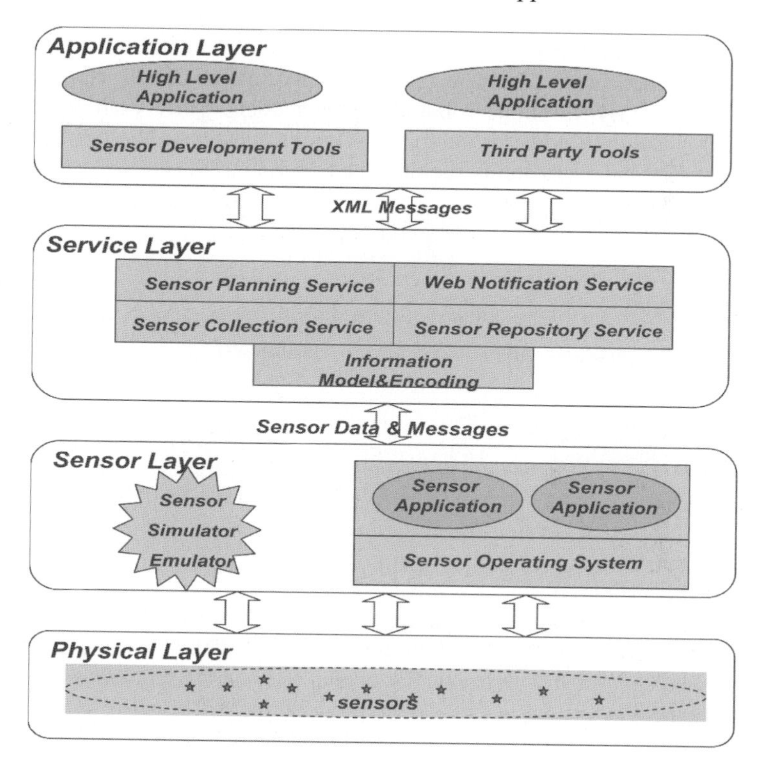

Fig. 3.6. A prototype instance of OSWA

SWE only provides the principle standard of how the Sensor Web looks, and does not have any reference implementation or working system available to the community; therefore, there are many design issues to consider, including all of the common issues faced by other distributed systems such as security, multi-threading, transactions, maintainability, performance, scalability and reusability, and the technical decisions that need to be made about which alternative technologies are best suitable to the system. Fig. 3.6 depicts a prototype instance of the OSWA, the implementation concentrates on the Service Layer and Sensor Layer as well as the XML encoding and the communication between the sensors and sensor networks. The following section will describe the key technologies that are relevant to different layers of the OSWA. In addition, the design and implementation of the core services are presented in this section. In order to better understand the whole Open Sensor Web Architecture including its design and implementation, several critical technologies are discussed briefly, which form the fundamental infrastructure across several layers of OSWA. The SOA is the essential infrastructure that supports the Service Layer and plays a very important role in presenting the core middleware components as services for client access. The main reason for Sensor Web relying heavily on SOA is because it simplifies integration of distributed heterogeneous systems which are loosely coupled. Moreover, SOA allows the services to be published, discovered and invoked by each other on the network dynamically. All the services communicate with each other through predefined protocols via a messaging mechanism which supports both synchronous and asynchronous communication models. Since each sensor network differs from each other, trying to put different sensors on the web, and providing discovery and accessibility requires the adoption of SOA.

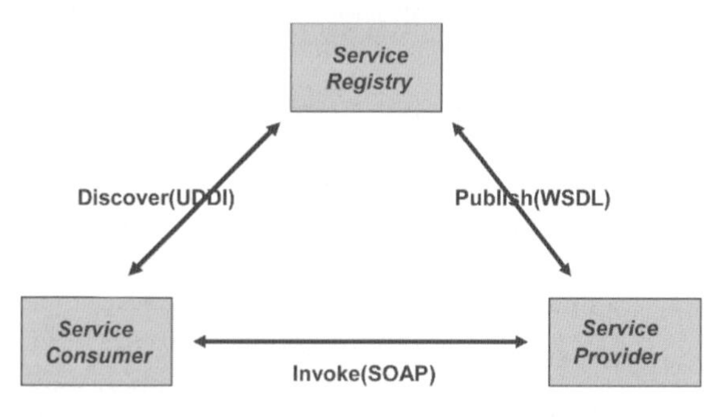

Fig. 3.7. Typical architecture of Web service

Web Services is one of the most popular implementations of SOA and is a language and platform neutral technology that can be implemented using any programming language in any platform. For example, a service written in C# can be accessed by a client written in Java. Web Services, technologically, depends on a group of standard specifications including HTTP, XML, Simple Object Application Protocol (SOAP), Universal Description Discovery and Integration (UDDI), Web Services Description Language (WSDL). XML is the key to Web Services technology, which is the standard format of the messages exchanged between services, and moreover almost every specifications used in Web Services are themselves XML data such as SOAP and WSDL. SOAP provides a unique framework that is used for packaging and transmitting XML messages over variety of network protocols, such as HTTP, FTP, SMTP, RMI/IIOP or proprietary messaging protocol[4]. WSDL describes the operations supported by web services and the structure of input and output messages required by these operations. It also de-

[4]http://www.w3.org/TR/2004/NOTE-ws-arch-20040211/wsa.pdf, Web Services Architecture, W3C, Feb 2004

scribes important information about the web services definition, support protocol, processing model and address. The typical architecture for Web Services is showed in Fig. 3.7. Service consumers may search the global registry (i.e. UDDI registry) about the WSDL address of a service that has been published by the service provider, and the consumers can invoke the relevant calls to the service once they obtain the WSDL for the service from the registry. As OSWA is primarily based on XML data model, Web Services provide a much better solution in terms of interoperability and flexibility.

3.4.1 Information Model and XML Data Binding

The information model of OSWA is based on Observation and Measurement and SensorML, both of them are built upon XML standards and are defined by an XML Schema. Transforming the data representation of the programming language to XML that conforms to an XML Schema refers to XML data binding, and is a very important and error-prone issue that may affect the performance and reliability of the system. In general, there are two common approaches to solve this problem. The first and most obvious way is to build the encoder/decoder directly by hand using the low-level SAX parser or DOM parse-tree API, however doing so is likely to be tedious and error-prone and require generating a lot of redundant codes that are hard to maintain. A better approach to deal with the transformation is to use an XML data binding mechanism that automatically generates the required code according to a DTD or an XML Schema. The data binding approach provides a simple and direct way to use XML in applications without being aware of the detailed structure of an XML document, and instead working directly with the data content of those documents. Moreover, the data binging approach makes access to data faster since it requires less memory than a document model approach like DOM or JDOM for working with documents in memory[5]. There are quite a few Java Data binding tools available such as JAXB, Castor, JBind, Quick, Zeus and Apache XMLBeans. Among those open source tools, XMLBeans seem to be the best choice not only because it provides full support for XML Schema, but also does it provide extra valuable features like XPath and XQuery supports, which directly enables performing queries on the XML documents.

3.4.2 Sensor Operating System

OSWA has the ability of dealing with heterogeneous sensor networks that may adopt quite different communication protocols including radio, blue tooth, and ZigBee/IEEE 802.11.4 protocols. As a results, it is quite desirable that the operating system level support for sensor networks can largely eliminate the work of developing device drivers and analyzing various protocol stacks directly in order to concentrate on higher-level issues related to the middleware development. TinyOS

[5]http://www-128.ibm.com/developerworks/xml/library/x-databdopt/

(Hill et al., 2000) is the de-facto standard and very mature Operating System for wireless sensor networks, which consists of a rich set of software components developed by nesC (Gay et. al., 2003) language, ranging from application level routing logic to low-level network stack. TinyOS provides a set of Java tools in order to communicate with sensor networks via a program called SerialForwarder. SerialForwarder runs as a server on the host machine and forwards the packets received from sensor networks to the local network, depicted by Fig. 3.8. Once the SerialForwarder program is running, the software located on the host machine can parse the raw packet and process the desired information. TinyDB (Maden 2003) is another useful component built on top of TinyOS, which constructs an illusion of distributed database running on each node of the sensor networks. SQL-like queries including simple and even grouped aggregating queries can be executed over the network to acquire data of sensors on each node.

Besides TinyOS, there are other Operating Systems existing as well. MANTIS (Abrach et. al. 2003) is a lightweight multithreaded sensor operating system, which supports C API enabling the cross-platform supports and reuse of C library. Moreover, it supports advanced sensor features including multi-model prototyping environment, dynamic reprogramming and remote shell. Contiki (Dunkels et. al., 2004), which is designed for memory constrained systems, is another event-driven sensor operating system like TinyOS with a highly dynamic nature that can be used to multiplex the hardware of a sensor network across multiple applications or even multiple users.

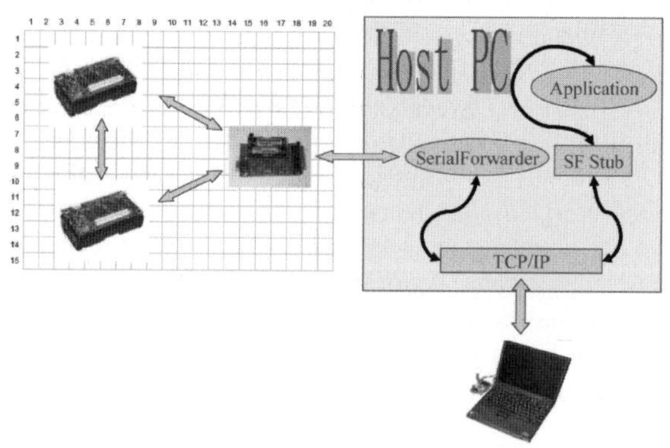

Fig. 3.8. TinyOS Serial forwarder architecture

3.4.4 Sensor Data Persistence

Persistence is one of the most important aspects for the purpose of manipulating the huge amount of data relevant to both sensor observation and sensor information. As the standard format for exchanging data between services is XML data which conforms to O&M and SensorML schema, transformations need to be done between different views of data including XML, Java object and relational database. In order to ease the operation of the transformation, the O/R mapping solution has been adopted to support the data persistence.

Java Data Objects (JDO) is one of the most popular O/R mapping solutions, which defines a high-level API that allows applications to store Java objects in a transactional data store by following defined standards. It supports transactions, queries, and the management of an object cache. JDO provides for transparent persistence for Java objects with an API that is independent of the underlying data store. JDO allows you to very easily store regular Java classes. JDOQL is used to query the database, which uses a Java-like syntax that can quickly be adopted by those familiar with Java. Together, these features improve developer productivity and no transformation codes need to be developed manually as JDO has done that complicated part underneath. To make use of JDO, the JDO Metadata is needed to describe persistence-capable classes. The information included is used at both enhancement time and runtime. The metadata assocoiated with each persistence-capable class must be contained within an XML file. In order to allow the JDO runtime to access it, the JDO metadata files must be located at paths that can be accessed by the Java classloader.

3.4.5 Design and Implementation

Currently, the primary design and implementation of OSWA focuses on its core services including SCS, WNS, and SPS (those extends from SWE) as well as the Sensor Repository Service that provides the persistent machanism for the sensor and the observation data. Fig. 3.9 illustrates an example of the client collection request and the invocations between relating services. As soon as the end user forwards an observation plan to the Planning Service, it checks the feasibility of the plan and submits it if feasible. The user will be registered in the Web Notification Service during this process and the user id will return to SPS. SPS is responsible for creating the observation request according to user's plan and retrieving the O&M data from the Sensor Collection Service. Once the O&M data is ready, the SPS will send an operation complete message to the WNS along with the user id and task id. The WNS will then notify the end user to collect the data via email or other protocols it supports.

3.4.6 Sensor Collection Service

Within those core services of OSWA, Sensor Collection Service (SCS) is one of the most important components residing in the service layer. The sensor collection service is the fundamental and unique component that needs to communicate directly with sensor networks, which is responsible for collecting real time sensing data and then translating the raw information into the XML encoding for other services to utilise and process. In other words, SCS is the gateway for entering into the sensor networks from outside clients and its design and implementation will affect the entire OSWA. The design of SCS provides an interface to both streaming data and query based sensor applications that are built on top of TinyOS and TinyDB respectively.

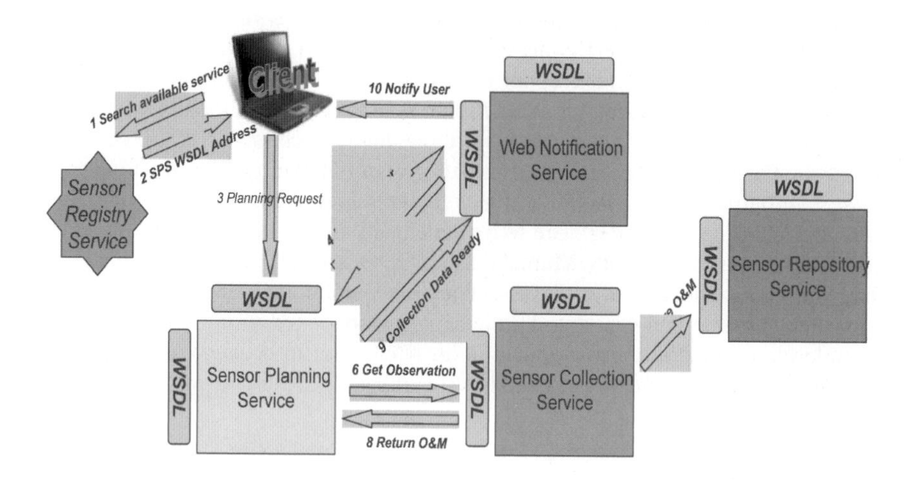

Fig. 3.9. A typical invocation for Sensor Web client

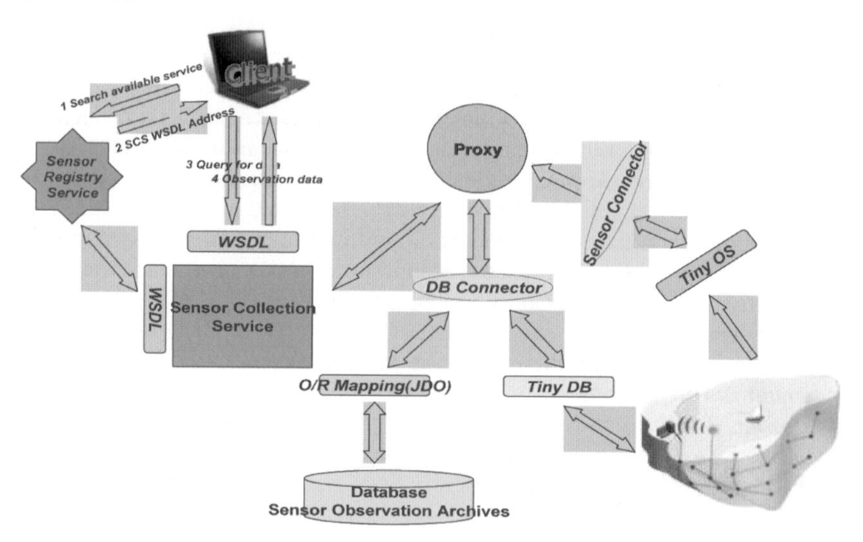

Fig. 3.10. Sensor collection service architecture

Fig. 3.10 illustrates the architecture of the Sensor Collection Service. It conforms to the interface definition that is described by the SCS specification and has been designed as web services that work with a proxy connecting to either real sensors or a remote database. Clients need to query the Sensor Registry Service about available SCS WSDL addresses according to their requirements and send a data query via SOAP to the SCS in order to obtain the observation data conforming to the O&M specification. The proxy acts as an agent working with various connectors that connect to the resources holding the information, and encode the raw observation into O&M compatible data. Different types of connectors have been designed to fit into different types of resources including sensor networks running on top of TinyOS or TinyDB and remote observation data archives.

The proxy needs to process the incoming messages from the client in order to determine what kind of connectors, either real-time based or archive based, to use. The design of the SCS is flexible and makes it easy to extend for further development if different sensor operating systems are adopted by the sensor networks such as MANTIS or Contiki. The only work is to implement a specific connector in order to connect to those resources and no modifications need to be made in the current system. The design of the proxy also encourages the implementation of a cache mechanism to improve the scalability and performance of the SCS. Load balancing mechanisms can be added to the system easily as well, by simply deploying the web service to different servers.

3.4.7 Sensor Planning Service

The design of the Sensor Planning Service (SPS) should consider the both short-term and long-term user's plan, which means that the SPS must provide response to the user immediately, rather than blocking to wait for the collection results.

Shown in the Fig. 3.11, SPS utilises a rule engine which reads a specific set of predefined rules in order to clarify the feasibility of the plan made by the user. The implementation of the rule engine can be quite complicated, expecting the system to accept rules within a configuration file as plain text, xml-based or other types of rule-based languages. . Currently, the rule engine is implemented as an abstract class that can be extended by the application developers to specify a set of boundary conditions that define the feasibility of the applications. For example, in a simple temperature application, a boundary condition for the temperature may be a range from 0 to 100.

The most important component that makes the SPS suitable for short or long term plan execution is the Scheduler which is implemented as a separate thread running in the background. The execution sequence of the Scheduler (i) composes the collection request according to user's plan and then invokes the getObservation of the SCS, (ii) asks the DataCollector to store the observation data in order for users to collect afterward, and (iii) sends notification to the WNS indicating the outcome of the collection request.

Notice that the time of the execution happened in the scheduler varies baesd on the requirements of the user's plan. The clients will get a response indicating that their plans will be processed right after they submit their plan to the SPS. The scheduler deals with the remaining time consuming activities. The clients may get the notification from the WNS as soon as the WNS receives the message from the scheduler, and collect the results from the DataCollector.

3.4.8 Web Notification Service

The current design of Web Notification Service is showed in Fig. 3.12, which contains two basic components: AccountManager and Notification. The SPS may request to register users via WNS, which asks the AccountManager to manage the user account in the DBMS in order to retrieve user information in the subsequent operations. The Notification is used to create a specific communication protocol and send the messages via the protocol to the user that has been registered in the DBMS. Currently, an EmailCommunicationProtocol has been implemented to send messages via email. Further implementations can be easily plugged into the existing architecture by implementing the CommunicationProtocol interface with a send method.

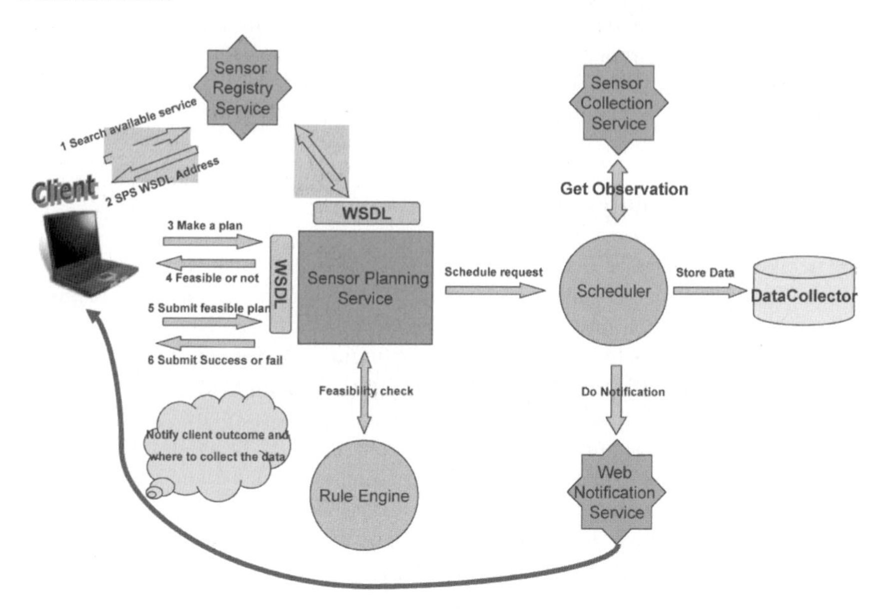

Fig. 3.11. Sensor planning service architecture

3.5 Experimentation and Evaluation

As the OWSA aims at providing a platform to serve numerous users globally through the Internet, it is quite important to test the services, and ensure that they are scalable and performing reasonably. The experiment platform for the services is built on TOSSIM (described by Levis et al., 2003 as a discrete event simulator that can simulate thousands of motes running complete sensor applications and allow a wide range of experimentation) and Crossbow's MOTE-KIT4x0 MICA2

Basic Kit[6] that consists of 3 Mica2 Radio board, 2 MTS300 Sensor Boards, a MIB510 programming and serial interface board. The experiment concentrates on the SCS, due to the fact that it is the gateway for other services to sensors, which would be the most heavily loaded service and possible bottleneck of the entire system. As can be seen in Fig. 3.13, SCS has been deployed on Apache Tomcat 5.0 on two different machines that run TinyDB application under TOSSIM and Temperature Monitoring Application under Crossbow's motes respectively. Meanwhile, a Sensor Registry Service is also configured on a separate machine that provides the functionality to access sensor registry and data repository.

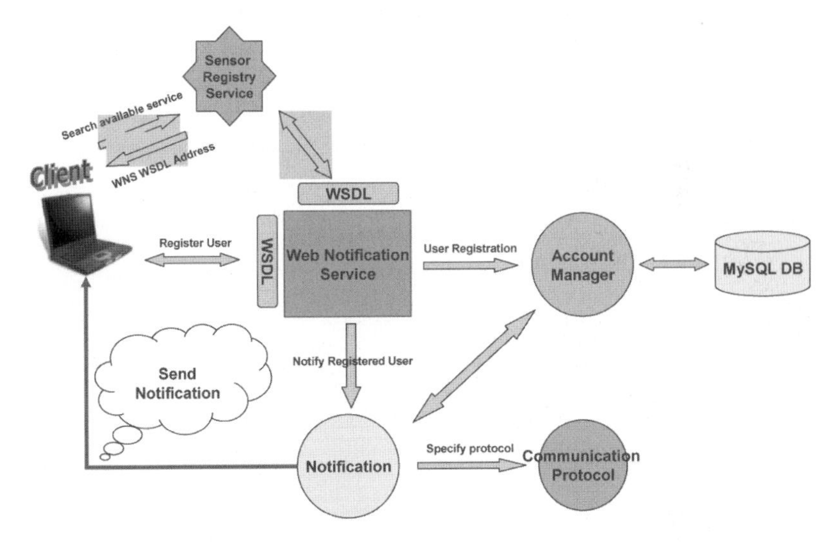

Fig. 3.12. Web notification service architecture

A simple temperature monitoring application has also been developed. The application is programmed using nesC and uses simple logic, which just broadcasts the sensing temperature, light and node address to the sensor network at regular intervals. The simple application does not consider any multi-hop routing and energy saving mechanism. Before installing the application to the Crossbow's mote, the functionality can be verified via the TOSSIM simulator. Fig. 3.14 demonstrates the simulation of the temperature application running under the TOSSIM visual GUI.

Once the application has been successfully installed onto each mote via the programming board, a wireless sensor network has been built with two nodes, and one base station connecting to the host machine via the serial cable. Besides installing the application itself, the SerialForwarder program also needs to run on the host machine in order to forward the data from the sensor network to the server. Fig. 3.15 demonstrates the list of results for a simple query "temp>200" to the sensors running TinyDB application under TOSSIM.

[6] http://www.xbow.com/Products/productsdetails.aspx?sid=67. Crossbow Technology Inc

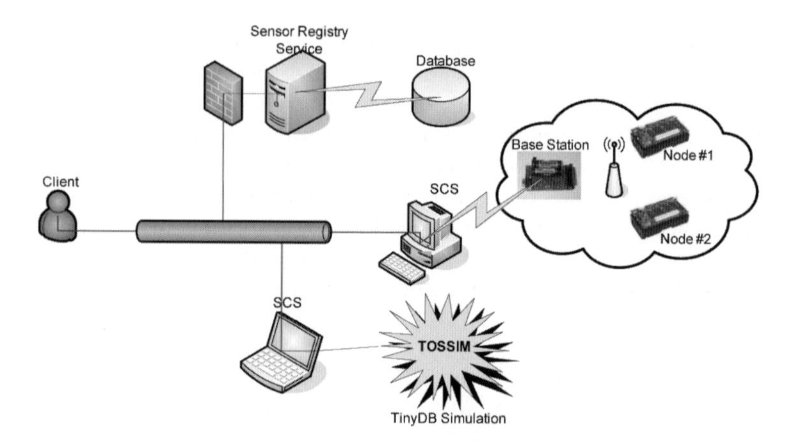

Fig. 3.13. Deployment of experiment

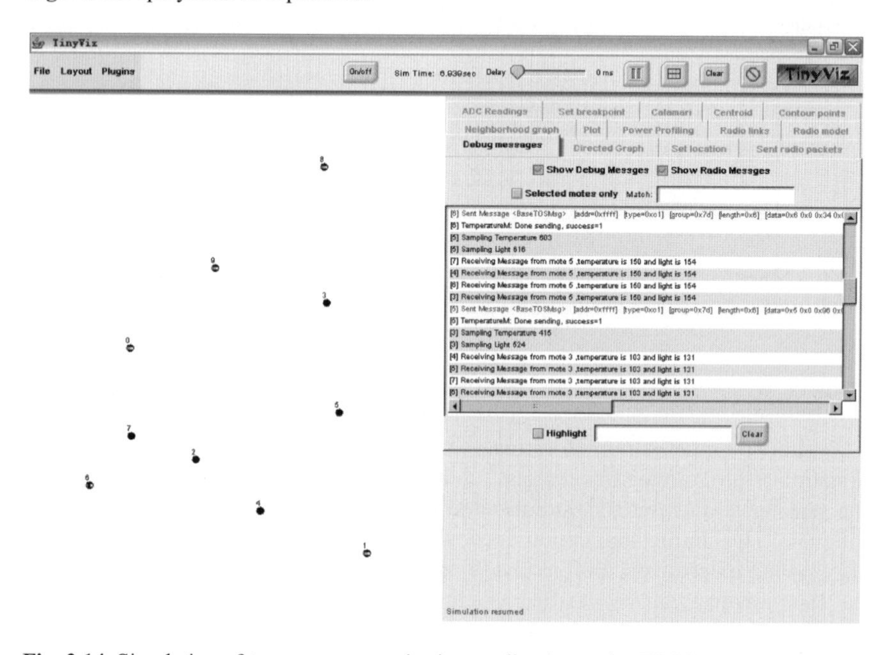

Fig. 3.14. Simulation of temperature monitoring application under TOSSIM

Regarding scalability, a simulation program that can stimulate different numbers of clients running at the same time has been used exclusively for the SCS. The performance measured by time variable (per second) for both auto-sending and query-based applications running on top of TinyOS is showed in the following figures. As can be seen from Fig. 3.16, the result of the auto-sending mode application is moderate when the number of clients who request the observation simultaneity is small. Even when the number of clients reaches 500; the response time for a small number of records is also acceptable. In contrast, the result showed in

Fig. 3.17 is fairly unacceptable as even just one client requesting a single observation takes 34 seconds. The response time increases near linearly when the number of clients and the number of records go up. The reason why the query-based approach has very poor performance is due to the execution mechanism of TinyDB. A lot of time has been spent on initializing each mote, and the application can only execute one query at one time, which means another query needs to wait until the current query has been stopped or has terminated. A solution to this problem may require the TinyDB application run a generic query for all clients, and the more specific query can be executed in-memory according to the observation data collected from the generic query.

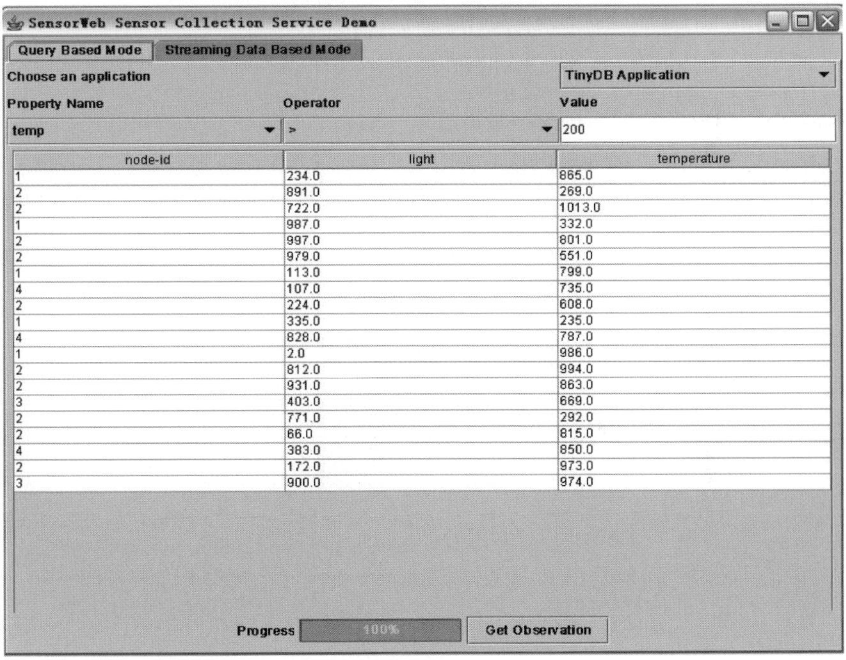
SensorWeb Sensor Collection Service Demo
Query Based Mode
Streaming Data Based Mode
Choose an application
TinyDB Application
Property Name
Operator
Value
temp
>
200
node-id
light
temperature
1 234.0 865.0
2 891.0 269.0
2 722.0 1013.0
1 987.0 332.0
2 997.0 801.0
2 979.0 551.0
1 113.0 799.0
4 107.0 735.0
2 224.0 608.0
1 335.0 235.0
4 828.0 787.0
1 2.0 986.0
2 812.0 994.0
2 931.0 863.0
3 403.0 669.0
2 771.0 292.0
2 66.0 815.0
4 383.0 850.0
2 172.0 973.0
3 900.0 974.0
Progress
100%
Get Observation

Fig. 3.15. Swing client showing query result for TinyDB application under TOSSIM

There are several possible ways to enhance the performance. A caching mechanism may be one of the possible approaches, the recent collected observation data can be cached in the proxy for a given period of time and the clients who request the same set of observation data can be read the observation data from the cache. However, as the data should be kept as close to real time as possible, it is quite hard to determine the period of time for the cache to be valid. A decision can be made according to the dynamic features of the information the application is targeting. For example, the temperature for a specific area may not change dynamically by minutes or by hours. Consequently, the period of time setting for the cache for each sensor application can vary based on the information the sensor is targeting. Another enhancement of query performance may be achieved by utilizing the query mechanism such as XQuery of the XML data directly other than asking the real sensor itself executing the query similar to TinyDB.

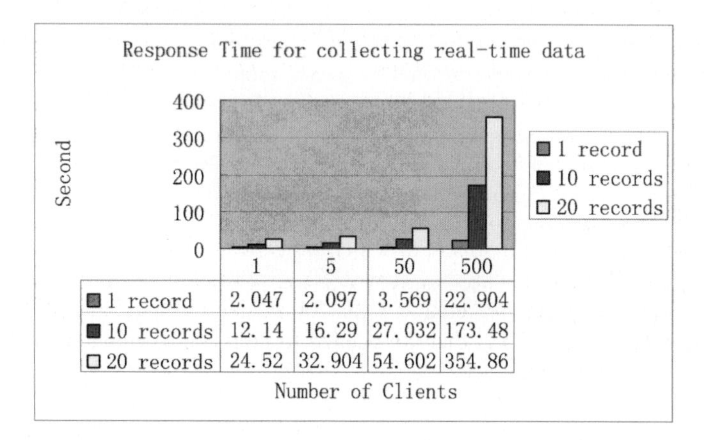

	1	5	50	500
1 record	2.047	2.097	3.569	22.904
10 records	12.14	16.29	27.032	173.48
20 records	24.52	32.904	54.602	354.86

Fig. 3.16. Performance for collecting auto-sending data

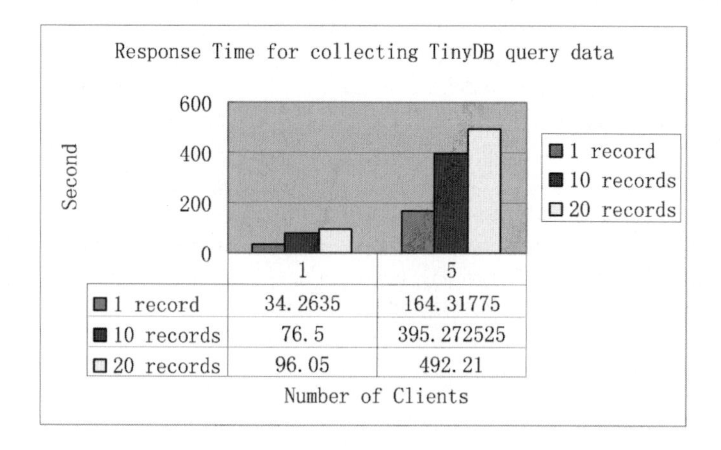

	1	5
1 record	34.2635	164.31775
10 records	76.5	395.272525
20 records	96.05	492.21

Fig. 3.17. Performance for collecting TinyDB query data

3.6 Summary and Future Works

In this chapter, we have introduced a new buzzword: Sensor Web in the research and academic community of sensor and sensor networks. There are a lot of efforts that aim to provide middleware support to the sensor development. Among those, the most important one is OGC's SWE standard that standardises the vision of Sensor Web. SensorML, O&M, SCS, SPS and WNS together, to create an integration platform to register, discover and access anonymous and heterogeneous sensors distributed all over the world through internet. A service oriented Sensor Web framework named Open Sensor Web Architecture (OSWA) has been discussed along with the design implementation of the core services targeting the sensor applications running on top of TinyOS. OSWA extends SWE and provides additional services to process the information collected from those resources accompanied by computational grids. In addition, the experiment of the scalability and performance of the prototype system is also presented.

Although, the services are all working properly with acceptable performance, we are still at an early stage of having the entire OSWA implemented. Even those services that we have implemented are not yet fully functional. The Sensor Collection Service is the key component of the entire OSWA, which affects the performance and reliability of the entire system. A lot of issues are left to future investigation, focusing on aspects of reliability, performance optimisation and scalability. There are a couple of efforts that are needed to be made to enhance the SCS and other services in the next stage.

- The query mechanism for the Sensor Collection Service will be enhanced to support in-memory XML document querying. XPath and XQuery technologies are planned to be adopted, as they are the standard way to query XML documents. The outcome of this enhancement is expected to improve the performance by moving the heavy workload of queries from the sensor network itself and onto the host machine instead.
- A caching mechanism may be implemented and placed into the Proxy to further enhance the performance and scalability.
- Other methods that are described in the specifications of the SWE services but are currently not available still need to be implemented.
- Other notification protocols need to be built for the WNS in the future.
- Sensor Registry via SensorML needs to be developed in order to support the worldwide sensor registration and discovery.
- Both XML-based configuration and rule-based configuration language may be developed in order to ease the deployment of the services.

3.7 References

Abrach H, Bhatti S, Carlson J, Dai H, Rose J, Sheth A, Shucker B, Deng J, and Han R (2003) MANTIS: system support for multi-modal networks of In-Situ sensors. Proceedings of the 2nd ACM Int. Conf. on WSN and applications, Sept. 19, San Diego, CA

Bonnet P, Gehrke J, and Seshadri P (2000) Querying the Physical World. IEEE personal Communications 7:10-15

Cardell-Oliver R, Smettern K, Kranz M, and Mayer K (2004) Field testing a WSN for reactive environmental monitoring, Int. Conf. on intelligent sensors, sensor networks and information processing, Dec. 14-17, Melbourne

Dunkels A, Gronvall B, and Voigt T (2004) Contiki: A lightweight and flexible operating system for tiny networked sensors, Proc. of the 29th Annual IEEE Int. Conf. on LAN Nov. 16-18, Tampa, Florida

Gay D, Levis P, von Behren R, Welsh M, Brewer E, and Culler D (2003) The nesC language: A holistic approach to networked embedded systems. Proc. of programming language design and implementation (PLDI), June 8-11, San Diego, CA

Gaynor M, Moulton SL, Welsh M, LaCombe E, Rowan A, and Wynne J (2004) Integrating WSN with the Grid, IEEE Internet Computing 8:32-39

Ghanem M, Guo Y, Hassard J, Osmond M, and Richards M (2004) Sensor grids for air pollution monitoring, Proc. of UK e-science all hands meeting, 31st Aug- 3rd Sept., Nottingham, UK

Gibbons PB, Karp B, Ke Y, Nath S, and Seshan S (2003) IrisNet: An architecture for a Worldwide sensor web, IEEE Pervasive Computing, 2: 22-33

Heinzelman W, Murphy A, Carvalho H, and Perillo M (2004) Middleware to support sensor network applications, IEEE Network Magazine 18: 6-14

Hu W, Tran VN, Bulusu N, Chou CT, Jha S, and Taylor A (2005) The design and evaluation of a hybrid sensor network for cane-toad monitoring, Proc. information processing in sensor networks, April 25-27, Los Angeles

Hill J, Szewczyk R, Woo A, Hollar S, Culler D, and Pister K (2000) System architecture directions for networked sensors, Proc. of architectural support for programming languages and operating systems, Nov. 12-15, Cambridge, MA

Liu T and Martonosi M (2003) Impala: A middleware system for managing autonomic, parallel sensor systems, Proc. of the 9th ACM SIGPLAN symposium on principles and practice of parallel programming, June 11-13, San Diego

Levis P, Lee N, Welsh M, Culler D (2003) TOSSIM: Accurate and scalable simulation of entire TinyOS applications., Proc. of the 1st Intl. Conf. on embedded networked sensor systems, Nov. 4-7, Los Angeles

Mainwaring A, Polastre J, Szewczyk R, Culler D, and Anderson J (2002) Wireless sensor networks for habitat monitoring, Proc. of the 1st ACM Int. workshop on WSN and applications, Sept. 28, Atlanta

Madden SR (2003) The design and evaluation of a query processing architecture for sensor networks, PhD thesis, UC Berkeley

Nickerson BG, Sun Z, and Arp JP (2005) A sensor web language for mesh architectures, 3rd annual communication networks and services Conf., May 16-18, Halifax, Canada.

Reichardt M (2005) Sensor Web Enablement: An OGC white paper, Open geospatial consortium (OCG), Inc.

Tao V, Liang SHL, Croitoru A, Haider Z, and Wang C (2004) GeoSWIFT: Open Geospatial Sensing Services for Sensor Web. In: Stefanidis A, Nittel S (eds), CRC Press, pp.267-274.

Soutoo E, Guimaraes G., Vasconcelos G., Vieira M, Rosa N, Ferraz C, and Freire L (2004) A message-oriented middleware for sensor networks, Proc. of 2nd Int. workshop on middleware for pervasive and Ad-hoc computing, Oct. 18-22, Toronto, Ontario

Tham CK and Buyya R (2005) SensorGrid: Integrating sensor networks and grid computing, CSI Communications 29:24-29

4 Cross-layer Designs

Dario Pompili, Mehmet Can Vuran and Tommaso Melodia

Broadband and Wireless Networking Laboratory, School of Electrical and Computer Engineering, Georgia Institute of Technology, Atlanta

4.1 Introduction

There exist exhaustive amount of research to enable efficient communication in wireless sensor networks (WSNs) (Akyildiz 2002). Most of the proposed communication protocols improve the energy efficiency to a certain extent by exploiting the collaborative nature of WSNs and its correlation characteristics. However, the main commonality of these protocols is that they follow the traditional layered protocol architecture. While these protocols may achieve very high performance in terms of the metrics related to each of these individual layers, they are not jointly optimised to maximise the overall network performance while minimising the energy expenditure. Considering the scarce energy and processing resources of WSNs, joint optimisation and design of networking layers, i.e., cross-layer design, stands as the most promising alternative to inefficient traditional layered protocol architectures.

Accordingly, an increasing number of recent works have focused on cross-layer development of wireless sensor network protocols. In fact, recent papers on WSNs (Fang 2004; Hoesel 2004; Vuran 2005) reveal that cross-layer integration and design techniques result in significant improvement in terms of energy conservation. Generally, there are three main reasons behind this improvement. First, the stringent energy, storage, and processing capabilities of wireless sensor nodes necessitate such an approach. The significant overhead of layered protocols results in high inefficiency. Moreover, recent empirical studies necessitate that the properties of low power radio transceivers and the wireless channel conditions be considered in protocol design (Ganesan 2002; Zuniga 2004). Finally, the event-centric paradigm of WSNs requires application-aware communication protocols.

Although a considerable amount of recent papers have focused on cross-layer design and improvement of protocols for WSNs, a systematic methodology to accurately model and leverage cross-layer interactions is still missing. With this respect, the design of networking protocols for multi-hop wireless ad hoc and sensor networks can be interpreted as the distributed solution of resource allocation problems at different layers. However, while most of the existing studies decompose the resource allocation problem at different layers, and consider allocation of resources at each layer separately, we review recent literature that has tried to establish sound cross-layer design methodologies based on the joint solution of resource allocation optimisation problems at different layers.

Several open research problems arise in the development of systematic techniques for cross-layer design of WSN protocols. In this chapter, we describe the performance improvement and the consequent risks of a cross-layer approach. We review literature proposing precautionary guidelines and principles for cross-layer design, and suggest some possible research directions. We also present some concerns and precautionary considerations regarding cross-layer design architectures. A cross-layer solution, in fact, generally decreases the level of modularity, which may loosen the decoupling between design and development process, making it more difficult to further design improvements and innovations. Moreover, it increases the risk of instability caused by unintended functional dependencies, which are not easily foreseen in a non-layered architecture.

This chapter is organised as follows. In next section, we overview the communication protocols devised for WSNs that focus on cross-layer design techniques. We classify these techniques based on the network layers they aim at replacing in the classical OSI (Open System Interconnection) network stack. In Section 3, a new communication paradigm, i.e., cross-layer module, is introduced. In Section 4, we discuss the resource allocation problems that relate to the cross-layer design and the proposed solutions in WSNs. Based on the experience in cross-layering in WSNs, in Section 5 we present the potential open problems that we foresee for WSNs. Then, we stress some reservations about cross-layer design by discussing its pros and cons in Section 6, and conclude the chapter in Section 7.

4.2 Pair-wise Cross Layer Protocols

In this section, we overview significant findings and representative communication protocols that are relevant to the cross-layering philosophy. So far, the term cross-layer has carried at least two meanings. In many papers, the cross-layer interaction is considered, where the traditional layered structure is preserved, while each layer is informed about the conditions of other layers. However, the mechanisms of each layer still stay intact. On the other hand, there is still much to be gained by rethinking the mechanisms of network layers in a unified way so as to provide a single communication module for efficient communication in WSNs. In this section, we only focus on the pair-wise cross-layer protocols and defer the discussion of cross-layer module design, where functionalities of multiple traditional layers are melted into a functional module, to Section 3.

The experience gained through both analytical studies and experimental work in WSNs revealed important interactions between different layers of the network stack. These interactions are especially important for the design of communication protocols for WSNs. As an example, in (Ganesan 2002), the effect of wireless channels on a simple communication protocol such as flooding is investigated through testbed experiments. Accordingly, the broadcast and asymmetric nature of the wireless channel results in a different performance than that predicted through the unit disk graph model (UGM). More specifically, the asymmetric nature of wireless channels introduces significant variance in the hop count between two nodes. Furthermore, the broadcast nature of the wireless channel results in significantly different flooding trees than predicted by the unit disk graph model (Ganesan 2002). Similarly, in (Zuniga 2004), the experimental studies reveal that the perfect-reception-within-range models can be misleading in performance evaluations due to the existence of a transitional region in low power links. The experiment results reported in (Zuniga 2004) and many others show that up to a certain threshold inter-node distance, two nodes can communicate with practically no errors. Moreover, nodes that are farther away from this threshold distance are also reachable with a certain probability. While this probability depends on the distance between nodes, it also varies with time due to the randomness in the wireless channel. Hence, protocols designed for WSNs need to capture this effect of low power wireless links. Moreover, in (Shih 2001), guidelines for physical-layer-driven protocol and algorithm design are investigated. These existing studies strongly advocate that communication protocols for WSNs need to be redesigned considering the wireless channel effects. Similarly, as pointed out in (Vuran 2005), the interdependency between local contention and end-to-end congestion is important to be considered during the phase of protocol design. The interdependency between these and other network layers calls for adaptive cross-layer mechanisms in order to achieve efficient data delivery in WSNs.

In addition to the wireless channel impact and cross-layer interactions, the content of the information sent by sensor nodes is also important in cross-layer protocol design. In fact, the spatial, temporal, and spatio-temporal correlation is another significant characteristic of WSNs. Dense deployment of sensor nodes results in the sensor observations being highly correlated in the space domain. Similarly, the nature of the energy-radiating physical phenomenon yields temporal correlation between each consecutive observation of a sensor node. Furthermore, the coupled effects of these two sources of correlation results in spatio-temporal correlation. Exploiting the spatial and temporal correlation further improves energy efficiency of communication in WSNs. In (Vuran 2004) and (Vuran 2006-2), the theory of spatial, temporal, and spatio-temporal correlation in WSNs is developed. The correlation between the observations of nodes are modelled by a correlation function based on two different source models, i.e., point and field sources. Based on this theory, the estimation error resulting in exploiting the correlation in the network can be calculated. This error is defined as distortion.

In Figs. 4.1 and 4.2, the effect of spatial and temporal correlation on the distortion in event reconstruction is shown, respectively. In general, lower distortion results in more accurate estimation of the event features. Hence, using more number of nodes in an event location as shown in Fig. 4.1 or sampling the physical locations in higher frequency as shown in Fig. 4.2 results in lower distortion. However, Fig. 4.1 reveals that, by using a small subset of nodes for reporting an event, e.g., 15 out of 50, the same distortion in event reconstruction can be achieved. Similarly, by reducing the sampling rate of sensor nodes, the same distortion level can be achieved as shown in Fig. 4.2 due to correlation between samples. As a result, the redundancy in the sensor readings can be removed. These results reveal that, significant energy savings are possible when the correlation in the content of information is exploited. Moreover, in Fig. 4.3, the feasible regions for number of nodes that are reporting an event and their reporting frequency tuple, (M,f), are shown for a given distortion constraint D_{max}. It is clearly shown that, using maximum values for both of these operation parameters may decrease distortion and that these parameters need to be collaboratively selected inside the feasible region using distributed protocols. In the following sections, two approaches in MAC and transport layers that exploit the spatial correlation in WSNs are described.

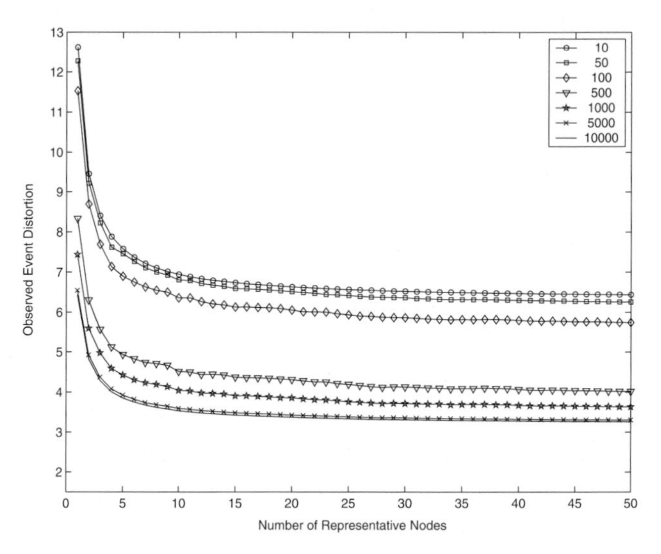

Fig. 4.1. Observed event distortion versus changing number of representative nodes

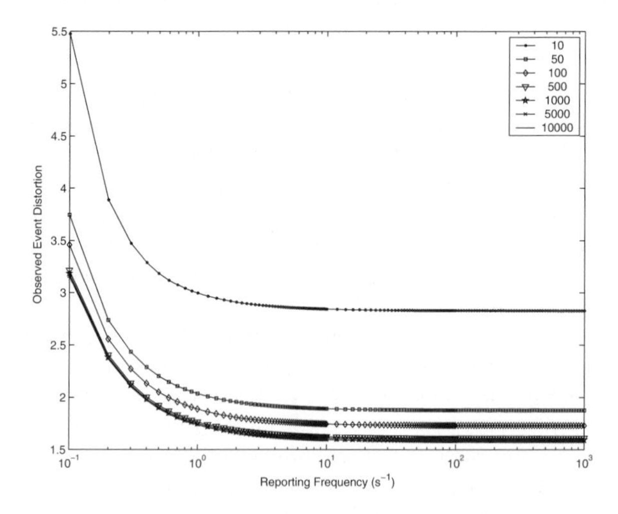

Fig. 4.2. Observed event distortion vs. varying normalised frequency (Vuran 2004)

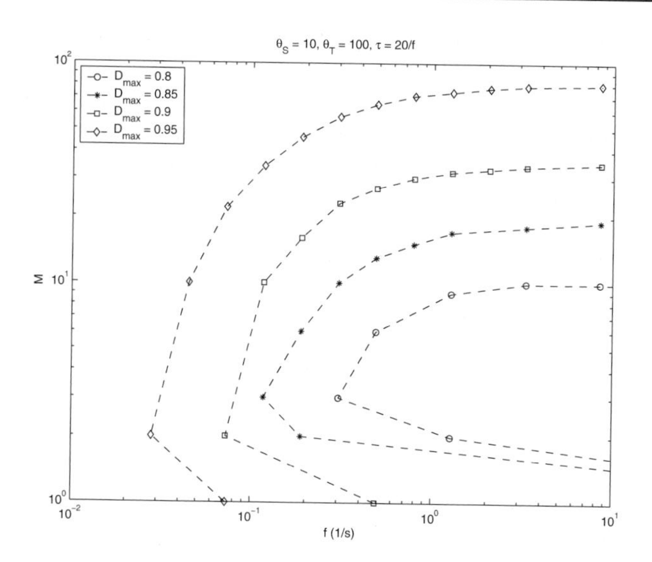

Fig. 4.3. No. of nodes vs. sampling rate, (M,f) tuples meeting various D_{max} constraints (Vuran 2006-2)

In the following, the literature of WSN protocols with cross-layer principles is surveyed. We classify these studies in terms of interactions or modularity among physical (PHY), medium access control (MAC), routing, and transport layers.

4.2.1 Transport and PHY Interactions

Transport layer functionalities, such as congestion control and reliability management, depend on the underlying physical properties of both the sensor transceiver and the physical phenomenon that is sensed. More specifically, the transmit power of the sensor nodes directly affects the one-hop reliability. This in effect improves end-to-end reliability. However, increasing the transmit power increases the interference range of a node and may cause increased contention in the wireless medium leading to overall network congestion (Akyildiz 2006). On the other hand, the spatial and the temporal correlation in the content of information enables energy efficient operation by definition of new reliability concepts (Akan 2005). In this section, we overview two representative solutions for pair-wise cross-layer protocols between transport and PHY layers.

In (Chiang 2005), a cross-layer optimisation solution for power control and congestion control is considered. More specifically, analysis of interactions between power control and congestion control is provided, and the trade-off between the layered and the cross-layer approach is presented, as further discussed in Section 4. In this analysis, a CDMA-based physical layer is assumed. Consequently, the received signal of a node is modelled as a global and nonlinear function of all the transmit powers of the neighbour nodes. Based on this framework, a cross-layer communication protocol is proposed, where the transmit power and the transmission rate are jointly controlled. The nodes control their transmit power based on the interference of other nodes and determine the transmission rate accordingly. However, the proposed solutions only apply to CDMA-based wireless multihop networks, which may not apply to a large class of WSNs where CDMA technology is not feasible. The spatial correlation between sensor nodes is exploited in (Akan 2005) with the definition of a new reliability notion. In conventional networks, since the information sent by different entities is independent of each other, a one-to-one and end-to-end reliability notion is used. In WSNs, however, the end user, e.g., sink, is often interested in the physical phenomenon in the vicinity of a group of sensors instead of the individual readings of each sensor. Consequently, in (Akan 2005), the event-to-sink reliability notion is defined for data traffic from sensors to the sink. This notion relies on the fact that the readings of a group of sensors in an event area are spatially correlated and the reporting rate of these sensors can be collectively controlled to ensure both reliability and prevent congestion. As a result, in event to sink reliable transport (ESRT) protocol, the transmission rate of sensors nodes is controlled by the sink iteratively through calculations during a decision interval.

4.2.2 Routing and PHY Interactions

Routing protocols are also affected by the transmit powers at PHY layer due to similar reasons explained in the previous section. While transmit power may improve the capacity of a link, the capacity of the whole network may degrade due to the increase in interference. Hence, power control can be performed in conjunction with route selections. In addition, the channel quality information resulting from a specific transmit power selection can also be exploited in route selection by preferring high quality links that can result in reliable transmission as well as minimum number of hops or minimum latency. In this section, we investigate the effects of power control and the channel quality on routing, and present two representative solutions, where a pair-wise cross-layer approach among routing and PHY layers is developed. A cross-layer optimisation of network throughput for multihop wireless networks is presented in (Yuan 2005). The authors split the throughput optimisation problem into two subproblems, i.e., multi-hop flow routing at the network layer and power allocation at the physical layer. The throughput is tied to the per-link data flow rates, which in turn depend on the link capacities and, consequently, on the per-node radio power level. On the other hand, the power allocation problem is tied to interference as well as to the link rate. Based on this solution, a CDMA/OFDM based solution is provided such that the power control and the routing are performed in a distributed manner. In (Saeda 2004), new forwarding strategies for geographic routing are proposed based on the results in (Zuniga 2004). Using a probabilistic model, the distribution for optimal forwarding distance for networks with automatic repeat request (ARQ) and without ARQ in a linear network is presented. Based on this analysis, new forwarding strategies are provided. The forwarding algorithms require the packet reception rate of each neighbour for determination of the next hop and construct routes accordingly. More specifically, a $PRR \times DIST$ method that selects the node with the maximum packet reception rate (PRR) and advancement (DIST) product is proposed. It is shown that this scheme is optimal for networks where ARQ is implemented. Moreover, two new blacklisting schemes are proposed such that nodes with very low packet reception rates are blacklisted for routing. These new forwarding metrics illustrate the advantages of cross-layer forwarding techniques in WSNs. On the other hand, the analysis for the distribution of optimal hop distance is based on a linear network structure. Hence, a more general analysis that considers networks with arbitrary topologies is required to prove the optimality of the proposed scheme in WSNs.

4.2.3 MAC and PHY Interactions

As explained above, physical layer properties of WSNs necessitate both channel-aware and physical phenomenon-aware design techniques. This necessity is also valid for medium access control (MAC) protocols. In this section, we present two major approaches in pair-wise interaction for MAC and PHY layers. The non-uniform properties of signal propagation in low power wireless channels need to be considered in MAC protocol design. MAC protocols aim at providing collision-free access to the wireless medium, and this collision can only be prevented by accurate knowledge of potential interfering nodes. Hence, an accurate wireless channel model is required for both evaluation and design of MAC protocols. In (Haapola 2005), the energy consumption analysis for physical and MAC layers is performed for MAC protocols. In this analysis, the energy consumption due to both processing and transmission is considered. Generally, in ad-hoc networks, multi-hop communication is preferred since transmission power is reduced. However, in WSNs, where processing and communication energy consumption are comparable, this preference is not that clear. Especially for low duty cycle networks, energy consumption due to processing may become comparable to energy consumption due to communication. In this analysis, this trade-off is investigated and it is concluded that single-hop communication can be more efficient when real radio models are used. This result necessitates new techniques for MAC protocols since the number of potential interferers increases significantly when single-hop communication is considered. Although this is an interesting result, the analysis in (Haapola 2005) is based on a linear network and it is necessary to generalise this result to networks with arbitrary topologies.

In addition to the characteristics of the wireless channel and the radio circuitry, the content of the information that will be sent by sensor nodes is also important in MAC design. The content of this information is closely related to the physical properties of the physical phenomenon since WSNs are primarily developed for sensing this phenomenon in the environment. As shown in Fig. 4.1, the spatial correlation between the information each sensor node gathers can be exploited for energy efficient operation. Furthermore, since the MAC layer coordinates interactions between closely located nodes, this layer is a perfect fit for exploiting spatial correlation. Consequently, a cross-layer solution among MAC layer, physical phenomenon, and the application layer for WSNs is proposed in (Vuran 2006). The main motivation behind this solution is illustrated in Fig. 4.4. Due to the spatial correlation between closely located nodes, in WSNs, a node may contain highly correlated sensor readings as its neighbours. Hence, any information sent by these neighbours may be redundant once this node sends its information. Based on the rate-distortion theory, in (Vuran 2006), it is shown that a sensor node can act as a representative node for several other sensor nodes as shown in Fig. 4.4. Accordingly, a distributed, spatial correlation-based collaborative medium access control (CC-MAC) protocol is proposed. In this protocol, using the statistical properties of the WSN topology, the maximum distance, d_{corr}, at which two nodes are still highly correlated, given a distortion requirement, is calculated at the sink. Each node then contends for the medium only if it does not hear any node in its correlation region transmitting information. This operation constructs correlation clusters as shown in Fig. 4.4. As a result, lower number of communication attempts are performed, which leads to lower contention, energy consumption, and latency while achieving acceptable distortion for reconstruction of event information at the sink. Simulation results in (Vuran 2006) show that this cross-layer interaction results in high performance in terms of energy, packet drop rate, and latency compared to MAC layer protocols designed for WSNs. The representative node transmits its record on behalf of the entire correlation region, while all correlation neighbours suppress their transmissions (Vuran 2006)

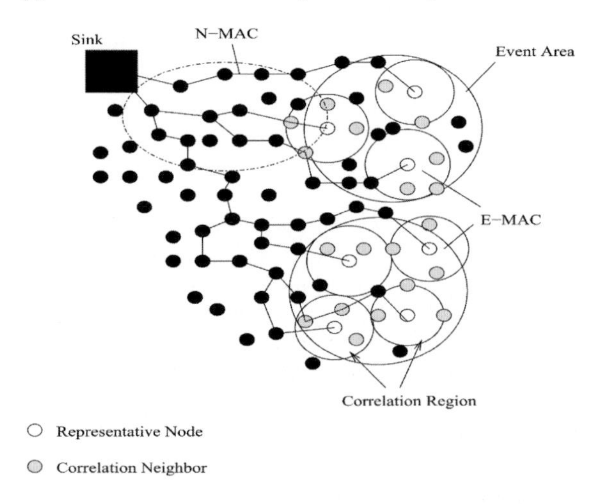

Fig. 4.4. CC-MAC protocol and its components Event-MAC (E-MAC) and Network-MAC

4.2.4 MAC and Routing Interactions

Recently, exploiting cross-layer interaction has gained much interest among MAC and routing layers. In this context, two main approaches are emerging. On one hand, the functions related to determining the next hop in a network is closely coupled with medium access. This concept is referred to as receiver-based routing. In this approach, the next hop is chosen as a result of the contention in the neighbourhood. Receiver-based routing has been independently proposed in (Skraba 2004), (Zorzi 2003), and (Zorzi 2003-2). Generally, the receiver-based routing can be described as shown in Fig. 3. Receiver-based routing couples the recent findings in geographical routing and channel-aware routing techniques with medium access procedures. When a node i has a packet to send, the neighbours of this node are notified by a broadcast message. Since the nodes that are closer to the sink than the node i are feasible nodes, the nodes in this feasible region, A(R,D), perform contention for routing. In order to provide minimum number of hops in routing, the feasible region is further divided into multiple priority zones. Consequently, the nodes closer to the sink, i.e., in D_1, perform backup for a smaller amount of time and contend for the packets.

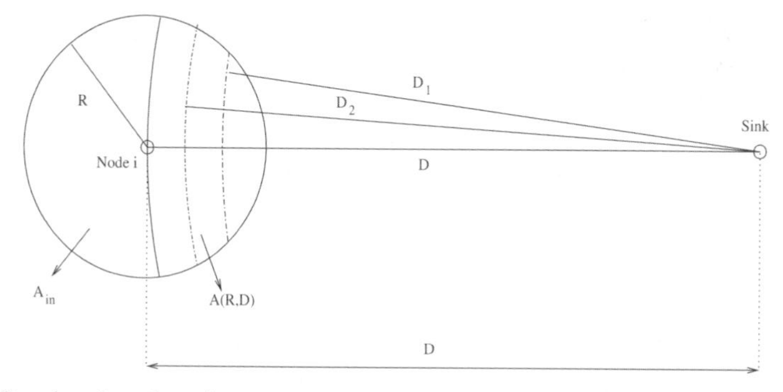

Fig. 4.5. Receiver-based routing

In (Zorzi 2003) and (Zorzi 2003-2), the energy efficiency, latency, and multi-hop performance of the receiver-based routing algorithm is discussed. In (Zorzi 2004), the work in (Zorzi 2003) and (Zorzi 2003-2) is extended for a single-radio node such that medium access is performed through a single channel. Furthermore, a new integrated MAC/routing solution is proposed in (Rossi 2005) for geographical routing in wireless sensor networks based on the results in (Zorzi 2004), (Zorzi 2003) and (Zorzi 2003-2). The proposed solution considers a realistic channel model including fading channel statistics. In (Skraba 2004), the receiver-based routing is also analysed based on a simple channel model and lossless links. Moreover, the latency performance of the protocol is presented based on different delay functions and collision rates. Although the authors provide insightful results for the receiver-based routing, the impact of physical layer is not considered in the protocol operation. Similarly, in (Ferrara 2005), the routing decision is performed as a result of successive competitions at the medium access level. More specifically, the next hop is selected based on a weighted progress factor, and the transmit power is increased successively until the most efficient node is found. The performance evaluations of all these propositions present the advantages of cross-layer approach at the routing and MAC layers.

Another aspect of cross-layer interaction of MAC and routing layers is through duty cycle operation. Since energy consumption is the major limiting factor in WSNs protocols, putting nodes into sleep as long as they are not required in the network is an energy-saving solution. However, this in effect results in degradation of the network connectivity leading to poor routes being constructed. Hence, routing algorithms need to consider the duty cycle operation at the MAC layer. As an example, a joint scheduling and routing scheme is proposed in (Sichitiu 2004) for periodic traffic in WSNs. In this scheme, the nodes form distributed on-off schedules for each flow in the network while the routes are established such that the nodes are only awake when necessary. Since the traffic is periodic, the schedules are then maintained to favor maximum efficiency. Furthermore, a comprehensive analysis of trade-off between on-off schedules and the connectivity of the network can be found in (Sichitiu 2004). The usage of on-off schedules in a cross-layer routing and MAC framework is also investigated in (van Hoesel 2004). In this work, a TDMA-based MAC scheme is devised, where nodes distributively select their appropriate time slots based on local topology information. The routing protocol also exploits this information for route establishment. The advantages of this approach are presented through comparative simulations with layered approach. Another approach in cross-layering MAC and routing layers is based on interference avoidance. WSNs are characterised by multiple flows from closely located nodes to a single sink. However, if this fact is not considered in route establishment, potential interfering routes can be established. In (Fang 2004), this effect of broadcast nature of MAC on routing is investigated. In this work, MAC interference between routes is minimised by constructing interference-aware routes. The routes are constructed using node codewords that indicate the interference level of nodes and each packet contains a route indicator for route establishment. As a result, the routes are constructed to minimise the interference among them.

4.3 Cross-layer Module Design

In addition to the proposed protocols that focus on pair-wise cross-layer interaction, more general cross-layer approaches among three protocol layers exist. In (Madan 2005), the optimisation of transmission power, transmission rate, and link schedule for TDMA-based WSNs is proposed. The optimisation is performed to maximise the network lifetime, instead of minimising the total average power consumption. In (Cui 2005), joint routing, MAC, and link layer optimisation is proposed. The authors consider a variable-length TDMA scheme and MQAM modulation. The optimisation problem considers energy consumption that includes both transmission energy and circuit processing energy. Based on this analysis, it is shown that single-hop communication may be optimal in some cases where the circuit energy dominates the energy consumption instead of transmission energy. Although the optimisation problems presented in this work are insightful, no communication protocol for practical implementation is proposed. Moreover, the transport layer issues such as congestion and flow control are not considered. A cross-layer approach, which considers routing, MAC, and PHY layers, is also proposed in (Kuruvila 2005). In this work, a MAC protocol is proposed such that the number of acknowledgements sent to the sender depends on the packet reception probability of the node. Moreover, the optimum hop distance to minimise the hop count is found to be less than the transmission range of a node, i.e., $0.72 \cdot R$, which motivates that nodes at the boundary of the transmission range should not be chosen as next hop. Finally, various combinations of greedy and progress-based routing algorithms are simulated showing the advantages of this cross-layer approach over well-known layered protocols. Although the existing solutions incorporate cross-layer interactions into protocol design, the layering concept still remains intact in these protocols. However, there is still much to be gained by rethinking the functionalities of each protocol layer and melting them into a single cross-layer module. The following section presents the overview of solution to cross-layer design in WSNs that incorporates transport, routing, MAC, and physical layer functionalities into a single cross-layer module.

4.3.1 XLM: Cross-layer Module

The cross-layer approach emerged recently still necessitates a unified cross-layer communication protocol for efficient and reliable event communication that considers transport, routing, and medium access functionalities with physical layer (wireless channel) effects for WSNs. Here, we overview a new communication paradigm, i.e., cross-layer module (XLM) for WSNs (Akyildiz 2006). XLM replaces the entire traditional layered protocol architecture that has been used so far in WSNs. The basis of communication in XLM is built on the *initiative* concept. The initiative concept constitutes the core of XLM and implicitly incorporates the intrinsic functionalities required for successful communication in WSN. A node initiates transmission by broadcasting an RTS packet to indicate its neighbours that it has a packet to send. Upon receiving an RTS packet, each neighbour of a node decides to participate in the communication through *initiative determination*. Denoting the initiative as I, it is determined as follows:

$$I = \begin{cases} 1, \; if \begin{cases} \xi_{RTS} \geq \xi_{Th} \\ \lambda_{relay} \leq \lambda_{relay}^{Th} \\ \beta \leq \beta^{\max} \\ E_{rem} \geq E_{rem}^{\min} \end{cases} \\ 0, \; otherwise \end{cases} \tag{4.1}$$

where, ξ_{RTS} is the received SNR value of the RTS packet, λ_{relay} is the rate of packets that are relayed by a node, β is the buffer occupancy of the node, and E_{rem} is the residual energy of the node, while the terms on the right side of the inequalities indicate the associated threshold values for these parameters, respectively. The initiative I is set to 1 if all four conditions in (Eq.4.1) are satisfied. The first condition ensures that reliable links be constructed for communication. The second and third conditions are used for local congestion control in XLM. The second condition prevents congestion by limiting the traffic a node can relay. The third condition ensures that the node does not experience any buffer overflow. The last condition ensures that the remaining energy of a node E_{rem} stays above a minimum value $E_{rem}^{\min}$. The cross-layer functionalities of XLM lie in these constraints that define the initiative of a node to participate in communication. Using the initiative concept, XLM performs local congestion control, hop-by-hop reliability, and distributed operation. For a successful communication, a node first initiates transmission by broadcasting an RTS packet, which serves as a link-quality indicator and also helps the potential destinations to perform receiver-based contention. Then, the nodes that hear this initiation perform initiative determination according to (Eq.4.1). The nodes that decide to participate in the communication contend for routing of the packet by transmitting CTS packets. The waiting time for the CTS packet transmission is determined based on the advancement of a node for routing (Akyildiz 2006). Moreover, the local congestion control component of XLM ensures energy efficient as well as reliable communication by a two-step congestion control. Analytical performance evaluation and simulation experiment results show that XLM significantly improves the communication performance and outperforms the traditional layered protocol architectures in terms of both network performance and implementation complexity.

4.4 Cross-layer Resource Allocation

Although a considerable amount of recent papers have focused on cross-layer design and improvement of protocols for WSNs, a systematic methodology to accurately model and leverage cross-layer interactions is still largely missing. With this respect, the design of networking protocols for multi-hop wireless ad hoc and sensor networks can be interpreted as the (possibly distributed) solution of resource allocation problems at different layers. From an engineering perspective, most networking problem can in fact be seen as resource allocation problem, where users (network nodes) are assigned resources (power, time slots, paths, rates, etc.) under some specified system constraints. Resource allocation in the context of multi-hop wireless networks has been extensively studied in the last few years, typically with the objectives of maximising the network lifetime (Chang 2000), minimising the energy consumption (Melodia 2005), or maximising the network throughput (Jain 2003). However, most of the existing studies decompose the resource allocation problem at different layers, and consider allocation of the resources at each layer separately. In most cases, resource allocation problems are treated either heuristically, or without considering cross-layer interdependencies, or by considering pair-wise interactions between isolated pairs of layers.

A typical example of the tight coupling between functionalities handled at different layers is the interaction between the congestion control and power control mechanisms (Chiang 2005). The congestion control regulates the allowed source rates so that the total traffic load on any link does not exceed the available capacity. In typical congestion control problems, the capacity of each link is assumed to be fixed and predetermined. However, in multi-hop wireless networks, the attainable capacity of each wireless link depends on the interference levels, which in turn depend on the power control policy. Hence, congestion control and power control are inherently coupled and should not be treated separately when efficient solutions are sought. Furthermore, the physical, medium access control (MAC), and routing layers together impact the contention for network resources. The physical layer has a direct impact on multiple access of nodes in wireless channels by affecting the interference at the receivers. The MAC layer determines the bandwidth allocated to each transmitter, which naturally affects the performance of the physical layer in terms of successfully detecting the desired signals. On the other hand, as a result of transmission schedules, high packet delays and/or low bandwidth can occur, forcing the routing layer to change its route decisions. Different routing decisions alter the set of links to be scheduled, and thereby influence the performance of the MAC layer. Several papers in the literature focus on the joint power control and MAC problem and/or power control and routing issues, although most of them study the interactions among different layers under restricted assumptions. In Section 1, we report a set of meaningful examples of papers considering pair-wise resource allocation problems. In particular, we report examples of joint scheduling and power control, joint routing and power control, and joint routing and scheduling. In Section 2, we describe previous work that dealt with cross-layer optimal resource allocation at the physical, MAC, and routing layer. In Section 3, we discuss recent work on cross-layer design techniques developed within the framework of network utility maximisation. Since these techniques often naturally lead to decompositions of the given problem and to distributed implementation, these can be considered promising results towards the development of systematic techniques for cross-layer design of sensor networks. Most of the papers described in this section consider general models of multi-hop wireless networks, and try to derive general methodologies for cross-layer design of wireless networks. Hence, unless otherwise specified, the techniques described here equally apply to the design of sensor networks and general purpose ad hoc networks.

4.4.1 Pair-wise Resource Allocation

Several recent papers have considered the problem of jointly optimised resource allocation at two layers of the protocol stack. Typical examples of pair-wise resource allocation are joint scheduling and power control, joint routing and power control, and joint routing and scheduling. Joint scheduling and power control is discussed in (ElBatt 2002), where the problem of scheduling the maximum number of links in the same time slot is studied. The objective of the paper is to develop a power control based multiple access algorithm for contention-based wireless ad hoc networks, so as to maximise the network per-hop throughput. To this end, the transmit powers are set to their minimum required levels such that all transmissions achieve a target signal to interference and noise ratio (SINR) threshold. In (Kozat 2004), the joint power control and scheduling problem is addressed under the assumption that the session paths are already given. This work aims at satisfying the rate requirements of the sessions not only in the long term, as considered in (Cruz 2003), but also in the short term, in order to prevent the sessions with low jitter or bounded delay requirement from suffering from the ambiguity of the long term guarantees. The main contribution in (Kozat 2004) is the formulation of a quality of service (QoS) framework that is able to capture both the different definitions of QoS from network layer to physical layer, and the general requirements of the individual sessions. The need for close interactions between these layers is demonstrated, and it is pointed out that independent decisions at different layers for achieving a local objective would deteriorate the performance of other layers. The impact of interference generated at the physical layer on the routes selected at the network layer has also been analysed in the literature. In (Jain 2003), the authors derive a methodology for computing bounds on the optimal throughput that can be supported by a multi-hop wireless network. The wireless interference generated among network links is modelled as a conflict graph, and interference-aware optimal routes are calculated with a centralised algorithm, and shown by ns-2 simulation to lead to a throughput improvement of a factor 2 with respect to interference-unaware shortest path routes. A different example of joint resource allocation at the physical and routing layers is discussed in (Melodia 2005). The analytical framework proposed allows analysing the relationship between the energy efficiency of the geographical routing functionality and the extension of the topology knowledge range for each node. A wider topology knowledge may improve the energy efficiency of the routing tasks but increases the cost of topology information due to signalling packets needed to acquire this information. The problem of determining the optimal topology knowledge range for each node to make energy efficient geographical routing decisions is tackled by integer linear programming. Finally, joint routing and scheduling for multi-hop wireless networks has been considered in (Kodialam 2003). The authors determine necessary and sufficient conditions for the achievability of a given rate vector between sources and sink, and develop polynomial-time algorithms for solving the routing and scheduling problem. However, only primary interference is considered, i.e., the only constraint is that each node can communicate with at most one other node at any given time.

4.4.2 Joint Routing, Scheduling, and Power Control

A few recent papers have considered the problem of jointly determining optimal strategies for routing, scheduling and power control, thus formulating joint resource allocation problems at three layers of the protocol stack. The problem of joint routing, link scheduling, and power control to support high data rates for broadband wireless multi-hop networks is analysed in (Cruz 2003). In particular, the work focuses on the minimisation of the total average transmission power, subject to given constraints on the minimum average data rate per link, as well as peak transmission power constraints per node. Interestingly, it is shown that, even though the focus is on minimising transmit power, the optimal joint allocation does not necessarily route traffic over minimum energy paths.

A joint scheduling, power control, and routing algorithm for multi-hop wireless networks has been proposed in (Li 2005). The authors assume a TDMA-based wireless ad hoc network and design a centralised algorithm for the joint solution. However, the algorithm is of limited practical interest as it is strictly centralised and suboptimal. In (Madan 2005), the problem of maximising the network lifetime of a wireless sensor network is tackled by defining a mixed integer convex problem that involves joint resource allocation at the physical, MAC, and routing layers. The problem is shown to be hard to solve, and the authors propose an approximate method that iteratively solves a series of convex optimisation problems. The algorithm is guaranteed to converge in a finite number of iterations, and the computation during each iteration can be decomposed into smaller subproblems and performed distributively over the network. However, this requires extensive communication among the nodes to communicate updated variables of each subproblem. Hence, although the proposed methodology is interesting, the practicality of the proposed method still needs to be demonstrated.

While most papers on cross layer design generally focus on CDMA- or TDMA-based wireless networks, a few recent efforts were concerned with the cross-layer design of Ultra-wideband (UWB) (Win 2000) networks. The jointly optimal power control, scheduling, and routing problem is formulated in (Radunovic 2004) for time-hopping Ultra-wideband networks, with the objective of maximising log-utility of flow rates subject to power constraint nodes. The problem is then solved with centralised algorithm and the focus is primarily on exposing peculiar features of networking with ultra-wideband. In particular, it is shown that, given the optimisation objective, power control is not needed, the design of the optimal MAC protocol is independent of the choice of the routing protocol, and that transmitting over minimum energy routes is always optimal even though the objective is maximising the rate.

In (Shi 2005), the authors consider the joint optimal power control, scheduling, and routing problem for Multi Carrier - Orthogonal Frequency Division Multiplexing Ultra-wideband (MC-OFDM UWB). A rate feasibility problem is considered, i.e., the problem of determining whether given a set of source sensor nodes generating data at a certain data rate, it is possible to relay this data successfully to the sink. The focus is more on developing an efficient centralised solution procedure for the problem. With respect to (Radunovic 2004), due to the different optimisation objectives, and to different assumptions on the form of the rate function (linear with the Signal-to-Interference ratio in (Radunovic 2004), logarithmic in (Shi 2005)), some of the results regarding power control and routing do not hold anymore.

4.4.3 Joint Resource Allocation Based on Dual Decomposition

Although some of the research efforts described above are extremely insightful, especially in exposing interdependencies among different layers, they mostly fail to clearly lay a unified foundation for the cross-layer design of sensor network protocols. In particular, desirable features of cross-layer design methodologies should allow transparently capturing and controlling the interdependencies of functionalities handled at different layers of the protocol stack. At the same time, it is desirable to somehow maintain some form logical separation. Desirable features are in fact vertical separation of functions, which is concerned with ease of design, and horizontal separation, which enables distributed implementation of network protocols, which is of fundamental importance for sensor networks.

An important step in the direction of the definition of a unified methodology for cross-layer design is constituted by the pioneering work by Low (Low 2003) and Chiang (Chiang 2005), who demonstrated the need to integrate various protocol layers into a coherent framework, to help provide a unified foundation for the analysis of resource allocation problems, and to develop systematic techniques for cross-layer design of multi-hop wireless networks. This can be accomplished within the framework of non-linear optimisation, by formulating a networking problem as a problem of maximising a utility function (Network Utility Maximisation (NUM)) or minimising a cost function over a set of variables confined within a constraint set. The objective of these efforts is twofold: first, to mathematically model the interactions of the resource allocation problem at different layers; second, to understand and control the interactions between quantities affecting the performance at different layers. This is aided by recent developments in nonlinear optimisation theory that provide powerful tools to solve even large-scale non-linear problems within reasonable time. Furthermore, advanced nonlinear optimisation techniques may naturally lead to decompositions based on Lagrange duality theory that, while maintaining some form of logical separation among the layers, may be suitable for distributed implementation. Hence, the development of these frameworks can lead to new theoretical results and to practical new design perspectives.

These results are built on recently developed non-linear optimisation theory for the design of communication systems, in particular convex optimisation (Boyd 2004) and geometric programming (Chiang 2005-2). The main technique used in these papers is the method of dual decomposition for convex optimisation problems. This technique is used by Low in (Low 2003), where the focus is on the design of TCP algorithms. The parameters describing congestion are interpreted as primal and dual optimisation variables, while the TCP protocol is interpreted as a distributed primal-dual algorithm solving an implicitly defined distributed network utility maximisation problem. This is in line with the work by Kelly et al. (Kelly 1998), where congestion control protocols are shown to approximate distributed algorithms that implicitly solve network utility maximisation problems. In (Chiang 2005), Chiang formulates the problem of joint optimisation of transmitted power levels and congestion window sizes. A multi-hop wireless network with interference-limited links is considered, and a delay-based congestion avoidance mechanism is modelled. The objective is to maximise a utility function of the source rates, hence to optimise the network throughput. The amount of bandwidth supplied to the upper layers is nonlinearly coupled to the bandwidth demanded by the congestion control through a dual variable. A quantitative framework for joint design of transport and physical layer protocols is provided, theorems of convergence are proved, and a sub-optimal version of the algorithm is proposed for scalable architectures. In (Lee 2006), price-based distributed algorithms are proposed for achieving optimal rate-reliability trade-offs (RRTO) in the framework of network utility maximisation. The utility of each user depends both on the achieved rate and on the reliability of the data stream, with a clear trade-off between the two. The authors consider networks where the rate-reliability trade-off is controlled by adapting channel code rates at the physical layer, and propose distributed algorithms to achieve the optimal trade-off. Moreover, they extend the framework to wireless MIMO multi-hop networks, in which diversity and multiplexing gains of each link are controlled to achieve the optimal RRTO.

In (Chen 2006), the cross-layer design of congestion-control, routing, and scheduling is jointly tackled by extending the framework of network utility maximisation. By dual decomposition, the resource allocation problem is decomposed into the three sub-problems (congestion control, routing, and scheduling) that interact through congestion prices. Based on this decomposition, a distributed subgradient algorithm is derived, which is shown to converge to the optimal solution, and to be implemented with low communication overhead. The convex optimisation approach guarantees global optimality of the proposed algorithm, and excludes the possibility of unintended cross-layer interaction. A joint source coding, routing, and channel coding problem for wireless sensor networks is formulated in (Yu 2005). Distributed source coding can be used to reduce the data rate of sensors observing spatially and temporally correlated phenomena. Hence, the proposed resource allocation framework jointly considers physical, network, and application layer. The joint optimisation problem in the dual domain leads to separation of the different sub problems at the different layers, and the interfaces between the different layers are the Lagrange multipliers or dual optimisation variables. A primal-dual method is proposed for distributed solution. In (Yuan 2005), an optimisation framework is proposed for the throughput optimisation problem in wireless networks, which jointly optimises multicast routing with network coding, and power allocation. A a primal-dual method is used to decompose the original problem into two sub-problems, data routing and power allocation, and the dual variables play the role of coordinating network layer demand for bandwidth and physical layer supply. An optimal primal-dual algorithm is developed to find the solution of the problem, suitable for distributed implementation.

4.5. Open Research Problems

As explained in Sections 2, 3 and 4, there exists remarkable effort on cross-layer design in order to develop new communication protocols. However, there is still much to be gained by rethinking the protocol functions of network layers in a unified way so as to provide a single communication module that limits the duplication of functions, which often characterises a layered design, and achieves global design objectives of sensor networks, such as minimal energy consumption and maximum network lifetime. In fact, research on cross-layer design and engineering is interdisciplinary in nature and it involves several research areas such as adaptive coding and modulation, channel modelling, traffic modelling, queuing theory, network protocol design, and optimisation techniques. There are several open research problems toward the development of systematic techniques for cross-layer design of wireless sensor network protocols. It is needed to acquire an improved understanding of energy consumption in WSNs. In fact, existing studies on cross-layer optimisation are mostly focused on jointly optimising functionalities at different layers, usually with the overall objective of maximising the network throughput. Conversely, in WSNs the ultimate objective is usually to minimise the energy consumption and/or to maximise the network lifetime. Hence, further study is needed to develop models and methodologies suitable to solve energy-oriented problems.

It is also necessary to develop sound models to include an accurate description of the end-to-end delay in the above framework as results from the interaction of the different layers. In particular, there is a need to develop mathematical models to accurately describe contention at the MAC layer. This would allow determining the set of feasible concurrent transmissions under different MAC strategies. This is particularly important for the design of sensor network protocols for monitoring applications that require real-time delivery of event data, such as those encountered in wireless sensor and actor networks (WSAN) (Akyildiz 2004). Moreover, characteristics of the physical layer communication, such as modulation and error control, that impact the overall resource allocation problem should be incorporated in the cross-layer design. For example, in future wireless communications, adaptive modulation could be applied to achieve better spectrum utilisation. To combat different levels of channel errors, adaptive forward error coding (FEC) is widely used in wireless transceivers. Further, joint consideration of adaptive modulation, adaptive FEC, and scheduling would provide each user with the ability to adjust the transmission rate and achieve the desired error protection level, thus facilitating the adaptation to various channel conditions (Liu 2005)(Cui 2005-2).

Another important open research issue is to study the network connectivity with realistic physical layer. Connectivity in wireless networks has been previously studied (Gupta 1998)(Bettstetter 2002), i.e., stochastic models have been developed to determine conditions under which a network is connected. These results, however, cannot be straightforwardly used, as they are based on the so-called unit disk graph communication model. Recent experimental studies, however, have demonstrated that the effects of the impairments of the wireless channel on higher-layer protocols are not negligible. In fact, the availability of links fluctuates because of channel fading phenomena that affect the wireless transmission medium. Furthermore, mobility of nodes is not considered. In fact, due to node mobility and node join and leave events, the network may be subject to frequent topological reconfigurations. Thus, links are continuously established and broken. For the above reasons, new analytical models are required to determine connectivity conditions that incorporate mobility and fading channels. Last but not least, new cross-layer network simulators need to be developed. Current discrete-event network simulators such as OPNET, ns-2, J-Sim, GloMoSim may be unsuitable to implement a cross-layer solution, since their inner structure is based on a layered architecture, and each implemented functionality run by the simulator engine is tightly tied to this architecture. Hence, implementing a cross-layer solution in one of these simulators may turn into a non-trivial task. For this reason, there is a need to develop new software simulators that are based on a new developing paradigm so as to ease the development and test of cross-layer algorithmic and protocol solutions.

4.6 Precautionary Guidelines in Cross-layer Design

In this section, we describe possible risks rising when a cross-layer approach is followed, and propose precautionary guidelines and principles for cross-layer design beyond the presented open research issues. As stressed in Sections 2 and 4, the increased interactions and dependencies across layers turn into an interesting optimisation opportunity that may be worth exploiting. Cross-layer design, in fact, makes the interfaces among different functionalities open and yields much more optimised solutions for resource-constrained devices. Following this intuition, many cross-layer design papers that explore a much richer interaction between parameters across layers have been proposed in the recent past. While, however, as an immediate outcome most of these cross-layer suggestions may yield a performance improvement in terms of throughput or delay, this result is often obtained by decreasing the architecture modularity, and by loosing the logical separation between designers and developers. This abstraction decoupling is needed to allow the former to understand the overall system, and the latter to realise a more efficient production. For these reasons, when a cross-layer solution is proposed, the system performance gain needs to be weighed against the possible longer-term downfalls raised by a diminished degree of modularity.

In (Kawadia 2005), the authors re-examine holistically the issue of cross-layer design and its architectural ramifications. They contend that a good architectural design leads to proliferation and longevity of a technology, and illustrate this with some historical examples. The first is John von Neumann's architecture for computer systems, which is at the origin of the separation of software and hardware; the second is represented by the layered OSI architecture for networking, which provides the base of the current Internet architecture success; another example is provided by the Shannon's architecture for communication systems, which motivated the non-obvious separation of source and channel coding; and, last but not least, the plant controller feedback paradigm in control systems, which provides universal principles common to human engineered systems as well as biological systems. Although the concerns and cautionary advice expressed in (Kawadia 2005) about cross-layer design are sound and well motivated, the layered-architecture, which turned to be a successful design choice for wired networks, may need to be carefully rethought for energy-constrained WSNs, where the concept itself of 'link' is liable and readily open to different interpretations, and many different effective transmission schemes and communication paradigms are conceivable. In fact, since the wireless medium is fundamentally different from the wired medium, the applicability of the OSI protocol stack to energy-constrained wireless networks such as WSNs needs to be carefully evaluated and possibly a completely different architecture following a cross-layer paradigm may be more suitable for wireless protocols.

This is also the conclusion drawn in (Toumpis 2003), where the pros and cons of cross-layer design approach are evaluated. In (Toumpis 2003), cross-layer design to improve reliability and optimise performance is advocated, although the design needs to be cautiously developed to provide long-term survivability of cross-layer architectures. In the following, we present some concerns and precautionary considerations, which need to be considered when a cross-layer design architecture is proposed, and suggest some possible research directions.

One of the most important concerns about cross-layer design is the degree of modularity that can be traded off for communication efficiency. In the classical layered design approach, system architecture is broken down into modular components, and the interactions and dependencies between these components are systematically specified. This design philosophy allows breaking complex problems into easier subproblems, which can then be solved in isolation, without considering all the details concerning the overall system. This approach guarantees the inter-operability of subsystems in the overall system once each subsystem is tested and standardised, leading to quick proliferation of technology and mass production. Conversely, a cross-layer design approach may loose the decoupling between design and development process, which may impair both the design and the implementation development and slow the innovation down.

A second concern involves system enhancement when a cross-layer approach is followed in the design of a communication system. In fact, design improvements and innovations may become difficult in a cross-layer design, since it will be hard to assess how a new modification will interact with the already existing solutions. Furthermore, a cross-layer architecture would be hard to upkeep, and the maintaining costs would be high. In the worst cases, rather than modifying just one subsystem, the entire system may need to be replaced. For these reasons, we advocate keeping some degree of modularity in the design of cross-layer solutions. This could be achieved by relying on functional entities - as opposed to layers in the classical design philosophy - that implement particular functions. This would also have the positive consequence of limiting the duplication of functions that often characterises a layered design. This functional redundancy is, in fact, one the causes for poor system performance.

Another important concern in cross-layer design is the risk of instability, which is worse than in layered architecture design. In cross-layer design, in fact, the effect of any single design choice may affect the whole system, leading to various negative consequences such as instability. This is a non trivial problem to solve, since it is well known from control theory that stability is a dominant issue in system design. Moreover, the fact that some interactions are not easily foreseen makes cross-layer design choices even trickier. Hence, great care should be paid to prevent design choices from negatively affecting the overall system performance. To this purpose, there is a need to integrate and further develop control theory techniques to study stability properties of system designed following a cross-layer approach. Dependency graphs, which may be used to capture the dependency relation between parameters, could be valuable means to prove stability, although it is hard to implement in some cases.

Besides stability, there is also the issue of robustness, which is related with the possible lack of modularity in cross-layer architectures. Robustness is the property of a system to be able to absorb parameter uncertainties, e.g., due to mismatch between estimated and actual statistics, and, in general, the degrading effect on the overall performance experienced by a system when unpredictable events occur such as transmission anomalies, channel impairments, loss of connectivity, or failures (Liu 2005). Techniques such as timescale separation and performance tracking and verification may need to be employed in a design phase to separate interactions and verify the system performance on-the-fly, as it is suggested in (Kawadia 2005). Moreover, an accompanying theoretical framework may be needed to fully support cross-layer design and study its robustness properties beforehand.

4.7 Conclusions

In this chapter, we reviewed and classified literature on cross-layer protocols, improvements, and design methodologies for wireless sensor networks (WSNs). We overviewed the communication protocols devised for WSNs that focus on cross-layer design techniques. We classified these techniques based on the network layers they aim at replacing in the classical OSI network stack. Furthermore, we discussed systematic methodologies for the design of cross-layer solution for sensor networks as resource allocation problems in the framework of non-linear optimisation. We outlined open research issues in the development of cross-layer methodologies for sensor networks and discussed possible research directions.

A cross-layer design methodology for energy-constrained wireless sensor networks is an appealing approach as long as cross-layer interactions are thoroughly studied and controlled. As pointed out in this chapter, in fact, no cross-layer dependency should be left unintended, since this may lead to poor performance of the entire system.

4.8 References

Akan OB and Akyildiz IF (2005) Event-to-sink reliable transport in wireless sensor networks. IEEE/ACM Transactions on Networking 13:5, pp 1003-1017

Akyildiz IF, Su W, Sankarasubramaniam Y, and Cayirci E (2002) Wireless sensor networks: a survey, Computer Networks (Elsevier) Journal 38:4, pp 393-422

Akyildiz IF and Kasimoglu IH (2004) Wireless sensor and actor networks: research challenges. Ad Hoc Networks Journal (Elsevier) 2:4, pp 351-367

Akyildiz IF, Vuran MC, and Akan OB (2006) Cross-layer module (XLM) for wireless sensor networks. In: Proc. CISS, Princeton, NJ

Bettstetter C (2002) On the minimum node degree and connectivity of a wireless multihop network. In: Proc. ACM MobiHoc 2002, Lausanne, Switzerland, pp 80-91

Boyd S and Vandenberghe L (2004) Convex optimisation. Cambridge University Press

Chang JH, Tassiulas L (2000) Energy conserving routing in wireless ad hoc networks. In: Proc. IEEE Infocom 2000, pp 22-31

Chen L, Low SH, Chiang M, and Doyle JC (2006) Jointly optimal congestion control, routing, and scheduling for wireless ad hoc networks. In: Proc. IEEE Infocom 2006, Barcelona

Chiang M (2005) Balancing transport and physical layers in wireless multihop networks: Jointly optimal congestion control and power control. IEEE Journal of Selected Areas in Communications, 23:1, pp 104-116

Chiang M (2005) Geometric programming for communication systems. Short monograph in Foundations and Trends in Communications and Information Theory, 2:1-2, pp 1-154

Cruz R and Santhanam A (2003) Optimal routing, link scheduling and power control in multi-hop wireless networks. In: Proc. IEEE Infocom 2003, San Francisco, pp 702-711

Cui S, Madan R, Goldsmith A, and Lall S (2005) Joint routing, MAC, and link layer optimisation in sensor networks with energy constraints. In: Proc. IEEE ICC 2005, pp 725-729

Cui S, Goldsmith AJ, and Bahai A (2005) Energy-constrained modulation optimisation. IEEE Trans. on Wireless Communications, 4:5, pp 2349-2360

ElBatt T and Ephremides A (2002) Joint scheduling and power control for wireless ad-hoc networks. In: Proc. IEEE Infocom 2002, New York, pp 976-984

Fang Y and McDonald B (2004) Dynamic codeword routing (DCR): a cross-layer approach for performance enhancement of general multi-hop wireless routing. In: Proc. IEEE SECON 2004, pp 255-263

Ferrara D, Galluccio L, Leonardi A, Morabito G, and Palazzo S (2005) MACRO: An integrated MAC/routing protocol for geographical forwarding in wireless sensor networks. In: Proc. IEEE Infocom 2005, pp 1770-1781

Ganesan D, Krishnamachari B, Woo A, Culler D, Estrin D, and Wicker S (2002) An empirical study of epidemic algorithms in large scale multihop wireless networks. Technical Report, Intel Research

Gupta P and Kumar PR(1998) Critical power for asymptotic connectivity in wireless networks. Stochastic Analysis, Control, Optimisation and Applications: a volume in honor of Fleming WH, McEneaney WH, Yin G, Zhang Q, Birkhauser, Boston, 1998

Haapola J, Shelby Z, Pomalaza-Raez C, and Mahonen P (2005) Cross-layer energy analysis of multi-hop wireless sensor networks. In: Proc. EWSN 2005, pp 33-44

van HL, Nieberg T, Wu J, and Havinga PJM (2004) Prolonging the lifetime of wireless sensor networks by cross-layer interaction. IEEE Wireless Communications, 11:6, pp 78-86

Jain K, Padhye J, Padmanabhan V, and Qiu L (2003) Impact of interference on multi-hop wireless network performance. In: Proc. ACM Mobicom 2003, San Diego, CA, USA

Kawadia V and Kumar PR (2005) A cautionary perspective on cross-layer design. IEEE Wireless Communications Magazine, 12:1, pp 3-11

Kelly FP, Maulloo A, and Tan D (1998) Rate control for communication networks: shadow prices, proportional fairness, and stability. J. of the Operations Research Society, 49:3, pp 237-252

Kodialam M and Nanagopal T (2003) Characterising achievable rates in multi-hop wireless networks: the joint routing and scheduling problem. In: Proc. ACM Mobicom 2003, San Diego

Kozat UC, Koutsopoulos I, and Tassiulas L (2004) A framework for cross-layer design of energy-efficient communication with QoS provisioning in multi-hop wireless networks. In: Proc. INFOCOM 2004, Honk Kong, pp 1446-1456

Kuruvila J, Nayak A, and Stojmenovic I (2005) Hop count optimal position based packet routing algorithms for ad hoc wireless networks with a realistic physical layer. IEEE Journal of Selected Areas in Communications, 23:6, pp 1267-1275

Lee JW, Chiang M, and Calderbank R (2006) Price-based distributed algorithms for optimal rate-reliability trade-off in network utility maximisation. To appear in IEEE Journal of Selected Areas in Communications

Li Y and Ephremides A (2005) Joint scheduling, power control, and routing algorithm for ad-hoc wireless networks. In: Proc. Hawaii Int. Conf. on System Sciences, Big Island, HI, USA

Liu Q, Zhou S, and Giannakis GB (2005) Cross-layer scheduling with prescribed QoS guarantees in adaptive wireless networks. Journal of Selected Areas in Communications, 23:5, pp 1056-1066

Low SH (2003) A duality model of TCP and queue management algorithms. IEEE/ACM Transactions on Networking, 11:4, pp 525-526

Madan R, Cui S, Lall S, and Goldsmith A (2005) Cross-layer design for lifetime maximisation in interference-limited wireless sensor networks. In: Proc. IEEE INFOCOM 2005, pp 1964-1975

Melodia T, Pompili D, and Akyildiz IF (2005) On the interdependence of distributed topology control and geographical routing in ad hoc and sensor networks. Journal of Selected Areas in Communications, 23:3, pp :520-532

Radunovic B and Le Boudec JY (2004) Optimal power control, scheduling, and routing in WWB networks. IEEE J. of Selected Areas in Communications, 22:7, pp 1252-1270

Rossi M and Zorzi M (2005) Cost efficient localised geographical forwarding strategies for wireless sensor networks. In: Proc. TIWDC 2005, Sorrento, Italy

Seada K, Zuniga M, Helmy A, and Krishnamachari B (2004) Energy-efficient forwarding strategies for geographic routing in lossy wireless sensor networks. In: Proc. ACM Sensys 2004, Baltimore, USA

Shi Y, Hou T, Sherali HD, and Midkiff SF (2005) Cross-layer optimisation for routing data traffic in UWB-based sensor networks. In: Proc. ACM Mobicom, Cologne, Germany

Shih E, Cho S, Ickes N, Min R, Sinha A, and Wang A, Chandrakasan A (2001) physical layer driven protocol and algorithm design for energy-efficient wireless sensor networks. In: Proc. ACM MOBICOM, Rome, Italy

Sichitiu ML (2004) Cross-layer scheduling for power efficiency in wireless sensor networks. In: Proc. IEEE Infocom 2004, Hong Kong, pp 1740-1750

Skraba P, Aghajan H, and Bahai A (2004) Cross-layer optimisation for high density sensor networks: distributed passive routing decisions. In: Proc. Ad-Hoc Now, Vancouver, Canada

Toumpis S and Goldsmith AJ (2003) Performance, optimisation, and cross-layer design of media access protocols for wireless ad hoc networks. In: Proc. ICC 2003, Seattle, WA, USA, pp 2234-2240

Vuran MC, Akan OB, and Akyildiz IF (2004) Spatio-Temporal Correlation: Theory and Applications for Wireless Sensor Networks. In: Computer Networks Journal (Elsevier), 45:3, pp 245-261

Vuran MC, Gungor VC, and Akan OB (2005) On the interdependency of congestion and contention in wireless sensor networks. In: Proc. SENMETRICS, San Diego, pp 136-147

Vuran MC and Akyildiz IF (2006) Spatial correlation-based collaborative medium access control in wireless sensor networks. To appear in IEEE/ACM Trans. on Networking

Vuran MC and Akan OB (2006) Spatio-temporal characteristics of point and field sources in wireless sensor networks. In: Proc. IEEE ICC 2006, Istanbul, Turkey

Win MZ and Scholtz RA (2000) Ultra-wide BW time-hopping spread-spectrum impulse radio for wireless multiple-access communications. IEEE Trans. Commun., 48:4, pp 679-691

Yu Wand Yuan J (2005) Joint source coding, routing, and resource allocation for wireless sensor networks. In: Proc. IEEE ICC 2005, Seoul, pp 737-741

Yuan J, Li Z, Yu W, and Li B (2005) A cross-layer optimisation framework for multicast in multi-hop wireless networks wireless internet. In: Proc. WICON 2005, Budapest, pp 47-54

Zorzi M and Rao R (2003) Geographic random forwarding (GeRaF) for ad hoc and sensor networks: multihop performance. IEEE Trans. Mobile Computing, 2:4, pp 337-348

Zorzi M and Rao R (2003) Geographic random forwarding (GeRaF) for ad hoc and sensor networks: energy and latency performance. IEEE Trans. Mobile Computing, 2:4, pp 349-365

Zorzi M (2004) A new contention-based MAC protocol for geographic forwarding in ad hoc and sensor networks. In Proc. IEEE ICC 2004, Paris, pp 3481-3485

Zuniga M and Krishnamachari B (2004) Analysing the transitional region in low power wireless links. In: Proc. IEEE SECON, Santa Clara, CA, pp 517-526

5 SensorGrid Architecture for Distributed Event Classification

Chen-Khong Tham

Dept of Electrical and Computer Engineering, National University of Singapore

5.1 Introduction

Recent advances in electronic circuit miniaturisation and micro-electromechanical systems (MEMS) have led to the creation of small sensor nodes which integrate several kinds of sensors, a central processing unit (CPU), memory and a wireless transceiver. A collection of these sensor nodes forms a *sensor network* which is easily deployable to provide a high degree of visibility into real-world physical processes as they happen, thus benefiting a variety of applications such as environmental monitoring, surveillance and target tracking. Some of these sensor nodes may also incorporate actuators such as buzzers and switches which can affect the environment directly. We shall simply use the generic term 'sensor node' to refer to these sensor-actuator nodes as well. A parallel development in the technology landscape is *grid computing*, which is essentially the federation of heterogeneous computational servers connected by high-speed network connections. Middleware technologies such as Globus (2006) and Gridbus (2005) enable secure and convenient sharing of resources such as CPU, memory, storage, content and databases by users and applications. This has caused grid computing to be referred to as 'computing on tap', utility computing and IBM's mantra, 'on demand' computing. Many countries have recognised the importance of grid computing for 'eScience' and the grid has a number of success stories from the fields of bioinformatics, drug design, engineering design, business, manufacturing, and logistics.

There is some existing work on the intersecting fields of sensor networks and grid computing which can be broadly grouped into three categories: (1) 'sensorwebs', (2) sensors to grid, and (3) sensor networks to grid. In the first category of 'sensorwebs', many different kinds of sensors are connected together through middleware to enable timely and secure access to sensor readings. Examples are the SensorNet effort by the Oak Ridge National Laboratories (ORNL) and NIST (USA) which aims to collect sensor data from different places to facilitate the operations of the emergency services; the IrisNet effort by Intel to create the 'seeing' Internet by enabling data collection and storage, and the support of rich queries over the Internet; and the Department of Defense (USA) ForceNet which integrates many kinds of sensor data to support air, sea and land military operations.

In the second category of sensors to grid, the aim is to connect sensors and instruments to the grid to facilitate collaborative scientific research and visualisation (Chiu and Frey 2005). Examples are the efforts by the Internet2 and eScience communities in areas such as the collaborative design of aircraft engines and environment monitoring; DiscoveryNet (Imperial College UK) which aims to perform knowledge discovery and air pollution monitoring; and the earthquake science efforts by the CrisisGrid team (Indiana University USA) and iSERVO (International Solid Earth Research Virtual Observatory). Finally, in the third category of sensor networks to grid, the aim is mainly to use grid web services to integrate sensor networks and enable queries on 'live' data. Examples are efforts by Gaynor et al (2004) to facilitate quicker medical response and supply chain management; and the SPRING framework proposed by Lim et al (2005).

Our focus and approach, which we refer to as *sensor-grid computing* executing on an integrated *sensor-grid architecture* or simply '*SensorGrid*' for short (Tham and Buyya 2005), see Fig. 5.1 builds on the three categories of existing work described above and aims to achieve more by exploiting the complementary strengths and characteristics of sensor networks and grid computing. Sensor-grid computing combines the real-time acquisition and processing of data about the environment by sensor networks with intensive distributed computations on the grid. This enables the construction of real-time models and databases of the environment and physical processes as they unfold, from which high-value computations such as analytics, data mining, decision-making, optimisation and prediction can be carried out to generate 'on-the-fly' results. This powerful combination would enable, for example, effective early warning of threats (such as missiles) and natural disasters, and real-time business process optimisation.

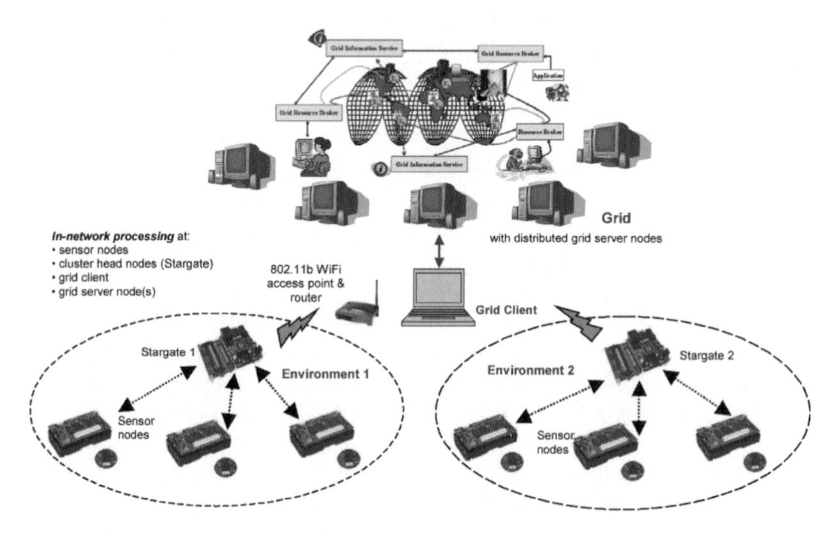

Fig. 5.1. SensorGrid architecture integrating sensor networks[1] and grid computing

[1]http://www.xbow.com

One other key aspect of sensor-grid computing is the emphasis on distributed and hierarchical *in-network processing* at several levels of the SensorGrid architecture. As will be explained later, the sensor-grid computing approach is more robust and scalable compared to other approaches in which computations are mainly performed on the grid itself. The sensor-grid computing approach together with the SensorGrid architecture enable more timely responses to be achieved and useful results to be available even in the presence of failures in some parts of the architecture. The organisation of this chapter is as follows. In Section 2, we describe a simple *centralised* approach to realise sensor-grid computing. We then point out its weaknesses and propose a *distributed* approach. In Section 3, we describe two applications of distributed sensor-grid computing which we have implemented. In Section 4, several challenges and research issues related to sensor-grid computing are discussed. Finally, we conclude in Section 5.

5.2 Approaches to Sensor-Grid Computing

One simple way to achieve sensor-grid computing is to simply connect and interface sensors and sensor networks to the grid and let all computations take place there. The grid will then issue commands to the appropriate actuators. In this case, all that is needed are high-speed communication links between the sensor-actuator nodes and the grid. We refer to this as the *centralised sensor-grid computing* approach executing on a centralised SensorGrid architecture.

However, the centralised approach has several serious drawbacks. Firstly, it leads to excessive communications in the sensor network which rapidly depletes its batteries since long range wireless communications is expensive compared to local computation. It also does not take advantage of the in-network processing capability of sensor networks, which permits computations to be carried out close to the source of the sensed data.

In the event of communication failure, such as when wireless communication in the sensor network is unavailable due to jamming, the entire system collapses. Other drawbacks include long latencies before results are available on the field since they have to be communicated back from the grid, and possible overloading of the grid (although this is unlikely).

The more robust and efficient alternative is the decentralised or *distributed sensor-grid computing* approach, which executes on a distributed *SensorGrid* architecture and alleviates most of the drawbacks of the centralised approach. The distributed sensor-grid computing approach involves in-network processing within the sensor network and at various levels of the SensorGrid architecture. We present specific realisations of the SensorGrid architecture with sensor nodes which are Crossbow motes, Stargate nodes, Hewlett Packard iPAQ nodes, grid clients and the grid comprising a number of grid server nodes. Fig. 5.2 shows several possible configurations of the SensorGrid architecture. The role and characteristics of each of the components and the features of the different configurations will be discussed later in this chapter.

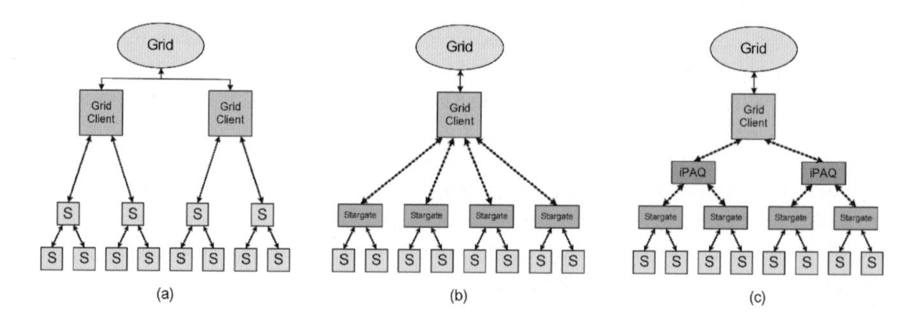

Fig. 5.2. Several possible configurations of the SensorGrid architecture (The squares with 'S' denote sensor nodes. Wired links - solid lines; wireless links - dotted lines.)

5.3 Distributed Sensor-Grid Computing Applications

The distributed sensor-grid computing approach is highly suitable for a number of distributed applications such as analytics, data mining, optimisation and prediction. In this chapter, we present a distributed information fusion, event detection and classification application, and a distributed autonomous decision-making application, as two examples of sensor-grid computing on the SensorGrid architecture.

5.3.1 Information fusion, event detection and classification

Since the nodes in a sensor network are independently sensing the environment, this gives rise to a high degree of redundant information. However, due to the severely resource-constrained nature of sensor nodes, some of these readings may be inaccurate. Information fusion algorithms, which compute the most probable sensor readings, have been studied extensively over the years, especially in relation to target detection and tracking. There are two methods for performing information fusion in the context of classification (Brooks et al 2003): (i) *decision fusion*, which is done when the different measurements are statistically independent, and (ii) *data* or *value fusion*, which is done when different measurements yield correlated information. In the case of decision fusion, classification is done at individual sensor nodes or sensing modalities and the classification results are combined or fused at the *fusion center* in a number of ways, such as by applying the product or sum rule on the likelihood estimates. In the case of data fusion, feature vectors from individual sensors are concatenated and a classification algorithm is applied on the concatenated feature vector at the fusion center. Although data fusion is likely to yield better classification performance, it is more expensive to implement from the communication and computational points of view. If there are M sensing modalities at each sensor node, K sensor nodes and n dimensions in each measurement, data fusion typically requires the transmission of $(K\text{-}1)Mn$ values to the fusion center (which is one of the K sensor nodes) and the classification algorithm there would need to operate on a KMn-dimensional concatenated feature vector. In contrast, decision fusion requires a classification operation to be applied on either an n or Mn-dimensional feature vector at each sensor node, followed by the communication of (K-1) decisions to the fusion center and the application of a decision fusion algorithm on the K component decisions there. In the subsequent parts of this sub-section, we will describe the design and implementation of a distributed decision fusion system for event detection and classification using the SensorGrid architecture and evaluate its performance.

5.3.2 Distributed Decision Fusion

To begin, we review basic statistical pattern classification theory and describe the *optimal decision fusion* (ODF) algorithm proposed by Duarte and Hu (2003). In statistical pattern classification, a feature vector x is assumed to be drawn from a probability distribution with probability density function $p(x)$. Due to a natural assignment process beyond our control, each sample x is assigned a class label C_n and we say that $x \in C_n$. The probability that any sample x belongs to a class C_n, denoted by $P(x \in C_n) = P(C_n)$, is known as the *prior probability*. The *likelihood* that a sample will assume a specific feature vector x given that it is drawn from a class C_n is denoted by the conditional probability $p(x \mid C_n) = p(x \mid x \in C_n)$. Using Bayes' rule, the probability that a particular sample belongs to class C_n, given that the sample assumes the value of x, is denoted by the *a posteriori* probability:

$$p(x \in C_n \mid x) = p(C_n \mid x) = \frac{p(x \mid C_n) P(C_n)}{p(x)} \tag{5.1}$$

A *decision rule* or *classifier* $d(x)$ maps a measured feature vector x to a class label C_n which is an element of the set of N class labels $\mathbf{C} = \{C_1, C_2, \ldots, C_N\}$, i.e. $d(x) \in \mathbf{C}$. If $x \in C_n$ and $d(x) = C_n$, then a correct classification has been made, otherwise it is a misclassification. The *maximum a posteriori* (MAP) *classifier* chooses the class label among the N class labels that yields the largest maximum a posteriori probability, i.e.

$$d(x) = \arg\max_n P(C_n \mid x) \tag{5.2}$$

where $d(x) = n$ means $d(x) = C_n$.

Let us consider a sensor network or SensorGrid configuration that consists of a fusion center and K distributed sensor nodes. Each sensor node observes feature vector x and applies a classification algorithm, such as the MAP classifier described above, to arrive at a local decision $d(x) \in \mathbf{C}$. The K sensor nodes forward their local decisions $d_k \in \mathbf{C}$ to the fusion center which forms a $K \times 1$ decision vector $\mathbf{d}(x) = [d_1 \ \ d_2 \ \ \cdots \ \ d_K]$. Duarte and Hu (2003) devised a decision fusion method which they referred to as *optimal decision fusion* (ODF). The decision fusion algorithm is itself a decision rule $l(\mathbf{d}(x)) \in \mathbf{C}$ that maps a feature vector $\mathbf{d}(x) = [d_1 \ \ d_2 \ \ \cdots \ \ d_K]$ to one of the class labels. The sample space of the decision fusion classifier is finite and countable: since each decision d_k has N possible values, the combined decision vector $\mathbf{d}(x)$ can have at most N^K different combinations. For each feature vector x, the set of K sensor nodes provide a unique decision vector $\mathbf{d}(x)$. The set of N^K decision vectors partition the feature space into N^K disjoint regions, denoted by $\{r_m; 1 \le m \le N^K\}$. Furthermore, let us denote the unique decision vector that every $x \in r_m$ maps to as $\mathbf{d}(m)$. Following the MAP principle at the fusion center, we have,

$$l(\mathbf{d}(m)) = C_{n*} \text{ if } P(C_{n*} \mid \mathbf{d}(m)) > P(C_n \mid \mathbf{d}(m)) \tag{5.3}$$

for $n \ne n^*$. Using Bayes' rule, if $P(x \in r_m) \ne 0$, then

$$\begin{aligned} P(C_n \mid \mathbf{d}(m)) &= \frac{P(\mathbf{d} = \mathbf{d}(m) \mid x \in C_n) \cdot P(x \in C_n)}{P(x \in r_m)} \\ &= \frac{P(\mathbf{d} = \mathbf{d}(m); x \in C_n)}{P(x \in r_m)} \\ &= \frac{\left|\{x \mid x \in r_m \cap C_n\}\right|}{\left|\{x \mid x \in r_m\}\right|} \end{aligned} \tag{5.4}$$

where $|\{\ldots\}|$ is the cardinal number of the set. Hence, the MAP classification label C_{n*} for r_m can be determined by,

$$n^* = \arg\max_n \left|\{x \mid x \in r_m \cap C_n\}\right| \tag{5.5}$$

when the feature space is discrete. This means that the class label of $\mathbf{d}(m)$ should be assigned according to a majority vote of class labels among all $x \in r_m$.

The ODF decision fusion algorithm is robust and produces high classification accuracy in the final classification even in the presence of faulty or noisy sensors or sensor node failures.

5.3.3 Event detection and classification on SensorGrid

The classification and decision fusion algorithms described above are implemented on the SensorGrid configuration shown in Fig. 5.2(a) which consists of leaf sensor nodes and several levels of fusion centers implemented in sensor nodes, one or more grid client(s) and one or more grid server node(s). The actual system can be seen in Fig. 5.3.

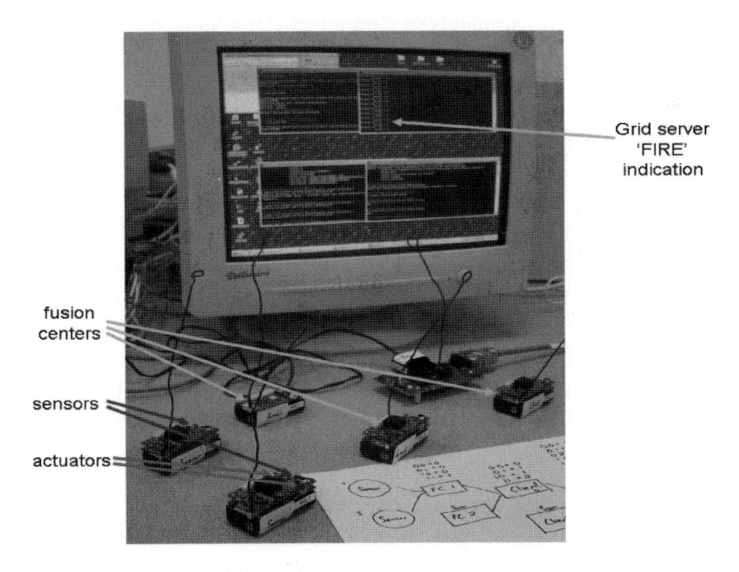

Fig. 5.3. SensorGrid for decision fusion

5.3.4 Sensor nodes

The sensor nodes are Crossbow MPR410 MICA2 nodes, each having a basic CPU, 433 MHz radio board and some SRAM, flash and non-volatile memory. The nodes run the TinyOS operating system which has a component-based architecture that allows modules to be 'wired' together for rapid development of applications, and an event-driven execution model designed for effective power management. The MTS310 sensor add-on board has sensing modalities such as 2-axis accelerometer, 2-axis magnetometer, light, temperature, acoustic and a sounder. We use the light and temperature sensing modalities in our application.

5.3.5 Grid client and grid server nodes

Components from the *Globus Toolkit* (Globus 2006) are implemented on the grid client and grid server nodes. The Globus Toolkit is an open source middleware platform, which allows the sharing of computational resources and the development of grid applications. There are several primary components which make up the Globus Toolkit:

- *Resource Management:* Globus Resource Allocation Manager (GRAM) provides a standardised interface for resource allocation to all local resource management tools and the Resource Specification Language (RSL) is used to exchange information between the resource management components.
- *Information Services:* The Globus Metacomputing Directory Service (MDS) makes use of the Lightweight Directory Access Protocol (LDAP) as a means to query information from system components. The Grid Resources Information Service (GRIS) also provides a standardised method of querying nodes in the grid for information on their configuration, capabilities and status.
- *Data Management:* GridFTP, based on the popular Internet-based File Transfer Protocol (FTP), is used to perform transfer of data or allow data access.
- *Grid Security Infrastructure (GSI):* Underlying the three above-mentioned modules is the GSI security protocol which provides secure communications and a 'single sign on' feature for users who use multiple resources.

The Open Grid Services Architecture (OGSA) in current versions of the Globus Toolkit enable the computational, data and storage resources on the grid to be packaged as services which are easily discovered and accessed by various users.

5.3.6 Implementation on SensorGrid

As mentioned earlier, our proposed sensor-grid computing approach involves more than simply connecting a sensor network and a grid computing system together. The judicious exploitation of the in-network processing capability of the SensorGrid architecture will lead to more efficient implementations of distributed algorithms, such as those for event detection and classification, in terms of the amount of processing and data transmission that needs to be carried out and the timeliness of the computed response.

The MAP classifier of Eqs. 5.1 and 5.2 is implemented on all the leaf sensor nodes and the optimal decision fusion (ODF) algorithm of Eqs. 5.4 and 5.5 is implemented on the sensor nodes which take on the role of fusion centers, grid clients and a grid server node.

The 'wiring' of TinyOS components in sensor nodes performing light and temperature sensing is shown in Fig. 5.4(a), while the 'wiring' of components which perform decision fusion at the sensor nodes which take on the role of fusion center is shown in Fig. 5.4(b).

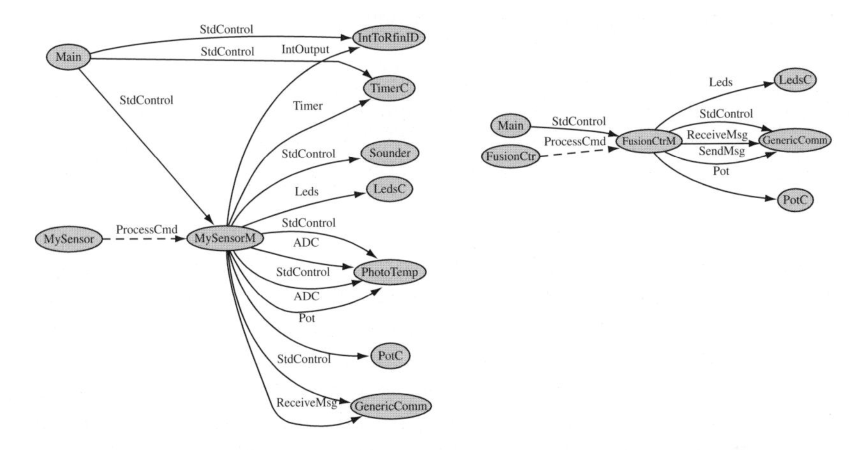

Fig. 5.4. (a) Sensor node component 'wiring'. (b) Fusion center component 'wiring'

5.3.7 Training phase and Operation phase

Before the SensorGrid architecture can be used for event detection and classification, the MAP classifiers and fusion centers running the ODF algorithm need to be trained. We developed a user client application, which initiates the training and operation phases, as well as specifies key parameters such as number of classes, details of the particular SensorGrid architecture configuration, number of samples for each class etc. Training is carried out in a natural manner one class at a time where a number of training samples of sensor readings under a variety of conditions corresponding to the same class are collected. A convenient method of finding the likelihood $p(x|C_n) = p(x|x \in C_n)$ is to use the Gaussian probability density function and determine the mean and covariance of the training feature vectors belonging to the same class. By using the 'equal priors' assumption, the MAP classifier shown in Eq. 5.2 becomes the Maximum Likelihood (ML) classifier where the decision rule is given by,

$$d(x) = \arg\max_n P(x|C_n) \tag{5.6}$$

Eq. 5.6 is implemented at every leaf sensor node to provide the local classification result. The local classification results for every sample of each known class, i.e. with the same class label C_n, are sent to the fusion center at the next higher level which then forms the decision vector $\mathbf{d}(x) = [d_1 \quad d_2 \quad \cdots \quad d_K]$ defining the region r_m, where K is the number of sensor nodes or lower level fusion centers reporting to a particular fusion center. As mentioned earlier, the fusion center can be a sensor node, grid client or grid server node. The counter corresponding to

$|\{x \mid x \in r_m \cap C_n\}|$ is incremented for each sample presented. This process is repeated for every class at various levels of the decision fusion system until the highest level, i.e. grid server node, is reached. At the end of the training phase, the decision fusion system is switched to the operation phase during which 'live' sensor measurements are made at regular intervals and a process similar to the training phase takes place, except that the counter corresponding to $|\{x \mid x \in r_m \cap C_n\}|$ is not incremented, but instead, Eq.5.5 is used to determine the classification outcome C_{n^*} of the entire decision fusion system.

5.4 Experimental Results

This sub-section describes a few experiments, which were conducted to validate the effectiveness of the decision fusion-based event detection and classification system implemented on the SensorGrid architecture. The environment is divided into four regions, each having a sensor node, which takes on the role of a fusion center, as shown in Fig. 5.2(a). The task is to accurately detect and classify different types and severities of fire occurring in the environment, as shown in Table 5.1. It is common practice to tabulate the results of classification experiments in an $N \times N$ *confusion matrix* whose elements $[\mathbf{CM}]_{ij} = n_{ij}$ are the number of feature vectors from class C_i classified as class C_j. Two performance metrics can be computed from the confusion matrix: (1) the probability of correct detection PD_m for class *m* is given by,

$$PD_m = \frac{n_{mm}}{\sum_{j=1}^{M} n_{mj}} \tag{5.7}$$

and (2) the probability of false alarm PFA_m for class *m*, which is the probability that an event is classified as class *m* when the true underlying class is different, is given by

$$PFA_m = \frac{\sum_{k=1,k\neq m}^{M} n_{km}}{\sum_{k=1,k\neq m}^{M} \sum_{j=1}^{M} n_{kj}} \tag{5.8}$$

From Table 5.2, we see that the ODF decision fusion method yields PD = 1.0, 0.9, 0.8, 0.8 (average PD = 0.875) and PFA = 0.067, 0.1, 0.0, 0.0 (average PFA = 0.042) for the four classes. The performance of the ODF decision fusion method is compared with the majority voting decision fusion method in which the final classification outcome is simply the highest frequency class label from among the classification decisions of local classifiers. From Table 5.3, we see that the majority voting method yields PD = 1.0, 0.85, 0.8, 0.75 (average PD = 0.850) and PFA = 0.067, 0.1, 0.033, 0.0 (average PFA = 0.050) for the four classes. Hence, the ODF decision fusion algorithm outperforms the majority voting method, and we have also validated the correct implementation of these algorithms in architecture.

Table 5.1: Fire event detection and classification

Class	Description
1	No fire
2	Fire in one region
3	Fire in two regions
4	Fire in all four regions

Table 5.2: Classification results of the ODF decision fusion method on SensorGrid

Class	Dec = 1	Dec = 2	Dec = 3	Dec = 4
Class = 1	20	0	0	0
Class = 2	2	18	0	0
Class = 3	2	2	16	0
Class = 4	0	4	0	16

Table 5.3: Classification results of the voting decision fusion method on SensorGrid

Class	Dec = 1	Dec = 2	Dec = 3	Dec = 4
Class = 1	20	0	0	0
Class = 2	3	17	0	0
Class = 3	1	3	16	0
Class = 4	0	3	2	15

5.4.1 Other Possible SensorGrid Configurations

In the previous section, the SensorGrid configuration shown in Fig. 5.2(a) have been used extensively. The drawback of this configuration is that, due to the limited radio range of the sensor nodes, the grid client has to be placed fairly close to the fusion centers and clusters of sensor nodes. Although multi-hop data forwarding over a series of sensor nodes can be done to extend the range of data communication, the latency between the occurrence of an event and its measurement reaching the 'sink', which is the grid and user applications in this case, is likely to be significant (Intanagonwiwat et al 2000). Furthermore, due to the limited energy available on sensor nodes, multi-hop data forwarding in a sensor network is likely to be unreliable with a high data loss rate. To overcome these difficulties and to facilitate more computationally intensive in-network processing close to the source of the sensor data, we introduce more powerful processing and communication nodes into the SensorGrid architecture, such as: (1) the Crossbow Stargate SPB400 single board computer, which is essentially a 400 MHz Linux computer with an Intel XScale processor with CompactFlash, PCMCIA, Ethernet and USB connectivity options (e.g. a CompactFlash WiFi network adapter can be connected for high speed wireless communications), and (2) Hewlett Packard iPAQ personal digital assistants (PDAs) with WiFi connectivity.

The resulting configurations of the SensorGrid architecture are shown in Fig. 5.2(b) and (c). In the SensorGrid configuration of Fig. 5.2(c), two additional levels in the form of Stargate and iPAQ cluster heads have been added between the sensor nodes and grid client levels to form a *hierarchical* SensorGrid architecture in which computational and communication resources, as well as the degree of data aggregation, increases when we move up the hierarchy. An implementation of the resulting system is shown in Fig. 5.5. This configuration also reduces the communication distances between the different levels, thus conserving power since radio communications over long distances consumes substantial amounts of energy in sensor networks. Instead of executing complex algorithms only at the grid, the presence of a larger amount of computational resources for in-network processing close to the source of sensor data improves the timeliness of the local and regional responses to sensed events.

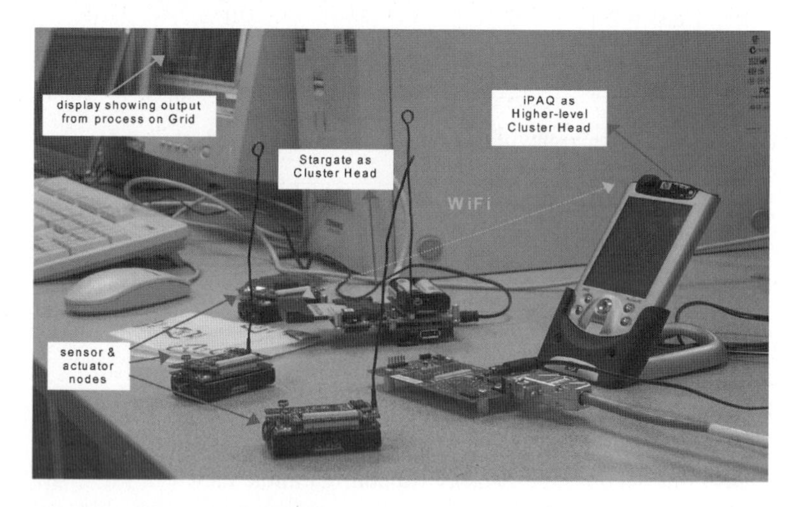

Fig. 5.5. Hierarchical SensorGrid architecture

5.4.2 Distributed Autonomous Decision-making

There are many cases in which a response is needed from the sensor-grid computing system, but the best action to take in different situations or *states* is not known in advance. This can be determined through an adaptive learning process, such as the Markov Decision Process (MDP) or reinforcement learning (RL) (Sutton and

Barto 1998) approach. MDP problems can be solved off-line using methods such as policy- or value-iteration, or on-line using RL or neuro-dynamic programming (NDP) (Bersekas and Tsitsiklis 1996) methods. We developed a multi-level distributed autonomous decision-making system and implemented it on the hierarchical SensorGrid architecture shown in Figure 5.5. Basic NDP agents were implemented in Crossbow motes (Tham and Renaud 2005) at the local or ground level, and more complex NDP agents at grid server nodes were implemented at the core of the grid. Each NDP agent is able to act autonomously and most parts of the sensor-grid computing system remain responsive even in the presence of communication failures due to radio jamming, router failures between some components.

5.5 Research Issues

Sensor networks is a relatively recent field and there are many research issues pertaining to sensor networks such as energy management, coverage, localisation, medium access control, routing and transport, security, as well as distributed information processing algorithms for target tracking, information fusion, inference, optimisation and data aggregation. Grid computing has been in existence longer, but nevertheless, still has a number of research challenges such as fair and efficient resource (i.e. CPU, network, storage) allocation to achieve quality of service (QoS) and high resource utilisation, workflow management, the development of grid and web services for ease of discovery and access of services on the grid, and security. Resource allocation itself involves a number of aspects such as scheduling at the grid and cluster or node-levels, Service Level Agreements (SLAs) and market-based mechanisms such as pricing. Apart from the afore-mentioned research issues in sensor networks and grid computing, sensor-grid computing and the SensorGrid architecture give rise to additional research challenges, especially when it is used in mission-critical situations. These research challenges are: web services and service discovery which work across both sensor networks and the grid, interconnection and networking, coordinated QoS mechanisms, robust and scalable distributed and hierarchical algorithms, efficient querying and self-organisation and adaptation. Each of these areas will be discussed in greater detail in the following sub-sections.

5.5.1 Web Services on Grid

The grid is rapidly advancing towards a utility computing paradigm and is increasingly based on web services standards. The Service-Oriented Architecture (SOA) approach has become a cornerstone in many recent grid efforts. It makes sense to adopt an SOA-approach as it enables the discovery, access and sharing of the services, data, computational and communication resources in the grid by many different users. On the SensorGrid architecture, the grid computing components are implemented as grid services using the Globus Toolkit 3 (GT3) shown in Fig. 5.6, which conforms to the Open Grid Services Infrastructure (OGSI) standard. A grid service is basically a form of web service that can be invoked remotely for Internet-based applications with loosely coupled clients and servers, but with improved characteristics suitable for grid-based applications. In particular, a grid service is a *stateful* web service, which remembers the state of operations invoked by grid clients. This is usually the case when the grid client has a chain of operations, where the result of one operation needs to be sent to the next operation.

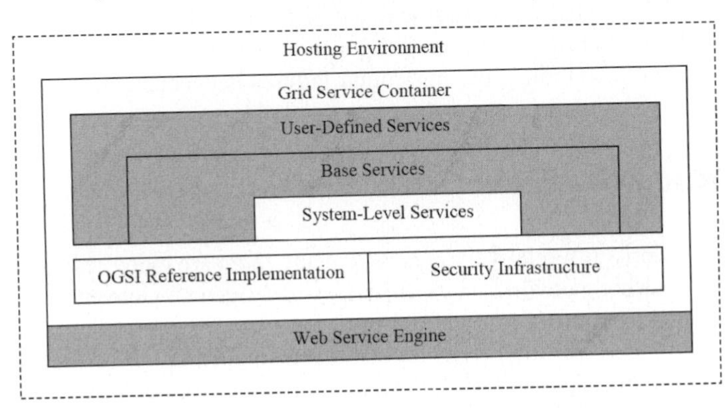

Fig. 5.6. GT3 core architecture

Another major improvement of grid services is the capability of being *transient*, as compared to normal *persistent* web services. Web services are referred to as *persistent* because their lifetime is bound to the web services container. After one client has finished using a web service, all the information stored in the web service can be accessed by subsequent clients. In fact, while one client is using the web service, another client can access the same service and potentially mix up the first client's operations. Grid services solve such problems by allowing programs to use a *factory* or *instance* approach to web services. When a client needs to create a new instance, it will communicate with a factory. These instances are *transient* since they have a limited lifetime, which is not bound to the lifetime of the container of the grid services. One can create and destroy instances at will whenever they are needed, instead of having one persistent service permanently available in the case of normal web services. The actual lifecycle of an instance can vary. In GT3, at the server-side, the main architecture components include:

- Web services engine. This is provided by the Apache AXIS framework and is used to deal with normal web services behaviours, SOAP message processing, JAX-RPC (Java API for XML-based Remote Procedure Call) handlers and Web Services configuration.
- Globus Container framework, which provides a software container to manage the web service through a unique instance handle, instance repository and life-cycle management, including service activation/passivation and soft-state management.

The GT3 container provides a pivot handler to the AXIS framework to pass the request messages to the Globus container (Fig. 5.). This container architecture is used to manage the stateful nature of web services and their life cycles.

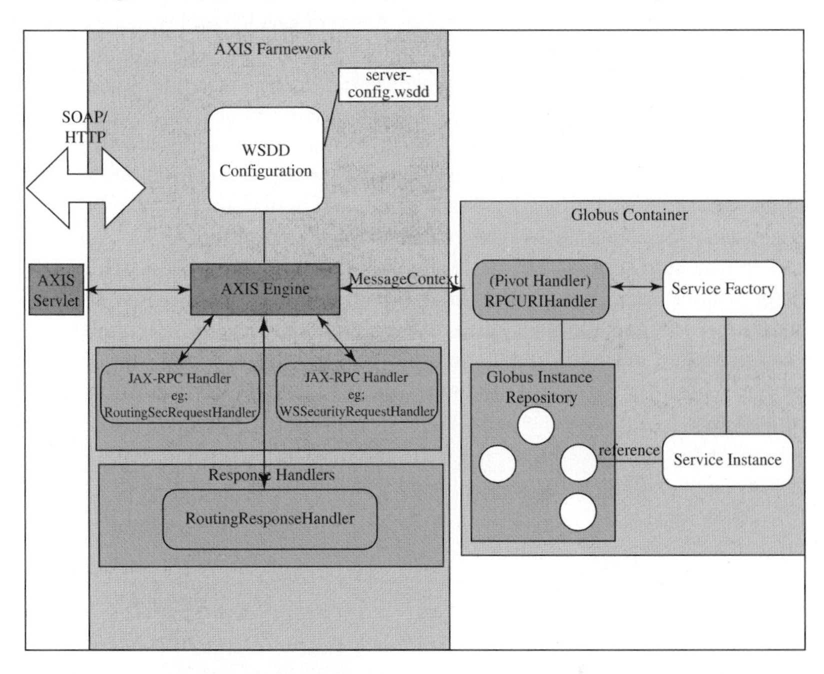

Fig. 5.7. Grid server-side framework for GT3

Referring to Fig. 5.7, once the service factory creates a grid service instance, the framework creates a unique grid service handle (GSH) for that instance, and that instance is registered with the container repository. This repository holds all of the stateful service instances and is contacted by the other framework components and handlers to perform services such as:

- Identifying services and invoke methods
- Getting/setting service properties (i.e. GSH and Grid Service Reference)
- Activating/passivating service
- Resolving grid service handles to reference and persist the service

At the client-side, Globus uses the normal JAX-RPC client-side programming model and AXIS client-side framework grid service clients. Essentially, the client AXIS engine communicates to the server via SOAP messages. In addition to the normal JAX-RPC programming model, Globus provides a number of helper classes at the client side to hide the details of the OGSI client-side programming model. Fig. 5.8 shows the client-side software framework used by Globus. On the SensorGrid architecture, such as the configuration shown in Fig. 5.2(b), the grid client forms the feature vector using the processed sensor readings from the Stargate embedded device. The Stargate acts as a gateway between the sensor network and the grid client and preprocesses the raw sensor readings from the sensor nodes. After that, the grid client invokes the grid service and provides it with the feature vector formed from the sensor network readings. Likewise, in sensor networks, it makes sense to share the sensor-actuator infrastructure among a number of different applications and users so that the environment is not swamped with an excessive number of sensor nodes, especially since these nodes are likely to interfere with one another when they communicate over the shared wireless medium and decrease the effectiveness of each node, and actuators may also take conflicting actions. There has been some recent work on adopting an SOA and web services approach to sensors and sensor networks. The OpenGeospatial Consortium's Sensor Model Language (SensorML) standard (OGC 2006) provides the XML schema for defining the geometric, dynamic and observational characteristics of sensors. We are currently developing a web services interface between the Stargate and grid client which will enable sensor network services to be discovered and accessed by grid applications as well as by user applications running on user client devices.

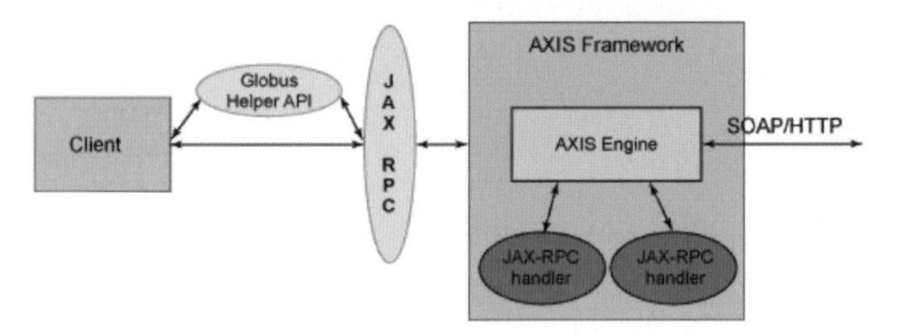

Fig. 5.8. Grid client-side framework of GT3

5.5.2 Interconnection and Networking

The communications and networking conditions in sensor networks and grid computing are worlds apart. In sensor networks, the emphasis is on low power wireless communications which has limited bandwidth and time-varying channel characteristics, while in grid computing, high-speed optical network interconnects are common. Thus, communications protocols for sensor-grids will have to be designed take into account this wide disparity. ZigBee has emerged as one of the first standards-based low power wireless communications technologies for sensor networks, e.g. Crossbow MICAz motes. Furthermore, a machine-to-machine (M2M) interface between ZigBee and GPRS has recently been announced, thus enabling sensor networks to be connected to the cellular network infrastructure. One other promising development is low-rate Ultra-Wide Band (UWB) wireless technology which has several characteristics suitable for sensor networks, i.e. extremely low power consumption, reasonable communication range, and integration with UWB-based localisation and positioning technology.

5.5.3 Coordinated QoS in Large Distributed System

The timeliness and correctness of computations have been studied extensively in the real-time systems community, while performance guarantees in terms of delay, loss, jitter and throughput in communication networks have been studied extensively by the networking research community. We shall refer to these as application-level and network-level QoS, respectively.

A number of QoS control mechanisms such as scheduling, admission control, buffer management and traffic regulation or shaping have been developed to achieve application-level and network-level QoS. However, all these QoS mechanisms usually relate to a particular attribute such as delay or loss, or operate at a particular router or server in the system. In order to bring about the desired system-level outcome such as meeting an end-to-end computational and communication delay requirement, these QoS mechanisms need to be coordinated instead of operating independently. We have developed a combined grid and network simulator (Sulistio et al 2005) to study these issues.

There are several methods to achieve coordinated QoS. For example, coordinated QoS can be viewed as a multi-agent Markov Decision Process (MDP) problem which can be solved using online stochastic optimal control techniques. Tham and Liu (2005) have shown that this technique can achieve end-to-end QoS in a multi-domain Differentiated Services network with multiple resource managers in a cost effective manner.

5.5.4 Robust and Scalable Distributed and Hierarchical Algorithms

In Section 3, we described the design and implementation of distributed information fusion and distributed autonomous decision-making algorithms on the SensorGrid architecture. Generally, it is more difficult to guarantee the optimality, correctness and convergence properties of distributed algorithms compared to their centralised counterparts, although the distributed versions are usually more attractive from an implementation point of view. Apart from distributed information fusion and decision-making, other current research efforts on distributed and hierarchical algorithms which are relevant to sensor-grid computing are distributed hierarchical target-tracking (Yeow et al 2005), distributed control and distributed optimisation (Rabbat and Nowak 2004).

5.6 Efficient Querying and Data Consistency

Another key area in sensor-grid computing is efficient querying of real-time information in sensor networks from grid applications and querying of grid databases by sensor network programs. It is expected that databases will be distributed and replicated at a number of places in the sensor-grid architecture to facilitate efficient storage and retrieval. Hence, the usual challenges of ensuring data consistency in distributed caches and databases would be present, with the added complexity of having to deal with a large amount of possibly redundant or erroneous real-time data from sensor networks.

5.6.1 Self-organisation, Adaptation, and Self-optimisation

The sensor-grid computing paradigm involves close interaction with the physical world, which is highly dynamic and event-driven. In addition, components in the SensorGrid architecture such as the sensor nodes are prone to failures arising from energy depletion, radio jamming and physical destruction due to bad weather conditions or tampering by people or animals. Self-organising and adaptive techniques are required to cope with these kind of failures as well as to respond to different unpredictable events, e.g. the computation and communication requirements at a certain part of the SensorGrid architecture may increase dramatically if an event such as a fire or an explosion happens in that region and the user demands higher resolution sensing. Since the dynamics of phenomena in the physical world are rarely known in advance, self-optimisation techniques are also needed to better manage system resources and improve application-level and platform-level QoS. Techniques from the field of *autonomic computing* address some of these issues and can be applied in the SensorGrid architecture.

5.7 Conclusions

In this chapter, we have provided an in-depth view of the potential and challenges in sensor-grid computing and described how it can be implemented on different configurations of the SensorGrid architecture. In the near future, as sensor systems, sentient computing and ambient intelligence applications become more widespread, we expect that the need for sensor-grid computing will grow as it offers an effective and scalable method for utilising and processing widespread sensor data in order to provide value in a number of domain areas. Last but not least, the success of the sensor-grid computing approach will depend on the ability of the sensor network and grid computing research communities to work together to tackle the research issues highlighted above, and to ensure compatibility in the techniques, algorithms and protocols that will be developed in these areas.

5.8 References

Bertsekas DP and Tsitsiklis JN (1996) Neuro-dynamic programming, Athena Scientific, USA

Brooks R, Ramanathan P, and Sayeed AK (2003) Distributed target classification and tracking in sensor networks, In: Proc. of IEEE, vol. 91, no. 8

Chiu K and Frey J (2005) Int. Workshop on scientific instruments and sensors on the grid, in conjunction with e-Science 2005, Melbourne, Australia

Duarte MF and Hu YH (2003) Optimal decision fusion for sensor network applications, In: Proceedings of First ACM Conf. on embedded networked sensor systems 2003 (SenSys 2003)

Gaynor M, Moulton S, Welsh M, LaCombe E, Rowan A, and Wynne J (2004) Integrating wireless sensor networks with the Grid, IEEE Internet Computing, July-August 2004

Globus (2006) The Globus Alliance, Online: http://www.globus.org

Gridbus (2005) The Gridbus Project, Online: http://www.gridbus.org

Intanagonwiwat C, Govindan R, and Estrin D (2000) Directed diffusion: A scalable and robust communication paradigm for sensor networks, In: Proceedings of MOBICOM 2000

Lim HB, Teo YM, Mukherjee P, Lam VT, Wong WF, and See S (2005), Sensor grid: Integration of WSN and the Grid, Proc. of IEEE Conf. on local computer networks (LCN'05), Nov.

OpenGeospatial Consortium (OGC) (2006) Sensor model language (SensorML), Online: http://vast.nsstc.uah.edu/SensorML

Rabbat M and Nowak R (2004) Distributed optimisation in sensor networks, Proc. of Information processing in sensor networks (IPSN)

Sulistio A, Poduval G, Buyya R, and Tham CK (2005) Constructing a grid simulation with differentiated network service using GridSim, In: ICOMP'05, The Int. Conf. on Internet Computing

Sutton R and Barto A (1998) Reinforcement learning: An introduction, MIT Press

Tham CK and Buyya R (2005), SensorGrid: Integrating sensor networks and grid computing, invited paper in: CSI Communications, Special Issue on Grid Computing, Computer Society of India, July

Tham CK and Liu Y (2005) Assured end-to-end QoS through adaptive marking in multi-domain differentiated services networks, Computer Communications, Special Issue on Current Areas of Interest in End-to-End QoS, Vol. 28, Issue 18, Elsevier, pp. 2009-2019.

Tham CK and Renaud JC (2005) Multi-agent systems on sensor networks: A distributed reinforcement learning approach, Invited Paper, Proc. of 2nd Int. Conf. on Intelligent Sensors, Sensor Networks and Information Processing - 2005 (ISSNIP 2005), Melbourne

Yeow WL, Tham CK, and Wong LWC (2005) A novel target movement model and efficient tracking in sensor networks, Proc. of IEEE VTC2005-Spring, Sweden, 29 May-1 June

6 Topology Control

Debashis Saha[1] and Arunava Saha Dalal[2]

[1]Indian Institute of Management, MIS & Computer Science Group, Kolkata India
[2]Nvidia Graphics Private Limited, Bangalore, India

6.1 Introduction

Networks of wireless units, communicating with each other via radio transceivers, typically along multihop paths, are generally known as wireless ad hoc networks (WANETs). WANETs can be used wherever a wired infrastructure is infeasible and/or economically inconvenient. For example, in order to provide communications during emergencies, special events (such as conferences, expos, concerts, etc.), or in hostile environments, WANETs are extremely useful (Santi 2005a). Wireless sensor networks (WSNs) are a special class of WANETs where network nodes are wireless sensors (which may be mobile in some cases). Sensor nodes collect the data of interest (e.g., temperature, pressure, soil makeup, etc.), and wirelessly transmit them, possibly compressed and/or aggregated with those of neighboring nodes, to other nodes. WSNs can be used to monitor remote and/or hostile geographical regions, to trace animal movement, to improve weather forecast, and so on. WSNs are envisioned to be an integral part of our lives in future, more so than the present day cell-phones or PCs (Akyildiz et al. 2002).

A protocol stack, used by the sensor nodes in WSNs, is suggested in Fig. 6.1 (Akyildiz et al. 2002). It follows the conventional 5-layer model of TCP/IP stack with some modifications. For example, this protocol stack combines power and routing awareness, integrates data with networking protocols, communicates power efficiently through the wireless medium, and promotes cooperative efforts of sensor nodes. Several energy-conserving methods have been proposed at various layers of the protocol stack of WSN. These range from design of low power CMOS circuitry to energy aware applications. All these energy-conserving methods together constitute the power management plane of the WSN architecture (Fig. 6.1), which manages how a sensor node uses its power,

Due to similarity in concept, realisation of WSN applications requires WANET techniques to be handy in the first place. But, although many protocols and algorithms have been proposed for traditional WANETs, they are not well suited for the unique features and application requirements of WSNs (Akyildiz et al. 2002). For instance, unlike wired networks, each node in a multi-hop WSN can potentially change its set of one-hop neighbors and subsequently the overall network topology, by changing its transmission power. Without proper TC algorithms in place, a randomly connected multi-hop WSN may suffer from poor network utilisation, high end-to-end delays, and short network lifetime. Interestingly, the topology also depends on several *uncontrollable* factors, such as node mobility, weather, interference, noise, along with *controllable* parameters, such as transmit power and antenna direction.

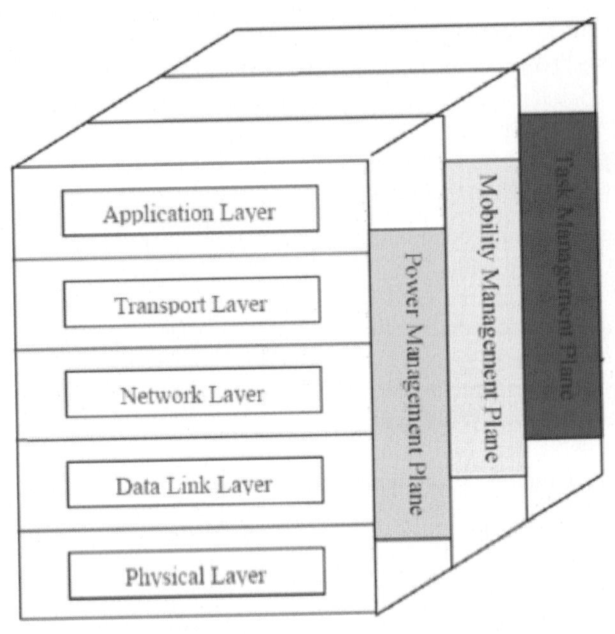

Fig. 6.1. A proposed protocol stack for WSNs

The primary goal of topology control is to design power-efficient algorithms that maintain network connectivity and optimise performance metrics, such as network lifetime and throughput by adjusting transmit and reception powers. Batteries are the primary source of energy for wireless sensor devices, and hence, a common thread restraining the true potential of WSNs is the limitation in energy supply. Consequently, energy aware system design has become an important tool in addressing this issue. In other words, a common approach to tackle the sensor network longevity problem is to develop energy-efficient algorithms and mechanisms that optimise the use of the battery power, while maintaining the necessary level of network connectivity. As pointed out by Chandrakasan et al., WANET protocols that minimise energy consumption are the key to the successful usage of control network optimisation problems and their solutions. Many researchers are currently engaged in developing algorithms that fulfill these requirements. The topology control network optimisation algorithms have many uses for the designing and implementation of WSNs: They can be used for:

- Efficient sensor network deployment,
- Determination of sensor density and transmit power for the nodes,
- Determination of the best transmission range,
- Facilitation of control message dissemination in an efficient way, and
- Better modeling for energy-aware broadcasting and multicasting

We concentrate on these energy-conserving schemes and specifically on topology control problems in this chapter. Our aim is to provide a clear understanding of the protocols being developed for topology control, and also to determine the significance of TC in the field of energy conservation methods. Before proceeding further, let us make some observations regarding terminology that is conventionally used in the related literature. The term topology control has been used with at least two different meanings in the WANET and WSN literature. For instance, several authors consider *topology control* techniques as superimposing a hierarchy on an otherwise flat network organisation in order to reduce, typically, energy consumption. This is the case, for instance, with clustering algorithms, which select some of the nodes in the network as cluster-heads, whose purpose is to optimise energy and communication efficiency in their cluster. Although, in a sense, clustering algorithms can be seen as a way of controlling the network topology, they cannot be classified purely as topology control mechanisms according to the informal definition presented above since the transmit power of the nodes is usually not modified by a clustering algorithm. Also, the terms *power control* and *topology control* are often confused with each other in the current literature. In our opinion, *power control* techniques are those, which by acting on the transmit power level of the nodes, aim at optimising a single wireless transmission. Although this transmission might, in general, be multihop, the focus of *power control* is on the efficiency of a single (possibly multihop) wireless channel. Again, this feature of *power control* does not fulfill our informal definition of *topology control* in which nodes adjust their transmitting range in order to achieve a certain network-wide target goal (e.g., network connectivity).

6.2 WANET Model and Topology Control (TC)

In this section, we introduce a simplified but widely accepted model of a WANET, which will be used in the definition of the various problems, related to TC considered in the literature. We are given a set of N possibly mobile nodes located in an m-dimensional space (m=1, 2, 3) The node configuration of the mobile WANET is represented by a pair $W_m = (N, C)$, where N is the set of nodes and $C : N \times T \rightarrow (0, l)^m$, for some $l > 0$, is the placement function. The placement function assigns to every element of N and to any time $t \in T$ a set of coordinates in the m-dimensional cube of side l, representing the node's physical position at time t. A node $u \in N$ is said to be stationary, if its physical placement does not vary with time. If all the nodes are stationary, the network is said to be stationary, and function C can be represented simply as $C : N \rightarrow (0, l)^m$. A range assignment for an m-dimensional node configuration $W_m = (N, C)$ is a function $RA : N \rightarrow (0, r_{max})$ that assigns to every element of N a value in $(0, r_{max})$, representing its transmitting range. Parameter r_{max} is called the maximum transmitting range of the nodes in the network and depends on the features of the radio transceivers equipping the mobile nodes. A common assumption is that all the nodes are equipped with transceivers having the same features; hence, we have a single value of r_{max}

for all the nodes in the network. Topology control (TC) is essentially a method in which a node can carefully select a set of neighbors to establish logical data links and dynamically adjust transmitting power for different links. In the protocol stack, TC methods are placed in the power management plane and implemented at the network layer. The TC problem can be formalised in the following way. Each node $u \in N$ has a power function p, where p(d) gives the minimum power needed to establish a communication link to a node $v \in N$ at distance d away from u. Assume that the maximum transmission power P is the same for every node, and the maximum distance for any two nodes to communicate directly is R, i.e., p(R) = P. If every node transmits with power P, then we have an induced graph $G_R = (V, E)$, where $E = \{(u,v)|d(u,v) < R; u,v \in N\}$ (where d(u,v) is the Euclidean distance between u and v). It is undesirable to have nodes transmitting with maximum power for two reasons. First, since the power required to transmit between nodes increases as the nth power of the distance between them (for some $n \geq 2$, Rappaport), it may require less power for a node u to relay messages through a series of intermediate nodes to v than to transmit directly to v. Second, the greater the power with which a node transmits, the greater is the likelihood of the transmission interfering with other transmissions. The goal in performing TC is to find a sub-graph G of G_R such that:

- G consists of all the nodes in G_R but has fewer edges,
- If u and v are connected in G_R, they are still connected in G, and
- A node u can transmit to all its neighbors in G using less power than is required to transmit to all its neighbors in G_R.

6.3 Overview of TC Algorithms

Various approaches to the TC problem, as they have appeared in the literature, can be broadly classified into two categories, depending on which kind of network they are suitable for. Between *stationary* and *mobile* networks, understandably, the latter poses a more complicated scenario wherein the nodes may change their positions with time. However, for each category of the networks, TC protocols may follow either the *homogeneous* or *non-homogeneous* approach (Fig. 6.2). In the former case, which is the simpler (and easier to analyse) type of TC, nodes are assumed to use the same transmitting range, and the TC problem reduces to the one of determining the minimum value of r (known as *critical transmitting range (CTR)*) such that a certain network-wide property is satisfied. Further, TC protocols for the homogeneous case can be subdivided into two categories, namely *practical* and *analytical*, based on the approach undertaken. In the analytical approach, mostly probabilistic, theoretical results are derived for TC, whereas, in the practical approach, real-world algorithms are depicted which may be put down to practice. In the non-homogenous case, nodes are allowed to choose different transmitting ranges (provided they do not exceed the maximum range). Non-homogeneous TC is classified into three categories (Fig. 6.2), namely *centralised*, *distributed*, and *hybrid*, depending on the strategy adopted by the algorithms in evaluating the optimum topology. Most of the earlier protocols were either purely centralised or purely distributed. In *centralised* TC algorithms, a network node is responsible for evaluating an optimum network topology based on the locations of the other network nodes. The location information along with the transmission range of the individual nodes is used to form a link between a node pair. This information is either exchanged between nodes and used to compute an "almost optimal" topology in a fully distributed manner or used globally by a centralised authority to compute a set of transmitting range assignments which optimises a certain measure (this is also known as the *range assignment problem* or variant of it). A centralised approach, although able to achieve strong connectivity (k-connectivity for k≥2), suffers from *scalability problems*. In contrast, a distributed approach, although scalable, lacks strong connectivity guarantees. Hence, a *hybrid* of these strategies is indeed essential to reap the benefits of both.

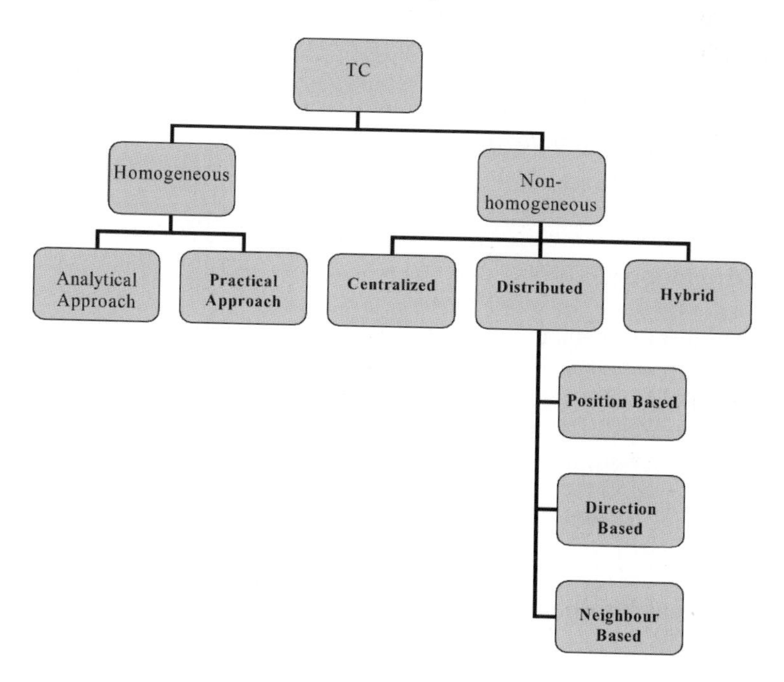

Fig. 6.2. A classification of TC protocols

The *distributed algorithms* are further categorised into three types (Fig. 6.2), based on the type of information that is used to compute the topology. In *position-based* approaches, exact node positions are known. In *direction-based* approaches, it is assumed that nodes do not know their position, but they can estimate the relative direction of each of their neighbors. Finally, in *neighbor-based* techniques, nodes are assumed to know only the ID of the neighbors and are able to order them according to some criterion (e.g., distance, or link quality). Besides classifying TC approaches based on the constraints we put on the range assignment (homogeneous or non-homogeneous), and on the type of information which is available to the network nodes, we can also distinguish the approaches, proposed in the literature, based on the properties of the network topology resulting from the application of TC techniques. Most of the approaches presented in the literature are concerned with building and maintaining a *connected* network topology, as network partitioning is highly undesirable. More recently, some authors have considered the problem of building a k-connected network topology (with $k > 1$), that is, a topology in which there exists at least k distinct paths between any two-network nodes. Guaranteeing k-connectivity of the communication graph is fundamental in all those applications in which a certain degree of *fault-tolerance* is needed: since there exist at least k paths between any two network nodes, network connectivity is guaranteed in the presence of up to $(k - 1)$ node failures. Other authors have recently also considered the TC problem, in which nodes alternate between active and sleeping times, and the goal is to build a network topology such that the subnetwork composed of the active nodes is connected at any time.

6.4 TC in Stationary WSNs

6.4.1 Homogenous TC

First we consider the problem concerning homogeneous range assignment: *given a set of N nodes in an m-dimensional space what is the minimum value of the transmitting range (r) to be assigned to all the nodes such that the placement results in a connected topology*? This minimum value of r such that homogenous range assignment is connecting is known as the CTR for connectivity. Two approaches have been followed in the literature to solve the CTR problem for stationary networks: *analytical* and *practical*. We will discuss the analytical approach first.

6.4.2 Analytical Approach

The solution to CTR depends on the information we have about the physical placement of nodes. If the node placement is known in advance, the CTR is the length of the longest edge of the Euclidean minimum spanning tree (MST) (Penrose 1997; Sanchez et al. 1999) built on the nodes (see Fig. 6.3). Unfortunately, in many realistic scenarios of WSNs, the node placement is not known in advance. For example, sensors could be spread from a moving vehicle (such as airplane, ship, or spacecraft). If node positions are not known, the minimum value of r ensuring connectivity in all possible cases is $r \approx l\sqrt{d}$, which accounts for the fact that nodes could be concentrated at opposite corners of the deployment region. However, this scenario is unlikely in most realistic situations. For this reason, CTR has been studied under the assumption that nodes are distributed in *R* according to some probability distribution. In this case, the goal is to characterise the minimum value of *r,* which provides connectivity with high probability i.e., with a probability that converges to 1 as the number of nodes (or the side *l* of the deployment region) increases.

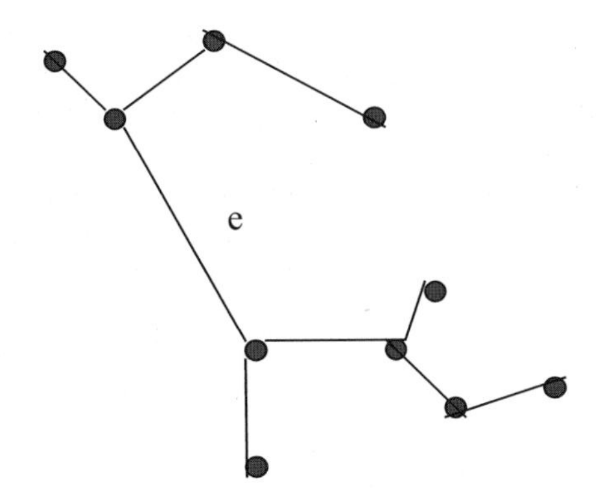

Fig. 6.3. CTR for connectivity: The longest edge length of the Euclidean MST (edge *e*).

For dense networks Panchapakesan and Manjunath (2001) have proved that, under the hypothesis that nodes are uniformly distributed in $(0, 1)^2$, the CTR for connectivity with high probability is $r = c\sqrt{((\log N)/ N)}$ for some constant $c > 0$. The characterisations of the critical range for connectivity in one- and three-dimensional networks can be obtained by combining some results derived in Dette and Henze, Holst, and Penrose are as follows. In one-dimensional networks, it is shown that if *N* nodes are distributed uniformly at random in (0, 1), then the critical range for connectivity with high probability is $r = (\log N)/N$. In three-dimensional networks, it is shown that if *N* nodes are distributed uniformly at random in $(0, 1)^3$, then the CTR is $r = \sqrt[3]{((\log N - \log\log N)/N\pi + 3/2(1.41 + g(N))/\pi N)}$, where $g(N)$ is an arbitrary function such that $\lim_{n\to\infty} g(N) = +\infty$.

The CTR for connectivity in sparse WANETs have been analysed in Santi et al. and Santi and Blough using the occupancy theory. It has been proven that, under the assumption that N nodes are distributed uniformly at random in $R = (0, l)^d$, the r-homogeneous range assignment is connecting with high probability if $r = l\sqrt[d]{c(\log l/N)}$ for some constant $c > 0$. The authors also prove that, if $r \in O(l\sqrt[d]{(1/N)})$, then the r-homogeneous range assignment is not connected with high probability. Bettsetter (2002a) analyses network connectivity for the homogenous case as well as non-homogeneous range assignment (N_1 nodes with r_1, N_2 with r_2 and so on). For homogeneous WANETs the authors prove that for N uniformly distributed nodes with node density ρ (=N/A, N>>1, A is the area of deployment of the N nodes) with range assignment r_0 ($A >> \pi r_0^2$) the probability of the network being k connected is approximately equal to the probability that it is connected when d_{min} is greater than equal to k , where d_{min} is the smallest node degree over all nodes in the network. For the more complicated non-homogeneous case with n nodes of J different node types; i.e., there are N_j nodes of type j with range r_j, the probability of the network being connected is an exponential function of the node density and square of a range termed as effective range by the authors.

6.4.3 Practical Approach

Besides analytical characterisation, the CTR has been investigated from a more practical viewpoint. We will now discuss each such viewpoint in fair amount of details. In Narayanaswamy et al. the authors present a distributed protocol, called COMPOW, which attempts to determine the *minimum common transmitting range* needed to: (1) ensure network connectivity, (2) maximise network capacity, (3) extend battery life, and (4) reduce collisions in the MAC layer. The results show that COMPOW, when implemented and tested on a real wireless test bed, reduces transmission power, redundancy and interference, and guarantees bi-directionality of links and connectivity. However, it has a major drawback in its significant message overhead, since each node runs multiple daemons, each of which has to exchange link state information with the counterparts at other nodes. Also, it tends to use higher power in the case of unevenly distributed nodes, and since the common power is collaboratively determined by the all nodes inside the network, global reconfiguration is required in the case of node joining/leaving.

Santi and Blough have presented experimental results presented in which show that, in sparse two and three-dimensional networks, the transmitting range can be reduced significantly if weaker requirements on connectivity are acceptable: halving the CTR, the largest connected component has an average size of approximately 0.9n. This means that a considerable amount of energy is spent to connect relatively few nodes. This behavior is not displayed in the case of one-dimensional networks in which a small decrease of the transmitting range with respect to the critical value split the network into at least two connected components of moderate size.

Another recent TC protocol is proposed by Li and Hou called FLSS, which is a fault tolerant version of LMST (Li et al. 2003). It is primarily developed for homogenous networks and aims to generate a k-connected topology and increase the level of routing redundancy. Two algorithms are described: a centralised greedy algorithm, known as Fault-tolerant Global Spanning Subgraph ($FGSS_k$), for fault-tolerant TC, and its localised version, known as Fault-tolerant Local Spanning Subgraph ($FLSS_k$). Compared to LMST, FLSS is k-1 failure tolerant, produces more robust topologies, yields better throughput and is more energy efficient. However on the other hand it produces longer links, higher average degree and consumes more total power than LMST.

The MobileGrid protocol of Liu and Li tries to keep the number of neighbours of a node within a low and high threshold centered around an optimal value. In order to identify one single parameter in controlling the network performance, the authors present a generic notion in WANETs, the contention index, which represents the number of contending nodes within the interference range. The authors contend that the contention index, rather than the transmission range on each node, is the primary and independent driving force that influences the network performance. MobileGrid attempts to maintain an optimal Contention Index (CI) and to maximise the network capacity and power efficiency. It is a distributed TC algorithm that attempts to achieve the best possible network capacity and power efficiency, by maintaining optimal contention index via dynamically adjusting the transmission range on each of the nodes in the network. It is hard for a node to compute the network contention index according to the provided equation accurately, due to the lack of global knowledge on either the number of nodes or the physical area of deployment.

By knowing how many neighbors a node has, the node can estimate the contention index. Based on this observation, MobileGrid, is implemented as a three-phase protocol. It introduces CI as a new parameter that influences network performance. The algorithm has no control message overhead and attempts to achieve best possible network capacity. But the accuracy of estimation of CI depends heavily on the network traffic and in the extreme case silent nodes may not be detected. Also CI checking is performed at least once which increases overhead. Finally connectivity has not been considered importantly in the protocol.

Loyd et al., (2002) have designed algorithms based on sorting and searching under several optimisation objectives: maintenance of desired graph properties and minimisation of maximum transmit power (MaxP) and total power (TotalP). The authors have provided a polynomial time algorithm for <DIR,P,MAXP>. Next an example of a non-monotone property was given for which the problem of minimising the maximum power is NP-complete.

Finally, it was shown that the additional requirement of minimising the number of nodes that use the maximum power also renders the problem NP- complete even for certain monotone properties. For a graph property P, problems (Undir/Dir,P,MaxP) have been solved and a general approach has been presented for developing approximation algorithms under total power minimisation objective. (Undir,2-NC,Total P) has also been solved and analogous result stated for (Undir,2-EC,Total P). (NC=Node Connected, EC=Edge Connected) However, the approach has some limitations. The graph properties in reality are mostly non-monotone and non-polynomial time testable. The requirement of minimising number of nodes that use maximum power renders the problem very hard. The methodology of determining a solution to a specific problem from the given general solution framework is not clear and explicit.

On the lines of Loyd et al. (2002), Krumke et al. (2003), solved the <UNDIR,DIAMETER,TOTALP> (graph induced by the power assignment has a diameter bound) and the <UNDIR,DEG LB,TOTALP> (node degree is bounded) problems under the symmetric power threshold model. For the first problem, they made extensive use of existing approximate algorithms for problem called MCTDC (minimum cost tree with a diameter constraint) (Charikar et al. 1999, Kortsarz and Peleg 1999; Marathe et al. 1998). Only results are provided for the second problem. They also studied the <UNDIR,CONNECTED,TOTALP> problem under the asymmetric threshold model where they used a combination of search and sort, and an algorithm for Steiner Trees (Guha and Khuller 1999) to find a connected dominating set.

6.5 Non Homogenous TC

6.5.1 Centralised Approach

Ramanathan and Rosales-Hain (2000) considered the problem of minimising the maximum of node transmitting ranges as well as total power while achieving global connectivity and bi-connectivity. The authors adopted a hybrid strategy of Minimum-spanning-tree and depth first search techniques for chalking out their algorithms. Two optimised centralised greedy polynomial time algorithms were described: one result in a connected and the other in a bi-connected static network. The former (CONNECT) is similar to Minimum-spanning-tree algorithm while the latter (BICONN-AUGMENT) uses a standard method based on depth first search. The static algorithms improve the throughput and battery life of portable networks and the heuristics increase capacity, battery life and connectivity of mobile networks (we will discuss the heuristics for mobile networks in a following section).In fact the optimum operating density can be calculated using the results. However, on the other hand the relative distance of all nodes has to be given as input to the centralised algorithm, which may be too demanding. Moreover there is no insight on how to select the appropriate number of neighbors for LINT (a distributed heuristic for mobile networks) and hence no guarantee of strong connectivity. The accuracy of the estimate of neighbour number given by overhearing depends on traffic present in the network. In the extreme case a node, which remains silent, will not be detected.

6.5.2 Distributed Approach

Position Based TC: In Rodoplu and Meng (1999), the authors presented a distributed TC algorithm that leverages on location information (provided by low-power GPS receivers) to build a topology that is proven to minimise the energy required to communicate with a given master node. Their primary aim is to design of a self-reconfiguring distributed network protocol that maintains strong network connectivity and minimises the total power from every sensor to a master site along a directed path. The distributed network protocol finds the minimum power topology for a stationary set of nodes with a master-site. The main idea in this protocol is that a node does not need to consider all the nodes in the network to find the global minimum power path to the master-site. By using a very localised search, it can eliminate any nodes in its relay regions from consideration and pick only those few links in its immediate neighborhood to be the only potential candidates. The proposed heuristic contains two phases:

- First, each node computes the relay set R such that the transmit power is minimised when reaching all nodes in N-R through R. This phase determines the transmit power for each node.

- Next, a directed path with minimum cost (the sum of both transmit and receive powers) to a designated master site from each node is computed by applying the distributed Bellman-Ford algorithm.

The algorithm by Rodoplu and Meng (1999) has primarily been developed for stationary networks. However, due to the localised nature of its search algorithm, it can be satisfactorily used for the mobile case as well. But there are a few problems with the protocol: It does not minimise over all energy consumption. The protocol does not give any clue on how to define the initial search region. The resultant topology can be significantly different from the energy optimal topology for the all-to-all communication scheme. Furthermore, no explicit reconfiguration protocol has been given and the assumption of existence of GPS may not be very practicable. The protocol of Rodoplu and Meng (1999) has been referred to as MECN (minimum energy communication network) by Li and Halpern (2001). The authors have identified the conditions that are necessary and sufficient for a graph to have the minimum energy property. They use this characterisation to improve upon the protocol of Rodoplu and Meng (1999) by proposing another protocol called SMECN (small minimum energy communication network). The subnetwork constructed by SMECN is demanded to be provably smaller than that constructed by MECN if broadcasts at a given power setting are able to reach all nodes in a circular region around the broadcaster and also authors demand SMECN to be computationally simpler than MECN. It is proved that a sub-graph G of G' has minimum energy property iff it contains G_{min} as a sub-graph where G_{min} is the smallest sub-graph of G' with the minimum energy property. The protocol constructs a communication network containing G_{min} as it is more power efficient than to construct G_{min} itself. The algorithm used was a variant of the one used by Rodoplu and Meng (1999) to compute the enclosure graph. The main difference between SMECN and MECN is the computation of a region called η. In MECN the nodes, which are not neighbors of transmitting node, are not considered whereas in SMECN all nodes that the transmitting node has so far found are considered. Li and Halpern extended their SMECN protocol in (2004) to incorporate a reconfiguration protocol whereby SMECN is run only when necessary i.e. it is only run when not running it may result in a network not satisfying the minimum-energy property. The reconfiguration protocol for SMECN can also be modified so that it works for MECN.

In Li et al. (2003), the authors introduced LMST, a fully distributed and localised protocol aimed at building an MST like topology such that network connectivity should be preserved with the use of minimal possible power. The algorithm should be less susceptible to the impact of mobility, and should depend only on the information collected locally. Desirable results of LMST are that the links in the topology derived under the algorithm should be bi-directional and the node degree should also be small.The topology is constructed by each node building its local MST independently (with the use of information locally collected) and only keeping one-hop on-tree nodes as neighbors. The proposed algorithm is composed of the following three phases: information collection, topology construction, and determination of transmission power, and an optional optimisation phase: construction of topology with only bidirectional edges. The LMST protocol generates a strongly connected communication graph. The node degree of any node in the generated topology is at most 6. Also the topology can be made symmetric by removing asymmetric links without impairing connectivity. In fact LMST outperforms CBTC (see the following sections) and the protocol of Rodoplu and Meng (1999) in terms of both average node degree and average node transmitting range. On the other hand it has some limitations also. LMST requires location information that can be provided only with a considerable hardware and/or message cost. There is a trade-off between the redundancy and the stability of the topology. The LMST TC algorithm has impact on other layers, especially MAC and routing layer. Hu (1993) designed a distributed novel TC (NTC) algorithm for packet radio networks (PRNs), based on the Voronoi diagram and the Delaunay triangulation, to control the transmitting power and logical neighbours in order to construct a reliable high-throughput topology. The TC algorithm ensures graceful degradation and reliability under topological changes and component failures. The algorithm first constructs a planar triangulation from locations of all nodes as a starting topology. Then, the minimum angles of all triangles in the planar triangulation are maximised by means of edge switching to improve connectivity and throughput. The resulting triangulation at this stage is called the Delaunay triangulation, and it can be determined locally at each node.

Finally, the topology is modified by negotiating among neighbors to satisfy a design requirement on the nodal degree parameter. Thus it maximises minimum angles of all triangles in the planar triangulation and hence minimises node degree. The author first proposes a centralised TC algorithm based on the Delaunay Triangulation and then extends it to the realistic distributed case. First we will discuss the centralised algorithm as it is needed to understand the heuristics behind the distributed approach. The distributed implementation of this algorithm relies on each node computing a part of the Delaunay triangulation containing all the nodes reachable within a radius R. The topology produced has better performance in terms of throughput and reliability, compared to topologies using a fixed R only or fixed Δ only (connecting to the nearest Δ nodes). The protocol produces reliable topology and a good throughput is achieved for PRNs. The degree parameter should be greater than or equal to six in order to maintain the underlying triangulation structure. The distributed heuristic is suitable for real-time applications. But the biggest drawback is that the protocol does not address the issue of assigning transmit powers to the nodes. The sorting operation in the last step of centralised NTC is the bottle-neck of the algorithm because the number of possible edges is $O(n^2)$ and hence the cost is in the order of $O(n^2 \log n)$.

6.5.3 Direction based TC protocols

In Wattenhofer et al. (2001), the authors introduced a distributed TC protocol based on directional information, called CBTC (Cone Based TC) that increases network lifetime, maintains strong 1-node, 2- node, k-node connectivity, minimises maximum transmit power and total power and thereby indirectly increases network lifetime. It aims to find a topology with small node degree, so that interference is minimal and hence throughput is sufficient. Each node in the multihop wireless network must use local information only for determining its transmission radius and hence its operational power. The local decisions must be made in such a way that they collectively guarantee the node connectivity in the global topology just as if all nodes were operating at full power. The algorithm has two phases. In the first phase the authors describe a decentralised scheme that builds a connected graph upon the node network by letting nodes find close neighbor nodes in different directions. The second phase improves the performance by eliminating non-efficient edges in the communication graph. The algorithm is simple and does not need any complicated operations. The algorithm is also distributed and without synchronisation. CBTC has been primarily designed for 2-D surface mainly for static networks. The protocol requires only local reachability information and reduces traffic interference. Moreover deployment region need not be specified. Since it uses the Angle-of-Arrival (AOA) technique GPS support is not required. But since it relies on directional information multiple directional antennae are needed. The initial power and its step increase required in the algorithm have not been discussed. Also no bound on the number of messages or on the energy expended in determining proper transmit power has been mentioned.

Li et al. (2001) extends the CBTC (Wattenhofer at al. 2001). The basic idea of this algorithm is that a node u transmits with the minimum power $p_{u,\alpha}$ required to ensure that in every cone of degree α around u, there is some node that u can reach with power $p_{u,\alpha}$. They showed that taking $\alpha = 5\pi/6$ is a necessary and sufficient condition to guarantee that network connectivity is preserved. More precisely, if there is a path from u to v when every node communicates at maximum power then, if $\alpha \leq 5\pi/6$, there is still a path in the smallest symmetric graph containing all edges (u,v) such that u can communicate with v using power $p_{u,\alpha}$. On the other hand, if $\alpha > 5\pi/6$, connectivity is not necessarily preserved. A set of optimisations was also provided that further reduces power consumption while retaining network connectivity. These optimisations are:

- Shrink back operation
- Asymmetric edge removal
- Pair-wise edge removal

The optimisations, however, rely on distance estimation. Also the authors improved the basic algorithm so that it could handle dynamic reconfigurations necessary to deal with failures and mobility. Huang et al. (2002), has implemented the CBTC (Wattenhofer at al. 2001) with directional antennas so as to adjust both the transmission power and the effective antenna direction(s) of each node. Bahramgiri et al, has extended the CBTC protocol to the three-dimensional case in (2002). They have also presented a modification of the protocol aimed at ensuring k-connectivity of the directed communication graph while being sufficiently fault tolerant. However, their method does not bind the node degree and does not optimally minimise the transmission power at each node. Borbash and Jennings (2002) gives another distributed algorithm for TC based on the local and directional information, which is similar to CBTC (Wattenhofer et al. 2001). The optimisation objectives considered were: minimisation of - node degrees, hop-diameter, maximum transmission radius, number of bi-connected components and the size of the largest bi-connected component, while guaranteeing connectivity. In the first phase of the algorithm, each node u grows it's transmit power i.e. its radius until a not-yet-covered node v is reached. In the second phase, u computes a cone, which spans the area covered by v. These two phases are repeated until cones jointly span the 2π surrounding area of u. This algorithm actually computes a Relative Nearest Graph (RNG). The topology maintained has good overall performance in terms of interference, reliability, and power usage.

6.5.4 Neighbour-based TC Protocols

Another class of TC protocols is based on the idea of connecting each node to its closest neighbours. The LINT/LILT(Local Information No/Link-State Topology) protocols of Ramanathan and Rosales-Hain (2000) try to keep the number of neighbors of a node within a low and high threshold centered around an optimal value. A significant shortcoming of LINT is its incognizance of network connectivity and the consequent danger of a network partition. In multihop wireless routing protocols based on the link state approach some amount of global connectivity is available locally at every node. This is available at no additional overhead to the topology control mechanism. The idea in LILT is to exploit such information for recognising and repairing network partitions. There are two main parts to LILT: the neighbour reduction protocol (NRP) – essentially same as LINT and the neighbour addition protocol (NAP) – Its purpose is to override the high threshold bounds and increase the power if the topology change indicated by the routing update results in undesirable connectivity. The main difference between these distributed algorithms is the nature of feedback information used and network property sought to be maintained.

- LINT uses locally available neighbor information and attempts to keep degree of each node bounded. It does not however guarantee global connectivity.
- LILT exploits global topology information also available from a suitable network layer routing protocol and then guarantees a connected network

However, no characterisation of the optimal value of the number of neighbors is given and, consequently, no guarantee on the connectivity of the resulting communication graph is provided. Another problem of the LINT protocols is that they estimate the number of neighbors by simply overhearing control and data messages at different layers. This approach has the advantage of generating no control message overhead but the accuracy of the resulting neighbor number estimate heavily depends on the traffic present in the network. In the extreme case, a node which remains silent is not detected by any of its actual neighbors. The problem of characterising the minimum value of k such that the resulting communication graph is connected (the Critical Neighbor Number) has been investigated in

Xue and Kumar (2004) where it is shown that k □ (log n) is a necessary and sufficient condition for connectivity with high probability Recently, Wan and Yi (2004) have improved the upper bound on the CNN for connectivity derived in Xue and Kumar (2004). Based on Xue and Kumar's (2004) theoretical result, Blough et al. (2003) proposed the *k*-NEIGH protocol which is a distributed, asynchronous, localised protocol for TC aimed at maintaining the number of neighbors of every node equal to or slightly below a specific value k and minimising of maximum transmit power to the symmetric neighbors of a node The algorithm is based on the computation of a symmetric sub-graph of the k nearest neighbors graph ordering the list of k nearest neighbors. Finally, a node sets its transmitting power such as to just transmit to the farthest node among its set of symmetric neighbors. The authors prove that the communication graph generated by k-NEIGH when $k \in \Theta(\log n)$ is connected with high probability. From a practical viewpoint, Blough et al. (2003) show through simulation that setting k = 9 is sufficient to obtain connected networks with high probability for networks with n ranging from 50 to 500.The k-neigh protocol does not require the knowledge of the exact number of nodes in the network. It is not influenced by a choice of specific transmission range. The authors show by simulation results that the topology generated by k-NEIGH is, on average, 20% more energy efficient than that generated by CBTC. However it does not deal with node mobility well and its objective function is not strong

A protocol that shares many similarities with *k*-NEIGH is the XTC protocol presented in Wattenhofer and Zollinger (2004): the neighbors of a node *u* are ordered according to some metric (e.g., distance or link quality), and *u* decides which nodes are kept as immediate neighbors in the final network topology based on a simple rule. Contrary to *k*-NEIGH, which achieves connectivity with high probability, XTC builds a topology which is connected whenever the maxpower communication graph is connected. To achieve this, the requirement of having an upper bound *k* on the number of neighbors of a node is dropped. Contrary to *k*-NEIGH, in XTC, a node can have as much as n−1 neighbors in the final topology.

Cheng et al. (2003), denotes the following problem as SMET (Strong Minimum Energy Topology) Problem: given a set of sensors in the plane, to compute the transmit power of each sensor, such that, there exists at least one bidirectional path between any pair of sensors and the sum of all the transmit powers is minimised. First SMET is proved to be NP-hard. Then two approximation heuristics are given to solve it. One of them is an incremental power heuristic and another is based on the MST. In the former case a globally connected topology is built from one node, selecting a node with minimum incremental power to add to the topology at each step. At any time in this heuristic, all bidirectional links in the partial topology form a connected graph. The incremental power heuristic takes time $O(n^3)$. For SMET, Cheng et al. (2003), proposed another algorithm based on minimum (edge-cost) spanning tree (MST). Power is assigned to each sensor such that it can reach the farthest neighbor in the tree. The topology generated by this power assignment (only bidirectional links are considered) is globally connected. The power assignment based on MST has performance ratio 2. The MST heuristic takes time $O(n^2 \log n)$, where n is the number of nodes in the network. The algorithms provide reference to WSN deployment. They can serve as a theoretical guide for determining strongly related key parameters such as sensor density and transmit power. These heuristics can provide the lower bound for the transmission range to maintain global connectivity, which is useful in network simulators. The authors demand SMET is a better model for energy-aware broadcasting and multicasting. However their heuristics have pretty high time complexity.

6.5.5 Hybrid TC

Apart from pure distributed or centralised techniques, Shen et al (2004) have described CLTC (Cluster Based TC), which is a hybrid of distributed and centralised approaches using clustering. It is a framework for the design of TC algorithms that guarantee strong 1-node, 2-node, k-node connectivity, while attempting to optimise a given function of the power specifications of the nodes. In fact, it minmises maximum transmit power and total power. To utilise the CLTC framework there are three algorithms that must be specified: the clustering algorithm in Phase 1; the intra cluster TC algorithm of Phase 2; and, the inter-cluster TC method of Phase 3. CLTC uses a centralised algorithm within a cluster and between adjacent clusters to achieve strong connectivity and yet achieves the scalability and adaptability of a distributed approach with localised information exchange between adjacent clusters. CLTC combines centralised and distributed schemes and achieves scalability as well as strong connectivity. It has lesser overhead than centralised schemes by at least a magnitude of one. But the direct and indirect paths cannot be simultaneously established and the clusters are not allowed to overlap which may give sub-optimal results.

6.6 Mobile WSNs

We discussed energy-efficient communication in stationary WSNs. In this kind of networks, a TC protocol is generally executed once at the beginning of the network operation, and then periodically to account for nodes' joining (or, leaving) the network. Thus, the message overhead of the protocol (usually expressed in the efficiency) has relatively little importance, and, hence, the emphasis is more on the quality of the produced topology. The situation will obviously be different in mobile WSNs, and, if deriving analytical results for stationary WSNs is difficult, deriving theoretical results regarding mobile WSNs is even more challenging. In this section, we will discuss how node mobility affects TC in general. To begin with, we note that the impact of mobility on TC is twofold:

- *Increased message overhead*- We know that any distributed TC protocol entails a certain amount of message overhead due to the fact that nodes need to exchange messages in order to set the transmitting range to an appropriate value. In the presence of mobility, a TC protocol must be executed frequently in order to account for the new positions of the nodes. Thus, reducing message overhead is fundamental, when implementing TC mechanisms in mobile WSNs (especially in the case of high mobility scenarios), even if reducing message overhead comes at the cost of a lower quality of the constructed topology.

- *Nonuniform spatial distribution*- As it will be discussed in detail later, some mobility patterns may cause a nonuniform spatial distribution of sensor nodes. This fact should be carefully taken into account in setting important network parameters (such as the CTR) at the design stage.

From the above discussion, it is clear that the impact of mobility on the effectiveness of TC techniques heavily depends on the mobility pattern. For this reason, we first present the mobility models which have been considered in the literature.

6.6.1 Mobility Models

The most widely used mobility model in the WANET community is the random waypoint model (Johnson and Maltz 1996). In this model, every node chooses at random a destination d (the *waypoint*) uniformly in $(0, l)$ and moves towards it along a straight line with a velocity chosen uniformly at random in the interval ($vmin$, $vmax$). When it reaches the destination, it remains stationary for a predefined pause time $tpause$, and then it starts moving again according to the same rule. A similar model is the random direction model (Bettstetter 2001; Royer et al. 2001) in which nodes move with a direction chosen uniformly in the interval $(0, 2\pi)$ and a velocity chosen uniformly in the interval ($vmin$, $vmax$). After a randomly chosen time taken usually from an exponential distribution, the node chooses a new direction. A similar procedure is used to change velocity, using an independent stochastic process.

Contrary to the case of the random waypoint and the random direction model which resemble (at least to some extent) intentional motion, the class of Brownian-like mobility models resembles non-intentional movement. For example, in the model used in Blough et al. (2002), mobility is modeled using parameters $pstat$, $pmove$, and m. Parameter $pstat$ represents the probability that a node remains stationary during the entire simulation time. Hence, only $(1 - pstat)*n$ nodes (on the average) will move. Introducing $pstat$ into the model accounts for those situations in which some nodes are not able to move. For example, this could be the case when sensors are spread from a moving vehicle, and some of them remain entangled, for example, in a bush or tree. This can also model a situation where two types of nodes are used, one type that is stationary and another type that is mobile. Parameter $pmove$ is the probability that a node moves at a given step. This parameter accounts for heterogeneous mobility patterns in which nodes may move at different times. Intuitively, the smaller the value of $pmove$, the more heterogeneous the mobility pattern is. However, values of $pmove$ close to 0 result in an almost stationary network. If a node is moving at step i, its position in step $i+1$ is chosen uniformly at random in the square of side $2m$ centered at the current node location. Parameter *m* models, to a certain extent, the velocity of the nodes: the larger *m* is, the more likely it is that a node moves far away from its position in the previous step. In the case of random direction or Brownian-like motion, nodes may, in principle, move out of the deployment region. Since a standard approach in simulations is to keep the number of network nodes constant, we need a so-called *border rule* (Bettstetter 2001) that defines what to do with nodes that are about to leave the deployment region. In this situation, a node can be:

(1) bounced back, *or*
(2) positioned at the point of intersection of the boundary with the line connecting the current and the desired next position, *or*
(3) wrapped to the other side of the region which is considered as a torus, *or*
(4) deleted, and a new node be initialised according to the initial distribution, *or*
(5) forced to choose another position until the chosen position is inside the boundaries of the deployment region.

Depending on the choice of the border rule, nonuniformity in the node spatial distribution can be produced. For example, the rule (2) places nodes exactly on the boundary of the region with higher probability than at other points. In fact, the only rules (3) and (5) do not appear to favor one part of the region over another. However, these rules appear quite unrealistic and are used mainly to artificially generate a more uniform node distribution. More exhaustive survey of mobility models is found in (Bettstetter 2001; Camp et al. 2002)

6.6.2 Homogeneous TC

When the range assignment is homogeneous, the message overhead is not an issue since the nodes' transmitting range is set at the design stage, and it cannot be changed dynamically. However, the node distribution generated by the mobility model could be an issue. For instance, it is known that the random waypoint model generates a node spatial distribution which is independent of the initial node positions and in which nodes are concentrated in the center of the deployment region (Bettstetter 2001; Bettstetter and Krause 2001; Bettstetter et al. 2003; Blough et al. 2002). This phenomenon, which is known as the *border effect*, is due to the fact that, in the random waypoint model, a node chooses a uniformly distributed destination point rather than a uniformly distributed angle. Therefore, nodes located at the border of the region are very likely to cross the center of the region on their way to the next waypoint. The intensity of the border effect mainly depends on the pause time *tpause*. In fact, a longer pause time tends to increase the percentage of nodes that are resting at any given time. Since the starting and destination points of a movement are chosen uniformly in $(0, l)^*d$, a relatively long pause time generates a more uniform node distribution. Consequently, the results concerning the CTR in stationary WSNs (which are based on the uniformity assumption) cannot be directly used. For this reason, the relationship between the CTR with and without mobility must be carefully investigated. Sanchez et al.

(1999) have analysed the probability distribution of the CTR in the presence of different mobility patterns (random waypoint, random direction, and Brownian-like) through simulation, and interestingly their results seem to indicate that the mobility pattern has little influence on the distribution of the CTR. Santi and Blough have also investigated the relationship between the CTR in stationary and in mobile networks through extensive simulation. They consider random waypoint and Brownian-like motion and analyse different critical values for the node transmitting range that are representative of different requirements on network connectivity (for instance, connectivity during 100% and 90% of the simulation time). Their results show that a relatively modest increase of the transmitting range with respect to the critical value in the stationary case is sufficient to ensure network connectivity during 100% of the simulation time. Further insights into the relationship between the stationary and mobile CTR can be derived from the statistical analysis of the node distribution of mobile networks reported in Blough et al. They considered random waypoint and Brownian-like mobility and perform several statistical tests on the node distribution generated by these models. The results of the tests show that the distribution generated by Brownian-like motion is virtually indistinguishable from the uniform distribution and confirm the occurrence of the border effect in random waypoint motion, whose intensity depends on the value of *tpause*. In the extreme case of *tpause*=0, the random waypoint model generates a node distribution which is considerably different from uniform. Overall, the analysis of Blough et al. indicate that Brownian-like mobility should have little influence on the value of the CTR, while the effect of random waypoint mobility on the CTR should depend on the settings of the mobility parameters.

6.6.3 Non-homogeneous TC

For non-homogenous cases, mobility introduces further problems because now the most important effect of node mobility is the message overhead generated to update every node's transmitting power. The amount of overhead depends on how frequently the network reconfiguration protocol is run to restore the desired network topology. In turn, this depends on several factors such as the mobility pattern and the properties of the topology generated by the protocol. To clarify this point, let us consider two TC protocols *P*1 and *P*2. Protocol *P*1 builds the MST in a distributed fashion and sets the nodes' transmitting range accordingly, while protocol *P*2 attempts to keep the number of neighbors of each node below a certain value *k* as in the *k*-NEIGH protocol of Blough et al. (2003). Protocol *P*1 is based on global and very precise information, since the MST can be built only if the exact position of every node in the network is known. In principle, *P*1 should be reconfigured every time the relative position of any two nodes in the network changes since this change could cause edge insertion/removal in the MST. On the other hand, *P*2 can be easily computed in a localised fashion and can be implemented using relatively inaccurate information such as distance estimation. In this case, the protocol should be re-executed only when the relative neighborhood relation of some node changes. It is quite intuitive that this occurs less frequently than edge insertion/removal in the MST. It should also be observed that having a topology that is not up-to-date is much more critical in the case of the MST than in case of the *k*-neighbors graph. In fact, a single edge removal in the MST is sufficient to disconnect the network, while several edges can, in general, be removed from the *k*-neighbors graph without impairing connectivity. Overall, we can reasonably state that *P*1 should be reexecuted much more frequently than *P*2. Further, we observe that the reconfiguration procedure needed to maintain the MST is more complicated than that required by the *k* neighbors graph since it relies on global information. So, we can conclude that protocol *P*1 is not suitable to be implemented in a mobile scenario; in other words, it is not *resilient to mobility*. From the previous discussion, it is clear that a mobility resilient TC protocol should be based on a topology which can be computed locally and which requires little maintenance in the presence of mobility. Many of the TC protocols presented in the literature meet this requirement. However, only some of them have been defined to explicitly deal with node mobility. In Li et al., an adaptation of the CBTC protocol to the case of mobile networks is discussed. It is shown that, if the topology ever stabilises and the reconfiguration protocol is executed, then the network topology remains connected. The reconfiguration procedure is adapted to the case of *k*-connectivity in Bahramgiri et al. In Rodoplu and Meng, the authors discuss how their protocol can be adapted to the mobile scenario and evaluate the protocol power consumption in the presence of a mobility pattern which resembles the random direction model. The MobileGrid (Liu and Li 2002) and LINT (Ramanathan and Rosales-Hain 2000) protocols, which are based on the *k*-neighbors graph, are explicitly designed to deal with node mobility. They are zero overhead protocols since the estimation of the number of neighbors is based on the overhearing of data and control traffic. However, no explicit guarantee on network connectivity is given except the simulation results. A more subtle effect of mobility on certain TC protocols is due to the possibly nonuniform node distribution generated by the mobility pattern. This fact is considered in setting fundamental protocol parameters such as the critical neighbor number in *k* neighbors graph-based protocols (Blough et al. 2002; Liu and Li 2002; Ramanathan and Hain 2000). The number of neighbors *k* needed to obtain connectivity with high probability in the presence of uniform node distribution is significantly different from the value *km* needed when the node distribution is nonuniform, such as in the presence of random waypoint mobility. So, if nodes are expected to move with random waypoint-like mobility, *km* must be used instead of *k*.

6.7 References

Akyildiz IF, Su W, Sankarasubramaniam Y, and Cayirci E (2002) Wireless sensor networks: a survey. Computer Networks 38, pp 393–422

Bahramgiri M, Hajjaghayi M, and Mirrokni VS (2002) Fault-tolerant and 3-dimensional distributed topology control algorithms in wireless multi-hop networks. In: Proc. IEEE Int. Conf on Computer Communications and Networks, pp 392-397

Bettstetter C (2001) Smooth is better than sharp: A random mobility model for simulation of wireless networks. In: Proceedings of the ACM Workshop on Modeling, Analysis and Simulation of Wireless and Mobile Systems (MSWiM) pp 19–27

Bettstetter C (2002a) On the connectivity of wireless multihop networks with homogeneous and inhomogeneous range assignment. In: Proceedings of the 56th IEEE Vehicular Technology Conf. (VTC), pp 1706–1710

Bettstetter C and Krause O (2001) On border effects in modeling and simulation of wireless ad hoc networks. In: Proceedings of the IEEE Int. Conf. on Mobile and Wireless Communication Network (MWCN)

Bettstetter C, Resta G, and Santi P (2003) The node distribution of the random waypoint mobility model for wireless ad hoc networks. IEEE Trans Mobile Comput 2:3, pp 257–269

Blough D, Leoncini M, Resta G, and Santi P (2002) On the symmetric range assignment problem in wireless ad hoc networks. In: Proc. of the IFIP Conf. on Theoretical Computer Science, pp 71–82

Blough D, Leoncini M, Resta G, and Santi P (2003) The k-neighbors protocol for symmetric topology control in ad hoc networks. In: Proc. of the ACM MobiHoc 03, pp 141–152

Blough D, Resta G, and Santi P (2002) A statistical analysis of the long-run node spatial distribution in mobile ad hoc networks. In: Proc. of the ACM Workshop on Modeling, Analysis, and Simulation of Wireless and Mobile Systems (MSWiM), pp 30–37

Borbash S and Jennings E (2002) Distributed topology control algorithm for multihop wireless networks. In: Proc. of the IEEE Int. Joint Conf. on Neural Networks pp 355–360

Camp T, Boleng J, and Davies V (2002) A survey of mobility models for ad hoc network research. Wirel Comm Mobile Comput 2:5, pp 483–502

Chandrakasan A., Amirtharajah R, Cho S H, Goodman J, Konduri G, Kulik J, Rabiner W, and Wang A (1999) Design considerations for distributed microsensor systems. In: Proceedings of IEEE Custom Integrated Circuits Conf. (CICC), May, pp 279 – 286

Charikar M, Chekuri C, Cheung T, Dai Z, Goel A, Guha S, and Li M (1999) Approximation algorithms for directed steiner problems. Journal of Algorithms, 33:1, pp 73-91

Cheng X, Narahari B, Simha R, Cheng M X, and Liu D (2003) Strong minimum energy topology in WSN: NP-completeness and heuristics. IEEE Trans on mobile computing, 2:3

Clementi A, Ferreira A , Penna P , Perennes S , and Silvestri R (2000a) The minimum range assignment problem on linear radio networks. In: Proceedings of the 8th European Symposium on Algorithms (ESA) pp 143–154

Cormen T H, Leiserrson C, Rivest R L, and Stein C (2001) Readings in introduction to algorithms. Second ed MIT Press

Dette H and Henze N (1989) The limit distribution of the largest nearest-neighbor link in the unit d-cube. J Appl Probab 26, PP 67–80

Guha and Khuller S (1999) Improved methods for approximating node weighted steiner trees and connected dominating sets information and computation, vol 150, pp 57-74

Holst L (1980) On multiple covering of a circle with random arcs. J Appl Probab 16, pp 284–290

Hu L (1993) Topology control for multihop packet radio networks. IEEE Trans Comm, vol 41, no 10, pp 1474-1481

Huang Z, Shen C, Srisathapornphat C, and Jaikaeo C (2002) Topology control for ad hoc networks with directional antennas. In: Proc. of the IEEE Int. Conf. on Computer Communications and Networks pp 16–21

Johnson D and Maltz D (1996) Dynamic source routing in ad hoc wireless networks. In Mobile Computing Kluwer Academic Publishers pp 153–181

Kortsarz G and Peleg D (1999) Approximating the weight of shallow light trees. Discrete Applied Mathematics, vol . 93, pp 265-285

Krumke S, Liu R, Loyd E L, Marathe M, Ramanathan R, ad Ravi SS (2003) Topology control problems under symmetric and asymmetric power thresholds. In: Proc. of the 2nd Int. Conf. on AD-HOC Networks and Wireless (ADHOC NOW 03), Lecture Notes in Computer Science, Vol. 2865, Montreal, Canada, Oct.-2003, pp 187-198

Li L and Halpern JY (2004) A minimum-energy path-preserving topology-control algorithm. IEEE Trans. on wireless communication, 3:3, pp 910 – 921

Li L and Halpern JY (2001) Minimum energy mobile wireless networks revisited. In: Proc. of IEEE Conf. on Communications (ICC'01), June 2001, pp 278-283

Li L, Halpern J, Bahl P, Wang Y, and Wattenhofer R (2001) Analysis of a cone based distributed topology control algorithm for wireless multi-hop networks. In: Proc. of ACM Principles of Distributed Company (PODC) pp 264–273

Li N and Hou J (2004) Flss: a fault-tolerant topology control algorithm for wireless networks. In: Proc. of ACM Mobicom pp 275–286

Li N, Hou J, and Sha L (2003) Design and analysis of an mst-based topology control algorithm. In: Proc. of the IEEE Infocom pp 1702–1712

Li N, Hou J, and Sha L (2003) Design and analysis of an mst-based topology control algorithm. In: Proc. of the IEEE Infocom pp 1702–1712

Liu J and Li B (2002) Mobilegrid: Capacity-aware topology control in mobile ad hoc networks. In: Proc. of the IEEE Int. Conf. on Computer Communications and Networks pp 570–574

Loyd E, Liu R, Marathe M, Ramanathan R, and Ravi S (2002) Algorithmic aspects of topology control problems for ad hoc networks. In: Proc. of ACM Mobihoc pp 123–134

Marathe MV, Ravi R, Sundaram R, Ravi SS, Rosenkrantz DJ, and Hunt HB (1998) Bicriteria Network Design Problems. Journal of Algorithms, 28:1, pp 142-171

Narayanaswamy S, Kawadia V, Sreenivas R, and Kumar P (2002) Power control in ad hoc networks: Theory, architecture, algorithm and implementation of the compow protocol. In: Proc. of European Wireless pp 156–162

Panchapakesan P and Majunath D (2001) On the transmission range in dense ad hoc radio networks. In: Proceedings of IEEE Signal Processing Communication (SPCOM).

Penrose M (1997) The longest edge of the random minimal spanning tree. Annals Appl Probab 7:2, pp 340–361

Penrose M (1999a) On k-connectivity for a geometric random graph. Rand Struct Algori 15: 2, pp 145–164

Penrose M (1999c) A strong law for the longest edge of the minimal spanning tree. The Annals Probab 27:1, pp 246–260

Pottie G and Kaiser W (2000) Wireless integrated network sensors. Comm ACM 43:5, pp 51–8

Ramanathan R and Rosales-hain R (2000) Topology control of multihop wireless networks using transmit power adjustment. In: Proceedings of IEEE Infocom'00 pp 404–413

Rappaport T 2002 Wireless communications: Principles and practice, 2nd Ed Prentice Hall, Upper Saddle River, NJ

Rodoplu V and Meng T (1999) Minimum energy mobile wireless networks. IEEE J Select Areas Comm 17: 8, pp 1333–1344

Royer E, Melliar SP, and Moser L (2001) An analysis of the optimum node density for ad hoc mobile networks. In: Proc. of the IEEE Int. Conf. on Communications pp 857–861

Sanchez M, Manzoni P, Haas Z 1999 Determination of critical transmitting range in ad hoc networks. In: Proc. of Multiaccess, Mobility and Teletraffic for Wireless Communications Conference

Santi P (2005) The critical transmitting range for connectivity in mobile ad hoc networks. IEEE Trans. Mobile Comput 4:3, pp 310–317

Santi P (2005a) Topology Control in wireless ad hoc and sensor networks .ACM Computing Surveys, 37:2, pp 164 – 194

Santi P and Blough D (2002) An evaluation of connectivity in mobile wireless ad hoc networks. In: Proc. of IEEE Dependable Systems and Networks (DSN) pp 89–98

Santi P and Blough D (2003) The critical transmitting range for connectivity in sparse wireless ad hoc networks. IEEE Trans Mobile Comput 2:1, pp 25–39

Santi P, Blough D, and Vainstein F (2001) A probabilistic analysis for the range assignment problem in ad hoc networks. In: Proc. of ACM Mobihoc pp 212–220

Shen C, Srisathapornphat C, Liu R, Hunag Z, Jaikaeo C, ad Lloyd EL (2004) CLTC: A cluster-based topology control framework for ad hoc networks. IEEE Trans. on Mobile Computing, 3:1, pp 18-32

Wan P and Yi C (2004) Asymptotical critical transmission radius and critical neighbor number for k-connectivity in wireless ad hoc networks. In: Proc. of ACM MobiHoc pp 1–8

Wattenhofer R, Li L, Bahl P, and Wang Y (2001) Distributed topology control for power efficient operation in multihop wireless ad hoc networks. In: Proc. of IEEE Infocom pp 1388–1397

Wattenhofer R and Zollinger A (2004) Xtc: A practical topology control algorithm for ad hoc networks. In: The 4th Int. Workshop on Algorithms for Wireless, Mobile, Ad Hoc and Sensor Networks (WMAN)

Xue F and Kumar P (2004) The number of neighbors needed for connectivity of wireless networks. ACM/Kluwer Wirel Netw 10:2, pp 169–181

7 Routing Protocols

Muhammad Umar, Usman Adeel and Sajid Qamar

Faculty of Computer Science & Engineering, Ghulam Ishaq Khan Institute of Engineering Sciences and Technology, Topi, Swabi, Pakistan

7.1 Introduction

A wireless sensor network (WSN) consists of hundreds or thousands of sensor nodes that are constrained in energy supply and bandwidth. Therefore, incorporating energy-awareness in all the layers of the protocol, stack is a basic requirement. Different sensor network applications generally require similar physical and link layers and the design of these layers is focused on system level power awareness such as dynamic voltage scaling, radio communication hardware, low duty cycle issues, system partitioning and energy aware MAC protocols (Heinzelman 2000; Min 2000; Woo 2001; Ye 2002; Eugene 2001). However, network layer gets influenced by the type of application besides ensuring energy-efficient route setup, reliable data-delivery and maximising network lifetime. Accordingly, routing in wireless sensor networks is nontrivial and should take into consideration its inherent features along with application and architectural requirements. The routing techniques are employed by contemporary communication and ad hoc networks cannot be applied directly to wireless sensor networks due to the following reasons:

- Traditional IP based protocols are inapplicable as it is infeasible to assign and maintain unique IDs for such a large number of sensor nodes
- As against a typical communication network nearly all the applications of WSNs require flow of data from several sources to one destination.
- Multiple sensor nodes within the vicinity of the sensed phenomenon might generate same data. In contrast to other networks, this fact can be exploited at the network layer to improve energy efficiency and bandwidth requirements.
- Sensor nodes have severe limitation on transmission power, energy reservoir, processing capability and storage that are not major issues in other networks.
- Sensor networks are application specific and the design requirement changes.
- Sensor networks are data-centric networks, i.e., they require attribute based addressing and queries are also formed in attribute-value pairs. For example, if a sensor network is queried with "temperature > 60" then only the sensors with temperature greater then 60 will respond.

7.2 Design Issues

In order to design an efficient routing protocol, several challenging factors that influence its design should be addressed carefully. These factors are discussed below:

- Network dynamics: Three main components of a sensor network, i.e., sink, sensor nodes and monitored events can be either stationary or mobile. Routing to a mobile node (sink or sensor node) is more challenging as compared to a stationary node. Similarly, keeping track of a dynamic event is more difficult.
- Node deployment: Node deployment can be either deterministic or random. Deterministic deployment is simple as sensors are manually placed according to a plan and data is routed along pre-determined paths. Whereas random deployment raises several issues as coverage, optimal clustering etc. which need to be addressed.
- Multi-hop vs. single-hop network: As the transmission energy varies directly with the square of distance therefore a multi-hop network is suitable for conserving energy. But a multi-hop network raises several issues regarding topology management and media access control.
- Data reporting models: In wireless sensor networks data reporting can be continuous, query-driven or event-driven. The data-delivery model effects the design of network layer, e.g., continuous data reporting generates a huge amount of data therefore, the routing protocol should be aware of data-aggregation.
- Node heterogeneity: Some applications of sensor networks might require a diverse mixture of sensor nodes with different types and capabilities to be deployed. Data from different sensors, can be generated at different rates, network can follow different data reporting models and can be subjected to different quality of service constraints. Such a heterogeneous environment makes routing more complex.
- Data Aggregation: Data aggregation is the combination of data from different sources according to a certain aggregation function, e.g., duplicates suppression, minima, maxima, and average. It is incorporated in routing protocols to reduce the amount of data coming from various sources and thus to achieve energy efficiency. But it adds to the complexity and makes the incorporation of security techniques in the protocol nearly impossible.

7.3 Flat Routing Protocols

Flat routing is a multi-hop routing in which all nodes are considered operational and same tasks are assigned to them. In flat network topology, all nodes collaborate with each other to perform sensing task. The number of nodes in a sensor network is usually very large therefore it is not feasible to assign each node a global identifier. This makes very hard to send a query to a particular node or set of nodes. Usually, data is transmitted from each node with quite redundancy, which results in high inefficiency in terms of energy consumption. This consideration has led to data centric routing in which data is named on the basis of some attributes, which specify its properties. The BS can send queries to a particular region about particular data and then waits for sensor nodes located in that region. It can also designate some sensor nodes to report data to it periodically whether any event occurs or not. Besides this the data reporting can follow an event-driven model i.e. sensor nodes report data to the base station only on the occurrence of an event. Fig. 7.1 depicts a typical scenario for multihop forwarding in flat routing.

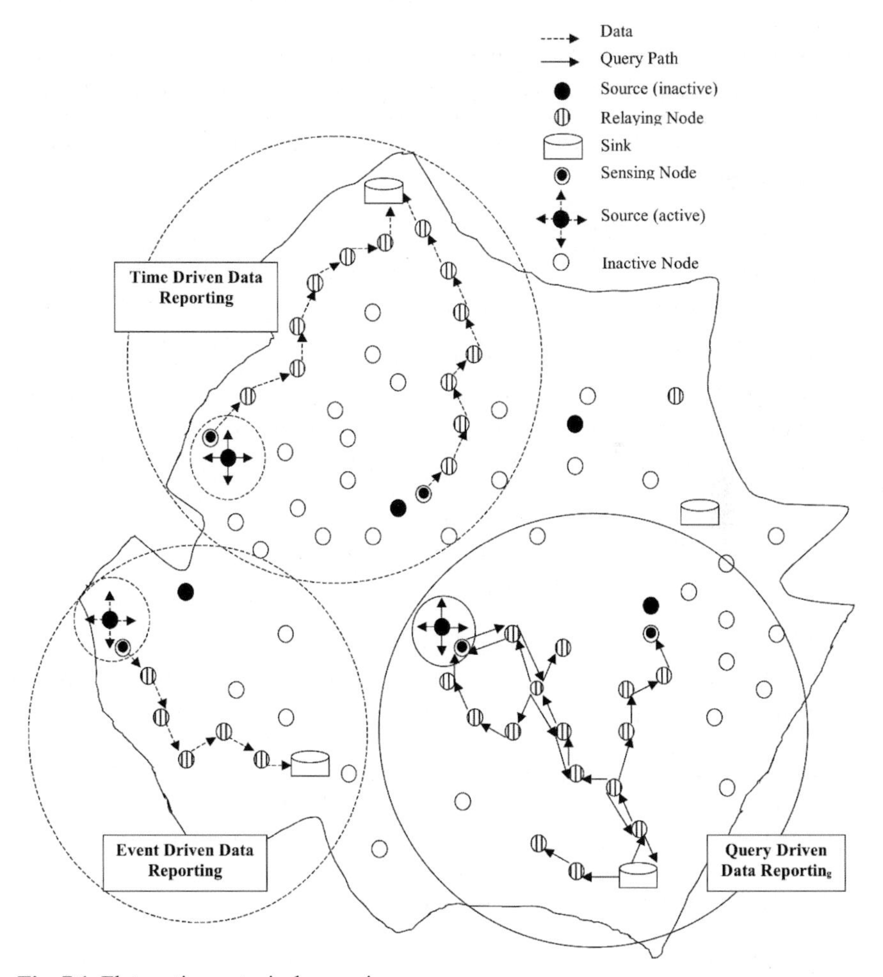

Fig. 7.1. Flat routing: a typical scenario

A brief introduction to some of flat routing schemes along with their advantages and drawbacks are discussed in the following subsections.

7.3.1 Sensor Protocol for Information via Negotiation (SPIN)

SPIN protocols proposed by Heinzelman (Heinzelman 1999) and Kulik (Kulik 2002) represents a family of routing protocols that achieve energy efficiency and elongate lifetime of the network through the use of negotiation based resource adaptive algorithms. The negotiation mechanism in SPIN is based on data advertisement and is implemented by the naming of sensed data using high-level descriptors or meta-data. The data is advertised to all the single-hop neighbours of the node using these descriptors. Any node interested in the data, i.e., a node not having the advertised data item, sends a request message back to the advertiser node, which then broadcast the data. In addition, the protocol has access to the current energy level of the node and adapts as to how much energy is remaining in the node. SPIN uses three stages for communication and has three types of messages accordingly, i.e., ADV, REQ and DATA. ADV messages are used to advertise the data, REQ messages are used to request the data and DATA messages are used to broadcast the requested data. SPIN represents a family of protocols, i.e., SPIN1, SPIN2, SPIN-BC, SPIN-PP, SPIN-EC, and SPIN-RL, each employing a three-stage mechanism for communication but with slight variations. SPIN1 is the initial proposal and SPIN2 extends it by incorporating threshold-based resource awareness in it, i.e., If the energy level in a node falls below a certain threshold, it can only participates in the protocol if it can complete all the three stages of the protocol. SPIN-BC is designed for broadcast channels, SPIN-PP for point-to-point communication, i.e., hop by hop routing, SPIN-EC is an extension to SPIN-PP but adds energy heuristic to it and SPIN-RL adapts the SPIN-PP protocol for lossy channels. The advantages of SPIN family of protocols include: localisation of topological changes, avoiding duplicate messages being sent to the same node, avoiding forwarding of same data from overlapped regions, elongation of the lifetime of network through resource adaptivity i.e. stopping the nodes with limited energy resource from forwarding. However, the data advertisement mechanism in SPIN cannot always guarantee the delivery of data, which is a key requirement of some critical applications.

7.3.2 Directed Diffusion

The main idea behind Directed Diffusion is to diffuse the sensed data through the network by using a naming scheme for it. Attribute-value pairs name all the data generated by the sensor nodes. This is done to get rid of the unnecessary operations of the network layer in order to save energy. This stored data is then retrieved by the base-station on an on-demand basis by the use of queries called interests. Interests represent a task to be done by the network and are created by the sink by using a list of attribute-value pairs. The interest is broadcast initially and each node receiving the interest can do caching for later use. The sensor nodes to compare the received data with the values in the interests later use these cached interests. Each sensor node that receives the interest set up a gradient towards the sensor nodes from which interests are received. An interest entry in the cache can contain several gradient fields and each gradient field is characterised by data-rate, duration and expiration time. Therefore, gradient is a reply link to a neighbour from which the interest was received. These gradients are setup to draw data satisfying the query towards the requesting node and this process continues until the gradients are setup from the source back to the base station. In general a gradient specifies an attribute value and a direction. Directed diffusion also achieves energy efficiency by allowing in-network consolidation of redundant data. It is a representative protocol of the data-centric paradigm, which is based on the idea of data-aggregation, i.e., combining the data from different sources. Unlike traditional end-to-end routing, data-centric routing finds route from multiple sources to a single destination, which facilitates the process of data-aggregation. The very first delivery of the requested data is made through several paths to the sink that resembles flooding but for subsequent deliveries the best path is reinforced according to a local rule at each intermediate node. In case of the failure of the data path, directed diffusion is equipped with path repair mechanism. In the post-failure repair approach, reinforcement is reinitiated to search among other paths. Whereas in the pre-failure approach, proposed in (Intanagonwiwat 2000), multiple paths are employed in advance so that in case of failure of one path another path is chosen without any cost for searching another path. There is an extra energy overhead in keeping multiple paths alive but more energy can be saved when a path fails and new path is to be chosen.

Some of the advantages of directed diffusion include energy efficiency, caching, data-aggregation, less delay in data transfer and no need of individual node addressing due to its data-centric approach. Whereas some of the weak points include extra overhead incurred in matching data and query, too much application dependence which requires the naming scheme to be defined before hand and finally it can only be applied to applications which support query-driven data delivery model. Directed Diffusion presented a breakthrough in data-centric routing therefore, its several variants are proposed in literature. Following three subsections present three such protocols that follow the same approach as Directed Diffusion but differ from it slightly.

7.3.3 Rumor Routing

Rumor routing (Braginsky 2002) achieves energy efficiency in the network through the use of controlled flooding of queries and is applicable in scenarios where geographic routing is infeasible. In routing protocols based on query-driven data delivery model, energy conservation can be achieved by allowing either queries or events to flood the network and not both. Former approach is suitable in scenarios when events are small as compared to the number of queries and the latter approach is suitable can be applied in scenarios where the queries are less. Rumor routing is based on the former approach and achieves this through the use of long-lived packets called *agents*. The algorithm starts with the detection of an event by a sensor node, which adds it to its locally maintained table called the event table and generates an agent. These agents travel through network to spread information about the event to distant nodes. When a query is generated in such a network, where the *rumour* about the event has been spread, the nodes can respond to it by checking their event-table and guiding it towards the source, thus saving the energy resource. Rumor routing is different from Directed Diffusion which uses flooding to spread query in the network and the data travels back to sink through multiple paths. The advantages of this approach are: significant energy saving through the avoidance of flooding and its capability to handle node failure whereas the limitations are: increased overhead of maintaining agents and event-tables if the number of events becomes large.

7.3.4 Gradient based Routing

Schurgers (Schurgers 2001) proposed the gradient based routing that is based on conserving energy in wireless sensor networks by the diffusion of data through selected nodes thus avoiding multiple paths. The technique is based on Directed Diffusion and uses a query-driven data delivery model. The protocol starts by memorising the hop number when the interest is flooded through the network. This hop number is used to calculate the height of each node, which is the minimum number of hops to reach the base-station. Each node also calculates another parameter called gradient, i.e., the difference between the height of current node and its neighbours. Forwarding of data packets occur in the direction of the largest gradient. The protocol also employs generic data-aggregation techniques to optimise energy consumption. Moreover, it discusses three data spreading techniques in order to achieve load-balancing and elongate lifetime of the network:

- Stochastic scheme picks one gradient at random when there are two or more neighbours at the same gradient.
- Energy based scheme increases the height of a node when its energy level falls below a certain threshold, so other sensors cannot sent data through it.
- Stream-based scheme avoids routing data through nodes that are on the path of other data streams.

The main advantage of the scheme is that it outperforms directed diffusion in the consumption of total communication energy.

7.3.5 Constrained Anisotropic Diffusion Routing (CADR)

CADR proposed by Chu (Chu 2002) seeks to improve the energy efficiency while minimising the latency and bandwidth. This routing scheme aims to be a general form of Directed Diffusion and is complemented by a routing scheme, i.e., information-driven sensor querying (IDSQ). It diffuses query to only selected sensors that get selected on the basis of an information criteria. For instance, the sensors can get selected on the basis of proximity to the occurring event thus adjusting the data routes dynamically. The nodes evaluate information versus cost objective in order to make forwarding decision where information utility is modelled through estimation theory. IDSQ complements it by specifying which node can provide most useful information and by performing load balancing. This scheme has proved to be more energy efficient than Directed Diffusion by incorporating intelligence in query diffusion.

7.3.6 Energy Aware Routing (EAR)

EAR proposed in (Rahul 2002) aims at increasing the network survivability through the use of a set of sub-optimal paths. It prevents forwarding of data through a single optimal path in order to avoid depletion of energy of the nodes on that path. A probability function based on energy consumption of each path is used to select a path from a set of sub-optimal paths. EAR makes an assumption that nodes in the network use a class-based addressing scheme, which can reveal the type and location of the node. The protocol operation can be divided into three phases. During the first phase each node creates a routing table with the help of local flooding. Each entry in the routing table contains a link cost for a particular neighbour and a probability that is inversely proportional to the cost of the link. The link cost for each neighbour, its associated probability and the average cost for reaching destination using neighbours in the routing table is calculated as:

$$LC_{i-j} = Cost(Node_i) + Energy(Link : Node_{i-j}) \tag{7.1}$$

where, $LC_{i\text{-}j}$ represents the link Cost between Node i and j, $Cost(Node_i)$ represents the average cost to reach destination using neighbours of node i and Energy ($Link$: $Node_{i\text{-}j}$) is a parameter which captures: cost of transmission, reception and residual energy of nodes i and j.

$$P_{Node_{i-j}} = (1/LC_{i-j})/(\sum_k 1/LC_{i-k}) \tag{7.2}$$

$$Cost(Node_i) = \sum_i (P_{j-i})(LC_{j-i}) \tag{7.3}$$

During the second phase data forwarding takes place in which a node uses the probabilities against each link for random selection and forwarding of packets. This last phase is maintenance phase that introduces a scheme based on infrequent flooding to keep the routing tables up to date. EAR resembles Directed Diffusion as both discover potential paths from source to destination. Energy Aware Routing always forwards data along a randomly selected single path but Directed Diffusion sends data along multiple paths at different data rates and reinforces one path. In this way, it outperforms Directed Diffusion in the consumption of energy but data delivery along single path fails in case of node failures. Furthermore, node addressing and location awareness complicates the protocol that should be addressed.

7.3.7 Minimum Cost Forwarding Algorithm (MCFA)

MCFA protocol proposed in (Ye 2001) conserves energy by always forwarding data along the least cost path towards the base-station. It assumes that data-forwarding always occur towards the base-station so no explicit addressing mechanism is required instead each node only needs to maintain a least cost estimate from itself to the base station. Cost is a general term that may represent energy, hops etc. Each node upon receiving the data only forwards it if it is on the least cost path from the source node to the base station, i-e the sum of consumed cost and remaining cost at the current node should be equal to the least cost estimate from that node to the base station.

During the setup phase each node calculates a least cost estimate from itself to the base station. This is done through a back-off based algorithm in which the base station initialises its cost with zero whereas every other node with infinity. The sink then broadcast its cost to its neighbours. Each neighbour upon the receipt of new cost value updates its cost only if it is greater then the received cost. If LC is the link cost from which the cost is received then each node will wait for a time period proportional to LC before broadcasting its cost. In this way, unnecessary broadcasts are avoided and energy is conserved. This process continues until each node in the network gets an estimate of its cost value.

Gradient broadcast (GRAB) proposed in (Ye 2005) is based on a similar cost field concept but it also addresses reliable data delivery. In this scheme, forwarding of data takes place along multiple paths forming a mesh. The width of the mesh and hence the reliability in data-delivery is controlled with a parameter named ‘credit’. GRAB proves its reliability by achieving over 90 % successful delivery of data packets in the presence of significant node failures.

MGRAB proposed in (Muhammad 2005) enhances the above-mentioned scheme by introducing a mobile sink without effecting reliable data-delivery thus enhancing the capability of previously proposed scheme. The only drawback of using a multi-path approach is that it increases energy consumption, which is a scarce resource in wireless sensor networks.

7.3.8 Cougar

Cougar, proposed in (Yao 2002), is based on data centric paradigm and views the sensor network as a distributed database system. In order to support this idea it adds a query layer above the network layer. In this scheme, one of the responsibilities of the base-station is to generate a query plan and to send it to the relevant nodes. This query plan incorporates information about data flow, in-network computation and selection criteria for the leader node. According to the architecture proposed in the scheme, the role of leader node is to perform data-aggregation from various nodes and to send the aggregated data back to the base-station. The protocol also supports in-network data aggregation by all the sensor nodes thus reducing the amount of data and saving energy. Although the scheme works well in a query driven data delivery application but it has some drawbacks. Firstly, query layer is an extra burden on each sensor node in terms of storage and layer is an extra burden on each sensor node in terms of storage and computation. Secondly data computation from several nodes requires synchronisation, which is a difficult task. Finally a dynamic mechanism is needed to maintain the leader node in order to avoid permanent failure, which further adds to complexity.

7.3.9 Active Query Forwarding (ACQUIRE)

Active Query Forwarding in Sensor Networks, proposed in (Sadagopan 2003) is another protocol that again views the sensor network as a distributed database system. The approach is based on a data-centric paradigm and is suitable for complex queries that can be divided into many sub-queries. The protocol starts from the sink by forwarding the query. Each node tries to respond to the query using information it has in its cache and then forwards the query to other sensors. This only occurs if the information in cache is up-to-date. If this is not the case then the sensor gathers information from its neighbours within a look-ahead of *d* hops. The query is then sent back to the sink where it is resolved completely either through reverse path or shortest path to the sink. The main motivation behind the proposal of ACQUIRE is to handle one-shot complex queries that can be answered by many nodes in the network. Other data-centric approaches based on Directed Diffusion propose a querying mechanism for continuous and aggregate queries whereas in the case of one-shot querying its use is not appropriate as it leads to considerable energy consumption. The value of the parameter *d* should be carefully manipulated as a very large *d* will make ACQUIRE resemble Directed Diffusion, thus consuming a lot of energy, and a very small *d* will make the query travel a large number of hops in order to get resolved, again consuming a lot of energy. The decision, as to which sensor node the query should get forwarded during its propagation, has been addressed in many protocols. CADR (Chu 2002) proposes the use of estimation theory to determine which node can provide most useful information. Rumor Routing recommends forwarding of the query to the node that knows about the event being searched because the event agent in it spread the information about the occurring event. Whereas ACQUIRE forwards the query randomly or on the basis of the fact that which node has the maximum potential for satisfying the query.

7.3.10 Routing based on Random Walks

The main idea behind protocols that are based on random walks (Servetto 2002) is to achieve load balancing statistically by employing multi-path routing. The scheme makes certain assumptions about the sensor network. Firstly, it requires the nodes to have very limited mobility. Secondly, the nodes can be turned on and off at any time. Thirdly, each node should have a unique identifier with no requirement of location information and lastly the nodes should get deployed in such a way that each node lie at the crossing point of a regular grid but the topology of the network can be irregular. To find the route between source and destination nodes; a distributed asynchronous version of bellman-ford algorithm is used, which computes the distance between the source and destination nodes. Each intermediate node forwards the data on the basis of probabilities assigned to each neighbour. These probabilities are based on the proximity of the neighbour to the destination and can be manipulated to use multiple paths thus balancing the load on the network. The major achievements in the scheme are its simplicity, as it requires little state information to be maintained in each node and its ability to choose different routes between same source and destination nodes at different time thus evenly distributing the load. The main limitation of this scheme is its assumption about an impracticable topology.

7.4 Hierarchical Routing Protocols

The factor of scalability is considered important while designing a sensor network. Single gateway architecture is not scalable when the number of sensors increases and is spread over a large area since usually sensor nodes cannot communicate to a very large distance. Moreover in single tier architecture, as the node density increases, there is more chance that gateway is overloaded. This overloading can cause inefficient communication and large latency. To cope with these problems cluster-based or hierarchical routing originally proposed for wired networks is utilised in wireless sensor networks to perform energy efficient routing. In hierarchical routing schemes, clusters are created and special tasks are assigned to the cluster-heads that increase the efficiency and scalability of the network. It increases efficiency by minimising energy consumption within a cluster by processing and aggregation of data, so that communication to the base-station can be reduced. There are usually two major phases in hierarchical routing. In the first phase clusters are formed and the second phase is usually related to routing. Some schemes use single layer clustering whereas others multi-layer clustering in which cluster-heads from one layer again form clusters at the next higher layer and so on. Fig. 7.2 depicts a typical scenario of clustering and data forwarding in hierarchical routing protocols. An introduction to some of the state of the art hierarchical routing schemes along with their advantages and drawbacks are discussed in the following subsections.

7.4.1 LEACH

Low energy adaptive clustering hierarchy proposed in (Heinzelman 2000) is a routing protocol, which includes distributed cluster formation. This cluster formation occurs by the selection of a certain percentage of sensor nodes as cluster-heads. Among other functions, the cluster-heads compress data arriving from various sensors within the cluster and send it to the base-station, hence reducing the amount of information that flows towards the base-station. Leach uses CDMA/TDMA to reduce the amount of inter- and intra-cluster collision. The protocol operates in two phases. The first phase is the setup phase that involves organisation of clusters and formation of cluster-heads. It starts by each node generating a random number between 0 and 1. If the generated number turns out to be less than a threshold value T(n) the node becomes a cluster head. The threshold is calculated as:

$$T(n) = \frac{p}{1 - p(r \bmod (1/p))} \quad if \;\; n \in G \tag{7.4}$$

where, p is the predetermined fraction of nodes that are planned to be cluster-heads and G is the set of all nodes that are participating in election process. The new elected cluster-heads send an advertisement message to all the nodes in the network. The nodes join the cluster-head from which they can receive maximum signal strength and cluster-heads calculate a TDMA schedule for the members of the cluster. The second phase involves sensing and transmitting data to the cluster-heads, which aggregate the data and send it to the base-station. After certain predetermined time the network again goes in the setup phase. There are certain limitations and assumptions in the protocol that should get addressed.

- Each node can transmit with enough power to reach base-station.
- Nodes always have data to send.
- Nodes located close to each other possess correlated data.

Whereas the drawbacks of the scheme are: the protocol can result in inefficient operation if the cluster-heads get concentrated in one part of the network and the idea of dynamic clustering brings extra overhead.

7.4.2 PEGASIS

Power efficient gathering in sensor information systems proposed in (Lindsey 2002) is a near optimal chain based protocol. It aims at reducing energy and elongate network lifetime by making nodes communicate only with nearest neighbours and they take turns in communicating with base-stations thus balancing the load. Also by allowing only local coordination between the nodes that are closer, communication bandwidth is reduced. Simulations have proved that the lifetime of networks using PEGASIS doubles as compared to LEACH because there is no overhead in dynamic cluster formation as in LEACH and data aggregation, reducing total transmission and reception but PEGASIS still requires overhead of dynamic topology management as in each node there is a requirement of knowing the energy status of each neighbour in order to make routing decision. During the setup phase each node uses signal strength to measure distance to the neighbouring nodes and then adjust its transmission power just to reach nearest neighbour. Thus members of the chain are only those nodes that are closer together and forming a multi hop connected path to the base station. PEGASIS make certain assumptions about sensor nodes that are hard to realise. It assumes that each node is capable of communicating with base-station, whereas generally sensor nodes use multi-hop communication. Each node is aware about the location of all other nodes but there is no method outlined for it. All nodes have equal energy and die at the same time. The limitations of this scheme include: Introduction of significant delay for distant nodes in the chain and the functionality of the protocol gets affected if the nodes move as there is no provision of node movement in it.

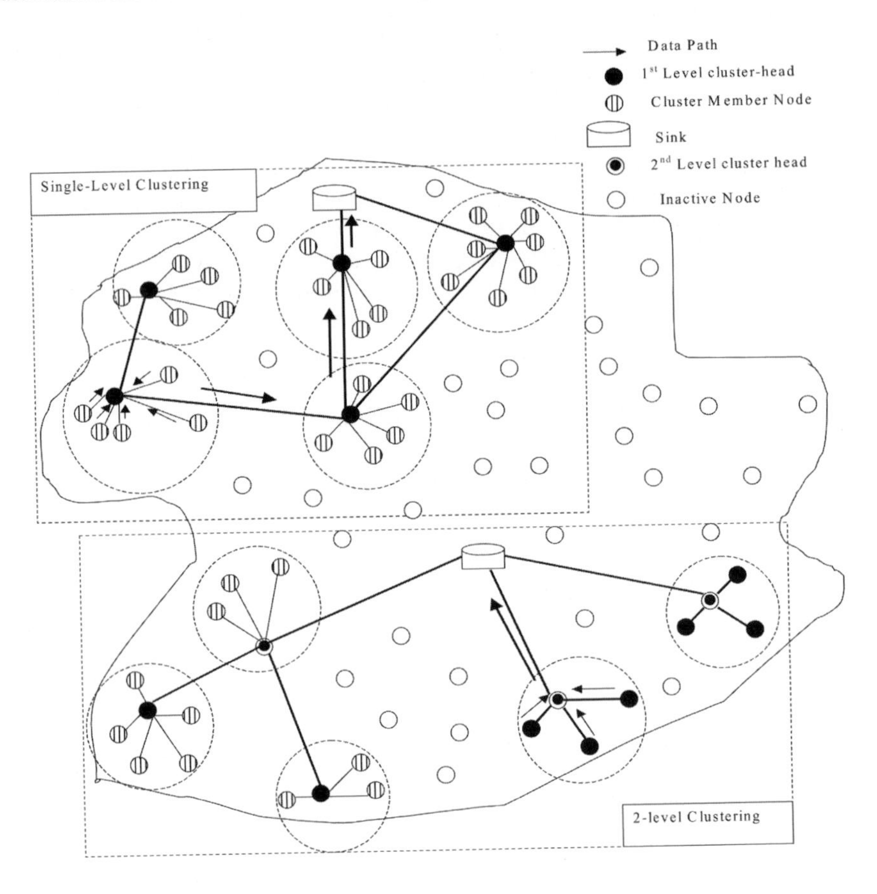

Fig. 7.2. Hierarchical routing: a typical scenario

7.5 Threshold Sensitive Energy Efficient Protocols: TEEN and APTEEN

TEEN and APTEEN proposed by (Manjeshwar 2001) and (Manjeshwar 2002) respectively, are two energy efficient protocols that also seek to elongate the lifetime of the network. These are most suitable choices for time critical sensing applications. TEEN protocol starts by the formation of cluster-heads. The cluster-heads inform the members about the attribute to be sensed and two threshold values, i.e., hard threshold (HT) and soft threshold (ST). HT indicates the threshold value that triggers the transmission of sensed attribute and ST indicates a change in sensed attribute that again triggers the transmitting process. When the value of sensed attribute, i.e., sensed value (SV) crosses HT for the first time, it is stored in an internal table and then transmitted. The next transmission only takes place when two conditions are true, i.e., current value again crosses HT and it differs from SV by an amount equal to or greater than ST. By carefully manipulating the values of HT and ST, number of transmission can be reduced but there is a trade-off between accuracy of sensed phenomenon and energy consumption. The values of HT and ST change with the change of cluster-head. APTEEN is an attempt to adopt the TEEN protocol according to different applications and user requirements. The cluster-heads are selected in a similar way as in TEEN but here they send two additional parameters to their respective members, i.e., TDMA schedule with a slot for each member and a count time (CT). Here again ST and HT govern the sensing and transmission but with little modification. Each node can only transmit within its assigned slot and a forced transmission can also take place if a node does not send data for a time equal to CT. In this way APTEEN implements a hybrid network by combining proactive and reactive policies. APTEEN provides more flexibility to the user and more control on energy consumption by choosing an appropriate count time (CT) but it adds complexity to the scheme. Both TEEN and APTEEN outperform LEACH in energy consumption and network lifetime.

7.5.1 Self-organising protocol

Self-organising protocol is proposed in (Subramanian 2000), which also gives taxonomy of sensor network applications. Moreover, based on this taxonomy, architectural components are proposed for the sensor application. The architecture defines three types of nodes:

- Sensors: sensor nodes can be stationary or mobile.
- Router nodes: stationary nodes forming the backbone of communication.
- Base-station nodes: Most powerful nodes to collect data from router nodes.

Addresses are assigned to sensing nodes on the basis of router node to which they are connected and thus forming a hierarchical routing architecture. Data can be retrieved from individual sensor nodes therefore the proposed algorithm is suitable for applications where communication to a particular node is required. The maintenance phase in the scheme employs Local Markov Loops (LML), which performs random walk on a spanning tree of a graph for fault tolerance. The advantage of the scheme include (i) Small cost associated with maintaining routing tables, (ii) Utilisation of broadcast trees to reduce the energy required for broadcasting a message, and (iii) Sophisticated fault tolerance by the use of LML. Some of its drawbacks include: Extra overhead introduced due to triggering of the organisation scheme on a not-on-demand basis and increased probability of reapplying the organisation phase if the network contains many cuts.

7.5.2 Virtual Grid Architecture Routing (VGA)

VGA routing proposed in (Jamal 2004) employs data aggregation and in-network processing to achieve energy efficiency and maximisation of network lifetime. The overall scheme can be divided into phases of clustering and routing of aggregated data. Clustering is addressed in (Xu 2001) where sensors are arranged in a fixed topology, as most of the applications require stationary sensors. In (Savvides 2001), another GPS-free approach is proposed which arranges sensors in clusters that are equal, fixed, adjacent, and non-overlapping having symmetric shapes. Square-clusters, proposed in (Jamal 2004), have a fixed rectilinear virtual topology. Inside each cluster a cluster-head, also known as local aggregator performs aggregation. A subset of these local aggregators (LA's) is selected to perform global or in-cluster aggregation and its members are known as master aggregator (MA's). However, the problem of optimal selection of local aggregators as master aggregators is N-P hard problem. The second phase is of routing with data aggregation (Jamal 2004; Al-Karaki 2004). In (Jamal 2004), besides the proposal of an exact linear programming based solution three heuristic schemes namely; GA-based heuristic, means heuristic and greedy-based heuristic are proposed which give simple, efficient, and near optimal solution. The assumption made in (Jamal 2004) is that LA nodes form groups, which may be overlapping. Hence the reading of members in a group can be correlated. However, each LA that exists in the overlapping region will be sending data to its respective MA for further aggregation. Another efficient heuristic namely; cluster based aggregation heuristic (CBAH) is proposed in (Al-Karaki 2004). These heuristics are fast, scalable, and produce near optimal result.

7.5.3 Two-tier Data Dissemination (TTDD)

Two-tier data dissemination (TTDD) presented in (Ye 2002) supports multiple mobile sinks in a field of stationary sensor nodes. The main theme behind this scheme is that each source node builds a virtual grid structure of dissemination points to supply data to mobile sinks. The protocol assumes that the sensor nodes are stationary and location aware but sinks are allowed to change their location dynamically. Once the event is sensed by nearby sensors, one of them becomes source to generate data reports. The next step is the building of virtual grid structure, which is initiated by source a node and choosing itself as a start crossing point of a gird. It sends a data announcement message to its four different adjacent crossing points using greedy geographical forwarding. The message only stops once it reaches to a node that is closest to the crossing point. This process continues until the message reaches boundary of the network. During this process, each node at the crossing point stores source's information and is chosen as a dissemination point, thus forming the grid structure. The sink to flood a query can use the grid structure thus formed. The query reaching the nearest dissemination point is instructed to retrieve data and send it to the sink. The data then follows the reverse path. Although a comparison with DD shows that TTDD out performs it by achieving a long network lifetime but the scheme has its limitations. The length of forwarding path in TTDD is longer than the length of the shortest path. The protocol doesn't study mobile sensor nodes and it is obvious that overhead of maintaining and recalculating grid is high as topology changes. Finally, the protocol requires very accurate positioning system for its functioning but doesn't specify algorithm to obtain location information.

7.5.4 Hierarchical Power Aware Routing (HPAR)

HPAR is another power aware routing protocol, which is proposed in (Li 2001) and divides the network into group of sensors called zones. Each zone is a group of geographically close sensor nodes and is treated as an entity. After the formation of clustered zones the function of routing scheme is to decide how a message is routed across other zones hierarchically so that battery lives of nodes in the system are maximised. For this, a message is routed along a path with a maximum power over all minimum remaining powers. This path is called max-min path. The idea behind making such a decision is that it may be possible that a path with high residual power has more energy consumption than the minimum energy consumption path. This scheme presents an approximation algorithm called max-min ZPmin algorithm. The algorithm first finds a path with least power consumption by applying Dijkstra algorithm. It then finds a second path that maximises the minimal residual power in the network. The protocol then tries to optimise both solution criteria. Li presents another solution for the routing problem called zone based routing (Li 2001). It divides the area into a small number of zones and a path from one zone to another is found in order to send messages across whole area. The sensors in a zone perform local routing and participate in estimating the power level of the zone. The zone power estimate is used to route message across zones. A global controller, which represents a node with highest energy level, performs global routing using the power level estimate of each zone. This algorithm is facilitated by a zone graph, in which a vertex represents each zone and there is an edge between two vertices if a path exists between two zones. The protocol introduces two algorithms for global and local path selection using zone graph.

7.6 Geographic Routing Protocols

Geographic Routing is a suitable routing scheme for wireless sensor networks. Unlike IP networks, it provides a mechanism for routing data to a geographic region instead of a destination node specified by an address. The idea of Geographic Routing is based on following reasons.

- Communication in sensor networks often needs to use physical locations as addresses. For example, “any node in the given region” or “node closest to a given location”. To meet these requirements a routing mechanism, which supports location awareness among the sensor nodes, is needed.
- Location information can be used to improve the performance of routing in wireless networks and provide new types of services.

Fig. 7.3 shows how location awareness in nodes can facilitate routing whereas the subsections following it review most of the geographic routing protocols.

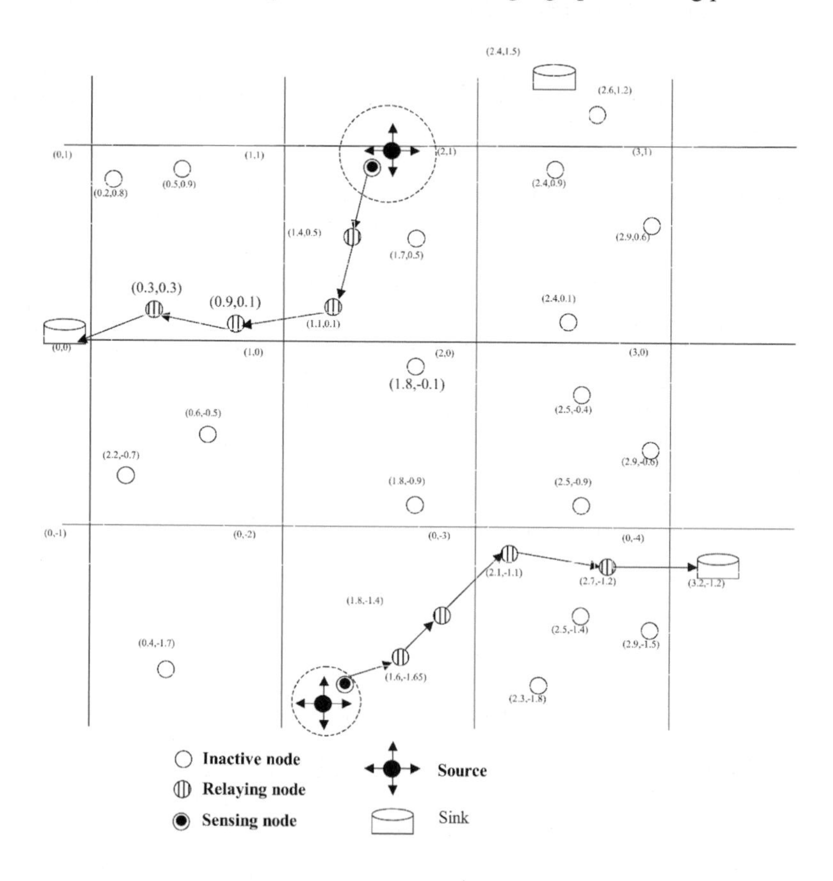

Fig. 7.3. Geographic routing: a typical scenario

7.6.1 Geographic and Energy Aware Routing (GEAR)

Yu et al. (Yan 2001) have suggested the use of geographic information while forwarding a packet to all the nodes inside a target region, which is a common primitive in data-centric sensor network applications (Chalermek 2000). This protocol is based on the knowledge of node's own as well as its neighbours' locations and energy information. It conserves the energy by minimising the number of interests in DD by only considering a certain region rather than sending the interests to the whole network. It also uses an energy-aware neighbour selection heuristic to forward packets. Nodes with more remaining energy are preferred over nodes with less energy to avoid nodes from becoming hotspot of the communication. Each node in GEAR keeps an estimated cost and a learning cost of reaching the destination through its neighbours. The estimated cost is a combination of residual energy and distance to the destination. The learned cost is a refinement of the estimated cost that accounts for routing around holes in the network. A hole occurs when a node does not have any closer neighbour to the target region than itself. If there are no holes, the estimated cost is equal to the learned cost. Whenever a packet reaches the destination the learned cost is communicated one hop back to update the route setup for next packet. There are two phases in the algorithm:

- **Forwarding packets towards the target region:** Upon receiving a packet, it is greedily forwarded to that neighbour node which is closest to the destination. If all the neighbouring nodes are farther than the node itself, this means there is a hole. In such a situation, a neighbour node is chosen on the basis of learning cost and packet is forwarded to that node.
- **Forwarding the packets within the region:** When a packet reaches the destination region, it can be routed to all nodes within that region by restricted flooding or recursive geographic forwarding. Restricted flooding is useful in the case of low-density networks but in case of high-density networks; recursive geographic forwarding is more useful. In recursive geographic forwarding, region is divided into sub regions and copies of packet are given to each sub region. This process continues until one node is left in the sub region.

GEAR helps in increasing the lifetime of the network and better connectivity is achieved after partition.

7.6.2 Greedy Perimeter Stateless Routing (GPSR)

GPSR (Karp 2000) is a geographic routing protocol that was originally designed for mobile ad-hoc networks but can be modified for use in Wireless Sensor Networks. It exploits the correspondence between geographic position and connectivity in a wireless network, by using the positions of nodes to make packet-forwarding decisions. Given the coordinates of the destination node, GPSR routes a packet to it using only the location information. This protocol assumes that each node knows its geographic coordinates and also possesses information about the geographical coordinates of its neighbouring nodes. GPSR uses two different algorithms for routing: greedy forwarding and perimeter forwarding. When a node receives a packet destined to a certain location, it selects the node among its neighbours that is closest to the destination than itself and forwards packet to that node by greedy forwarding. If there is no node among the neighbours that is closer to destination than the node itself invokes perimeter routing. Fig. 7.4 depicts a situation where a packet destined for node D arrives at a node A. No neighbour of A is closer to D than A itself. When such situation arises, we say that the packet has encountered a void or hole. Voids can result due to irregular deployment of nodes, or due to the presence of radio-opaque obstacles. Perimeter forwarding is used to route packet around a void by right hand rule. The packet travels along the perimeter around the void and when it reaches a node, closer to the destination than the node that invokes the perimeter routing, greedy forwarding resumes.

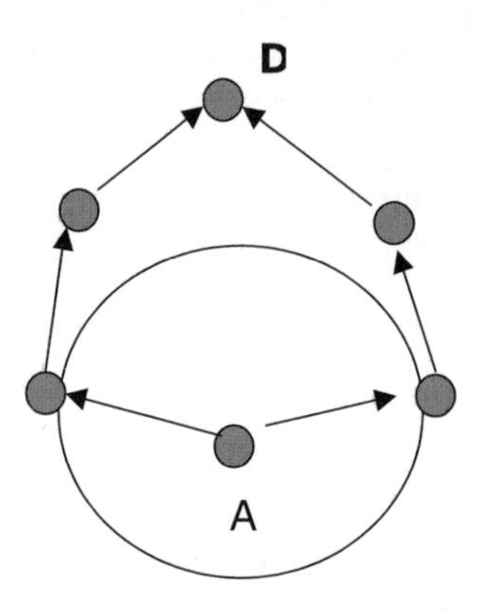

Fig. 7.4. Perimeter routing

By keeping state only about local topology and using it in the packet forwarding, GPSR scales better than other ad-hoc routing protocols. There is more traffic on the nodes on the perimeter of the void causing them to be depleted quickly.

7.6.3 Graph Embedding for Routing (GEM)

This protocol (Newsome 2003) is built on principles of geographic routing without location information. The idea is to assign labels to the sensor nodes uniquely in a distributed manner so that nodes can route messages knowing only the labels of their immediate neighbours. In GEM, virtual coordinates are used instead of actual physical coordinates. This algorithm consists of two parts, construction of virtual polar coordinates and routing using polar coordinate system. For the construction of Virtual Polar Coordinate Space (VPCS), a node in the centre of the network, usually sink, is chosen as root and spanning tree is created. Each node is assigned an *angle range*, which can be used to assign angles to its sub-trees. Each node splits its *angle range* into its children based on the size of the sub-tree of each child. If the children know their position they can be easily sorted and proper ranges can be assigned to them, but the authors (Newsome 2003) proposed a method, which can work without any physical information. For each sub-tree its centre-of-mass and average position of all the nodes are computed and propagated to the parent of that tree. This helps to determine the ordering of the child nodes that leads to a fairly accurate VPCS. When a message is to be sent to a particular node by another node, the message is passed up in the tree until it reaches the common parent of source and destination nodes (can be easily computed by checking if the *angle range* of that node is the super set of the *angle range* of the destination). From the parent node then it is routed down in the tree to the destination. This scheme is a little inefficient because it does not take advantage of cross-links that are shortcuts between two neighbouring nodes belonging to different sub-trees. The proposed routing scheme, Virtual Polar Coordinate Routing (VPCR) first checks if there is any neighbouring node whose *angle range* is closer to the destination than the current node. If such a node exists, the packet is routed to that node instead of routing up in the tree. This scheme works well in the absence of information about geographical location and also when there is a difference between geographical location and network topology. Moreover, it also supports data-centric routing and storage.

7.6.4 Location Aided Routing (LAR)

Location Aided Routing (LAR) (Ko 1998) provides location based routing using restrained/directed flooding. The first phase of LAR protocol is the route discovery using flooding. Whenever a node needs to find a route to another node, it sends a route request to all its neighbours. On receipt of this route request, the neighbours check whether the request is meant for them. If not, they broadcast the route request again to all their neighbours only once, discarding any more route requests for the same combination of sender and receiver. At every hop, a node is added to the route that is contained in the route request. Once a route request reaches the destination node it responds by sending a route reply. This route reply follows a path obtained by reversing the path contained in the route request. LAR assumes that the mobile nodes are constantly moving and that a node's location at two different times will most probably be different. The expected-zone of a receiving-node from the viewpoint of a sending-node is the zone in which the receiver is expected to be present based on the prior information of the receiver's location and its velocity of movement. If no such information is available, then the expected-zone may potentially be the entire network, leading to the algorithm being reduced to flooding. Another terminology used is the request-zone that is the zone in which a forwarding node must lie. An intermediate node may only forward a route request if it is within the request-zone.

The main thrust of LAR is the methodology used to determine whether a node is in the request-zone or not and this can be accomplished with two schemes. Scheme-1 assigns the request zone to be a rectangle with its sides being parallel/perpendicular to the X-Y axes. This rectangle is cornered at one side by the sending node. The other corner of the rectangle is formed by the intersection of the tangents to the expected-zone (usually a circle) of the destination. In case of the sender being located within the expected-zone, the request-zone is specified to be a rectangle enclosing the expected-zone. The request-zone is not specified explicitly in Scheme-2 as was done in Scheme-1. Instead, the route request contains two parameters. One is the distance of sender from the last known position of the destination. The last known co-ordinates of the destination are also specified in the route message. On receipt of the message, an intermediate node calculates its own distance from the destination. If this distance is less than the distance contained in the message, and it is at least some specific distance away from the previous hop's node, the node will accept the route request, else it will drop it.

LAR is another protocol that is designed for the wireless networks in general, but does not account for the unique and stringent characteristics of sensor networks. Thus expensive routing tables need to be maintained. This protocol is similar to DSR in operation but differs in the aspect of route building. However, an added disadvantage of this protocol is that the route is found out using flooding. This gives an O(n) complexity to each route discovery. However, each node receives same route request from each of its neighbours making it to process O(n) requests for propagation. These control messages are numerous. The LAR protocol chooses a route that is of the smallest length. However, on receipt of multiple routes of same length resulting from the route request, LAR is unclear about which route to accept and store.

7.6.5 Geographic Adaptive Fidelity (GAF)

GAF (Xu 2001) is an energy-aware location-based routing algorithm designed primarily for mobile ad-hoc networks but is applicable to sensor networks as well. The network area is first divided into fixed zones forming a virtual grid. Inside each zone, nodes collaborate with each other to play different roles. For example, nodes elect one sensor node to stay awake for a certain period of time while they go to sleep. This node is responsible for monitoring and reporting data to the BS on behalf of the nodes in the zone. Hence, GAF conserves energy by turning off unnecessary nodes in the network without affecting the level of routing fidelity.

Each node uses its GPS-indicated location to associate itself with a point in the virtual grid. Nodes associated with the same point on the grid are considered equivalent in terms of the cost of packet routing. Such equivalence is exploited in keeping some nodes located in a particular grid area in sleeping state in order to save energy.

Thus, GAF can substantially increase the network lifetime as the number of nodes increases. There are three states defined in GAF. These states are discovery (for determining the neighbours in the grid, active reflecting participation in routing) and sleep (when the radio is turned off). In order to handle the mobility, each node in the grid estimates its leaving-time of the grid and sends this to its neighbours. The sleeping neighbours adjust their sleeping-time accordingly in order to keep the routing fidelity. Before the leaving-time of the active node expires, sleeping nodes wake up and one of them becomes active. GAF is implemented both for non-mobility (GAF-basic) and mobility (GAF-mobility adaptation) of nodes. The fixed clusters are selected to be equal and square in shape. The selection of the square size is dependent on the required transmitting power and the communication direction.

The issue is how to schedule roles for the nodes to act as cluster-heads. A cluster-head can ask the sensor nodes in its cluster to switch on and start gathering data if it senses an object. The cluster-head is then responsible for receiving raw data from other nodes in its cluster and forward it to the BS. It is assumed (Xu 2001) that sensor nodes can know their locations using GPS cards, which is inconceivable with the current technology. GAF strives to keep the network connected by keeping a representative node always in active mode for each region on its virtual grid. Simulation results show that GAF performs at least as well as a normal ad-hoc routing protocol in terms of latency and packet loss and increases the lifetime of the network by saving energy. Although GAF is a location-based protocol, it may also be considered as a hierarchical protocol, where the clusters are based on geographic location. For each particular grid area, a representative node acts as the leader to transmit the data to other nodes. The leader node however, does not do any aggregation or fusion.

7.6.6 GeoTORA

GeoTORA (Ko 2000, 2003) is another geocast protocol for ad-hoc networks. It is based on TORA (Temporally Ordered Routing Algorithm) (Park 1997, 1998), which is a unicast routing algorithm for ad-hoc networks. In TORA a directed acyclic graph (DAG) is maintained for each destination. The DAG shows for each node the direction to the destination node; hence it can be used for forwarding a packet to a destination starting at any node. The GeoTORA algorithm is based on an anycast modification of TORA. First a DAG is maintained for each anycast group. Between members of the anycast group there is no direction in the DAG, that is, they are all possible destinations. The directions within the DAG are defined by assigning a height to each node. A packet is always forwarded to a neighbour with lower height. Basically, the height is the distance to the destination region. Two Members of the geocast group are assigned height 0. The initial DAG is created as follows. When a node first requires a route to a geocast group it broadcasts a query to all neighbours. The query is rebroadcast until a member of the DAG is found (the neighbour nodes of the destination region are already members). A flag on each node helps to identify duplicates, which are discarded. On receiving a query, a member of the DAG responds by broadcasting its height to its neighbours. A node that waits for a connection to the DAG sets its own height to the minimal height of all neighbours increased by one and broadcasts its height. Since nodes are moving, the DAG is not stable. However, maintaining the DAG is achieved without flooding, therefore GeoTORA is classified as a routing protocol without flooding. GeoTORA reacts to changes in the DAG if a node no longer has outgoing links. Then the direction of one or more links is changed, which is called link reversal. The neighbouring nodes are only affected by this measure if their last outgoing link has changed to an ingoing link, which means that they have to repeat the link reversal process. If the directed links of the DAG are followed to forward an anycast packet, it is finally delivered to a random node of the anycast group. Geocasting with this algorithm works as follows. It starts with an anycast to a random member of the geocast group using the approach described above. Upon receiving the first geocast packet the geocast group member floods the packet within the geocast region.

7.6.7 Location Based Multicast (LBM)

A recent research area is geocast in ad-hoc networks, that is, spontaneous constituting networks without a fixed infrastructure. In wireless ad hoc environments, Ko and Vaidya (Ko 1999, 2002) identified two approaches: modified multicast flooding and the modified multicast tree-based approach. For efficiency reasons, multicasting in traditional networks is mainly based on tree-based approaches. However, as tree-based approaches require frequent reconfigurations in ad-hoc network environments, they are considered unsuitable in (Ko 1999; Ko 2002) to solve the geocast problem in ad-hoc networks. Therefore, two schemes that improve multicast flooding with position information are presented, which are both derived from Location Aided Routing (LAR) (Ko 1998), a protocol for unicast routing in ad-hoc networks. Defining a forwarding-zone that includes at least the destination region and a path between the sender and the destination region as described above modifies simple flooding. An intermediate node forwards a packet only if it belongs to the forwarding zone. By increasing the forwarding zone, the probability for reception of a geocast packet at all destination nodes can be increased; however, overhead is also increased. Similar to their unicast routing protocol, the forwarding zone can be the smallest rectangular shape that includes the sender and the destination region possibly increased by a parameter δ to increase the probability for message reception. The forwarding zone is included in each geocast packet in order to allow each node to determine whether it belongs to the forwarding zone. A second scheme defines the forwarding zone by the coordinates of the sender, the destination region, and the distance of a node to the centre of the destination region. A node receiving a geocast packet determines whether it belongs to the forwarding zone by calculating its own geographic distance to the centre of the destination region. If its distance, decreased by δ, is not larger than the distance stored in the geocast packet, which is initially the sender's distance, the geocast packet is forwarded to all neighbours and the packet sender's distance is replaced by the calculated own distance. In other words, a node forwards a packet if it is not farther away from the destination region than the one-hop predecessor of the packet increased by δ. Finally, a geocast packet is forwarded to all neighbours if the one-hop predecessor is located inside the destination region.

7.7 References

Al-Karaki, Kamal AE (2004) On the correlated data gathering problem in WSN. In: 9th IEEE Symp. on Computers and Communications, Alexandria, Egypt

Braginsky and Estrin D (2002) Rumor routing algorithm for sensor networks. In: 1st Workshop on Sensor Networks and Applications (WSNA), Atlanta

Chalermek I, Govindan R, and Deborah Estrin (2000) Directed diffusion: A scalable and robust communication paradigm for sensor networks. In: ACM/IEEE Int. Conf. on Mobile Computing and Networking (Mobicom), pages 56–67, Boston

Chu, Haussecker H, and Zhao F (2002) Scalable information-driven sensor querying and routing for ad hoc heterogeneous sensor networks, J High Performance Computing Applications 16:3

Eugene Shih, et al. (2001) Physical layer driven protocol and algorithm design for energy-efficient WSNs. In: 7th Annual ACM/IEEE Int. Conf. on Mobile Computing and Networking (Mobicom'01), Rome

Ganesan, et al. (2002) Highly resilient, energy efficient multipath routing in WSA, J Mobile Computing and Communications Review (MC2R) 1:2

Heinzelman, Kulik J, and Balakrishnan H (1999) Adaptive protocols for information dissemination in WSN, In: 5th ACM/IEEE Mobicom Conf. (MobiCom '99), Seattle, pp. 174-85.

Heinzelman, Chandrakasan A and Balakrishnan H (2000) Energy-efficient communication protocol for wireless micro-sensor networks. In: 33rd Hawaii Int. Conf. on System Sciences (HICSS '00)

Heinzelman et al. (2000) Energy-scalable algorithms and protocols for WSN, In: Int. Conf. on Acoustics, Speech, and Signal Processing (ICASSP '00), Turkey

Intanagonwiwat, Govindan R, and Estrin D (2000) Directed diffusion: a scalable and robust communication paradigm for sensor networks. In: ACM MobiCom '00, Boston. pp. 56-67

Jamal NAK, Raza Ul-Mustafa, and Kamal AE(2004) Data aggregation in WSN exact and approximate algorithms. In: IEEE Workshop on High Performance Switching and Routing (HPSR), Phoenix, Arizona

Karp, Brad and Kung HT (2000). GPSR: Greedy perimeter stateless routing for wireless networks. In: Proc. of ACM/IEEE Int. Conf. on Mobile Computing and Networking, pages 243-254, Boston

Ko and Vaidya NH (1998) Location-aided routing (LAR) in mobile ad hoc networks. In: 4th ACM/IEEE Int.. Conf. Mobile Comp. and Net. (MobiCom '98), Dallas

Ko and N. H. Vaidya (2000) GeoTORA: A protocol for geocasting in mobile ad hoc networks. In: 8th Int. Conf. Network Protocols (ICNP), Osaka, Japan, pp. 240–50

Ko and Vaidya NH (2002) Flooding-based geocasting protocols for mobile ad hoc networks. J Mobile Networks and Applications 7: pp. 471–80

Ko and Vaidya NH (2003) Anycasting-based protocol for geocast service in mobile ad hoc networks. J Comp. Net. 41: pp. 743–60

Kulik, Heinzelman WR, and Balakrishnan H (2002) Negotiation-based protocols for disseminating information in WSN, J Wireless Networks, 8: 169-185

Li and Aslam J, and Rus D (2001) Hierarchical power-aware routing in sensor networks. In: DIMACS Workshop on Pervasive Networking

Lindsey and Raghavendra C (2002) PEGASIS: Power-efficient Gathering in sensor information systems. In: IEEE Aerospace Conf. Proc. vol. 3, 9-16 pp. 1125-1130

Manjeshwar and Agarwal DP (2001) TEEN: a routing protocol for enhanced efficiency in WSNs. In: 1st Int. Workshop on Parallel and Distributed Computing Issues in Wireless Networks and Mobile Computing

Manjeshwar and Agarwal DP (2002) APTEEN: A hybrid protocol for efficient routing and comprehensive information retrieval in WSNs. In: Parallel and Distributed Processing Symp., Proc. Int., IPDPS.pp. 195-202

Min et al. (2000) An architecture for a power aware distributed micro sensor node. In: IEEE Workshop on signal processing systems (SIPS'00)

Muhammad U and Qamar S (2005) Robust data delivery to mobile Sink in WSN. In: 4the WSEAS Int. Conf. on Information Security, Communications and Computers (ISCOCO 2005), Tenerife, Canary Islands, Spain

Newsome and Song D (2003) GEM: Graph embedding for routing and data-centric storage in sensor networks without geographic information. In: Int. Conf. on Embedded Networked Sensor Systems (SenSys'03)

Park and Corson MS (1997) A highly adaptive distributed routing algorithm for mobile wireless networks. In: IEEE INFOCOM, Kobe, Japan, pp. 1405–13

Park and M. Corson (1998) A performance comparison of the temporally-ordered routing algorithm and ideal link-state routing. In: IEEE Symp. Comp. and Commun., Athens

Rahul, Rabaey J (2002) Energy aware routing for low energy ad hoc sensor etworks. In: IEEE Wireless Communications and Networking Conf. (WCNC), vol.1, Orlando, pp. 350-355

Sadagopan et al. (2003) The ACQUIRE mechanism for efficient querying in sensor networks. In: 1st Int. Workshop on Sensor Network Protocol and Applications, Anchorage, Alaska

Savvides, Han CC, and Srivastava M (2001) Dynamic fine-grained localisation in ad-hoc networks of sensors. In: 7th ACM Annual Int. Conf. on Mobile Computing and Networking (MobiCom). pp. 166-179

Schurgers and Srivastava MB (2001) Energy efficient routing in WSNs. In: MILCOM Proc. on Communications for Network-Centric Operations: Creating the Information Force, McLean

Servetto and Barrenechea G (2002) Constrained random walks on random graphs: routing algorithms for large scale WSNs. In: 1st ACM Int. Workshop on WSNs and Applications, Atlanta, Georgia

Subramanian and Katz RH (2000) An architecture for building self-configurable systems. In IEEE/ACM Workshop on Mobile Ad Hoc Networking and Computing, Boston

Woo and D. Culler (2001) A transmission control scheme for media access in sensor networks. In: 7th Annual ACM/IEEE Int. Conf. on Mobile Computing and Networking (Mobicom'01), Rome

Xu, Heidemann J, and Estrin D (2001) Geography-informed energy conservation for ad-hoc routing. In: 7th Annual ACM/IEEE Int. Conf. on Mobile Computing and Networking, pp. 70-84.

Yan Y, Govindan R, and Deborah Estrin (2001) Geographical and energy aware routing: A recursive data dissemination protocol for WSNs. Technical report. University of California, Los Angeles

Yao and J. Gehrke (2002) The cougar approach to in-network query processing in sensor networks, SIGMOD Record

Ye, Chen A, Liu S, and Zhang L (2001) A scalable solution to minimum cost forwarding in large sensor networks. In: 10th Int. Conf. on Computer Communications and Networks (ICCCN), pp. 304-309

Ye, Zhong G, Lu S, and Zhang L (2005) Gradient broadcast: A robust data delivery protocol for large-scale sensor networks. J. ACM Wireless Networks. (WINET) 11:pp. 99-115

Ye, Luo H, Cheng J, Lu S, and Zhang L (2002) A two-tier data dissemination model for large-scale WSNs. In: ACM/IEEE MOBICOM

Ye, Heidemann J, and Estrin D (2002) An energy-efficient MAC protocol for WSNs. In: IEEE Infocom, New York

8 Energy Efficient Routing

Ibrahim Korpeoglu

Department of Computer Engineering, Bilkent University, Ankara, Turkey

8.1 Introduction

The design of routing protocols for wireless sensor networks (WSNs) requires new approaches to be followed due to the different characteristics and applications. The primary objective to optimise WSNs is energy efficiency, and this is also the case for routing protocols for WSNs. This chapter introduces some of the routing protocols proposed in literature for WSNs. The communication needs in WSNs exhibit many differences from *ad hoc* networks or Internet. Besides unicast delivery of data packets from one node to another, there is also significant need for broadcasting data to a subset nodes, gathering of data from all nodes to a single or multiple locations, or dissemination of data to a set of nodes where the set is determined dynamically during dissemination. We call the mechanisms that a WSN implements to perform these types of delivery as routing protocols, data gathering protocols, or data dissemination protocols. We will use these three terms interchangeably in text to refer to those mechanisms.

In this chapter, first the differences between WSNs and ad hoc networks as far as routing problem is concerned are introduced. Ad hoc networks are the type of networks that have the most similar features with WSNs; however there are still many important differences that require new routing protocols to be designed and adapted for WSNs. Later in the chapter, new design approaches for WSN protocols are discussed and cross-layer design is emphasised. Then some wireless link layer technologies that can support realisation of WSNs and that can affect the design of routing protocols are discussed. The descriptions of some of the routing protocols proposed for WSNs available in the literature are included afterwards. The chapter ends with a summary.

8.2 Need for New Routing Protocols

There are already routing protocols designed for the dynamic and mobile environment of wireless ad hoc networks. But since there are differences between ad hoc networks and WSNs, we need to design new routing protocols for WSNs. The differences of WSNs from ad hoc networks, as far as routing is concerned, are many. Since WSNs are most of the time application specific, the classical layered architecture, where each network layer is independent from each other, is not a very efficient paradigm for WSNs. Therefore, cross-layer design, where a couple of layers designed together or layers interact with each other to use resources in a more efficient manner, can be a more favorable paradigm for WSNs. As a result of this, the routing layer used in WSNs is usually *application aware* (Hac 2003). In other type of networks, the routing layer cannot look inside and process the content of received packets. But in WSNs, the routing layer can look inside the received packets and can process the payload (data). This facilitates *in-network processing* of data in WSNs. For example, *aggregation* inside network is possible which reduces the amount of data transported and the number of packets transmitted. This saves energy. Moreover, since sensor nodes carry data belonging to a single application as opposed to ad hoc network nodes, which carry traffic of different types of applications at the same time, application specific policies can be applied. In WSNs, nodes can be assumed to be *not mobile*. But in wireless ad hoc networks, mobility is a common case for most of the scenarios. Therefore routing protocols for WSNs can be designed assuming there will be no mobility. This assumption facilitates more efficient design.

Nodes in both ad hoc networks and WSNs are expected to be powered by batteries and therefore they are expected to be energy constrained. But batteries used in ad hoc networks are usually rechargeable batteries and also can be replaced if they are no longer functional. In WSNs, however, we should be ready for the fact that *batteries cannot be replaceable* or rechargeable. This may not possible, for example, because of vast number of nodes or because of harsh deployment environments of sensor nodes. Therefore, the *energy becomes more critical resource* in WSNs than wireless ad hoc networks. We should design routing and related protocols so that the energy is used very efficiently and batteries can be functional for months or even years.

In ad hoc networks and in Internet, data packets are routed based on destination addresses. The destinations in these networks are specified by numerical addresses. IP addresses (unicast or multicast), MAC addresses, or some other addressing scheme can be used for this purpose, and routing is done accordingly. This is called *node-centric* routing. In WSNs the destination of query or data packets can be more suitably specified by the attributes of the data carried in the packets. Hence the routing decisions can be based on the data carried inside the packets. This is called *data-centric* routing. Moreover, WSN applications are usually interested from which location data arrives instead of caring from which sensor node data arrives. Therefore, sometimes it makes more sense to route packets based on the position of the destination. Hence, *geo-centric* (position based) or data-centric routing approaches may be more natural to take for routing than node-centric approaches.

The *traffic pattern* in WSNs is also different than in ad hoc networks. In ad hoc networks, we can assume that the traffic is distributed uniformly among all pairs of nodes of an ad hoc network. In WSNs, this may not be a very realistic assumption. In WSNs, the traffic flow is more likely to be between one or more sinks (base stations) and a set of or all of sensor nodes. For example, we can conceive a lot of WSN applications where the majority of traffic is directed from all sensor nodes to a single sink node. This traffic flow characteristics can provide opportunity to design more efficient routing protocols for WSNs.

A lot of WSN applications require sensor nodes to be active only for a fraction of time; the sensor nodes can be asleep most of the time. Hence the *duty cycle* can be very low, less than 1 percent. In ad hoc networks this is not the case. A node participating in an ad hoc network is usually active most of the time until it is explicitly put into stand-by mode. The fact that most sensor nodes can be put into *sleep mode* provides an additional power efficient design opportunity for WSN routing protocols.

Scalability becomes a more important objective for WSNs than ad hoc networks. We expect an ad hoc network to have number of nodes in the order of 100s, or 1000s at most. But the number of nodes in a WSN is expected to be much more than this. For example, a WSN consisting of 1 million nodes is possible. Therefore a routing protocol designed for such a network should be very scalable.

Since nodes in a WSN are expected to be operated with irreplaceable batteries and in harsh environments, it is highly possible that some nodes will become non-functional at random times after they got deployed to the field. The routing protocols should be *adaptive* to those cases where some nodes fail and can not take part in routing anymore. Therefore we need routing solutions that are *robust* against node failures and resource lack.

The sensor nodes are also expected to be much more *resource-limited* than nodes in an ad hoc network in terms of *energy*, *processing power*, *memory* and *link bandwidth*. The routing protocols designed for WSNs should consider this fact. For example, if we have a WSN consisting of 1 million nodes, we can not afford to implement a table-driven routing protocol requiring each node to store a routing table of size 1 million entries.

New metrics are important for the performance evaluation of WSNs, which may not make sense for wireless ad hoc networks. For example, *network lifetime* (*survivability*) is a metric that is very commonly used for WSNs (although it may be used sometimes for ad hoc networks as well). It means how long it will take before a WSN that is installed into a region becomes no more functional.

8.3 Sources of Energy Consumption

One of the most important objectives of routing protocols for WSNs is energy efficiency. A routing protocol designed for WSNs should decrease the energy usage and try to maximise the network lifetime. Selecting always the lowest energy paths may not increase the network lifetime since the nodes on those paths will quickly deplete their energy, and WSN lifetime will be negatively affected. Therefore protocols that distribute the load and energy consumption evenly to nodes are more favorable (Chang 2000). There are sources of energy consumption in WSNs (Ye 2002), and those sources need to be attacked in order to save energy at the routing layer. We can list some of the sources of energy consumption related to communication as below:

- *Packet transmissions*. Each packet transmission causes energy consumption and energy consumed is proportional to the number of packets transmitted. Therefore a routing scheme that reduces the number of packet transmissions can save energy.
- *Packet receptions*. Each packet that is received by a node is processed and therefore causes energy consumption. The processing includes demodulation, copying of bytes, etc. Therefore routing protocols should try to reduce the packet receptions. A node sometimes receives not only the packets destined to itself, but also the packets destined to other nodes in the vicinity. This is called overhearing of packets. Each such packet is also received and processed at the transceiver electronics before actually dropped at the MAC layer. Therefore each such packet also causes energy waste. The number of overheard packets should be reduced.
- *Idle listening*: When a sensor node is in idle listening mode, not sending, or receiving data, it can still consume a substantial amount of energy. Therefore, a sensor node that is not sending or a receiving data should not stay in idle listening mode, but should go into sleep (be powered off). Depending on the communication technology, there might be several low-power modes available. Those modes should be used whenever possible.
- *Packet size*: Size of a packet determines how long a transmission will last. Therefore it is effective in energy consumption. We have to reduce the packet sizes if possible. This can be achieved, for example, by combining several packets into one large packet, or by compression.
- *Distance*: The distance between the transmitter and receiver affects how much output power is required at the transmitter to send the packets to the receiver; this in turn affects the energy consumption. Routing algorithms can select paths that use shorter distances between nodes and in this way can reduce energy consumption.

These sources of energy consumption can be attacked by *node-level* or *network-level* schemes. Each node can apply a node-level scheme independently; a network-level scheme, on the other hand, should be applied by cooperation of several nodes or all nodes in the network.

8.4 New Approaches

Many new approaches related to routing can be used to improve the performance of WSNs in many dimensions, including energy and bandwidth efficiency, delay, network lifetime, and adaptation to changes. These approaches include, but are not limited to, multi-path routing (Ganesan 2002), load balancing (Schurgers 2001), and cross-layer design (Shakkottai 2003) (Madan 2005) (Goldsmith 2002). In this section we will briefly look to some cross-layer strategies that also include the routing protocols and that improves energy efficiency.

Cross-layer design means designing the network layers and functions together or to be interdependent to each other to serve an objective in a better manner. For example, for energy efficiency purposes, MAC layer and routing layer, or physical layer and routing layer can be designed together and operated in an interdependent manner. There are various studies proposing cross-layer solutions that involve the routing layer (Madan 2005)(Van Hoesel 2004) (Chang 2000)(Conti 2004) (Madan 2005).

The physical and MAC layers can provide status information to the routing layer so that the routing protocols can adapt better to the changing conditions. Alternatively, the physical and MAC layers can be designed or optimised with the information available about the routing layer properties or information provided by the routing layer. For example, the LEACH protocol's routing layer requires a clustered topology to be formed, and each member in the cluster sends its data to the clusterhead (Heinzelman 2000). This property of the topology implied by the routing strategy can be used to design a CDMA and TDMA based MAC layer, where CDMA is used among clusters, and TDMA is used among members of a cluster. A clusterhead can easily coordinate a TDMA channel. This is indeed the approach followed by LEACH. LEACH can also utilise the interaction between routing layer and physical layers. After clusters are formed, each member node transmits its data to a clusterhead. Here, there are two options: 1) transmitting with a fixed power (no interaction between routing and physical layers) and interfering other nodes' transmissions and also consuming extra energy; 2) transmission with a power level that is just enough to reach to the clusterhead (requires interaction between routing and MAC/PHY layers). This second option is better, since it causes less energy consumption and less interference (which indirectly affects again the energy consumption).

If cross-layer design approach is not used, there is possibility that some functions can be executed at more than one layer, causing energy waste. For example, neighbor discovery can be done at the network layer by using network layer packets. But some link layers also provide a similar functionality. For example, in Bluetooth, a piconet has a master and slaves that are neighbors to the master. When the Bluetooth link layer is formed, the master knows its slaves and the slaves know their master. There is no need for neighbor discovery at the network layer of the master in this case. So, the routing layer can use the information about neighbors maintained at the link layer, if cross-layer interaction is allowed.

For cross-layer interaction, the routing layer should be able to receive information from the link layer and physical layer. Similarly, it may need to pass information to the link and physical layers. There are a number of ways of doing this. One is piggybacking the cross-layer information to the control or data packets that are already passed between those layers (Van Hoesel 2004). Another method is to design new messages just to pass cross-layer information. A third method can be to use a lightweight database which stores all the metrics and counters maintained at all layers of the protocol stack and to which each layer can access (Conti 2004). The information stored in such a database may include the RSSI (received signal strength indicator) value obtained from the physical layer, the transmit power level received again from the physical layer, the number of re-transmissions obtained from the MAC layer, the maximum link layer packet size defined at the MAC layer, and the identities of the neighbors obtained from the MAC or routing layer.

8.5 Effects of Wireless Technology on Routing

While WSN protocols were being developed, there have been also efforts to develop wireless communication standards that can support construction of WSNs. Two such efforts are IEEE 802.15.4 and Bluetooth standards. As a fairly new standard, IEEE 802.15.4 (IEEE 802.15.4 2003) (Gutierrez 2001) defines the physical and MAC layers of a wireless technology that can be used in WSNs to connect sensor nodes to each other and to the base stations. While 802.15.4 standard defines the physical and MAC layers, the ZigBee standard (ZigBee 2004), to be placed on top of 802.15.4, defines the higher layer protocols including the network layer protocols and application layer interfaces. Besides this, Bluetooth, a more mature short-range wireless technology (Bluetooth 2004), is also a candidate for being used as the underlying communication technology for WSNs. The lower layers of Bluetooth technology are defined under IEEE 802.15.1 standardisation activity and include the specification of physical and MAC layer protocols (IEEE 802.15.1 2005). The link layer technology (technology defining the physical and MAC layers) has an impact on the design of efficient routing protocols (Fig. 8.1). For example, use of Bluetooth technology for WSNs restricts the design space for possible routing algorithms. Bluetooth range is limited and therefore some routing schemes that assume arbitrary power control capability will not work. Bluetooth uses star topology and there is a limit on the number of slaves that can be connected to a master so the nodes will have degree constraint. IEEE 802.15.4, on the other hand, can support mesh, peer-to-peer and star topologies.

Bluetooth can be used as the link layer for moderate data-rate WSN applications, and can enable development of QoS based routing protocols for the network. ZigBee also provides support for real-time applications and QoS based traffic, but the data rate provided is much less compared to Bluetooth. In IEEE 802.15.4, the nodes constituting the network can be classified into two categories (reduced functional device and full functional device). This provides opportunity to design *asymmetric* protocols where less functionality will be given to simple devices, and sophisticated functionality will be given to more powerful devices. This also provides a natural foundation for clustering-based routing protocols.

Depending on the MAC layer properties, the routing scheme can be tailored to provide efficient transfer of information both in terms of energy and delay. Most MAC layers for WSNs are expected to be CSMA/CA or TDMA based. For example, the 802.15.4 technology uses a CSMA/CA based MAC protocol. This has a number of advantages including easy self-configuration and maintenance at the link layer, and possibility of efficient peer-to-peer communication between nodes in the network. The MAC layer affects the decisions that will be given at the routing layer. For example, if the MAC layer is TDMA based, coordinated by a coordinator, peer-to-peer communication between any pair nodes in the range of each other may not be very efficient. In Bluetooth, for example, all communication between two slaves has to pass through their master node. Such a TDMA based scheme, however, combined with a tree-based routing protocol can provide an efficient joint MAC/routing solution (Zaruba 2001).

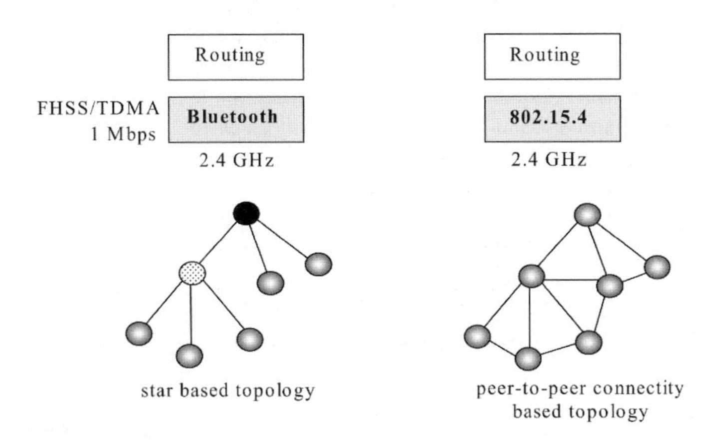

Fig. 8.1. Link layer technology and its properties affect the design of the routing protocols

Use of CSMA/CA and peer-to-peer topology at the link layer also helps matching the physical topology (defined by the range, i.e. reach ability of signals) to the logical topology (defined by the link layer connectivity). In Bluetooth, however, the physical topology can be very different than the logical topology. For example, all nodes in a Bluetooth network can be in range of each other, allowing a full mesh physical topology among nodes; but the logical topology defined by the scatternet covering all these nodes will not be a full mesh topology. It will be a topology defined by the inter-connection star-shaped small networks (piconets). What is important for the routing layer, however, is the logical topology, since it is the topology that defines who can send to whom directly.

Bluetooth networks can be built as scatternets. A scatternet is nothing but a collection of piconets inter-connected together using nodes taking roles as bridges. Therefore, most of the time, we first need to form a Bluetooth scatternet before running the routing protocol on top of it to set up the routes. The routing structures (routes or paths) can be tightly coupled with the structure of the scatternet. There are lots of scatternet formation algorithms defined in the literature that result with different shapes for the scatternets (Zaruba 2001) (Tan 2001) (Salonidis 2001) (Petrioli 2004) (Wang 2002). However, tree-based scatternets are quite common. Some scatternet formation algorithms prefer generating tree-based scatternets because of easy routing on a tree. Hence tree-based routing is a natural fit for WSNs that are based on Bluetooth technology. Tree-based routing, however, has its own deficiencies.

Another affect of Bluetooth and 802.15.4 protocols on the routing design is visible when the addressing scheme is considered. Bluetooth link layer packets carry only a local 3-bit address, which is not enough to address all the Bluetooth nodes in a large Bluetooth network. Therefore, we need an additional addressing scheme to be used by the routing protocol designed for a large Bluetooth network. In 802.15.4, however, the link layer address range, which is also used as part of transmitted packets, is much bigger, and therefore allows these addresses to be used by the routing layer, if desired. The routing layer, however, can also use some other addressing scheme, such as IP addresses. Of course this issue is something to be concerned if only node centric communication is used in the WSN. For data centric communication it is not an important issue since forwarding is not done based on addresses.

Both technologies support sleep modes and therefore higher layers, including the routing layer, can be designed so that only a subset of all the nodes in the network can be used as active nodes and participate in routing; and the rest can be put into sleep mode or can take less responsibility. The set of active nodes can be rotated according to an algorithm and depending on the conditions.

A good approach for developing routing schemes for Bluetooth based WSNs can be considering scatternet construction and routing at the same time. Some studies follow this approach (Zhang 2005). There are also studies which focuses on a scatternet topology over which routing will be implicit and *efficient.* (Sain 2005) proposes a scatternet formation scheme, which produces tree-based scatternets rooted at the sink and suitable for energy efficient routing in WSNs. The constructed scatternet aims having nodes reaching to the sink over shortest paths, but also tries to balance the energy load on the nodes that consumes most power. In a tree-based topology, these are the nodes that are directly connected to the sink. Hence the formation algorithm tries to assign equal number of descendants to nodes that are directly connected to the sink. The scheme assumes there is no data aggregation applied. Therefore, the number of descendants of a node determines the energy load of the node. The scheme also considers the limited range of Bluetooth nodes and does not make an assumption requiring every node to be reachable from every other node. The piconet properties of Bluetooth are considered and the degrees of nodes are constrained.

8.6 Routing Classification

The design space for routing algorithms for WSNs is quite large and we can classify the routing algorithms for WSNs in many different ways. Table x.1 shows some possible classification criteria for WSN routing protocols together with some example protocols meeting the criteria. Those protocols, which are given as examples, are described in the next section.

Classify the protocols based on the way the destinations are specified. Node-centric, location-based (geo-centric), or data-centric specification of destinations is possible in WSNs. Most ad hoc network routing protocols are node-centric protocols where destinations are specified based on the numerical addresses (or identifiers) of nodes. In WSNs, node-centric communication is not a commonly expected communication type (Niculescu 2005). Therefore, routing protocols designed for WSNs are more data-centric or geo-centric. There are routing protocols for WSNs, like LEACH (Heinzelman 2000) and PEDAP (Tan 2003), where there is a single destination, which is the sink node. Those protocols can be considered as a special case of node-centric communication. But a significant number of routing protocols designed for WSNs are data-centric. Directed Diffusion (Intanagonwiwat 2000), ACQUIRE (Sadagopan 2003), Epidemic Algorithms (Akdere 2006), Cougar (Yao 2002) are example protocols that follow data-centric communication paradigm. Specification of a destination based on geographical coordinates (location) or geographical attributes is also common. GPSR (Karp 2000), GAF (Xu 2001), and GEAR (Yu 2001) are example protocols for location-based routing.

Classify the protocols based on whether they are reactive or proactive. A proactive protocol sets up routing paths and states before there is a demand for routing traffic. Paths are maintained even there is no traffic flow at that time. LEACH and PEDAP, for example, are proactive protocols. SPIN (Heinzelman 1999), on the other hand, is a reactive protocol. The routing protocol actions are triggered when there is data to be sent and disseminated to other nodes. Directed Diffusion and ACQUIRE can be considered as reactive protocols, since paths are setup on demand when queries are initiated. Otherwise, no path setup is performed.

Classify the protocols based on whether they are destination-initiated or source-initiated. A source-initiated protocol sets up the routing paths upon the demand of the source node, and starting from the source node. A destination-initiated protocol, on the other hand, initiates path setup from a destination node. We can consider LEACH as a destination initiated protocol since the sink triggers the setup process. SPIN, on the other hand is a source-initiated protocol. The source advertises the data when available and initiates the data delivery. Push-based Epidemic Algorithms are also source-initiated. Directed Diffusion, on the other hand, is a destination (sink) initiated protocol.

Classify protocols based on sensor network architecture. Routing protocols depend on the architecture of the WSNs and vice versa. Some WSNs consist of homogenous nodes, whereas some consist of heterogeneous nodes. A hierarchical routing protocol is a natural approach to take for heterogeneous networks where some of the nodes are more powerful than the other ones. Therefore we can classify the protocols based on whether they are operating on a flat topology or on a hierarchical topology. The hierarchy does not always depend on the power of nodes. Hierarchy can be defined also in a physically homogenous network by assigning different responsibilities to nodes. The responsibility assignment can change dynamically during the lifetime of the network. The responsibility to a node can be given as a clusterhead, leader, gateway, or as an ordinary node. LEACH and TEEN (Manjeshwar 2001) are examples for hierarchical protocols. PEDAP and PEGASIS (Lindsey 2002), on the other hand, are not hierarchical; each node has equal responsibility. Directed diffusion is also a non-hierarchical protocol.

Table 8.1. Some example routing protocols that have those features

Property	Example Protocol(s)
Node-centric	LEACH, PEDAP, PEGASIS
Data-centric	Directed Diffusion, ACQUIRE, SPIN
Geo-centric	GEAR
Proactive	LEACH, PEDAP
Reactive/On-demand	SPIN, Directed Diffusion, Epidemic Algorithms
Dst-initiated	Directed Diffusion, LEACH
Src-initiated	SPIN, Epidemic Algorithms
Flat topology	Directed Diffusion, SPIN
Hierarchical topology	LEACH
Regular Structure	PEDAP
Irregular structure	Directed Diffusion
No structure	Epidemic Algorithms, SPIN
Centralised/ Distributed	PEDAP/Directed Diffusion, LEACH
Multipath-based	EAR, Directed Diffusion
Singlepath-based	PEGASIS, LEACH

Classify the protocols based on whether they use a regular routing structure. Some protocols build a regular structure over which routing will take place prior to actually routing the packets. LEACH, PEDAP, PEGAS are examples of such protocols. Some other protocols do not build a regular structure for routing; for example, Directed Diffusion. On the other hand, some protocols do not have any structure at all: they do not maintain state information. For example, SPIN disseminates data just based on local interactions and do not establish paths from sources to destinations. Similarly, Epidemic Algorithms, Rumor Routing (Braginsky 2002), and Gossiping (Hedetniemi 1988) protocols can be considered as having no structure, i.e. maintaining no end-to-end paths. *Classify the routing protocols based on protocol operation.* Multipath routing and QoS based routing are example categories that are based on the operation of the protocol. The QoS based protocols operate considering the QoS that can be supported by the paths. Multipath routing protocols use more than one path to route the flows and packets. For example, the Energy Aware Routing (EAR) protocol proposed in (Shah 2002) uses multiple paths to route packets towards the sink. Although this causes some sub-optimal paths to be used, it distributes the load more evenly to sensor nodes and prolongs the network lifetime. *Classify the protocols based on whether they are centralised or distributed.* Distributed computation of routing paths and states is preferable for WSNs, which are highly dynamic and self-organising. But computing the routing paths at a central location has also some advantages like providing optimal paths, easy implementation, etc. Therefore, we can design routing protocols so that they compute the paths in a distributed manner or in a central location. For example, PEDAP computes the paths at a central location and then informs the sensor nodes to follow the computed structure. Direction Diffusion, SPIN, and LEACH, on the other hand, are example protocols that compute the paths in a distributed fashion.

8.7 Routing Schemes

Table 8.2. There are different applications for WSNs. Each application has its own traffic characteristics and this affects the selection of the most suitable routing protocol

Application	Traffic Properties	Suitable Routing Protocols
Environmental Monitoring – 1	All nodes send data periodically to sink	LEACH, PEGASIS, PEDAP
Environmental Monitoring – 2	Nodes send data when an event occurs	Directed Diffusion, Epidemic Algorithms, SPIN
Sink issues single-shot queries and some nodes reply	A query has to be disseminated; data has to be transported to the sink from nodes that can answer	Cougar, Acquire, GEAR
Sink issues long-running queries	A query has to be disseminated. Data has to be transported to the sink from nodes that can answer; this happens for a long duration.	Flooding, Directed Diffusion, LEACH, PEGASIS, PEDAP

There are plenty of routing algorithms proposed in the literature for WSNs. The choice of which routing protocol to use for a WSN depends on the application to run and the traffic properties of the application. Table x.2 shows the description of some application types, their traffic properties and some suitable WNS routing protocols. In this section, some widely known routing protocols will be described briefly.

8.7.1 LEACH

LEACH is a clustering-based hierarchical routing protocol designed for WSNs. The protocol requires a set of sensor nodes to be elected as clusterheads according to a randomised algorithm. The set of nodes that are clusterheads are rotated during the lifetime of the network. Each rotation is called a round. In this way, the energy load is more evenly distributed to sensor nodes, since being a clusterhead is much more energy consuming than being a member node (Fig. 8.2). LEACH is designed for WSN applications where there is a single sink node and all sensor nodes are required to sense the environment periodically and send the data to the sink. Hence the data communication is directed towards the sink; the only destination. The protocol is not suitable for applications that require some sensor nodes to be destinations as well. The protocol, in each round, first elects a percentage (5%) of sensor nodes as clusterheads. Each sensor node is assigned to a clusterhead. Then, at each sensing period, all sensor nodes send their data to their clusterheads.

A clusterhead aggregates the data received from member nodes into a single packet and transmits it to the base station (sink). Hence a data packet originating at a sensor node reaches to the sink after two hops. This provides quite fast delivery of data to the sink in an energy efficient manner. But being able to reach to the base station in two hops requires the nodes to have power control capability so that they can adjust their power level to reach to arbitrary distances. If they cannot do this, then the coverage becomes limited and LEACH cannot be used for WSNs deployed over large areas in this case. The percentage of clusterheads does not have to be fixed to 5%. It can be increased or decreased. If it is increased, it means more clusterheads will be on the field and therefore the average distance that sensor nodes will transmit to reach clusterheads will be reduced. On the other hand, more packets will be transmitted from the clusterhead to the sink. The clusterhead selection algorithm promotes uniform distribution of clusterheads to the field. Some of these clusterheads, which are far away from the base station, will spend more energy than the other ones. The clustering based architecture of LEACH, the property that each clusterhead sends its data directly to the base station, and rotating the clusterheads provide an important advantage which is distributing the load more evenly to sensor nodes. In multi-path based routing protocols, which may be operating on a tree-like topology, we have the problem of uneven distribution of load to sensor nodes that are close to the root (sink), since all traffic has to go through those nodes. In LEACH, however, this is not a problem since the clusterheads are distributed evenly to the sensor field and they are rotated at every round.

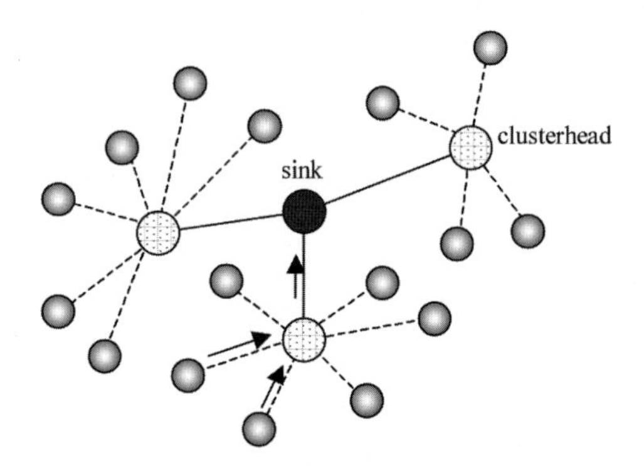

Fig. 8.2. LEACH routes packets in two hops. Clusters are formed, and each node sends its data packet to its clusterhead. Each clusterhead aggregates the data received from the cluster members into a single packet and sends it to the sink.

8.7.2 PEDAP

PEDAP (Tan 2003) is another routing scheme that is suitable for WSN applications that require periodic sensing of environment by each sensor node and sending of the sensed information to a sink node. It is assumed that data aggregation at intermediate nodes is possible for the applications. The data flow is again directed from all sensor nodes to a single point, which is the sink node. Again, power control capability at each node is needed for PEDAP to operate efficiently. PEDAP is based on the idea that at each round of communication (during which data from all sensor nodes is delivered to the sink once) the network should consume minimum amount of energy and the energy load should be balanced among the nodes so that a longer lifetime can be achieved. To spend less energy per round, a minimum spanning tree (MST) spanning all the sensor nodes and rooted at the base station is constructed, and routing is performed over the edges of this tree (Fig. 8.3). Accordingly, each node will receive data from one or more sensor nodes (children in MST), aggregate the received data, and then will send the aggregated data to the next sensor node (parent in MST) on the way towards the sink. The cost of each link (edge) is the energy consumed for sending a packet over that link. This energy depends on the size of the packet, the distance between sender and receiver, and the transceiver electronics energy consumption constants (Heinzelman 2000). Hence, routing over MST implies minimum total energy consumption per round.

If the distance between a node i and j is d_{ij}, then the cost of the edge connecting those nodes depending on the packet size k is computed as follows:

$$C_{ij}(k) = 2E_{elec}k + E_{amp}kd_{ij}^2 \tag{8.1}$$

Similarly, the cost of an edge connecting a sensor node i to the sink (which is mains powered) is computed as follows (again for a packet of size k):

$$C'_{i}(k) = E_{elec}k + E_{amp}kd_{ib}^2 \tag{8.2}$$

Here, E_{elec} and E_{amp} are radio transceiver dependent constants. This cost model is based on the radio model proposed by (Heinzelman 2000).

PEDAP also tries to distribute the energy load evenly to sensor nodes by increasing the cost of edges connected to nodes that have very little energy left. For that, it modifies the edge cost model by dividing it with the remaining energy of the node.

This implies that, the cost of an edge connected to the node increases, while the remaining energy of the node decreases. Hence, a node that has very little energy left will have a very high cost for its edges and therefore those edges will be excluded from the minimum spanning tree and will not be used for routing in the next phase. In this way, traffic load on this node will be reduced. PEDAP reconstructs the MST periodically (at each phase), which means the remaining energy levels of nodes are checked and taken into account periodically.

The route computation in PEDAP is done at a central location and the position of sensor nodes are assumed to be known. Performing route computation in a centralised manner has the advantage of obtaining better routes and preventing routing loops. But it has also some drawbacks. Maintenance of routes is more difficult when nodes fail or deplete their energy or when links break.

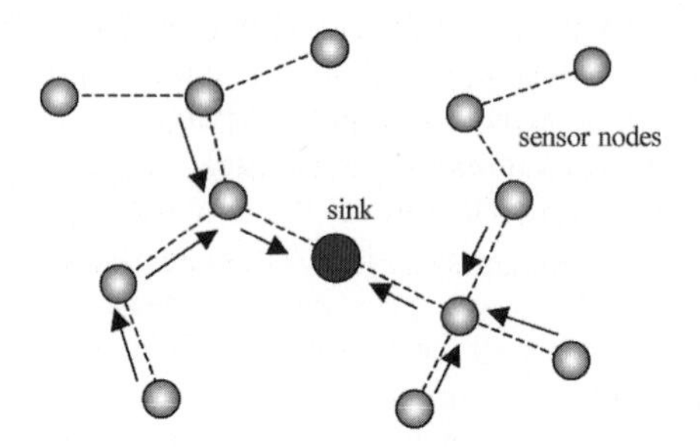

Fig. 8.3. A sensor network, where routing paths are established by the PEDAP protocol. PEDAP builds an MST spanning the sensor nodes and rooted at the sink. The packets are routing over the edges of the MST

8.7.3 PEGASIS

PEGASIS (Lindsey 2002) is a routing scheme that is based on forming a chain passing through all sensor nodes, and packets are routed over this chain (Fig. 8.4). The chain is formed before data gathering process starts. Each node on the path aggregates the received data with its own data and sends the aggregated data to the next node on the chain. One node on the chain is selected as the node that will finally forward the aggregated data to the base station. This node partitions the chain into two. The data is collected at this node starting from each end of the chain. The responsibility of sending the aggregated data to the sink is rotated among the sensor nodes, and in this way energy load is more evenly distributed.

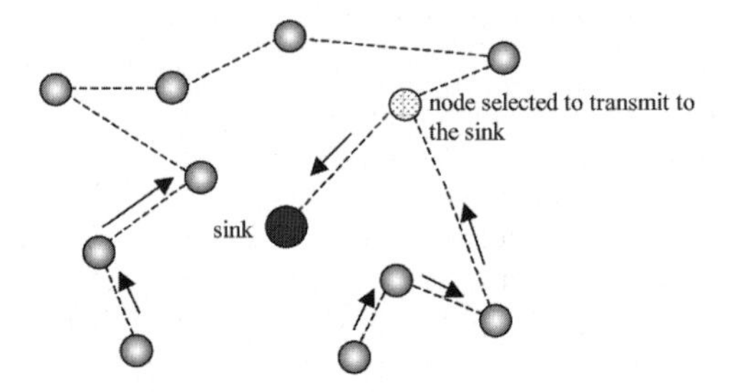

Fig. 8. 4. The PEDAP protocol establishes a chain passing through all sensor nodes. A node on the chain is selected to transmit the data to the base station. Data is transported on the chain after getting aggregated at each node

A greedy algorithm can form the chain. Starting at a node, each node selects a nearest neighbor to be the next node on the chain. In this way, the transmission energy cost over the chain is reduced, since the distances between adjacent nodes are reduced. This greedy approach, however, does not give an optimal solution. The optimal solution is the solution of the TSP (traveling salesman problem) problem. The disadvantage of PEGASIS is the long delays for packets sent by the nodes that are close to the ends of the chain. The distance of these nodes to the node that is transmitting to the sink is O(n) where n is the number of nodes in the network. PEGASIS can work in environments where range is limited, provided that there are neighbors of each node that are in the range. But this does not guarantee forming a chain. PEGASIS is more scalable compared to LEACH in terms of the area of the region is concerned, since it does not have two-hops limitation.

8.7.4 Directed Diffusion

Directed Diffusion is a data-centric routing protocol (Intanagonwiwat 2000) designed for WSNs. The routing and path setup decisions are made according to the data that packets deliver. Directed diffusion is developed with the aim of being a robust, scalable, and energy efficient protocol, all of which are very important objectives for the design of protocols for WSNs. In data-centric routing, the destinations are identified with the content of the packets. The content is organised in a format that consists of sequence of attribute-value pairs. A setting of those attributes specifies the destinations that will process the packet. Directed diffusion is a distributed protocol and uses local interactions to setup paths and route packets. There is no central coordination required. It is also scalable since the state information stored at intermediate nodes is independent of the number of nodes in the network. It can work for both mono-sink and multiple-sink scenarios. In Directed Diffusion, the sinks via messages called interests activate sensor nodes. A sink that is interested in obtaining information about a region in the sensor field builds up an interest message and disseminates the message through the network. An interest message consists of attribute-value pairs. Those attributes specify what kind of information the sink would like to receive about the network. The attributes may also include the coordinates of the region from where data will collected, the rate of data collection, and the duration of the data gathering process. The interest message can also be considered as a query message or a task description message.

The dissemination of interest is realised as follows. The sink node, that originates the interest, broadcasts the interest to its neighboring sensor nodes. Each sensor node receiving an interest creates an entry in its local table for the interest and creates gradients (pointers to the next node) in the entry pointing to the nodes from which the interest is received. These gradients will be used during data delivery to forward the data packets to the next nodes towards the sink. After gradients are setup at the node, the node also broadcasts the interest to its neighbors. This process continues and the interest gets diffused through the network. Each sensor node receiving the interest checks the attributes of the interest, and if required, processes the interest. For example, if the node is inside the region specified by the target coordinates stored in the interest, then the node activates its sensors and becomes ready to send data towards the sink. When the interest is disseminated in the network, some sensor nodes that decide to process the interest will be the sources of data (and events) to be sent to the sink. When a source sensor node would like to send data to the sink, it sends the data packet to the directions (neighbors) shown by the gradients stored in the corresponding interest entry. Each intermediate node that is receiving a data packet checks its interest table and finds the matching interest entry. The matching interest entry will have gradients showing the next hops (neighbors) the data packet has to be forwarded. The node then forwards the packet to those neighbors. In this way, the data packet is routed back to the sink node, possibly over multiple paths.

It is not desirable to receive the same information over multiple paths. It causes energy waste. Therefore, Directed Diffusion has a path-reinforcement mechanism through which a single path among several paths is selected and only that one is used for further information sent from the source to the sink. This is achieved by reinforcing the neighbors along the selected path, starting from the sink node. Reinforcement is done by sending a higher rate interest message starting at the sink node and along the selected path towards the source node. The gradients corresponding to the links of this selected path will be involved in activity. Each gradient setup in Directed Diffusion protocol has a timer associated with it. If no packet is sent over that gradient for the time period for which a timer has been setup, the gradient is deleted. In this way, the gradients corresponding to the paths that are undesired will be deleted with time and only one path will be used to deliver packets from source to the sink.

8.7.5 SPIN

SPIN (Heinzelman 1999) is also a data-centric data dissemination scheme that is distributed, negotiation-based, and resource-aware. It can be considered as reactive (on-demand) protocol where data delivery is initiated by the source node. It is more suitable for applications where data will be available at unpredictable times. It is an energy and bandwidth efficient protocol. SPIN is a negotiation-based protocol. Neighboring sensor nodes negotiate with each other for transfer of information. For this, meta-data based approach is used to describe the data that is to be disseminated to other nodes. It is assumed that the size of the meta-data will be much smaller compared to the size of the corresponding data for most applications. Therefore, meta-data is exchanged between neighbors before the actual data is delivered. A node that has new data first advertises the availability of the data to its neighbor (by using an ADV – advertisement – message). Any neighbor that does not have the data and is interested in receiving the data, requests the data explicitly (by using a REQ – data request – message) upon receiving the advertisement. Then the data is transmitted (by using a DATA – payload – message). If there is no neighbor that would like to receive the data, then the data is not transmitted. Any node receiving a new data from a neighbor applies the same procedure: it advertises the new data to its neighbors and the interested neighbors request the data. This process continues as long as there are nodes that are interested in receiving the data that is advertised by some neighbor. In this way, transmission of redundant information is eliminated. SPIN is also resource-aware and resource-adaptive. It requires each node to monitor its own resources, such as remaining battery energy, traffic load, etc., and adjust its behavior and activities accordingly.

SPIN aims to solve the following problems that can be seen in flooding-based data dissemination schemes: 1) implosion; 2) overlap; and 3) resource blindness. Implosion means that a node receives the same data packet more than once, possibly through different paths. Overlap means that several data items that is received by a node correspond or include the same information. For example, two sensor nodes may sense overlapping regions and therefore produce data packets that include overlapping information. Resource blindness is acting of sensor nodes without caring about the resources consumed and the level of available resources. SPIN attacks the first two problems by use of meta-data and negotiations. It attacks the third problem by being resource-aware. It adapts the activities of a node based on the level of resources the node has.

8.7.6 Epidemic Algorithms

Epidemic Algorithms are based on the theory of epidemics that is about the models for spread of diseases. Epidemic Algorithms follow these natural models to disseminate information in networks and distributed environments (Demers 1987). These models are using interaction between neighbors to disseminate information, as in SPIN. The interaction can be pull-based or push-based. In *pull-based* interaction and delivery, a node requests information from a neighbor and gets the information if it is available at the neighbor. In *push-based* interaction, a node that has some new information sends it to a selected neighbor. In this way the information is delivered to the neighbor and the neighbor can execute the same steps to disseminate the information further. The information that a node has can be cached in the node for a while even though it is sent to a neighbor. There are different policies defining when to delete an information message from the cache. This approach for delivering information in a distributed environment can also be used in WSNs (Akdere 2006). Sensor nodes can push data to their neighbors by local interactions. Similarly a sink node, or another sensor node can pull data from neighbors by local exchange of messages. An important property of Epidemic Algorithms is that they are robust to changes in the network. Links can break, nodes can move or fail, but the algorithms still work and deliver data. Epidemic Algorithms are realised by distributed protocols requiring local interactions. There is no need for a central control. There is also no path setup needed to deliver messages between nodes. This is something different than Directed Diffusion. Directed Diffusion is also a distributed protocol and is robust, but it first establishes paths between sources and destinations via gradients set up at intermediate nodes.

8.7.7 ACQUIRE

ACQUIRE is a data gathering scheme that is more suitable for WSN database applications (Sadagopan 2003). In ACQUIRE, a query is resolved while it is being disseminated. In other words, a query is processed inside the network, as opposed to being processed at a central location after all data is available at the central location. ACQUIRE aims improving energy efficiency by not sending a query to all nodes and by not requiring all sensor data to be collected first at the center before processing. Therefore it is more efficient than flooding based approaches, and can be used efficiently for one-shot queries for which flooding the whole network would cause too much overhead. It is also suitable for complex queries that contain several interests and variables, and that can be progressively resolved at intermediate nodes. In this way, the amount of data packets (and their sizes) that should be transported to the sink is reduced.

ACQUIRE works as follows. The sink injects an active query, which can be a one-shot or complex query, into the network. Any node receiving the query tries to answer (resolve) the query as much as it can. This is achieved by the cached information at the node that is received from *d*-hop neighbors of the node. If the cached information is out-of-date (has been received long time ago), the node can initiate a data request from its *d*-hop neighbors upon receiving the query. In this way, the node will have information about its local neighborhood and this information is used to resolve the query as much as possible. If the query is completely resolved, we are done and the query is propagated back to the sink.

If the query is not completely resolved, it is sent to another sensor node in the network which will further process the query. That node can be determined in a random manner, or by an intelligent decision. That node receiving the query will execute a similar procedure to process the query as much as possible. If the query is still not completely resolved, it is sent to another node. In this way, the query is processed step-by-step inside the network. Hopefully, most of the time, the query will be completely resolved without needing to travel along all the sensor nodes.

8.7.8 GEAR

GEAR – Geographical and Energy Aware Routing - (Yu 2001) is a routing protocol that is location-based and energy-aware. It makes routing decisions based on the coordinates of a destination region that packets are destined to. Therefore, a packet has to include information about the region. The region can be specified as a rectangle and its coordinates can be included in the packet. Assume such a packet is to be delivered to all nodes in the rectangular region located somewhere in the field where WSN is deployed. GEAR performs the delivery in two steps. 1) First it delivers the packet from the source node (which can be the base station) to a node in that region. For this, it uses location-based and energy-aware routing. 2) Then, when the packet arrives to a node inside that region, that node delivers the packet to all nodes in the region by Recursive Geographic Routing. If this may not work for some scenarios, then restricted flooding can be used as well.

Forwarding of a packet towards the destination region is achieved by a combination of geographical and energy-aware route selection techniques. GEAR assumes that each node knows its neighbors, their positions, and their remaining energy levels. This can be achieved by infrequent message exchanges between neighbors. Using these values at a node, GEAR computes a cost value for each neighbor of the node. Using these costs, and the costs of sending packets from the node to each neighbor, GEAR selects a neighbor to forward the packet next. As a result of this algorithm, when all neighbor nodes have equal level of energy, for example, the protocol selects the neighbor that is closest in distance to the destination region. On the other hand, when all neighbors have equal distance to the destination region, then the protocol selects the neighbor that has the largest remaining energy. In this way, GEAR tries to both balance energy load on neighbors and route the packets to the destination over shorter distances.

When a packet arrives to a destination region, it is delivered to all nodes in the region using Recursive Geographic Routing. For this, the destination region is first divided into four sub-regions and a copy of the packet is sent to each sub-region using location-based routing. Then, when a copy of the packet arrives to a sub-region, the same procedure is applied in that sub-region as well: it is further divided into four other sub-regions and a copy of packet is sent to each sub-region. This continues until a sub-region has just one node. As a result of this Recursive Geographic Routing, all nodes in the destination region can receive the packet sent from the source.

GEAR enables efficient routing of a packet to a destination region. This is usually required, for example, when a base station issues a query and the query has to be forwarded to all nodes in a destination region. GEAR provides an energy-efficient and location-based routing protocol for delivering the query. The same protocol can be used, however, for sending the data packets from those nodes to the sink back. All sensor nodes can know the location information of the sink, and the data packets generated by these sensor nodes can be targeted to that location. That delivery can also be achieved by the location-based and energy-aware forwarding mechanism of GEAR.

8.8 Summary

This chapter describes the routing problem in WSNs. The routing problem is also identified as data dissemination or data gathering problem in WSNs. Although there are some differences between data dissemination, data gathering, and data routing, what all aim is efficient delivery of data packets from one or more sources to one or more destinations. Therefore, here in this chapter, routing, data dissemination, and data gathering are considered as the same problem in WSNs, and they are unified under a single topic: routing and data dissemination protocols for WSNs.

WSNs are sometimes considered as a sub-class of ad hoc networks. But there are actually very important differences. WSNs consist of unmanned nodes which are powered by irreplaceable batteries. Therefore energy is the most important resource. The number of nodes in a WSN can be much larger than the number of nodes in an ad hoc network; therefore scalability of protocols is very important. The traffic is usually directed from or to a base station in WSNs. This provides opportunity to design special routing protocols tailored for this property. WSNs are also application-specific networks and therefore it is possible to design routing protocols that are application-aware and application-specific. There are a number of wireless technologies that can be used as the link layer technology in WSNs. Bluetooth and IEEE 802.15.4 are two such technologies. The properties of the link layer technology affect the design of routing protocols. The topologies that can be generated by the link layer, the MAC protocol, the bit-rate, data range, and whether it is possible to adjust the transmission power are all important factors that can affect the design and optimisation of the routing protocols for WSNs.

Several routing protocols specifically designed for WSNs are introduced in this chapter. Those routing protocols can be classified in many different ways: 1) based on whether they are node-centric, geo-centric, or data-centric; 2) based on whether they operate on a flat topology or on a hierarchical topology; 3) based on the whether they are reactive or proactive; etc. Most routing protocols designed for WSNs follow a data-centric approach where the destinations are specified not by numerical addresses, but by values of the attributes of the data included in the packets. Data-centric routing is a new approach that is proposed for WSNs. Since WSNs are application-specific, the design space for routing algorithms for WSNs is very large. Each WSN and application may require its own routing protocol to be designed; but still there are some common approaches and mechanisms that can be followed by routing protocols designed for many different applications. The chapter therefore has provided the descriptions of some of the widely known routing protocols developed for WSNs. These protocols encompass many of the important design approaches and techniques for routing in WSNs.

8.9 References

Akdere M, Bilgin C, Gerdaneri O, Korpeoglu I, Ulusoy O, and Cetintemel U (2006) A comparison of epidemic algorithms in WSNs. Computer Communications (to appear)

Akkaya K and Younis M (2005) A survey on routing protocols for WSNs. Ad hoc networks 3:325-349

Al-Karaki JN and Kamal AE (2004) Routing techniques in WSNs: a survey, IEEE Wireless Communications 11:6-28

Bluetooth Special Interest Group (2004) Bluetooth core specifications. Available at http://bluetooth.com/Bluetooth/Learn/Technology/Specifications/

Braginsky D and Estrin D (2002) Rumor routing algorithm for sensor Networks. In: Proc. of 1st Workshop on Sensor Networks and Applications, Atlanta

Chang JH and Tassiulas L (200) Energy conserving routing in wireless ad-hoc networks, In: Proc. of IEEE INFOCOM. pp 22-31

C. –K. Toh (2001) Maximum battery life routing to support ubiquitous mobile computing in wireless ad hoc networks. IEEE Communications Magazine 39:138-147

Conti M, Maselli G, Turi G, and Giordano S (2004) Cross-Layering in mobile ad hoc network design. IEEE Computer 37:48-51

Demers A, Greene D, Hauser C, Irish W, Larson J, Shenker S, Sturgins H, Swinchart D, and Terry D (1987) Epidemic algorithms for replicated database maintenance. In: The Proc. of the ACM Symp. on principles of distributed computing. pp 1-12

Ganesan D, Govindan R, Shenker S, and Estrin D (2002) Highly resilient, energy efficient multipath routing in WSNs. Mobile Computing and Communications Review 5:11-25

Ganesan D, Krishnamachari B, Woo A, Culler D, Estrin D, and Wicker S (2002) An emprical study of epidemic algorithms in large-scale multihop wireless networks. (Technical report UCLA/CSD-TR-02-0013 by the Department of Computer Science of UCLA)

Goldsmith AJ and Wicker WB (2002) Design challenges for energy-constrained ad hoc wireless networks. IEEE Wireless Communications 9:8-27

Goussevskaia O, Machado MV, Mini RAF, Loureiro AAF, Mateus GR, and Nogueira JM (2005) Data dissemination on the energy map. IEEE Communications 43:134-143

Gutierrez JA, Naeve M, Callaway E, Bourgeois M, Mitter V, and Heile B (2001) IEEE 802.15.4: a developing standard for low-power low-cost wireless personal area networks. IEEE Network 15:12-19

Hac A (2003) WSNs designs. Wiley

Hedetniemi S and Liestman A (1988) A survey of gossiping and broadcasting in communication networks. IEEE Network 18:319-349

Heinzelman WR, Chandrakasan A, and Balakrishnan H (2000) Energy-efficient communication protocol for wireless microsensor networks, In: Proc. of the Hawaii Int. Conf. on System Sciences

Heinzelman WR, Kulik J, and Balakrishnan H (1999) Adaptive protocols for information dissemination in WSNs. In: Proc. of ACM/IEEE, Int. Conf. on mobile computing and networking. Seattle, pp 174-185

IEEE 802.15.1 (2005) IEEE 802.15.1 Standard Specification. Available from http://standards.ieee.org/getieee802/download/802.15.1-2005.pdf

IEEE 802.15.4 (2003) Standard Specification. Available from http://www.ieee802.org/15/pub/TG4.html

Intanagonwiwat C, Govindan R, and Estrin D (2000) Directed diffusion: A scalable and robust communication paradigm for sensor networks, In: Proc. of the ACM/IEEE Int. Conf. on mobile computing and networking. Boston, pp 56-67

Karp B and Kung HT (2000) GPSR: Greedy perimeter stateless routing for wireless networks. In: Proc. of the 6th Annual Int. Conf. on mobile computing and networking. pp 243 – 254

Lindsey S and Raghavendra CS (2002) PEGASIS: power-efficient gathering in sensor information systems, In: Proc. of IEEE Aerospace Conf.. pp 1125-1130

Madan R, Chui S, Lall S, and Goldsmith A (2005) Cross-Layer design for lifetime maximisation in interference-limited WSNs. In: Proc. of IEEE INFOCOM. pp 1964-1975

Manjeshwar A and Agrawal DP (2001) TEEN: A routing protocol for enhanced efficiency in WSNs. In: Proc. of the Int. parallel and distributed processing Symp. workshops

Niculescu D (2005) Communication paradigms for sensor networks. IEEE Communication Magazine 43:116-122

Petrioli C, Basagni S, and Chlamtac I (2004) BlueMesh: degree-constrained multi-hop scatternet formation for Bluetooth networks. Mobile Networks and Applications 9:33-47

Sadagopan N, Krishnamachari B, and Hemly A (2003) The ACQUIRE mechanism for efficient querying in sensor networks, In: Proc. of the IEEE Int. Workshop on Sensor Network Protocols and Applications. pp 149-155

Saginbekov S and Korpeoglu I (2005) An energy efficient scatternet formation algorithm for Bluetooth based sensor networks. In: Proc. of the European Workshop on WSNs. Istanbul

Salonidis T, Bhagwat P, Tassiulas L, and LaMaire R (2001) Distributed topology construction of Bluetooth personal area networks. In: Proc. of IEEE INFOCOM. pp. 1577-1586.

Schurgers C and Srivastava MB (2001) Energy efficient routing in WSNs, In: Proc. of MILCOM Conf.. Vienna. pp. 357-361

Shah RC and Rabaey JM (2002) Energy aware routing for low energy ad hoc sensor networks, In: Proc. of WCNC, pp 350-355

Shakkottai S, Rappaport TS, and Karlsson PC (2003) Cross-layer design for wireless networks. IEEE Communications 41:74-80

Tan G, Miu A, Guttag J, and Balakrishnan H (2001) Forming scatternets from Bluetooth personal area networks. (Technical report MIT-LCS-TR-826 by MIT)

Tan HO and Korpeoglu I (2003) Power efficient data gathering and aggregation in WSNs, ACM SIGMOD Record 32:66-71

Van Hoesel L, Nieberg T., Wu J, and Havinga PJM (2004) Prolonging the lifetime of WSNs by cross-layer interaction. IEEE Wireless Communications 11:78 - 86

Wang Z, Thomas RJ, and Haas Z (2002) Bluenet – a new scatternet formation scheme. In: Proc. of the 35th Hawaii Int. Conf. on System Sciences

Xu Y, Heidemann J, and Estrin D (2001) Geography-informed energy conservation for ad hoc routing. In: Proc. of the ACM/IEEE Int. Conf. on mobile computing and networking. Rome. pp 70-84.

Yao Y and Gehrke J (2002) The Cougar approach to in-network query processing in sensor networks. ACM SIGMOD Record 31:9-18

Ye W, Heidemann J, and Estrin D (2002) An energy efficient MAC protocol for WSNs. In: Proc. of IEEE Infocom Conf.. New York, pp 1567-1576

Yu Y, Govindan R, and Estrin D (2001) Geographical and energy aware routing: a recursive data dissemination protocol for WSNs. (Technical Report UCLA/CSD-TR-01-0023 by UCLA, 2001.

Zaruba G, Basagni S, and and Chlamtac I (2001) BlueTrees - scatternet formation to enable Bluetooth-based personal area network. In: Proc. of the IEEE Int. Conf. on communications. pp 273-277

Zhang X and Riley GF (2005) Enery aware on-demand scatternet formation and routing for Bluetooth-based WSNs, IEEE Communications 43:126-133

Zhao F, Guibas L (2004) WSNs: an information processing approach. Elsevier-Morgan Kaufmann, Boston

ZigBee Alliance (2004) ZigBee standard specification. Available at: http://www.zigbee.org/en/spec_download/download_request.as

9 Quality Assurances of Probabilistic Queries

Reynold Cheng[1], Edward Chan[2] and Kam-Yiu Lam[2]

[1]Department of Computing, Hong Kong Polytechnic University, Hong Kong,
[2]Department of Computer Science, City University of Hong Kong, Hong Kong

9.1 Introduction

Many applications use sensors extensively to capture and monitor the status of physical entities. In a habitat-monitoring system, for example, the temperature values of birds' nests are investigated through the use of wireless sensor networks (Deshpande 2004). Sensors are also installed in different parts of the building, so that the temperature offices can be adjusted by an internal air-conditioning system. In fact, sensors are increasingly used in various applications, which deliver monitoring and querying services based on various attribute values of physical environments, such as location data, temperature, pressure, rainfall, wind speed, and UV-index. In this kind of systems, one common problem is that the reading of a sensor can be uncertain, noisy and error-prone (Elnahrawy 2003, Cheng 2003). A sensor's value can be contaminated by measurement errors. Moreover, the environment being monitored by sensors can change continuously with time, but due to limited battery power and network bandwidth, the state of the environment is only sampled periodically.

As a result, the data received from the sensor can be uncertain and stale. If the system uses these data, it may yield incorrect information to users and make false decisions. In order to solve this problem, the *uncertainty* of the sensor data must be taken into account in order to process a query (Cheng 2003). In order to consider uncertain data during query execution, the concept of *probabilistic queries* has been studied extensively in recent years (Wolfson 1999; Pfoser 2001; Cheng 2003; Desphande 2004). In these works, uncertain data are modeled as a range of possible values with a certain probability distribution (e.g., Gaussian and uniform distribution). Probabilistic queries process these uncertain data and produce "imprecise answers". These answers are those that are augmented with probabilistic guarantees to indicate the confidence of answers. As an example, consider a query asking "which area yields a wind speed over 1 km/hr", where the wind speed values of some regions are reported by sensors. By modeling each wind speed value as an uncertain data item, a probabilistic query returns a list of areas, together with their probabilities, which indicate the chance that the area yields a wind speed higher than 1 km/hr. Notice that although probabilistic queries do not return exact answers, the probability values reflect the degree of confidence for the answer, rather than a completely wrong answer when uncertainty is not considered. In fact, the probability values augmented to a probabilistic query answer can serve as some indicators for the *quality* of query answers. Consider a MAX query, which returns all objects which have non-zero probability of giving a maximum value. This query is executed over two different sets of data, namely {A, B, C} and {D, E, F}, and yields the following answers:

Answer set I: (A, 95%), (B, 4%), (C, 1%)
Answer set II: (D, 40%), (E, 30%), (F, 30%)

Items A and D have the highest chance of satisfying the MAX query among their corresponding answer sets. However, it is clear that A (with a 95% chance) can be placed with a much higher confidence of satisfying the query than D (with a 40% chance). We say that Answer set I has a *higher quality* than Answer set II since the former is less ambiguous than the latter.

A user who submits a query may only want to accept answers with a high quality. For example, the user may specify that he only accepts an answer if its member with the highest probability exceeds 80%. In the previous example, only Answer set I satisfies the user's requirement. Now, a natural question to ask is: can the quality of a probabilistic answer be improved in order to satisfy the user's quality requirements?

To answer this question, notice that the answer quality is directly related to the uncertainty of sensor data. In general, if the data items have a higher degree of uncertainty/error, the answer is less precise and this results in a lower quality. Hence a natural solution is to reduce data uncertainty in order to improve the answer quality. For example, one may request all sensors to increase their sampling rates so that the uncertainty due to sampling is reduced. This solution, however, exhausts the limited resources of a wireless sensor network easily. Clearly, we need solutions that can balance between query quality and resource utilisation.

In this chapter, we present two methods that are aimed at improving the query quality towards the user's quality requirement. The first method, known as *query-centric probing policies*, are heuristics developed to reduce the effect of uncertainty due to sampling. The main idea is to selectively increase the sampling rate of sensors that have the highest effect on query quality. The second method, called *sensor selection techniques*, exploits the fact that nowadays many low-cost and redundant sensors can be deployed to monitor the same region of interest. Multiple sensors are selected to get an average reading in order to reduce data uncertainty due to measurement error. The redundancy of sensors is exploited carefully, in order to attain an average value with a small variance, with low energy and communication costs.

The rest of this chapter is organized as follows. Section 2 presents related works. In Section 3 we describe a classification scheme of probabilistic queries and the basic evaluation methods of several representative queries. Next, Section 4 describes the important topic of query quality metrics for different probabilistic queries. Section 5 and Section 6 then explain the essence of query-centric probing policies and sensor selection techniques. We conclude the chapter in Section 7.

9.2 Review

There is a large body of research in querying sensor data in wireless sensor networks. However, it is only recently that researchers have started to consider the effect of data uncertainty as it becomes increasingly evident that the noisy nature of sensor readings must be addressed before users can be confident about the accuracy of sensor-based monitoring. Indeed, researchers have pointed out that there is a *lower bound* of uncertainty in location measurements in wireless sensor networks (Wang 2004).

A number of researchers have adopted uncertainty reduction approaches for specific types of errors. Eiman and Badri (Elnahrawy 2003) proposed a Bayesian approach for reducing one important type of data uncertainty – the random noise. They also proposed algorithms for answering queries over uncertain data. However, their proposed method does not take into consideration the quality requirement of probabilistic queries. Moreover, their noise cleaning technique aims at cleaning random error of individual sensors, while our methods probe the right sensors in order to obtain more precise data values to meet the quality requirement of probabilistic queries with a low data collection cost. The hierarchical sensor network model described in this chapter, where a coordinator handles a number of sensors, is similar to that proposed in a number of other works like the microsensor net (Kumar 2003) which provides authentication and Denial of Services (DoS) protection. Other researchers proposed overlaying a data gathering tree to perform aggregate queries efficiently (Madden 2002), including a multiple sink version (Dubois-Ferriere 2004). Although we also use intermediate nodes to collect data and reduce communication cost in evaluating aggregate queries, we consider *probabilistic* aggregate queries and sensor uncertainty which none of these works does. The problem of selecting appropriate sensors in a wireless environment has been studied by researchers, but so far only in the context of improving accuracy in location tracking. In (Etrin 2003, Liu 2003), mutual information between the distribution of an object's location and the predicted location observed by a sensor is used to compute the information gain due to the sensor. The sensor with the highest information gain is selected to reduce the uncertainty of the sensor reading. Another approach, based on entropy-based selection heuristics, is claimed to be computationally more efficient than the above methods based on mutual information (Wang 2004). Note, however, that these schemes are designed primarily for location tracking, but not for the common types of continuous queries (such as minimum/maximum query) considered in this chapter.

9.3 Data Uncertainty and Probabilistic Queries

In this section, we present the concept of sensor uncertainty and probabilistic queries that provide us a foundation of further discussion on query quality maintenance. The discussions in this section are based on our previous work (Cheng 2003).

9. 3.1 Probabilistic Uncertainty Model

To capture data uncertainty, a data scheme known as probabilistic uncertainty model was described in previous literature (Cheng 2003, Pfoser 2001, Wolfson 1999). This model assumes that each data item can be represented by a range of possible values and their distributions. Formally, suppose a database consists of n tuples, namely Oi (where i = 1,.., n). Each tuple Oi consists of an uncertain value a, characterized by two elements – uncertainty interval and uncertainty pdf:

- **Definition 1** An uncertainty interval, denoted by $a.U$, is an interval $[a.l, a.r]$ where $a.l, a.r \in \Re$, $a.r \geq a.l$ and $a \in a.U$.
- **Definition 2** An uncertainty pdf of a, denoted by a.f(x), is a probability distribution function of a, such that $\int_{a.l}^{a.r} a.f(x)dx = 1$ and $a.f(x)=0$ if $x \notin a.U$.

As an example, $a.U$ can be a fixed bound d, which is a result of negotiation between the database system and the sensor (Wolfson 1999). In a more general model $a.U$ is a function of t, where the uncertainty grows with time until an update with time (Cheng 2003, Cheng 2004). An example uncertainty pdf is the Gaussian distribution, which models the measurement inaccuracy of location data (Wolfson 1999) and data from sensor network (Desphande 2004). It can also be a uniform distribution, which represents the worst-case uncertainty within a given uncertainty interval. Thus, the uncertainty model provides flexibility and is able to model different types of uncertainty. We investigate both how sampling uncertainty (i.e., uncertainty interval is modeled as a function of time) can be reduced by probing policies. We also study how measurement uncertainty (i.e., a time-independent Gaussian distribution with infinitely long uncertainty intervals) can be handled by sensor selection policies. Next, let us study how data uncertainty is managed by queries known as probabilistic queries.

9.3.2 Classification of Probabilistic Queries

We now describe a classification scheme for probabilistic queries (Cheng 2003). There are two reasons to justify the need for such a classification. First, queries in the same class have similar evaluation algorithms. Another reason is that probabilistic queries produce probabilistic (or inexact) answers. The vagueness of a probabilistic answer is captured by its quality metrics, which are useful to decide whether the answer is too ambiguous and any action needs to be done to reduce data uncertainty. Quality metrics differ according to the query class. To illustrate, denote (O_i, p_i) be the probability p_i that O_i is the answer. In a range query, an answer $(O_1, 0.9)$ is better than $(O_1, 0.1)$ since there is a higher level of confidence that the former answer is within a user-specified range, compared with the latter one, which has only a half chance of satisfying the query. A simple metric such as $\frac{|p_i - 0.5|}{0.5}$ can indicate the quality of the answer. For a maximum query, the answer

$\{(O_1, 0.8), (O_2, 0.2)\}$ is better than $\{(O_1, 0.5), (O_2, 0.5)\}$, since from the first answer we are more confident that O_1 gives the maximum value. To capture this, an entropy-based metric is needed. Probabilistic queries can be classified in two ways. First, we can classify them according to the forms of answers required. An *entity-based query* returns a set of objects, whereas a *value-based* query returns a single numeric value. Another criterion is based on whether an aggregate operator is used to produce results. An *aggregate query* is one which involves operators like MAX, AVG – for these operators, an interplay between objects determines the results. In contrast, for a *non-aggregate query*, the suitability of an object as the result to an answer is independent of other objects. A range query is a typical example. Based on this classification, we obtain four query classes.

1. Value-based Non-Aggregate Class. An example of this class is to return the uncertain attribute values a which are larger than a constant.
2. Entity-based Non-Aggregate Class. One example query is the range query: given a closed interval $[l,u]$, a list of tuples (O_i,p_i) are returned, where p_i is the non-zero probability that $O_i.a \in [l,u]$. Another example is a join over two tables R and S. It returns a pair of tuples (Ri, S_j, p_{ij}), where p_{ij} is the probability that the two tuples R_i and S_j join (using comparison operators such as $=$, $\neq$, $>$, and $<$).
3. Entity-based Aggregate Class. An example is the minimum query: a set of tuples (O_i,p_i) are returned, where p_i is the non-zero probability that $O_i.a$ is the minimum among all items in the database. Other examples are maximum and nearest-neighbor queries.
4. Value-based Aggregate Class. Queries for this class include any aggregate operators (e.g., addition, subtraction, multiplication, division, minimum value, SUM, AVG) that involves two or more values, and produce a single uncertain value. A SUM query, for example, yields a bound $[l,u] \in \Re$ and $\{p(x) \mid x \in [l,u]\}$, where X is a random variable for the sum of values of a for all objects in the database, and $p(x)$ is a *pdf* of X such that $\int_l^u p(x)dx = 1$.

9.3.3 Evaluation of Probabilistic Queries

Let us now investigate how the imprecise answers for probabilistic queries can be investigated. We have selected three representative queries: range query, minimum/maximum query and SUM/AVG query for detailed discussions. A comprehensive presentation of efficient evaluation algorithms for different query classes is described in (Cheng 2003, Cheng 2005).

(a) Probabilistic Range Query. Given a closed interval $[l,u]$, where $l,u \in \Re$ and $l \leq u$, a Range Query returns a set of tuples (O_i,p_i), where p_i is the non-zero probability that $O_i.a \in [l,u]$.

Using the classification scheme, a range query is a entity-based non-aggregate query. The query can be used in sensor networks and location monitoring. For example, one may ask which sensor(s) return temperature values that are within a user-specified interval.

To evaluate a probabilistic range query, each object's p_i is computed as follows: first, the amount of overlap OI between the uncertainty interval of O_i and $[l,u]$ is found (i.e., $OI = O_i.a.U \cap [l,u]$). If OI has zero width, we are assured that $O_i.a$ does not lie in $[l,u]$, and thus $p_i = 0$ and O_i will not be included in the result. Otherwise, we calculate the probability that i:a is inside $[l,u]$ by integrating $O_i.a.f(x)$ over OI (i.e., $p_i = \int_{OI} O_i.a.f(x)dx$), and return the result (O_i,p_i) if $p_i \neq 0$.

Next, we illustrate the evaluation of *aggregate queries* i.e., those that generate answers based on an aggregate function on $O_1,O_2,\ldots, O_n$.

(b) Probabilistic Minimum/Maximum Query. This query returns the objects that contains the minimum (maximum) value among objects $O_1,O_2,\ldots, O_n$, together with their probability values. This query is a entity-based aggregate query. This kind of query has wide application in aggregating information in data streams sensor networks. For example, one may want to acquire the identity of the sensor that yields the maximum temperature value over a region being monitored by sensors (Desphande 2004). Notice that the condition $\sum_{O_i \in R} p_i = 1$ holds.

To evaluate this query, consider two data values O_1 and O_2. The probability of O_1 being the maximum data object, i.e. the value of O_1 is larger than that of O_2, is:

$$p_1 = \int_{-\infty}^{+\infty} f_1(s) \cdot \left(\int_{-\infty}^{s} f_2(t)dt \right) ds \tag{9.1}$$

where $f_1(s)$ and $f_2(t)$ are the probability density functions for O_1 and O_2 respectively. Furthermore, if we suppose the sampled data values are independent random variables satisfying the Gaussian distribution (which is often true for modeling measurement uncertainty of sensor instruments), the calculation of p_1 can be simplified to:

$$p_1 = P\{O_1 - O_2 \geq 0\} = \int_{0}^{+\infty} g_{1,2}(s)ds \tag{9.2}$$

where $g_{1,2}(s)$ is the probability density function for $O_1 - O_2$ which also satisfies Gaussian distribution. It can be seen from Fig. 9.1 that the variance decreases with increases in p_1. It is consistent with the fact that O_1 is more likely to be the maximum. For the case of multiple data values, suppose the size of the calculation set is N, the probability of O_i being the maximum is:

$$p_i = \int_{-\infty}^{+\infty} f_i(s) \cdot \left(\prod_{j=1 \wedge j \neq i}^{N} \int_{-\infty}^{s} f_j(t)dt \right) ds \tag{9.3}$$

where $f_i(s)=\frac{1}{\sqrt{2\pi}\sigma_i}e^{\frac{(s-\mu_i)^2}{2\sigma_i^2}}$ and $f_j(t)=\frac{1}{\sqrt{2\pi}\sigma_j}e^{\frac{(t-\mu_j)^2}{2\sigma_j^2}}$

Similarly, the calculation could be simplified given the assumption that the data values are pairwise independent:

$$p_i=\prod_{j=1}^{n}P\{O_i-O_j\geq 0\ \ (j\neq i)\}=\prod_{j=1}^{n}\int_0^{+\infty}g_{i,j}(s)ds$$

$$where\ \ g_{i,j}(s)=\frac{1}{\sqrt{2\pi(\sigma_i^2+\sigma_j^2)}}e^{\frac{(s-(\mu_i-\mu_j))^2}{2(\sigma_i^2+\sigma_j^2)}} \tag{9.4}$$

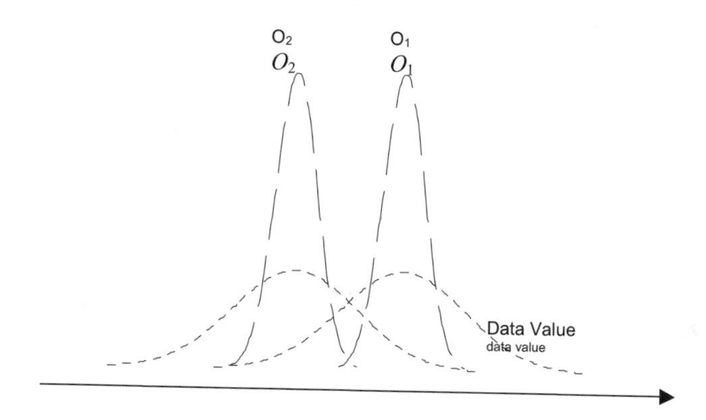

Fig. 9.1. MAX and MIN Queries

We remark that the above algorithm does not take into account that in some scenarios the uncertainty intervals are bounded (Definition 1). In those cases the integration bounds have to be modified accordingly. In (Cheng, 2003), we have also investigated how to exploit the ordering of those bounds to compute probabilistic queries more efficiently.

(c) SUM/AVG queries. The final query described in this section, an AVG/SUM query returns the probability density function of the average value (sum) of objects. One typical example of this query to inquire about the average pressure value among all the sensors in a monitoring sensor network (Desphande 2004). It return answers in the form of a probabilistic distribution $p(x)$ in an interval $[l,u]$, such that $\int_l^u p(x)dx=1$. We also remark that this is a value-based aggregate query. The pdf of the resultant SUM is the convolution of the pdf of the addends. The convolution operator ⊗ is defined $(f \otimes g)(z) := \int_{-\infty}^{+\infty} f(s)g(z-s)ds$. Then the pdf of the sum of the values of $O_{1_1}, O_2, \ldots, O_n$, whose pdfs are $f_1(t), f_2(t), \ldots, f_n(t)$, is $f_S = f_1 \otimes f_2 \otimes \cdots \otimes f_n$. The pdf of the average is simply the result of the SUM scaled by the reciprocal of the number of addends. If all the pdfs under consideration is normal, the sum is a parameterized function with $\sigma_1, \sigma_2, \ldots, \sigma_n$ (the variance of $f_i(t)$) as parameters. Furthermore, according to probability theory, the sum of a finite number of normally-distributed random variables still follows a normal distribution. More specifically, for our case, the sum follows a normal distribution

$$N(\sum_{j=1}^{n} \mu_j, \sum_{j=1}^{n} \sigma_j^2) \tag{9.5}$$

This is an important result, which we will use extensively in the discussion of sensor selection policies. We are now ready to describe the quality metrics for measuring the degree of imprecision of these queries.

9.4 Query Quality Metrics

We now discuss how quality of the results is measured for probabilistic queries. The query quality reflects the degree of ambiguity of answers due to data imprecision. For each type of probabilistic query, we present different quality metrics according to (Cheng, 2003) and (Lam 2004).We identify two major types of metrics for measuring query quality:

- Probability-based quality metric: this metric is based on the user's specification on the minimum value P for the maximum probability or variance of the answer.
- Entropy-based quality metric: this class of metric measures the answer uncertainty using the concept of entropy (Shannon, 1949).

We will discuss how these two kinds of quality metrics are defined for the probabilistic queries that we have described. As illustrated next, entropy-based metrics make use of information about the query answer and are thus more informative than probability-based metrics.

9.4.1 Entity-based Non-aggregate Queries

For queries that belong to the entity-based non-aggregate query class, it is necessary to define the quality metric for each (T_i, p_i) individually, independent of other tuples in the result. This is because whether an object satisfies the query is independent of the presence of other objects. We illustrate this point by explaining how the metric of *range query* is defined.

(a) Probability-based Metric: For a range query with query range $[l,u]$, the result is the best if we are sure either an uncertain value is completely inside or outside $[l,u]$. Uncertainty arises when we are less than 100% sure whether the value is inside $[l,u]$. We are confident that O_i's value is inside $[l,u]$ if a large part of its uncertainty U_i overlaps $[l,u]$ i.e., p_i is large. Likewise, we are also confident that the value is outside $[l,u]$ if only a very small portion of Ui overlaps $[l,u]$ i.e., p_i is small. The worst case happens when p_i is 0.5, where we cannot tell if O_i satisfies the range query or not. Hence a reasonable metric for the quality of p_i is:

$$\frac{|p_i - 0.5|}{0.5} \tag{9.6}$$

In Eq. 9.1 we measure the difference between p_i and 0.5. Its highest value, which equals 1, is obtained when p_i equals 0 or 1, and its lowest value, which equals 0, occurs when p_i equals 0.5. Hence the value of Eq.9.6 varies between 0 and 1, and a large value represents good quality. Let us now define the probability-based quality of a range query: Probability-based quality of a range query becomes:

$$\frac{1}{|R|}\sum_{T_i \in R} \frac{|p_i - 0.5|}{0.5} \tag{9.7}$$

where R is the set of tuples (T_i, p_i) returned by the range query. Essentially, Eq.9.7 evaluates the average over all tuples in R. Notice that in defining the metric of the range query, Eq.9.7 is defined for each O_i, disregarding other objects. In general, to define quality metrics for the entity-based non-aggregate query class, we can define the quality of each object individually. The overall score can then be obtained by averaging the quality value for each object.

(b) Entropy-based Metric. The second quality metric for range query makes use of the uncertainty of the query answer in terms of *entropy*. Before we describe the metric, it is useful to have a brief review of entropy (Shannon, 1949):

Let $X_1,\ldots,X_n$ be all possible messages, with respective non-zero probabilities $p(X_1),\ldots, p(X_n)$ such that $\sum_{i=1}^{n} p(X_i) = 1$. The entropy of a message $X \in \{X_1,\ldots,X_n\}$ is:

$$H(X) = \sum_{i=1}^{n} p(X_i)\log_2 \frac{1}{p(X_i)} \tag{9.8}$$

The entropy, $H(X)$, measures the average number of bits required to encode X, or the amount of information carried in X. If $H(X)$ equals 0, there exists some i

such that $p(X_i) = 1$, and we are certain that X_i is the message, and there is no uncertainty associated with X. On the other hand, $H(X)$ attains the maximum value ($\log_2 n$) when all the messages are equally likely. For each object O_i that satisfies a range query with a non-zero probability p_i, it has a probability $q_i = 1-p_i$ of failing it. We can then consider the set R_i consisting of two events for which O_i satisfies and fails the query as two *messages*, such that $p_i + q_i = 1$. The *entropy* of R_i, or $H(R_i)$, is equal to:

$$p_i \log_2 \frac{1}{p_i} + q_i \log_2 \frac{1}{q_i} \tag{9.9}$$

The above equation is equivalent to $-(p_i \log_2 p_i + q_i \log_2 q_i)$. Similar to the probability-based query, we can also define the entropy-based score of the range query, which is the average over the answer set R of the entropy for each individual object in R, as follows. Th entropy-based quality of a range query becomes:

$$-\frac{1}{|R|} \sum_{T_i \in R} (p_i \log_2 p_i + q_i \log_2 q_i) \tag{9.10}$$

Notice that the validity of Eq.9.10 is based on the assumption that the effect of each object on the query answer is independent of each other. Next, we investigate how quality metrics are defined for entity-based aggregate queries.

9.4.2 Entity-based Aggregate Queries

Contrary to an entity-based non-aggregate query, we observe that for an entity-based aggregate query, whether an object appears in the result depends on the existence of other objects. Therefore we cannot just define the quality of an object and then compute the average of all the objects that satisfy the query.

(a) Probability-based Metric. One simple way to define the quality of this type of query is to use the maximum of the probability values of all the objects that satisfy the query. Consider, for example, the result of a maximum query:

$$\{(O_1,0.7), (O_2,0.2), (O_3,0.1)\} \tag{9.11}$$

The quality is then equal to 0.7, the maximum of the probabilities for the answer above. This metric is based on the intuition that the user usually only accepts answers that are higher than a certain threshold. We will use this definition to illustrate our sensor selection scheme. Although this metric is easy to understand and use, it does not make use of all the probability information. For example, consider the following two sets of answers to a minimum query: $\{(O_i,0.6), (O_2,0.4)\}$ and $\{(O_1,0.6), (O_2,0.3), (O_3,0.1)\}$. The quality of both answer sets is 0.6, and we cannot tell which answer is better. Next, we illustrate how entropy can be used to quantify this difference.

(b) Entropy-based Metric. Recall that the result to the queries we defined in this class is returned in a set *R* consisting of tuples (T_i, p_i). We can view *R* as a set of messages, each of which has a probability p_i. Moreover, the property that $\sum_{i=1}^{n} p_i = 1$ holds. Then, the entropy-based quality of entity-based aggregate query becomes, $-H(R)$. Notice that $H(R)$ measures the uncertainty of the answer to these queries; the lower the value of $H(R)$, the less uncertainty is associated with *R*, which produces a higher value. In particular, if $H(R)$ is zero, then the maximum quality value is 0. Compared with the probability-based metric, this metric makes use of *all* the probability information for all the objects, and is thus more informative than the former. We describe probing policies in the next section to illustrate the use of this metric.

9.4.3 Value-based Queries

Probability-based Metric. The first metric uses the *variance* of the probability distribution function of the query answer. Naturally, the higher the variance, the more uncertain is the answer. The metric is simply defined as probability-based quality of value-based query, which is:

$$-1 * \text{Variance of Query Answer} \tag{9.12}$$

Entropy-based Metric. The second metric uses the concept of entropy again. It assumes that the result of a value-based query is in the form $(l, u, \{p(x) \mid x \in [l, u]\})$, i.e., a probability distribution of values in interval $[l, u]$. To measure the quality of such queries, we can use the concept of differential entropy, defined as follows:

$$\widehat{H}(X) = -\int_{l}^{u} p(x) \log_2(x) dx \tag{9.13}$$

where $\widehat{H}(X)$ is the differential entropy of a continuous random variable *X* with probability density function $p(x)$ defined in the interval $[l, u]$ (Shannon, 1949). Similar to the notion of entropy, $\widehat{H}(X)$ measures the uncertainty associated with the value of *X*. Moreover, $\widehat{H}(X)$ attains the maximum value, $\log_2(u-l)$ when *X* is uniformly distributed in $[l, u]$. When $u-l = 1$, $\widehat{H}(X) = 0$. Therefore, if a random variable has more uncertainty than a uniform distribution in [0,1], it will have a positive entropy value; otherwise, it will have a negative entropy value. We use differential entropy to measure the quality of value-based queries. Specifically, we apply Eq.9.13 to $p(x)$ to measure the uncertainty inherent to the answer. The lower the differential entropy value, the better quality is the answer. In particular, if there is a value *y* in $[l, u]$ such that the value of $p(y)$ is high, then the entropy will be low. We can now define the entropy-based quality metric of a probabilistic value-based query:

$$\text{Score of a Value-Based Query} = -\widehat{H}(X) \tag{9.14}$$

The quality of a value-based query can thus be measured by the uncertainty associated with its result: the lower the uncertainty, the higher score can be obtained as indicated by Eq. 9.14. We conclude this section by summarising all the quality metrics that we have presented. Table x.1 lists the quality metrics defined for each type of query. In the rest of this chapter, we describe probing and sensor selection policies for satisfying these quality metrics.

Table x.1. Quality Metrics of Probabilistic Queries

Probabilistic Query Class	**Probability-based Metric**	**Entropy-based Metric**
Entity-based non-aggregate	$\frac{1}{\lvert R \rvert}\sum_{T_i \in R}\frac{\lvert p_i - 0.5 \rvert}{0.5}$	$-\frac{1}{\lvert R \rvert}\sum_{T_i \in R}(p_i \log_2 p_i + q_i \log_2 q_i)$
Entity-based aggregate	$\max(p_i \mid T_i \in R)$	$\sum_{T_i \in R} p_i \log_2 \frac{1}{p_i}$
Value-based	-1 * Variance of Answer	$\int_l^u p(x)\log_2(x)dx$

9.5 Query-centric Probing Policies

In this section, we discuss several update policies that can be used to improve the entropy-based quality of probabilistic queries defined in the last section. We assume that the sensors cooperate with the central server i.e., a sensor can respond to update requests from the sensor by sending the newest value to the server, as in the system model described in the literature (Olston, 2002). As explained, one major source of data uncertainty arises from the fact that the sensor data, while continuously evolving, is only sampled at some discrete point of time. Once a data value is received at the server, the uncertainty interval will grow with time, until another update is applied. Hence, the more frequently the data is sampled, the lower the uncertainty for each data item.

This results in a smaller effect on query answer correctness. In other words, one simple way for the server to improve the query quality is by requesting updates from sensors, so that the uncertainty intervals of some sensor data are reduced, resulting in an improvement of the answer quality. Ideally, a system can demand updates from all sensors involved in the query; however, this is not practical in a limited-bandwidth environment. The issue is, therefore, to improve the quality with as few updates as possible. Here we propose a number of update policies for some query types we identified earlier.

- *Improving the Quality of Range Query.* The policy for choosing objects to update for a range query is simple: choose the object with the minimum entropy, with an attempt to improve the overall quality of the range query.
- *Improving the Quality of Other Queries.* There are two classes of policies. The first type of policies, called global update policy, focus on the freshness of the whole database:
- *Glb_RR.* In this policy, the server picks a sensor to update in a round-robin fashion. The policy ensures that each item gets a fair chance of being refreshed.

The second batches of policies, called query-centric policies, refresh data that are only required by the queries active in the system. The following policies are examples in this group:

- **Loc_RR**. This policy is basically the same as Glb_RR except that only objects that are relevant to the query are involved. As the set of relevant object changes, it needs to tell the irrelevant objects to stop sending their updates, and inform the new relevant objects to report.
- **MinMin**. This policy is specially designed for minimum queries, where an object with the lowest lower bound of the uncertainty interval among the objects interest to query is chosen for update. The rationale of choosing this object is that this object possibly carries the minimum value; refreshing this object has a better chance of separating it from other candidate objects, thereby rendering a higher quality output. A symmetric policy (**MaxMax**) can be defined for maximum queries where the candidate object with the highest upper bound is chosen.
- **MaxUnc**. This heuristic simply chooses the uncertainty interval with the maximum width to update, with an attempt to reduce the overlapping of the uncertainty intervals.
- **MinExpEntropy**. The final heuristic is to check, for all the objects whose uncertainty intervals overlap, the effect to the overall entropy if we choose to update the value of an object. Recall that once the object is updated, its uncertainty interval will shrink to a single value. The new uncertainty is then a point in the uncertainty interval before the update is applied. For each value in the uncertainty interval before the update, we evaluate the entropy, assuming that the uncertainty shrinks to that value after the update. The mean of these entropy values is then computed. The object that yields the minimum expected entropy is then selected for probing.

In (Cheng 2005), we have conducted experimental simulations and investigated the performance of these few techniques. Generally, for Minimum and Maximum queries, MinMin and MaxMax perform the best. We have also studied the tradeoff between the precision of MinExpEntropy and the execution time. Readers are referred to (Cheng 2005) for more details.

9.6 Sensor Selection Techniques

An important issue with sensors is that their readings are noisy and error-prone (Niculescu 2004). This translates to a large data uncertainty and can have large impact to the quality of queries. In order to alleviate this problem, we have investigated the possibility of obtaining a more accurate sensor reading (and better query quality) by getting the average of multiple low-cost sensors that monitor the same region. The major problem of this approach is that the limited resource of a sensor network (e.g., network bandwidth and battery power) can be exhausted easily. Moreover, since some sensors may not work properly, they may generate abnormal readings that skew the average value. In the rest of this chapter, we will investigate two key issues that need to be solved: (i) What is the smallest number of sensors that can achieve a guaranteed query quality?, and (ii) Which sensor should be selected?

9.6.1 System Model

In order to understand our method, we briefly describe the underlying system model and the query model. The wireless sensor system model consists of a *base station* (BS) and a collection of *sensor nodes*. It is assumed that the system environment is divided into a number of *regions*, each of which consists of a node with high computational capability, called the *coordinator node* that manages nodes in the same region. The base station is responsible for communication between the coordinator nodes and the users of the system, i.e., transmitting PQs to the coordinator nodes and returning results to the users. The base station communicates with the coordinator nodes through a low bandwidth wireless network and may require the relay of other sensor nodes and coordinator nodes. We assume that the base station knows the distribution and connections of the coordinator nodes and which sensor node monitors which region. Fig. 9.2 illustrates the overall system architecture. A two-level architecture is shown, but this processing model can be applied to a multiple-level architecture and the regions managed by different coordinators can overlap.

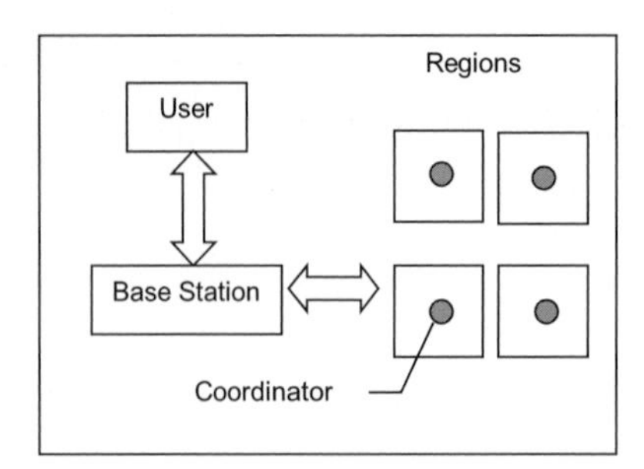

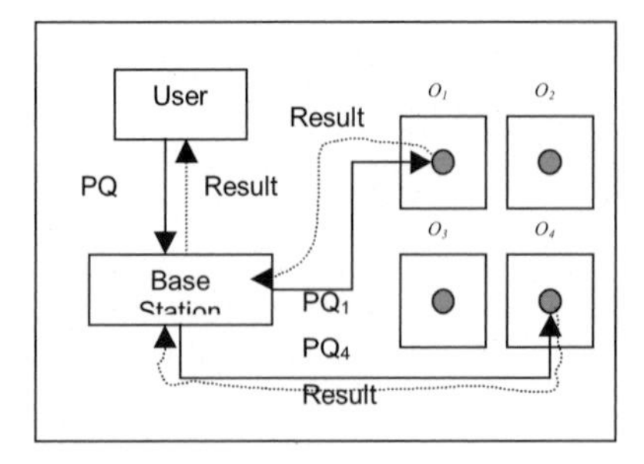

Fig. 9.2. (a) System architecture, (b) Probabilistic sub-queries

9.6.2 Probabilistic Sub-queries

A probabilistic query (PQ) is submitted by a user to the sensor network system, executed over a set R of n objects $O_1, O_2, \ldots, O_n$. Here we define *probabilistic sub-query*, PQ_i, as follows:

Definition 1: A Probabilistic Sub-query i, denoted as PQ_i, is a sub-query of PQ, which accesses item O_i in the list of objects specified by PQ. It returns to the PQ a Gaussian distribution (u_i, σ_i) of O_i.

When the base station receives a PQ, it determines the set of data items required by the PQ according to the required regions of the query and which coordinator nodes are responsible for generating the required data items. The base station then breaks down the PQ into sub-queries $\{PQ_1, PQ_2, \ldots, PQ_n\}$. Each subquery PQ_i is then sent to the coordinator node, which is responsible for reading O_i and generating a Gaussian distribution for the reading of O_i to describe the distribution of O_i. Notice that the Gaussian distribution has been used for modelling measurement error and noisy data (Wolfson 1999, Pfoser 2001). Each coordinator sends its results back to the base station, which then computes the final result and sends it back to the user. Fig. 9.2(b) illustrates an example of a PQ executing on objects O_1 and O_4. The PQ submitted by the user is broken down into two subqueries, PQ_1 and PQ_4, which access regions O_1 and O_4 respectively. The results from coordinators for O_1 and O_4 are sent to the base station, which subsequently returns the result to the user.

9.6.3 Sensor Data and the Role of the Coordinator Node

We assume that the sampled values from a sensor follow a *continuous* function of the system environment. Once a sensor data value is generated, it is associated with two time-stamps, called the lower time stamp (*lts, the current sampling time*) and upper time stamp (*uts, the next sampling time*), to indicate its validity interval. A data value is invalid at the sampling time of the next value. The time-stamps can be used for ensuring temporal consistency in query execution (Lam 2004, Sharaf 2003) such that all the data values are valid at the same time interval. The synchronisation of the clocks at different sensor nodes for generating the time-stamps can be done by the coordinator node of the region. Note that the generated sensor data values from a sensor node may contain error due to noises from the surrounding environment. The sensor nodes in a region are handled by a coordinator node. As shown in Fig. 9.3 the coordinator node collects readings from individual sensors, computes the average value of these sensors and the statistical information (e.g., Gaussian distribution) of these sensor values, and passes them to the base station for generating query results. Note also that *not* all the sensors are involved in sending their values to the coordinator. This is because reading data values from all sensors can result in a heavy network load. The issues of sensor selection will be discussed in more detail in the next section.

The processing of a PQ is divided into two levels: (1) Aggregation of multiple sensors responsible for generating a value for a data item required by the query at the coordinator, and (2) Aggregation of the data items required by the query to generate the query results to be performed at the base station. The aggregation at the coordinator node will follow the time-stamps of the data values of different sensor data streams such that they are *relatively consistent* with each other as shown in Fig. 9.3 (Lam 2004). Although the sampling periods of the sensor nodes within the same region may be the same (i.e., the sensor values have the same validity period length), the sensor data values from different streams may not be synchronized with each other since they may be activated at different time points as shown in Fig. 9.3 in which an aggregate function, *max*, is performed on two sensor streams x_i and x_k. The results are obtained from the maximum of the two streams for the same time points, and they are presented as a function of time.

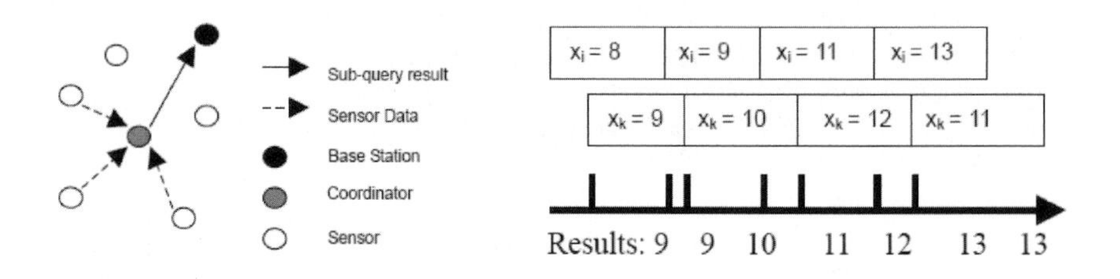

Fig. 9.3. (a) Coordinator Model, (b) Accessing sensor data value with relative consistency requirement

9.6.4 Sensor Selection Framework

Accessing more sensor nodes can improve the reliability of query results, at the expense of an increased aggregation workload. Our goal is to meet the quality requirement of a query using the minimum number of sensor nodes for generating the value of a data item required by a query. Specifically, we want to determine the set of sensor nodes to participate in sampling for aggregation of the values at the time when the reading from a certain region is required. Since different types of probabilistic queries have various forms, the derivation of the probabilistic query result, in the form of probability density function (*pdf*), can be different. This in turn affects how many regions and how many sensors are selected to report data. Fig. 9.4 summarizes our approach. In the figure Seps 1 to 3 initialize the environment for the sensor selection algorithm. Step 4 is the main part of the algorithm. It consists of a conditional loop, which only stops when he quality requirement of the PQ is met. In Step 4, the quality of the PQ is checked, the maximum allowed variance for each region is derived, and the optimum number of sensors selected is computed. Finally, in Step 4(d) these sensors are used to re-compute the corresponding probabilistic sub-queries and quality value. In the rest of this section, we will investigate some key steps. In Fig. 9.4 sensor statistics initialisation, maximum allowed variance derivation and sensor selection. For simplicity, the algorithms described here utilize probability-based quality metrics discussed.

However, the framework in Fig. 9.4. can also be used to satisfy entropy-based quality metrics.

1. Let R be the set of regions (objects) involved in a PQ with a quality requirement.
2. Let M be the maximum number of times that PQ is re-evaluated.
3. The coordinator of each region computes the region's statistic properties.
4. **while** (the base station decides that the quality requirement of the PQ is not satisfied)
   (a) For each region in R, BS derives the maximum allowed variance (MAV) of the sample data for meeting the query quality requirement.
   (b) BS sends the information about MAV to each region in R.
   (c) The coordinator of each region in R computes the sample size and the set of sensor nodes for sending the data.
   (d) Each selected sensor sends its updated value to the coordinators, which then re-evaluate the probabilistic sub-query and sends the result back to the base station.
5. Return the computed result of the PQ to the user.

Fig. 9.4. Outline of the sensor node selection algorithm for PQ

9.6.5 Computing the Region's Initial Statistic Properties

In Step 3 of Fig. 9.4 the initial statistical properties of each region, including the initial expected value and the estimated population variance, are computed by the coordinator. Specifically, for each region O_i required by a sub-query PQ_i, the coordinator node identifies the set of sensor nodes S_i that are responsible for generating values for O_i. Then it sends out data request messages to all these sensor nodes. Each sensor responds to the request message by returning its latest sampled data value of O_i to the coordinator. The received sensor data values from each sensor node are first buffered and the mean values are calculated by the coordinator after a pre-determined waiting time has expired. To improve accuracy in calculating the mean value, each node may send multiple sensor values which cover the period from (*current time - some time interval*) to *current time*. Then, the mean value calculation is performed in two steps: (1) the mean value for the set of data values from a sensor node, and (2) the mean value over all sensor nodes. If the variance of the values from a sensor node is too high, (e.g., higher than a predefined threshold), it is assumed that the node is either currently located at a high-noise environment or it is currently in an abnormal state. The node will be marked as abnormal and the coordinator will not consider the node for further processing. Based on the variances of the values from all the sensor nodes selected, σ_i, the population variance for the region O_i can be estimated as the average of the variances. The uncertain value (pdf) represented by each region is then shipped to the base station for computing the probabilistic query result.

We now briefly describe how the base station, based on the distribution of the probabilistic results from the regions selected and the probabilistic requirements of the queries, derives the variance of the sampled data being queried for different types of probabilistic queries. This corresponds to Step 4(a) of the selection algorithm. The impact of errors in sampled data values on the query result depends on the query type. Here we consider two types of aggregate queries: (1) MAX and MIN, (3) AVG and SUM. The data being queried are aggregated from multiple sensors from the same regions by the coordinator nodes. Thus it is reasonable to assume that the data values follow normal distributions with specific means and variances. In essence, the sub-query PQ_i executed at each coordinator returns a normal distribution of its sensor reading to the base station. We now describe how the maximum variance allowed for each region is derived.

9.6.6 MAX and MIN Queries

We have already examined how the probabilistic results of MAX and MIN queries are computed. One important observation about MAX and MIN queries is that as the variance of the sampled data values decreases, the conclusion about the maximum and the minimum value is less ambiguous. It can be seen from the Fig. 9.1 that the variance decreases with an increase in p_1 (that is, the probability that O_1 satisfies the query). It is consistent with the fact that O_1 is more likely to be the maximum. The algorithm in Fig. 9.5 finds the maximum allowed variance for each region to satisfy the probability-based quality metric: the highest probability of the region holding the maximum or minimum value is larger than P_T. As shown in Fig. 9.5, Step 1 finds the highest probability value P of all the objects satisfying the MAX or MIN query. Step 2 then searches for the sensor (k_{max}) whose variation has the highest impact on P. For this sensor, we reduce its variance by $\Delta\sigma$ (Step 3), in order to increase the chance that P will be higher than P_T when the PQ is re-evaluated again (Step 4). The final maximum allowed variance value for each object (such that $P > P_T$) is returned in step 5.

1. **if** (query type is MAX) **then**

$$P(\sigma_1,\sigma_2,...,\sigma_n) = \max_{i=1,...,n}\left(\int_{-\infty}^{+\infty} f_i(s)\cdot\left(\prod_{j=1\wedge j\neq i}^{N} \int_{-\infty}^{s} f_j(t)dt\right)ds\right)$$

   if (query type is MIN) **then**

$$P(\sigma_1,\sigma_2,...,\sigma_n) = \max_{i=1,...,n}\left(\int_{-\infty}^{+\infty} f_i(s)\cdot\left(\prod_{j=1\wedge j\neq i}^{N} (1-\int_{-\infty}^{s} f_j(t)dt)\right)ds\right)$$

2. Find k_{max}, such that $\frac{\partial}{\partial\sigma_{k\max}}P(\sigma_1,\sigma_2,...,\sigma_n) = \max_{k=1,...,n}\left(\frac{\partial}{\partial\sigma_k}P(\sigma_1,\sigma_2,...,\sigma_n)\right)$ i.e., the sensor that has the greatest impact on P.
3. Adjust variance requirement of the k_{max} sensor (i.e., O_{kmax}), as $\sigma_{k_{max}} = \sigma_{k_{max}} - \Delta\sigma$
4. Repeat 1 to 3 until $P(\sigma_1,\sigma_2,...,\sigma_n) \geq P_T$
5. Return $\sigma_1,\sigma_2,...,\sigma_n$

Fig. 9.5. Finding the maximum allowed variance for MAX and MIN queries

9.6.7 SUM and AVG Queries

We have discussed how the probabilistic results of SUM and AVG query can be found. Notice that since we have assumed the data uncertainty follows Gaussian distribution, the sum should also follow a normal distribution, i.e.,

$$N(\sum_{j=1}^{n} \mu_j, \sum_{j=1}^{n} \sigma_j^2) \tag{9.15}$$

Hence, our task is to find out $\sigma_1, \sigma_2, \ldots, \sigma_n$ such that the probability-based quality metric is met, i.e., σ_s, given below, is smaller than a given threshold σ_T. The AVG query is just the result of SUM divided by m.

$$\sigma_S = (\sum_{j=1}^{n} \sigma_j^2)^{1/2} \tag{9.16}$$

Thus, it follows the same rationale except that its distribution is given by:

$$N(\frac{1}{n}\sum_{j=1}^{n} \mu_j, \frac{1}{n^2}\sum_{j=1}^{n} \sigma_j^2) \tag{9.17}$$

Fig. 9.6 illustrates how the MAV for each region is found for SUM and AVG queries.

---

1. **Find k_{max}, the sensor that has the greatest impact on σ_S:**
   if (query type is SUM) then $\frac{\partial}{\partial \sigma_{k_{max}}}\sigma_S = \max_{k=1,\ldots,n}(\frac{\partial}{\partial \sigma_k}\sigma_S) = \max_{k=1,\ldots,n}(\sigma_k(\sum_{j=1}^{n}\sigma_j^2)^{-1/2})$
   if (query type is AVG) $\frac{\partial}{\partial \sigma_{k_{max}}}\sigma_s = \max_{k=1,\ldots,n}(\frac{\partial}{\partial \sigma_k}\sigma_s) = \max_{k=1,\ldots,n}(\frac{\sigma_k}{n}(\sum_{j=1}^{n}\sigma_j^2)^{-1/2})$
2. **Adjust variance requirement of the k_{max}th sensor:** $\sigma_{k_{max}} = \sigma_{k_{max}} - \Delta\sigma$
3. **Repeat 2 to 4 until $\sigma_S \le \sigma_T$**
4. **Return $\sigma_1, \sigma_2, \ldots, \sigma_n$**

Fig. 9.6. Algorithm for determining the Maximum Allowed Variance in AVG and SUM queries

As we can see, Step 1 finds the sensor k_{max} such that it has the maximum impact on the deviation of SUM or AVG. Then the variance of that sensor is reduced, and this is repeated until the quality requirement, i.e., $\sigma_S \le \sigma_T$, is reached (Step 3). The resulting variance values for the objects are returned in Step 4.

9.7 Determining Sample Size and the Set of Sensors

Based on the information about the variance values transmitted from the base station, the coordinator node in each region is now ready to determine the sample size and the set of sensor nodes to be sampled in order to meet the confidence requirement of data being queried. Notice that this corresponds to Step 4(c) of Fig. 9.4. We first determine the sample size, i.e. how many sensors need to be included in the aggregation process to get the average value. Suppose the sample size is n_s and the approximate mean value $\overline{S} = \frac{1}{n_s}\sum_{i=1}^{n_s} s_{k_i}$, where $1 \leq k_i \leq n$ and $k_i \neq k_j$ for all $i \neq j$. If all s_{k_i} 's follow an identical distribution $N(\mu,\sigma^2)$, then $\overline{S}$ must follow the normal distribution $N(\mu,\sigma^2/n_s)$, where μ is the expected value and σ is the region's estimated population variance calculated. To satisfy the accuracy requirement, we need to choose a value of n_s that satisfies the constraint $\sigma/\sqrt{n_s} \leq \sigma_i$ (where σ_i is the MAV for object O_i computed in the last section). Hence, the sample size n_s is given by $\lceil \sigma^2/\sigma_i^2 \rceil$. Next, we determine the set of sensor nodes to be sampled. We have shown how to calculate the mean value for each region, by using the data from all the sensors residing in the region. For each sensor, we calculate the difference d_i between a sensor data item s_i and the expected value for the region, and use it as the selection criterion for sensor selection: $d_i = s_i - E(s)$. We sort the sensors in ascending order of d_i. At each sampling time, with a certain variance σ_i, the coordinator will calculate the sampling size n_s and select the top n_s sensors to sample. Since the selected sensor may be erroneous during the sampling period, at each sampling time, when the coordinator collects all the possible sensor data with a small delay (to handle the disconnection case), it will calculate the value of d_i again to check whether a sensor's value exceeds the expectation a pre-fixed threshold. If the threshold is exceeded, we assume that the sensor is in error at the sampling time, and the coordinator will send a new request for information to other sensor candidates in the sorted list and it will also re-sort the sensor list as well. Once the sensors are selected, they will be resampled again and submitted to re-evaluate the probabilistic query (Fig. 9.4, Step 4(d)), and the process will go on until the desired quality is reached. We have conducted experiments and they illustrate that our sensor selection scheme can reduce the use of network bandwidth while maintaining quality. Interested readers can refer to our previous paper (Han 2005) for more details.

9.8 Conclusions

In this chapter we investigated the problem of achieving high quality probabilistic results over uncertain data in sensor networks. We discussed two methods, one of which tackles the problem of sampling uncertainty by probing appropriate sensors, and the other one exploits the abundance of low-cost sensors. Both methods improve the reliability in sensor readings, and are aware of the limited resources of a sensor network. We also study how the two methods can be applied to some of the probabilistic queries that we have described. With prices of sensors dropping continuously, we expect that more applications will deploy large sensor networks for monitoring purposes. Managing uncertainty in such a resource-limited system in order to attain high-quality results will be an important issue. One particularly interesting extension of the work described in this chapter is to study how entropy-based quality metrics can be used to aid sensor-selection.

9.9 References

Cheng R, Kalashnikov D, and Prabhakar S (2005) Evaluation of probabilistic queries over imprecise data in constantly-evolving environments. Accepted in Information Systems, Elsevier

Cheng R, Kalashnikov D, and Prabhakar S (2004) Querying imprecise data in moving object environments. In IEEE Trans. on Knowledge and Data Engineering (IEEE TKDE), 16:9, pp. 1112-1127, Sept.

Cheng R, Kalashnikov D, and Prabhakar S (2003) Evaluating probabilistic queries over imprecise data. In Proc. of the ACM SIGMOD Intl. Conf. on Management of Data, June 2003.

Cheng R, Prabhakar S, and Kalashnikov D (2003) Querying imprecise data in moving object environments. In: Proc. of the Intl. Conf. on Data Engineering (IEEE ICDE 2003), pp. 723-725, Bangalore

Deshpande A, Guestrin C, Madden S, Hellerstein J, and Hong W (2004) Model-driven data acquisition in sensor networks. In: Proc. of the 30th VLDB Conference, Toronto, Canada

Elnahrawy E and Nath B (2003) Cleaning and querying noisy sensors. In: ACM WSNA'03, Sept. 2003, San Diego

Ertin E, Fisher J, and Potter L (2003). Maximum mutual information principle for dynamic sensor query problems. In: Proc. IPSN'03, Palo Alto, April

Dubois-Ferriere H, and Estrin D (2004) Efficient and practical query scooping in sensor networks. CENS Technical Report #39, April

Han S, Chan E, Cheng R, and Lam K (2005) A statistics-based sensor selection scheme for continuous probabilistic queries in sensor networks. Accepted in the 11th IEEE Intl. Conf. on Embedded and Real-Time Computing Systems and Applications (IEEE RTCSA 2005), Hong Kong, Aug.

Madden S, Franklin M, Hellerstein J, and Hong W (2002) Tiny aggregate queries in ad-hoc sensor networks. In: Proc. of the 5th Symp.on Operating Systems Design and Implementation (OSDI), Boston, USA

Kumar V (2003) Sensor: The atomic computing particle. In SIGMOD Record, 32:4, Dec.

Krishnamachari B and Iyengar S (2004) Distributed Bayesian algorithms for fault-tolerant event region detection in WSN, IEEE Trans. on Computers, 53:3, pp. 241-250, March

Lam K, Cheng R, Liang B, and Chau J (2004) Sensor node selection for execution of continuous probabilistic queries in WSN. In: Proc. of ACM 2nd Int. Workshop on Video Surveillance and Sensor Networks, Oct. New York

Lam K and Pang H (2004) Correct execution of continuous monitoring queries in wireless sensor systems. In: Proc. of the Second Int. Workshop on Mobile Distributed Computing (MDC'2004), March, Tokyo

Liu J, Reich J, and Zhao F (2003) Collaborative in-network processing for target tracking. In EURASIP JASP: Special Issues on Sensor Networks, March, vol. 4, 378-391

The National Institute of Standards and Technology. Wireless ad hoc networks: smart sensor networks. URL: http://w3.antd.nist.gov/wahn_ssn.shtml

Niculescu D and Nath B (2004) Error characteristics of ad hoc positioning systems. In Proc. of the ACM Mobihoc, May Tokyo

Olston C and Widom J (2002) Best-effort cache synchronisation with source cooperation. In Proc. of the ACM SIGMOD Int. Conf. on Management of Data, pp. 73-84

Pfoser D and Jensen C (2001) Querying the trajectories of on-line mobile objects. In: Proc. of the ACM Int. Workshop on Data Engineering for Wireless and Mobile Access (MobiDE)

Shannon C (1949) The mathematical theory of communication, University of Illinois Press, Champaign, 1949

Sharaf M, Beaver J, Labrinidis A, and Chrysanthis P (2003) TiNA: A scheme for temporal coherency-aware in-network aggregation. In: Proc. of 2003 Int. Workshop in Mobile Data Engineering

Wang H, Yao K, Pottie G, and Estrin D (2004) Entropy-based sensor selection heuristic for localisation. In: Proc. 3rd Int. Workshop on Information Processing in Sensor Networks (IPSN'04), April

Wang H, Yip L, Yao K, and Estrin D (2004) Lower bounds of localisation uncertainty in sensor networks. In: Proc. of IEEE Int. Conf. on Acoustics, Speech, May

Wolfson O, Sistla P, Chamberlain S, and Yesha Y (2005) Updating and querying databases that track mobile units. Distributed and Parallel Databases, 7:3

10 Resilient Aggregations: Statistical Approach

Levente Buttyán, Péter Schaffer and István Vajda

Budapest University of Technology and Economics, Laboratory of Cryptography and Systems Security

10.1 Introduction

Sensor networks are distributed systems, consisting of hundreds or thousands of tiny, low-cost, low-power sensor nodes and one (or a few) more powerful base station(s). These networks are designed to interact with the physical environment. Typically, sensors measure some physical phenomena (e.g., temperature and humidity) and send their measurements to the base station using wireless communications. The base station performs data processing functions and provides gateway services to other networks (e.g., the Internet). Sensor nodes are able to transmit messages only within a short communication range, therefore, it is envisioned that the sensors form a multi-hop network in which the nodes forward messages on behalf of other nodes toward the base station. The typical toplogy of a sensor network is a tree with the base station at the root (see Fig. 10.1 for illustration). In order to reduce the total number of messages sent by the sensors, in-network processing may be employed, whereby some sensor nodes perform data aggregation functions. Aggregator nodes collect data from surrounding sensors, process the collected data locally, and transmit only a single, aggregated message toward the base station. Finally, the base station computes a single aggregated value (e.g., average, minimum, or maximum) from the data received from the network.

After deployment, sensors are typically left unattended for a long period of time. In order to keep their cost acceptable, common sensor nodes are not tamper resistant. This means that they can be captured and compromised at a reasonable cost. Therefore, we cannot assume that common sensors attacked by a determined adversary are able to protect any secret cryptographique elements (e.g., secret keys). Once a sensor is compromised, it can send authentique messages to other nodes and to the base station, but those messages may contain arbitrary data created by the adversray (e.g., bogus measurments).

Note that even if we assumed that the adversary is less powerful or that the nodes are tamper resistant, the adversary can still perform *input based attacks*, meaning that it can directly manipulate the physical environment monitored by some of the sensors, and in this way, it can distort their measurements and the output of the aggregation mechanism at the base station.

In this chapter, we will focus on this problem. More precisely, we assume that a small fraction of the sensor nodes are compromised (or their input is directly manipulated) by an adversary. In general, we do not make assumptions about how the compromised nodes behave (i.e., Byzantine fault model is assumed). We do assume, however, that the objective of the attacker is to maximally distort the aggregated value computed by the base station, while at the same time, the adversary does not want to be discovered (stealthy attack).

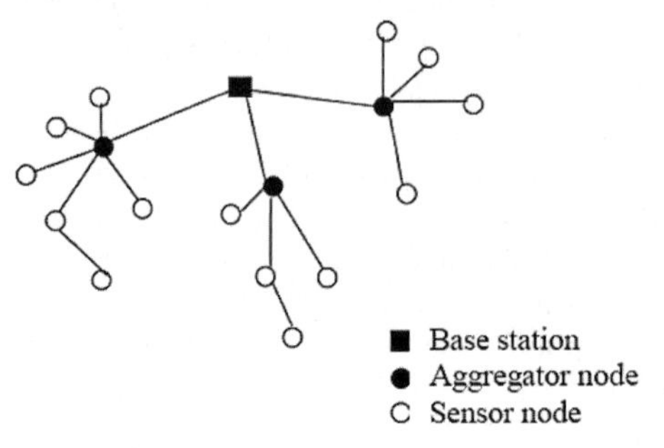

Fig. 10.1. Sensor network architecture

The adversary may not limit its attack to simple sensor nodes, but it may try to compromise aggregator nodes too. Aggregator nodes are, obviously, more valuable targets for attack. The reason is that via gaining control over such nodes, the adversary might manipulate the aggregated data collected from a larger set of common sensors. Therefore, we cannot assume that aggregation points are trustworthy. We will assume, however, that the base stations cannot be compromised, and they always behave correctly, since there are only a few of them, they can be placed in a protected environment, and they can be monitored continuously.

The use of wireless communication channels means additional vulnerabilities. For instance, data can be easily eavesdropped or disrupted by jamming. Jamming may lead to temporary link failures, which may be viewed as a kind of Denial of Service (DoS) attack. We note, however, that in general, the adversary can be assumed to be more or less aware of the data measured in the network. Consequently, its primary goal may not be eavesdropping, but instead, it wants to modify the data delivered to the base station. Similarly, we can assume that it is more valuable for an adversary to manipulate the data collected from a given part of the network, than completely eliminate the corresponding subset of data via applying DoS attacks. On the other hand, if degradation of the communication graph endangers the fulfillment of the mission of the network, then techniques ensuring reliable communications (e.g., redundant paths, diversification techniques) must also be applied.

While cryptographique techniques (e.g., message integrity protection) and reliability methods (e.g., path diversification) are also important defense measures for resilient aggregation, in this chapter, we put our emphasis on *statistical approaches* to detect deliberate attacks stemming from the insertion of bogus data by compromised sensors or by the manipulation of the physical environment of some uncompromised sensors. Note that such attacks simply cannot be detected by cryptographique techniques, that is why we need statistical methods, such as *extreme value detection* and *robust statistics*. These approaches help to detect and eliminate unusual, strange data values from the pool of data before the aggregation is performed, or they can considerably suppress the effect of bogus data on the aggregated value. The strength of these methods depends on the accuracy of the statistical model assumed for the data as statistical sample, and the accuracy of our assumptions about the capabilities of the adversary (e.g., the percentage of the compromised sensors and/or aggregators).

The rest of the chapter is organised as follows: In section 2, we give an overview of the state-of-the-art in resilient aggregation in sensor networks. In section 3, we summarise the main results of outlier tests and robust statistics; these tools can be used to design data aggregation schemes that are resilient to attacks stemming from the insertion of bogus data by compromised sensors or by the manipulation of the physical environment of some uncompromised sensors. In section 4, we give a detailed example of a statistical approach to resilient data aggregation. This example is based on an approach called RAndom SAmple Consensus (RANSAC). We also present some simulation results that show the efficiency of the RANSAC approach.

10.2 State-of-the-art

Resilient aggregation is a recent topic in the world of sensor networks; that is why the list of related papers is relatively short. In this section, we give an overview of these related papers. We begin, in subsection 10.2.1, with some proposed data aggregation schemes that are resilient to node and/or link failures. As we mentioned before, such failures can be the result of jamming attacks, therefore, these schemes can be viewed as data aggregation algorithms that would work in the presence of a jamming adversary. Then, in subsection 10.2.2, we give an overview of those papers that assume an adversary that can compromise aggregator nodes in the network and modify the data that they pass on toward the base station. As we will see, such attacks can be detected by appropriate cryptographique techniques. Finally, in subsection 10.2.3, we deal with papers that address the problem of input based attacks and bogus data introduced by compromised sensors. The countermesaures for this latter class of attack are based on statistical approaches.

10.2.1 Data Aggregation in case of Node or Link Failures

Shrivastava *et al.* present a new approach for computing complicated statistics like the histogram of the sensor readings (Shrivastava et al. 2004). The topology of the sensor networks here is assumed to be a typical tree structure, where each intermediary node performs aggregation (i.e., compression). No attacker is assumed and even no link failures are considered.

Each sensor maintains a q-digest which comprises the summary of data measured by it. A q-digest is a binary tree built on the value space in the following way. The leaf nodes represent buckets corresponding to optionally overlapping intervals of the value space and store the count of readings that came from the corresponding interval. At the beginning, the intermediary nodes do not store any counts, they are needed for later processing. Note, that here the term 'node' is not corresponding to a sensor, but reflects to an egde in the graph of the q-digest.

To construct a memory-effective q-digest we need to hierarchically merge and reduce the number of buckets. This means going through all nodes bottom-up and checking if any node violates the digest property, which declares that none of the nodes should have a high count (this is needed to prove error bounds on the q-digest) and no parents-children triple should have low count. (The upper and the lower bounds for this count are determined by a compression parameter *k*.) This latter property is responsible for the compression, because if two sibling buckets have low counts then we do not want to have two separate counters for them. They will be merged into its parent and the empty buckets will be deleted. Doing this recursively bottom-up in the q-digest, we archieve a degree of compression. This emphasises a key feature of q-digest: detailed information concerning data values which occur frequently are preserved in the digest, while less frequently occuring values are lumped into larger buckets resulting in compression, but also in information loss. This technique will always produce only an approximate result, but it is aware of the limited memory of the sensors.

Q-digests can be constructed in a distributed fashion too, that makes them available for calculating aggregates in sensor networks. If two sensors send their q-digests to their parent sensor (i.e., parent in the rooting tree), it can merge the received q-digests together and add its own q-digest to produce a common q-digest. The idea in merging of two q-digests is to take the union of the two q-digests and then run the above described compression algorithm on the union q-digest.

Q-digests are best deployable at such queries, where the output of the query can be represented with a single node - the error derived by the compression becomes minimal in this case. Such queries are the quantiles or specially the median. By a given memory *m*, the error of a quantile query can be upper bounded by $3\log(\sigma)/m$, where σ stands for the size of the value space which is assumed to be $[1\ldots\sigma]$. However, other queries can be executed with the help of q-digests too (like inverse quantile, range query, consensus query or histogram), but their included error will possibly grow above this limit.

Chen *et al.* propose a method to calculate a statistic in a distributed fashion (Chen et al. 2005). The sensor network is abstracted as a connected undirected graph, where the sensor nodes are the vertices and the bi-directional wireless communication links are the edges. This underlying graph can be of any kind depending on the placement of the sensor nodes. The chapter does not consider attackers, but handles link failures.

The method proposed in (Chen et al. 2005) is called Distributed Random Grouping (DRG), and it can be used to calculate the average or the min/max of the sensor readings in a distributed fashion. The main idea of DRG is to randomly generate groups of nodes which locally synchronise their aggregate values to the common aggregate value of the group. Sooner or later this leads to a global consensus about the aggregate value in the sensor network. The detailed algorithm can be described as consecutive steps in a round: In the first step, each node in idle mode independently originates to form a group and become the group leader with probability p. A node i which decides to be the group leader enters the leader mode and broadcasts a group call message containing its id (i.e., i). After this, it waits for the response of its neighbours. In the second step, a neighbouring node j, at the idle mode that successfully received the group call message, responds to the group leader i with a joining acknowledgement containing its reading v_j. Node j then enters to member mode and waits for the group assignment message from group leader i. In the third step of the algorithm the group leader i gathers the received joining acknowledgements and computes the number of the group members and the desired statistic on the readings (e.g., the average). Then it broadcasts the group assignment message comprising the calculated statistic and returns to idle mode. The neighbouring nodes at member mode of group i which receive the assignment message, update their value to the received one and return to idle mode.

Although this algorithm is natural, it is not straightforward to give bounds on the running time (i.e., the number of rounds needed until the whole network owns roughly the same value for the desired statistic). The authors assumed that the nodes run DRG in synchronous rounds to evaluate the performance of the algorithm. Based on theoretical calculations, they succeed to upper bound the number of rounds needed as a function of the properties of the graph, the grouping probability, the desired accuracy and the grand variance of the initial value distribution. An other way of evaluation was the simulation of the algorithm on hyphotetical sensor networks deployed according to Poisson random geometric graphs. The authors also compared DRG with known distributed localised algorithms, like the Flooding algorithm and the Uniform Gossip algorithm. In each comparison scenario, the DRG outperformed these two algorithms in the total number of transmissions, which is closely related to the energy consumption of the sensor network.

10.2.2 Attacking Aggregators: Cryptographique Countermeasures

In (Anand et al. 2005), Anand *et al.* consider an eavesdropping attacker who wants to eavesdrop the network-wide aggregatum. The topology is the same as in (Shrivastava et al. 2004), each sensor senses and aggregates (except the leaf nodes, they have no aggregation task).

The adversary has the capability only for eavesdropping the communication between some sensor nodes. The goal of the adversary is to determine the aggregatum as precisely as possible. Two types of eavesdropping attacks are distinguished in the chapter. There are passive eavesdroppers (who only listens to the broadcast medium) and active eavesdroppers (who have the power to send queries to the sensor network).

The authors of (Anand et al. 2005) show a way how the probability of a meaningful eavesdrop can be calculated, where meaningful means that the information obtained by the eavesdropper helps him to calculate a good estimate of the real aggregatum. The function that measures this probability is called eavesdropping vulnerability and it depends on the set of eavesdropped nodes, on the adversary's error tolerance and on the aggregation function used to calculate the aggregatum. Eavesdropping vulnerability can be calculated for a group containing only one aggregation point or for hierarchical groups, here the goal is to consider how close the adversary gets to the aggregatum higher in the tree when he eavesdrops on data in the lower level comprising that group (this latter kind of analysis is called eavesdropping vulnerability over a hierarchy).

In (Hu et al. 2003), Hu and Evans consider large sensor networks where the sensor nodes organise themselves into a tree for the purpose of routing data packets to a single base station represented by the root of the tree. Intermediate sensor nodes (i.e., nodes that are not leaves) perform aggregation on the data received from their children before forwarding the data to their parent. The details of the aggregation are not given, presumably it can be the computation of the minimum, the maximum, and the sum (for computing the average at the base station).

Two types of adversaries are considered: adversaries that deploy intruder nodes into the network, and adversaries that can compromise a *single* node in the network. Intruder nodes try to defeat the system by introducing bogus data into the network or altering the contents of the packets sent by the legitimate nodes. However, they do not have access to any key material. On the other hand, adversaries that can compromise a node are assumed to gain access to all the secrets of the compromised node. These two types of adversaries are considered separately, meaning that the key material of the compromised node is not distributed to the intruder nodes.

The goal of the adversary is to distort the final aggregated value by either modifying the contents of the data packets using the intruder nodes or forwarding a bogus aggregated value to the parent of a single compromised node.

The authors make the following system assumptions. The base station is powerful enough to send packets to all sensor nodes directly (i.e., in a single hop). On the other hand, sensor nodes can only communicate with other nodes in their vicinity. Message delivery between two neighboring nodes is reliable. The size of the network is large enough so that most of the nodes are several hops away from the base station (and the other nodes). However, the network is dense enough so that every node has several neighbors. A shared secret is established between the base station and each node in a safe environment before the deployment of the network.

The proposed solution is based on two mechanisms: delayed aggregation and delayed authentication. Delayed aggregation means that instead of performing the aggregation at the parent node, messages are forwarded unchanged to the grandparent and aggregation is performed there. This increases the overall transmission cost but it allows the detection of a compromised parent node if the grandparent is not compromised (and by assumption it cannot be compromised if the parent is compromised). Delayed authentication does, in fact, refer to the μTESLA protocol which allows broadcast authentication with using purely symmetric cryptography. During the operation phase, leaf nodes send sensor readings to their parents. The integrity of these messages is protected by a MAC. Parent nodes compute the aggregated value of the data received from their children, but do not forward this aggregated value. Instead a parent node forwards the messages (together with their MACs) received from its children to its own parent (i.e., the grandparent of its children). In addition, the parent also sends a MAC which is computed over the aggregated value. Thus, intermediate nodes receive messages that contain values authenticated by their grandchildren, and a MAC on the aggregated value computed by their child. An intermediate node verifies the MACs of its grandchildren, computes the aggregated value (which should be the same as the one computed by its child) and then verifies the MAC of its child. If any of these verification fails, then the node raises an alarm. The idea is that since there is only a single compromised node in the network, if the child is compromised then the grandchildren are honest. Thus, if the child sends a MAC on a bogus aggregated value, then the node will detect this because it also receives the correct input values from the honest grandchildren.

MACs are computed with TESLA keys that are shared by the sender of the MAC and the base station but not yet known to the other nodes at the time of sending. These keys are disclosed after some time delay (specified by the TESLA protocol) by the base station. Thus messages and their MACs are stored and verified later, when the corresponding keys are revealed. The base station authenticates the disclosed keys by its own current TESLA key, which is also disclosed in a delayed manner. The authors use informal arguments to illustrate that the scheme is resistant to the type of adversaries considered. In addition some cost analysis is performed too, which focuses on the communication costs of the scheme as sending and receiving messages are the most energy consuming operations. Finally some scalability issues are discussed.

In (Przydatek et al. 2003), Przydatek *et al.* present cryptography based countermeasures against the attacker, who wants to distort the aggregatum. The topology assumed is the following. A distant home server collects statistics of the measured data, which are the median, the average and the minimum/maximum of the sample. Intermediary nodes, aggregators with enhanced computation and communication power are used to optimise the communication need by calculating partial statistics on spot over subset of sensors (leaf-nodes in this tree structured topology). The communication complexity between the home server and the aggregator is a central issue for the optimisation of the approach, because it is assumed that in real application this is an expensive long-distance communication link. The case of a single aggregator is considered in (Przydatek et al. 2003). Each sensor shares a separate secret key with the home server and the aggregator. The data sent from the nodes to the aggregator is authenticated and encrypted (using both keys).

The adversary can corrupt a small fraction of the sensors, and by using the compromised key(s) it is able to change the measured values arbitrarily. No model is assumed about the statistical properties (like probability distribution) of the measured data. Consequently the approach does not use outlier tests to detect and filter out corrupted data. While the median is the most common and simplest robust statistics, the other two statistics calculated (the average and the min/max) are not robust at all, so the protection in this chapter is basically not statistical, and the countermeasures deployed at the level of the aggregator node(s) are based on cryptographic approaches. The main attack is launched against the aggregator (stealthy attack), by corrupting it and modifying the aggregated value.

The secure aggregation approach proposed in this chapter is based on cryptographic commitment and interactive proof techniques. The aggregator sends the aggregate statistics to the home server together with a Merkle hash tree based commitment. In the proof step the home server asks the aggregator for a randomly selected sub-sample. In this step the aggregator sends the wanted sub-sample in the form protected by the keys shared between the nodes and the home server. The home server checks elements of this sub-sample against the commitment by interacting with the aggregator. If this check is successful, i.e., the home server is convinced that the sub-sample is really comes from the sample used for the calculation of the commitment and sent by the corresponding sensors, it calculates the actual statistics for this sub-sample and compares the result to the value sent previously by the aggregator and calculated for the whole sample. If the distance between these two values are slight enough, the home server accepts the statistics authentic.

The quality of the approach is measured by a pair (ε, δ), where δ is an upper bound on the probability of not detecting a cheating aggregator, where the cheating means that the reported value sent by the aggregator is not within ε bounds.

10.2.3 Input based Attacks: Statistical Countermeasures

One of the most important papers in the field of resilient aggregation in sensor networks is from Wagner (Wagner 2004). Wagner assumes an abstract sensor network topology, where the inessential underlying physical structures are abstracted away. All the sensor nodes are connected with a single base station each over its own separate link. The base station is assumed to be trusted and the links are secure and independent, meaning that capturing one sensor node might compromise the contents of that node's channel but it has no influence on the other node's channel. The sensors send their packets directly to the base station via their channel, every packet contains a measurement value. The base station collects the measurements from all the sensor nodes and then it computes the desired aggregatum with the aggregation function *f*. The overall question is: Which aggregation functions can be securely and meaningfully computed in the presence of a few compromised nodes?

The main threat considered is that of malicious data. The adversary is able to inject malicious data to the list of sensor measurements by capturing a few nodes and modify their readings, e.g., by altering the environment around the sensor or by reverse-engineering the sensor. The modification of the sensor's environment can be considered for example as lighting a lighter near to a temperature sensor or as flashing with a flashlight to a photometer sensor. The adversary is only able to compromise a small percent of the sensor nodes (at least fewer than the half of the network), but he can inject arbitrary values in place of the original readings. Thus, the adversary is modelled by the Byzantine fault model. The goal of the adversary is to skew the computed aggregate as much as possible.

The term *resilient aggregation* has been coined in this chapter and it refers to those aggregation function that are reliable even in the case if some of the sensor nodes are compromised. The resiliency of an aggregation function f is measured by the root-mean-square (r.m.s.) error value of the function for the best possible k-node attack, where the best possible k-node attack is that one which skewes the aggregatum at most. The mathematical framework beyond the root-mean-square error is based on independent and identically distributed sample elements and on the knowledge of the parametrised distribution. An aggregation function is declared to be resilient if its r.m.s. error for the best k-node attack is beyond the critical value corresponding to α times the root-mean-square error in the uncompromised state, concisely expressed if it is (k,α)-resilient for some α that is not too large. The analysis of the chapter shows that the commonly used aggregation functions are inherently insecure. The min/max, the sum and the average are totally insecure, because only one compromised reading can mislead the base station (or generally the aggregator) without an upper bound – this means that α becomes infinity. The aggregation functions that are secure are the median and the count – these have acceptable α values. The tools proposed in (Wagner 2004) to archieve better resiliency is truncation (i.e., if we know that valid sensor readings will usually be in the interval $[l,u]$, then we can truncate every input to be within this range) and trimming (i.e., throwing away the highest and the lowest part of the input).

The chapter is based on strong statistical background and contains a lot of simple proofs which help to understand the theory of the root-mean-square error calculation. There were no simulations made to demonstrate the viability of the approach and no real-life experience are shown. However, this type of investigation does not need any simulation because it is purely mathematical.

The spectrum of the potencial application of resilient aggregation is sensor networks is very wide. Resilient aggregation is best-suited to settings where the data is highly redundant, so that one can cross-check sensor readings for consistency. None the less there are many possible applications of such robust aggregation functions, the paper does not mention any of them. But there are scenarios where aggregation should not be applied, for instance in fire-detection or in any systems searching for a needle in the haystack.

Wagner's idea has been developed further by Buttyán *et al.* in (Buttyán et al. 2006). In (Buttyán et al. 2006), the abstract topology is the same as in the previous paper. The sensor nodes perform some measurements and send their readings to the base station. The base station runs the aggregation procedure (after an analysis phase) and it is assumed to be reliable. The communication between the nodes and the base station is assumed to be secure – there are effective cryptographic techniques to protect it and resilient aggregation is not concerned with this problem. Nonetheless, some percentage of the sensor readings can be compromised by artificially altering the measured phenomena.

The adversary is allowed to modify the sensor readings before they are submitted to the aggregation function. The affected sensors are chosen randomly without knowing their readings (myopic attacker). The adversary's goal is to maximise the distortion of the aggregation function, defined as the expected value of the difference between the real parameter and the computed one. In addition, the adversary does not want to be detected, or more precisely, he wants to keep the probability of successful detection of an attack under a given small threshold value. This criterion upper bounds the possibilities of the adversary, even for aggregation functions that were considered to be insecure earlier.

The novel data aggregation model in (Buttyán et al. 2006) consists of an aggregator function and of a detection algorithm. The detection algorithm analyses the input data before the aggregation function is called and tries to detect unexpected deviations in the received sensor readings. In fact, trimming (proposed by Wagner in (Wagner 2004)) is a special case of this more general idea. The detection algorithm uses the technique of sample halving, i.e., it first halves the sample and computes the sum of the halves separately. Then it subtracts the two sums from each other and indicates attack if the result is above a limit. The concrete value of this limit can be calculated from the desired false positive probability (i.e., when there is no attack but the detection algorithm indicates attack). As it can be seen, (Buttyán et al. 2006) uses a natural statistical tool to analyse the skewness of the distribution of the sensor readings and to filter out unusual samples. However, to filter out something that is unusual, we need some a priori knowledge about what is usual. Here the assumption is that the sensor readings are independent and identically distributed and that the variance of the sample is 1.

The efficiency of this novel data model and the sample halving technique is measured by the probability of successful detecting an attack. This probability is shown as a function of the distortion archieved by the adversary for different number of compromised sensors. Thanks to the detection algorithm, the probability of an attack detection grows steeply even for small ammount of compromised sensor readings and thus the archievable distortion is strictly upper bounded.

10.3 Outliers and Robust Statistics

By modifying the readings of the compromised sensors, or by manipultaing the sensors' physical environment (in order to influence their measurements), the adversary contaminates the samples received by the aggregation function with bogus data. Methods for eliminating the effects of bogus data have been extensively studied in statistics (although not with a determined adversary in mind), and we find it useful to summarise the main achievements in this field here. Thus, after a brief introduction, we give an overview of *outlier tests* and *robust statistics* in this section.

An outlier is an observation, which is not consistent with the data, meaning that we have some assumptions about the statistical characteristics of the data (e.g., the type of distribution) and some data points do not fit this assumption.

If an outlier is really an error in the measurement, it will distort the interpretation of the data, having undue influence on many summary statistics. Many statistical techniques are sensitive to the presence of outliers. For example, simple calculations of the mean and standard deviation may be distorted by a single grossly false data point. The most frequently cited common example is the estimation of linear regression from sample contaminated with outliers: Because of the way in which the regression line is determined (especially the fact that it is based on minimising not the sum of simple distances but the sum of squares of distances of data points from the line), outliers have a dangerous influence on the slope of the regression line and consequently on the value of the correlation coefficient. A single outlier is capable of considerably changing the slope of the regression line and, consequently, the value of the correlation. Therefore it is tempting to remove atypical values automatically from a data set. However, we have to be very careful: if an "outlier" is a genuine result (a genuine extreme value), such a data point is important because it might indicate an extreme behavior of the process under study. No matter how extreme a value is in a set of data, the suspect value could nonetheless be a correct piece of information. Only with experience or the identification of a particular cause can data be declared 'wrong' and removed.

Accordingly outlier tests and elimination of detected outliers are appropriate if we are confident about the distribution of the data set. If we are not sure in that, then robust statistics and/or non-parametric (distribution independent) tests can be applied to the data.

If we know extreme values represent a certain segment of the population, then we must decide between biasing the results (by removing them) or using a nonparametric test that can deal with them. In statistics, classical least squares regression relies on model assumptions (the Gauss-Markov hypothesis) which are often not met in practice. 'Nonparametric' tests make few or no assumptions about the distributions, and do not rely on distribution parameters. Their chief advantage is improved reliability when the distribution is unknown. However, non-parametric models give very imprecise results, compared to their parametric counterparts. Therefore, a compromise between parametric and non-parametric methods was created: robust statistics.

The aim of robust statistics is to create statistical methods which are resistant to departure from model assumptions, i.e., outliers. If there is no reason to believe that the outlying point is an error, it should not be deleted without careful consideration. However, the use of more robust techniques may be warranted.

Robust techniques will often downweight the effect of outlying points without deleting them. Robust statistics include methods that are largely unaffected by the presence of extreme values. Therefore the three approaches handling data seemingly "contaminated" with "atypical" points are the following:

- Outlier tests
- Robust estimates
- Non-parametric methods

Below the main approaches in outlier testing and robust statistics are summarised.

10.3.1 Outlier Tests

There are outlier tests which perform quite good across a wide array of distributions, like the Box Plot which is a traditional graphical method, however can also be automated. We mark the largest data point that is less than or equal to the value that is 1.5 times the interquartile range (*IQR*) above the 3rd quartile. Similarly we mark the smallest data point that is less than or equal to the value that is 1.5 times the interquartile range (*IQR*) below the 1st quartile. Data point above or below this upper and lower marks (so called whiskers) is considered an outlier. Why 1.5 *IQR*s? John Tukey, the inventor of Box Plot, commented: "1 is too small and 2 is too large." Really, in practice this outlier rule is quite good across a wide array of distributions. For instance, consider the following ranked data set of daily temperatures in winter season:

$$-15, -7, -4, -3, -2, +1, +2, +4, +5$$

Here is an example, using the first set of numbers above using Tukey's method of determining *Q1* and *Q3*. From $Q1 = -4$, $Q3 = +2$, we get $IQR = 6$. Now 1.5 times 6 equals 9. Subtract 9 from the first quartile: $-4 - 9 = -13$. Note that -15 is an outlier, and the whisker should be drawn to -7, which is the smallest value that is not an outlier. Add 9 to the third quartile: $+2 + 9 = 11$. Any value larger than 11 is an outlier, so in this side we have no outlier. Draw the whisker to the largest value in the dataset that is not an outlier, in this case $+5$. Since this value is the 3rd quartile, we draw no whisker at all. Mark -15 as outlier.

Tests are more sharp if we know something about the underlying distribution of data. Assuming the knowledge of the mean and the standard deviation some researchers use simple quantitative technique to exclude outliers. They simply exclude observations that are outside the range of ±2 standard deviations (or even ±1.5 sd's) around the mean. Refining this method we can draw confidence intervals also for the mean, which give us a range of values around the mean where we expect the "true" (population) mean is located (with a given level of certainty). For example, if the (sample) mean is 12, and the lower and upper limits of the $p = 0.05$ confidence interval are 8 and 16 respectively, then we can conclude that there is a 95 percent probability that the population mean is greater than 8 and lower than 16. The width of the confidence interval depends on the sample size and on the variation of data values. The larger the sample size, the more reliable its mean. The larger the variation, the less reliable the mean. The standard calculation of confidence intervals is based on the assumption that the variable is normally distributed in the population. The estimate may not be valid if normality assumption is not met, unless the sample size is large, say $n = 100$ or more, which is not unusual in case of sensor networks.

There are outlier tests which are explicitly based on the assumption of normality. A typical such test is the Grubbs' test, which is also known as the maximum normed residual test. Grubbs' test detects one outlier at a time. This outlier is expunged from the set of data and the test is iterated until no outliers are detected. More formally, Grubbs' test is defined for the hypothesis:

H_0: There are no outliers in the data set.
H_1: There is at least one outlier in the data set.

The Grubbs' test statistic is defined as:

$$G_1 = \max_i \left|\bar{x} - x\right| / s \tag{10.1}$$

with x and s denoting the sample mean and standard deviation, respectively. In words, the Grubbs' test statistic is the largest absolute deviation from the sample mean in units of the sample standard deviation. This is the two-sided version of the test. The Grubbs' test can also be defined as one of the following one-sided tests: we can test if the minimum value is an outlier

$$G_2 = \left|\bar{x} - x_{\min}\right| / s \tag{10.2}$$

with x_{min} denoting the minimum value. Similarly we can test if the maximum value is an outlier,

$$G_3 = \left|\bar{x} - x_{\max}\right| / s \tag{10.3}$$

with x_{max} denoting the maximum value. The critical values are calculated according to Student t-distribution with appropriate degree of freedom (Grubbs 1969).

There are approaches, like the Random Sample Consensus approach (RANSAC) (Fischler et al. 1981), which rely on random sampling selection to search for the best fit. The model parameters are computed for each randomly selected subset of points. Then the points within some error tolerance are called the consensus set of the model, and if the cardinality of this set exceeds a pre-specified threshold, the model is accepted and its parameters are recomputed based on the whole consensus set. Otherwise, the random sampling and validation is repeated as in the above. Hence, RANSAC can be considered to seek the best model that maximises the number of inliers. The problem with this approach is that it requires the prior specification of a tolerance threshold limit which is actually related to the inlier bound. In section 4, we will present a possible application of the RANSAC approach. The Minimum Probability of Randomness (MINPRAN) (Stewart 1995) is a similar approach, however it relies on the assumption that the noise comes from a well known distribution. As in RANSAC, this approach uses random sampling to search for the fit and the inliers to this fit that are least likely to come from the known noise distribution.

10.3.2 Robust Statistics

As it was mentioned above, robust techniques will often downweight the effect of outlying points without deleting them. Several questions arise: How many outliers can a given algorithm tolerate? How can we describe the influence of outliers on the algorithm? What are the properties desirable for robust statistical procedures? Accordingly, three basic tools are used in robust statistics to describe robustness:

1. The breakdown point,
2. The influence function,
3. The sensitivity curve.

Intuitively, the breakdown point of an estimator is the maximum amount of outliers it can handle. The higher the breakdown point of an estimator, the more robust it is. The finite sample breakdown point ε_n^* of an estimator T_n at the sample $(x_1,\dots,x_n)$ is given by:

$$\varepsilon_n^*(T_n) = \frac{1}{n} \max_{m \geq 0} \left\{ m : \max_{i_1,\dots,i_m} \sup_{y_1,\dots,y_m} \left| T_n(z_1,\dots,z_n) \right| < +\infty \right\} \tag{10.4}$$

where $(z_1,\dots,z_n)$ is obtained by replacing the m data points $(x_{i_1},\dots,x_{i_m})$ by arbitrary values $(y_1,\dots,y_n)$. The maximum breakdown point is 0.5 and there are estimators which achieve such a breakdown point. For example, the most commonly used robust statistics, the median, has a breakdown point of 0.5.

As for the influence function we distinguish the empirical influence function and the (theoretical) influence function (Hampel et al. 1986). The empirical influence function gives us an idea of how an estimator behaves when we change one point in the sample and no model assumptions is made. The definition of the empirical influence function EIF_i at observation i is defined by replacing the i-th value x_i in the sample by an arbitrary value x and looking at the output of the estimator (i.e., considering it as a function in variable x):

$$T_n(x_1,\ldots,x_{i-1},x,x_{i+1},\ldots,x_n) \tag{10.5}$$

The influence function tells us what happens to an estimator when we change the distribution of the data slightly. It describes the effect of an infinitesimal "contamination" at the point x on the estimate. Let Δ_x be the probability measure which gives probability 1 to x. The influence function is then defined by:

$$IF(x,T,F) = \lim_{t\to 0}\frac{T(t\Delta_x + (1-t)F) - T(F)}{t} \tag{10.6}$$

A good influence function is qualified by the following properties: finite rejection point, small gross-error sensitivity and small local-shift sensitivity. The rejection point is defined as the point beyond which function IF becomes zero (Goodall 1983), formally,

$$\rho^* = \inf_{r>0}\{r : IF(x,T,F) = 0, |x| > r\} \tag{10.7}$$

Observations beyond the rejection point have zero influence. Hence they make no contribution to the final estimate. However, a finite rejection point may result in the underestimation of scale. This is because when the samples near the tails of a distribution are ignored, too little of the samples may remain for the estimation process (Goodall 1983). The Gross Error Sensitivity expresses asymptotically the maximum effect a contaminated observation can have on the estimator. It is the maximum absolute value of the IF when x is varied. The Local Shift Sensitivity (l.s.s.) measures the effect of the removal of a mass at x and its reintroduction at y. For a continuous and differentiable IF, l.s.s. is given by the maximum absolute value of the slope of IF at any point:

$$\lambda^*(T,F) = \sup_{x\neq y}\left|\frac{IF(y,T,F) - IF(x,T,F)}{y-x}\right| \tag{10.8}$$

General constructions for robust estimators are the M-estimators (Rey 1983). A motivation behind M-estimators can be a generalisation of maximum likelihood estimators (MLE). MLE are therefore a special case of M-estimators (hence the name: "generalised Maximum likelihood estimators"). The M-estimate, $T(x_1,\ldots,x_n)$ for the function ρ and the sample $x_1,\ldots,x_n$ is the value that minimises the following objective function,

$$\min_t \sum_{j=1}^{n} \rho(x_j;t) \tag{10.9}$$

i.e., the estimate T of the parameter is determined by solving,

$$\sum_{j=1}^{n} \psi(x_j;t) = 0 \tag{10.10}$$

where,

$$\psi(x_j;t) = \frac{\partial \rho(x_j;t)}{\partial t} \tag{10.11}$$

One of the main reasons for studying M-estimators in robust statistics is that their influence function is proportional to $\psi : IF(x,T,F) = b\psi(x;T(F))$. When the M-estimator is equivariant, i.e., $T(x_1 + a,\ldots,x_n + a) = T(x_1,\ldots,x_n) + a$, for any real constant a then we can write functions ψ and ρ in terms of residuals $x - t$. If additionally scale estimate S is used we obtain the, so called, scaled residuals $r = (x-t)/S$, and we can write $\psi(r) = \psi((x-t)/S)$ and $\rho(r) = \rho((x-t)/S)$. For the purpose of scale estimate most M-estimators use the MAD (Median Absolute Deviation). For example $\rho(r) = r^2/2$ (least-squares) estimators are not robust because their influence function is not bounded; if no scale estimate (i.e., formally, $S = 1$) is used we get back the mean as estimate. $\rho(r) = |r|$ estimators reduce the influence of large errors, but they still have an influence because the influence function has no cut off point; using this estimator we get the median. For more details about M-estimators and their robustness we refer the reader to (Huber 1981) and (Rey 1983).

The W-estimators (Goodall 1983) represent an alternative form of M-estimators. The L-estimators are also known as trimmed means for the case of location estimation (Koenker et al. 1978). All the above estimators are either obliged to perform an exhaustive search or assume a known value for the amount of noise present in the data set (contamination rate). When faced with more noise than assumed, these estimators will lack robustness. And when the amount of noise is less than the assumed level, they will lack efficiency, i.e., the parameter estimates suffer in terms of accuracy, since not all the good data points are taken into account. Now that we have got an inside view of robust statistics, in section 4, we will present an example how to use the above mentioned RANSAC paradigm to build a filtering and model fitting tool that can be applied in resilient aggregation in sensor networks.

10.4 An Example of the RANSAC Approach

Usually it is impossible to protect the sensor nodes from malicious mishandling, therefore, we need resilient aggregation techniques to treat the situation of receiving some amount of bogus data. The RANSAC paradigm is capable of handling data containing a significant percentage of gross errors by using random sampling. That makes it suitable for environments such as sensor networks where the sensors can be affected by compromising their measurements and that is why it can be convenient as a building block in robust statistical tools.

RANSAC is the abbreviation of Random Sample Consensus and it defines a principle how non-consistent data can be filtered from a sample, with other words, how a model can be fitted to experimental data (smoothing). The principle is the opposite to that of conventional smoothing techniques: Rather than using as much of the data as possible to obtain an initial solution and then attempting to eliminate the non-consistent data elements, RANSAC uses as few of the data as feasible to determine a possible model and then tries to enlarge the initial datum set with the consistent data. Algorithm 1 (below) shows how the RANSAC principle can work as the heart of an algorithm.

Algorithm 1 RANSAC Pseudo-Algorithm

1: **while** $No.\ of\ trials \leq Max\ trials$ **do**
2: Randomly select s data elements (S)
3: Instantiate the model M
4: Depute all data elements within some error tolerance of M (S^*)
5: **if** $\#(S^*) > threshold$ **then**
6: Instantiate the model M^* based on S^*
7: Quit the loop
8: **end if**
9: **end while**
10: **if** $\#(S^*) < threshold$ **then**
11: Compute M^* on the largest S^* or terminate in failure
12: **end if**

For example, if the task is to fit a circle to a set of points with two-dimensional coordinates, then the above algorithm would randomly choose three points for the initial set S (since three points are required to determine a circle), and would fit a circle to this three points (this circle would be model M). With this the algorithm would enlarge the initial set with all the points that are not too far from the arc of the circle (this would be S^*, called also the consensus set of S). If the size of the consensus set is above a threshold, then the algorithm would finish by fitting a final circle by considering all the points within the consensus set. If the consensus set is too small, then the algorithm would drop S and would retry to establish a suitable S^* by picking another three points and running the algorithm again. If after some number of trials the algorithm would not find a suitable consensus set, it would finish with the best possible fit (that would include more errors than desired) or would return with an error message.

10.4.1 The Applied RANSAC Algorithm

To show how to use the RANSAC paradigm in resilient aggregation, we have implemented a filtering function based on this paradigm that filters out outlier measurements supposing that all the samples are independent and normally distributed (with the same parameters) in the uncompromised case. For sake of simplicity we have called it the RANSAC algorithm. Notice, that we assumed nothing about the expected value or the standard deviation of the distribution.

Our RANSAC algorithm is applied against an attacker who can distort some sensor measurements. The assumptions we have made on the attacker are the following. The attacker has limited power resources, but he has full control over some part of the sensor network, so he knows the concrete values measured by the compromised nodes and he can arbitrarily modify the values sent to the base station.

The attacker knows the applied RANSAC algorithm in detail, but he cannot horn in it since it runs on the base station which is assumed to be secure. The attacker also knows that the sample is normally distributed. The sensors communicate with the base station by cryptographically protected messages so the attacker cannot modify the messages after they were encrypted. However, there is no need for it, because the attacker can arbitrary modify the messages of the compromised nodes just before the encryption. The attackers object is to cause as high distortion in the aggregation result as possible while remaining undetected (stealthy attack). We investigated a naive attacker who simply modifies all the compromised sensor measurement values to one common value. For example, if the attacker wants the cooling system to turn on, then he lights a lighter just beside one (or more) temperature sensors. That implies a peak in the histogram of the sample received by the base station. The attacker is able to set this peak to an arbitrary place in the histogram, with other words, he can modify the measured values to an arbitrary common one.

The working of the RANSAC algorithms is as follows (see Fig. 10.1). The base station receives the sample compromised previously by the attacker. The sample is the input of the RANSAC algorithm along with an upper estimation of the percentage of compromised nodes κ, and the required confidence level α. At first, a bag S of minimum size will be randomly chosen to establish a preliminary model. Here we use bag instead of set because the sample may contain repetitive elements. The size of bag S is s, the model M is the theoretical histogram of the empirical Gaussian distribution with the expected value of:

$$\hat{\theta} = \frac{1}{s}\sum_{i=1}^{s} S_i \tag{10.12}$$

and with the standard deviation of

$$\hat{\sigma}^2 = \frac{1}{s-1}\sum_{i=1}^{s}(S_i - \overline{S})^2 = \frac{1}{s-1}\sum_{i=1}^{s}(S_i - \hat{\theta})^2 \tag{10.13}$$

with the restriction that $\hat{\sigma}^2$ cannot be 0. After the histogram is imagined, the algorithm checks whether the sample is consistent with the histogram or not. The algorithm collects those elements from the sample that can be fit to the theoretical histogram and temporally drops the other elements. This phase is denoted by *Depute S** on Fig. 10.2 and can be described as:

$$y_x = \min(c_x, n\phi(x \mid \hat{\theta}, \hat{\sigma})) = \min\left(c_x, \frac{n}{\sqrt{2\pi}\hat{\sigma}} e^{-\frac{(x-\hat{\theta})^2}{2\hat{\sigma}^2}}\right) \tag{10.14}$$

where y_x is the number of elements with a value of x that should not be dropped, n is the size of the sample, c_x is the number of elements in the sample with a value of x and $\phi(x \mid \hat{\theta}, \hat{\sigma})$ is the probability density function of the Gaussian distribution. The bag

$$S^* = \bigcup_i y_i \tag{10.15}$$

is the consensus bag of S. If the size of S^* is smaller than a required size t then the algorithm starts again from the beginning, otherwise S^* will be forwarded to the aggregator. There is an upper bound on the maximum number of retrials denoted by f. If there were more iterations than f, the algorithm ends with failure. The aggregator can be of any kind, here we used the average to estimate the expected value of the distribution of the sample. The value produced by the aggregator is M^*.

The RANSAC algorithm consists of three parameters that have not been defined yet. These are the size s of the initial bag S, the required size t of the consensus bag S^* and the maximum permitted number f of iterations. The size s of the initial bag is desired to be as small as possible according to the RANSAC paradigm. For the RANSAC algorithm, we need to establish a theoretical histogram according to a Gaussian distribution. The Gaussian distribution has two parameters, the expected value and the standard deviation. The expected value can come from only one element, but for the standard deviation we need at least two elements. That was the motivation by the choice of $s = 2$. The required size t of the consensus bag is the most important parameter in the algorithm, however, the

RANSAC paradigm does not gives us any hint about the correct choice of its value. If t is small, then the algorithm has a higher probability to succeed, but the aggregatum at the end will contain a high level of error caused by the attacker. If t is too big, the algorithm cannot work because of the high expectations on the number of elements in the final bag. Nevertheless, we required the algorithm to be as effective as possible by filtering all the attacked elements:

$$t = (1-\kappa)n \tag{10.16}$$

where κ is the upper estimation for the number of attacked sensor nodes and n is the total number of sensor nodes in the network.

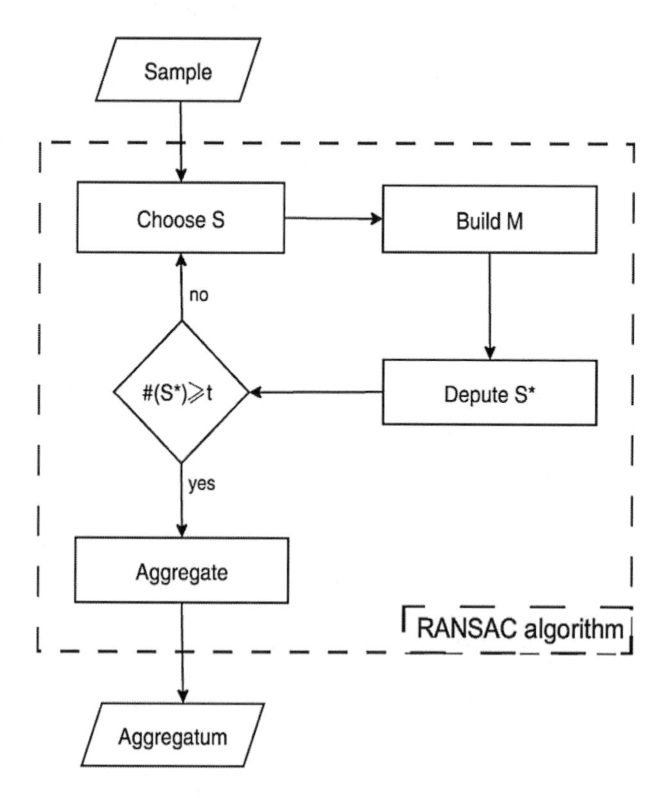

Fig. 10.2. The RANSAC algorithm

The maximum number f of iterations can be figured on probabilistic analysis. Assuming that we need only one S bag that satisfies the criterion that it does not contain compromised elements, we can try to select s elements for S as long as the probability of finding a correct S is beyond a given value $1-\alpha$, where α is the confidence level. Since the elements are selected without replacement, the probability of taking a good S is

$$P_S = \frac{\binom{n-\kappa n}{s}}{\binom{n}{s}} = \prod_{i=0}^{\kappa n-1} \frac{n-s-i}{n-i} \qquad (10.17)$$

The probability of finding at least one correct S out of f trials is,

$$P_S^f = 1-(1-P_S)^f = 1-\left(1-\frac{\binom{n-\kappa n}{s}}{\binom{n}{s}}\right)^f \qquad (10.18)$$

and we require

$$P_S^f \geq 1-\alpha \qquad (10.19)$$

This leads to

$$f \geq \log_{1-P_S} \alpha \qquad (10.20)$$

We have chosen,

$$f = \log_{1-P_S} \alpha + 1 \qquad (10.21)$$

because there is no need for a higher perturbation. The value of f is related to the punctuality of the algorithm as well as the confidence level α. Since the confidence level is an input of the algorithm, the value f can be controlled through the proper choice of α.

10.4.2 Simulations Results

To validate the RANSAC algorithm, we have implemented it in a simulation environment. The parameters for the simulation are included in Table 1. The sample for the simulations has been generated using the randn function of Matlab. After the attacker has compromised κn nodes (i.e., he set the value of κn nodes to a fixed value), the RANSAC algorithm analyses the modified sample by trying to establish a model that fits at least t readings. If the algorithm can not make it in f

iterations, it stops in failure. In this case either α was too low or κ was too high and so the algorithm could not estimate under the given criteria.

Table 10.1: Simulation parameters

Parameter name	Value
n (sample size)	100
Sample distribution	$N(\mu,\sigma)$
μ (expected value, unknown to the algorithm)	0
σ (standard deviation, unknown to the algorithm)	1
s (size of random set)	2
α (confidence level)	0.01

Fig. 10.3 shows a typical results for the maximum distortion achieved by the attacker as a function of the place of the peak (i.e., the value to which it sets the readings of the compromised sensors), where the maximum is taken over 50 simulation runs, and $\kappa = 0.05$ (i.e., the proportion of compromised nodes is 0.05 which means that 5 nodes were compromised). The distortion is defined as the absolute value of the difference between the real average and the estimated average produced by the RANSAC algorithm.

Notice, that the expected value of the test data is 0, therefore Fig. 10..3 is nearly symmetrical to $x = 0$. The slight asymmetry comes from the randomness of the input data. As it can be seen on the figure, the adversary can never achieve a distorion more than 0.6. This means that the estimated average will be always in the range $(-0.6, 0.6)$ surrounding the real average given that 5 nodes out of 100 are compromised. In other words, the adversary can perturb the compromised readings with an arbitrary value, but this will not significantly change the estimate. Moreover, this $(-0.6, 0.6)$ bound is independent of the expected value of the sample, thus for a higher expected value, the relative distortion (i.e., distortion divived by the expected value) can be arbitrarily small.

Fig. 10..3 tells us even more about the algorithm. The steeply ascending lines composed of points correspond to accepting some elements from the peak after the filtering phase. For some x values, it looks like there would be two distinct values for the same x value. For example at $x = -22$, the two lines composed of points appear as if they both had a value for this value of x. As a matter of fact, there is only one y value for every x, but in some cases the algorithm is uncertain to accept a compromised reading. In a small range surrounding $x = -22$, the algorithm accepts a compromised reading in some cases and drops it in another cases. This implies points that look like if they belonged to the same x. Some additional bursts on the figure stem from the inaccuracy of the approximation of the Gaussian distribution and from the random property of the algorithm. As it can be seen, it is meaningless for the adversary to push the peak too far from the average because the highest distortion occurs when the peak is about 10 units far from the real average. Above the distance of approximately 42 units the peak will have no influence on the estimated average (i.e., it will be always filtered out by the algorithm). The unit can be of any kind, for example degree in case of a thermometer sensor or lumen in case of a photometer sensor.

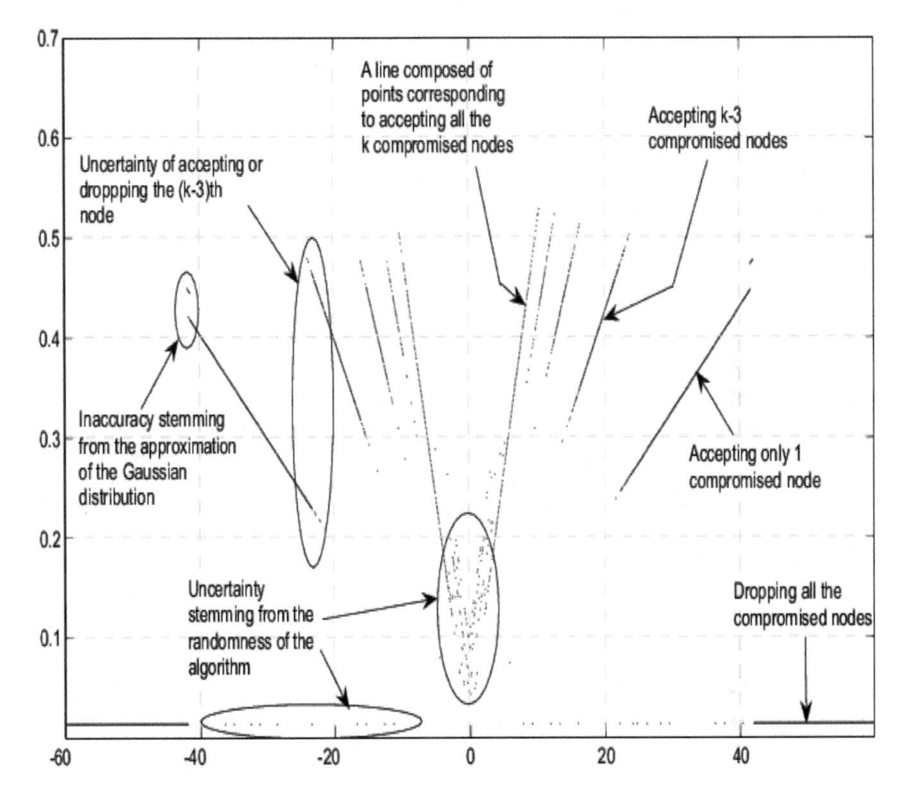

Fig. 10.3. A typical simulation result and its explanation

Another demonstration on the performance of the RANSAC algorithm is Fig. 10.4. This shows the highest reacheable distortion (y axis) as a function of different values of κ (x axis). We made 50 simulation runs with the same parameter set for each values of κ and plotted out the one with the maximum distortion. As one can see, the RANSAC algorithm reaches the theoretical maximum of the breakdown point, since the distortion never goes to infinity for $\kappa \leq 0.5$ (which means that at most 50% of the nodes are compromised). Notice, that the theoretical maximum for the breakdown point is 0.5. Another important issue is that the RANSAC algorithm strictly upper bounds the reacheable distortion even for high values of κ. For example, for a node compromisation ratio of 45% ($\kappa = 0.45$) the error is upper bounded in 3 units. This means, that even if almost the half of the nodes send bogus information to the base station, the RANSAC algorithm is capable to produce a result with at most 3 units of perturbation.

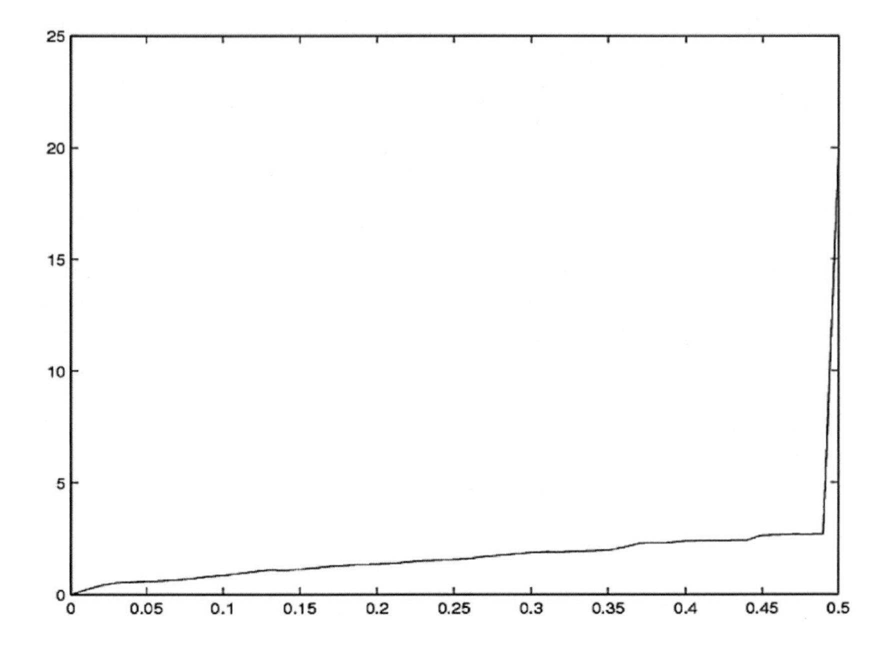

Fig. 10.4. Demonstration of the breakdown point

10.5 Conclusions

In typical sensor network applications, the sensors are left unattended for a long period of time once they have been deployed. In addition, due to cost reasons, sensor nodes are usually not tamper resistant. Consequently, sensors can be easily captured and compromised by an adversary. Moreover, when a sensor is compromised, it can send authentique messages to other nodes and to the base station, but those messages may contain arbitrary data created by the adversray (e.g., bogus measurments). Even if we assumed that the adversary is less powerful or that the nodes are tamper resistant, the adversary can still perform *input based attacks*, meaning that it can directly manipulate the physical environment monitored by some of the sensors, and in this way, it can distort their measurements and the output of the aggregation mechanism at the base station. The task of resisilient aggregation is to perform the aggregation despite the possibility of the above mentioned attacks, essentially by supressing their effect.

Resilient aggregation cannot be based on pure cryptographic countermeasures (e.g., message authentication) and/or reliability mechanisms (e.g., diversification techniques). Statistical methods, such as extreme value detection and robust statistics, are relevant and inevitable additional elements in protecting data aggregation mechanisms in sensor networks. However, their efficiency basically depends on the accuracy of the model describing the characteristics of the data as a statistical sample. This means, for instance, that we have some a priori information about the probability distribution of data, the normal range of time-variability of the main statistical characteristics, the statistical dependence between measurements of certain groups of sensors etc. Many statistical techniques are available, and the task of the designer of the resilient aggregation scheme is to find the appropriate and valid adversary model, distortion measure, and description of the sample, as well as to appropriately apply the toolkit provided by statisticians. Besides giving an overview of the state-of-the-art in resilient aggregation in sensor networks, and a brief summary of the relevant techniques in the field of mathematical statistics, we also introduced a particular approach for resilient aggregation in somewhat more details. This approach is based on RANSAC (RAndom SAmple Consensus), which we adopted for our purposes. We presented some initial simulation results about the performance of our resilient data aggregation scheme, where we assumed a priori knowledge of the type of the distribution of the clean sample, and a particular type of adversary. These results show that the our RANSAC based approach is extremely efficient in this particular case, as it can cope with a very high fraction of compromised nodes.

10.6 References

Anand M, Ives Z, and Lee I (2005) Quantifying eavesdropping vulnerability in sensor networks. In: Proc. of the Second VLDB Workshop on Data Management for Sensor Networks (DMSN), pp 3-9

Buttyán L, Schaffer P, and Vajda I (2006) Resilient aggregation with attack detection in sensor networks. In: Proc. of the Second IEEE Workshop on Sensor Networks and Systems for Pervasive Computing (PerSeNS), accepted for publication

Bychkovskiy V, Megerian S, Estrin D, and Potkonjak M (2003) A collaborative approach to in-place sensor calibration. In: Proc. of the 2nd Int. Workshop on Information Processing in Sensor Networks (IPSN), pp 301-316

Chen J-Y, Pandurangan G, and Xu D (2005) Robust computation of aggregates in WSN: distributed randomised algorithms and analysis. In: Proc. of the 4th Int. Symp. on Information Processing in Sensor Networks (IPSN), pp 348-355

Chum O, Matas J (2002) Randomised RANSAC with $T_{d,d}$ test. In: Image and Vision Computing, 22: 837-843

Fischler MA and Bolles RC (1981) Random sample consensus for model fitting with applications to image analysis and automated cartography. In: Communications of the ACM 24: 381-395

Goodall C (1983) M-estimators of location: An outline of the theory. In: Hoaglin DC, Mosteller F, Tukey JW (eds) Understanding robust and exploratory data analysis, Wiley, New York, pp 339-403

Grubbs F (1969) Procedures for detecting outlying observations in samples. In: Technometrics 11: 1-21

Hampel FR, Ronchetti EM, Rousseeuw PJ, and Stahel WA (1986) Robust statistics: The approach based on influence functions. Wiley, New York

Hu L and Evans D (2003) Secure aggregation for wireless networks. In: Proc. of the 2003 Symp. on Applications and the Internet Workshops (SAINT), pp 384-394

Huber PJ (1981) Robust Statistics. Wiley, New York

Koenker R and Basset Jr G (1978) Regression quantiles. In: Econometrica 36: 33-50

Lacey AJ, Pinitkarn N, Thacker NA (2000) An evaluation of the performance of RANSAC Algorithms for Stereo Camera Calibration. In: Proc. of the British Machine Vision Conference (BMVC)

Li Z, Trappe W, Zhang Y, and Nath B (2005) Robust statistical methods for securing wireless localisation in sensor networks. In: Proc. of the 4th Int. Symp. on Information Processing in Sensor Networks (IPSN)

Przydatek B, Song D, and Perrig A (2003) SIA: Secure information aggregation in sensor networks. In: Proc. of the ACM Conf. on Embedded Networked Sensor Systems (SenSys), ACM Press, New York, pp 255-265

Rey WJJ (1983) Introduction to Robust and quasi-robust statistical methods. Springer, Berlin, Heidelberg

Shrivastava N, Buragohain C, Agrawal D, and Suri S (2004) Medians and beyond: New aggregation techniques for sensor networks. In: Proc. of theACM Conf. on Embedded Networked Sensor Systems (SenSys), ACM Press, Baltimore, pp 239-249

Stewart CV (1995) MINPRAN: A new robust estimator for computer vision. In: IEEE Transactions on Pattern Analysis and Machine Intelligence, 17: 925-938

Wagner D (2004) Resilient aggregation in aensor networks. In: Proc. of the ACM Workshop on Security in Ad Hoc and Sensor Networks, ACM Press, pp 78-87

Yan J and Pollefeys M (2005) Articulated motion segmentation using RANSAC with priors. In: Proc. of the ICCV Workshop on Dynamical Vision

Zuliani M, Kenney CS, and Manjunath BS (2005) The MultiRANSAC algorithm and its application to Detect Planar Homographies. In: Proc. of the IEEE Int. Conf. on Image Processing, pp 153-156

11 Communication Performance Study

Dario Rossi, Claudio Casetti and Carla-Fabiana Chiasserini

Dipartimento di Elettronica, Politecnico di Torino, Italy

11.1 Introduction

Wireless sensor networks (WSNs) applications are characterised by different requirements in terms of reliability and timeliness of the information to be delivered. It is of fundamental importance to develop efficient sensor technologies that can meet the requirements of the desired applications. In this chapter we investigate through experimental measurements the performance that can be achieved by two well-known technologies: Mica2 (Mica2 2002) and MicaZ (MicaZ 2004) sensors. Our work furthers the existing studies on motes communication by (i) considering both the Mica2 and MicaZ technologies, (ii) conducting different experiments with stationary and mobile sensors, (iii) developing a measurement framework to collect and analyse large amounts of empirical data. More specifically, our objective is to get a better understanding of the actual behaviour of Mica2 and MicaZ sensors, as well as of the quality of the communication links. The insights that we provide help in identifying which application requirements can be satisfied (and under which conditions), and in designing protocols at the higher layers that mitigate the inefficiencies of the sensor nodes. To derive meaningful results in spite of the difficulties in conducting empirical measurement, we developed a measurement framework and followed a rigorous methodology in the experimental setup. We focus on the performance metrics that better enable us to understand and characterise the communication dynamics in sensor networks, that is: packet loss probability and average length of a packet loss burst. Furthermore, we carry out experiments under some set-ups that help most in highlighting the effects of channel conditions and interfering transmissions. We hope that the description of the methodology used for our experiments and the measurement tool that we developed can be useful to other researchers.

11.2 Review

Several studies have been presented on sensor networks but only few of them rely on experimental measurements, though their currently increasing number (Ganesan 2002; Zhao 2003; Woo 2003; Lal 2003; Zhou 2004; Reijers 2004; Cerpa 2005; Whitehouse 2005; Anastasi 2005; Malesci 2006; Lymberopoulos 2006) testify the growing interest of the whole research community in this field. Recent experimental studies number (Ganesan 2002; Zhao 2003; Woo 2003), identify the existence of three distinct reception regions in the wireless link: connected, transitional, and disconnected. Therefore, the real channel behaviour deviates to a large extent from the idealised disc-shape model used in most published results, and it is now well recognised that the anisotropy of the wireless channel is so important that the use of simple radio models may lead to wrong simulation results at the upper layers. In particular (Ganesan 2002), one of the earliest works presents empirical results on flooding in a dense sensor network composed of Mica motes radio, studying different effects at the link, MAC, and application layers. Results show that under realistic communication, the flooding structure exhibits a high clustering behaviour, in contrast to the more uniformly distributed structure that would be obtained with a disc-shape model. The work in (Zhao 2003), reports measurements of packet delivery for a sixty-node test-bed in different indoor and outdoor environments, studying the impact of the wireless link on packet delivery at the physical and at the MAC layer by testing different encoding schemes (physical layer) and traffic loads (MAC layer). The effect of link connectivity on distance-vector based routing in sensor networks is the object of investigation in (Woo 2003), where authors show that cost-based routing using a minimum expected transmission metric shows good performance. More complex cost metrics are investigated in (Lal 2003), where the authors also address the problem of measuring such metrics in an energy-efficient way.

Building over the rationale that for large-scale networks on-site testing may be unfeasible and models for simulators will be needed, many tools and models have also been proposed. As an example, (Woo 2003) also derives a packet loss model based on aggregate statistical measures, such as mean and standard deviation of packet reception rate, which is simplistically assumed to be gaussian distributed for a given distance. Using the SCALE (Cerpa 2003) tool, the work in (Cerpa 2005) identifies other important factors for link modelling providing a spectrum of models of increasing complexity and accuracy. The work in (Zhou 2004) is based on the experimental investigations of the non-isotropic behavior of the Mica2 motes radio and the effects of the heterogeneity in hardware calibration of these devices, whereas the study in (Whitehouse 2005) addresses the interesting aspect of sensor capture capability and its effect on MAC layer performance.

Closest to our studies are (Reijers 2004), (Anastasi 2005). The empirical experiments in (Reijers 2004) are carried out using the EYES sensors (Eyes 2002) and aim at assessing the link quality in different environments, in presence of interfering transmissions and varying the sensors orientation. In (Anastasi 2005) the maximum throughput and the communication distance that can be achieved with Mica2 sensors are investigated, though the study is not entirely devoted to wireless sensor networks as it also focuses on 802.11b stations. Finally, recent work has been devoted to experimentally evaluate interactions between several protocols at the MAC and network layers (Malesci 2006), which have usually been studied in isolation, in terms of throughput and packet loss. Other works, such as

(Lymberopoulos 2006), provide a detailed characterisation of signal strength properties and link asymmetries for covering the most recent architectures, such as the MicaZ motes used in this study. To the best of our knowledge, however, no experimental work provided a systematic, rigorous investigation of the impact of mobility on communication performance of wireless sensor networks, which is among the main objectives of this study.

11.3 Experimental Setup

To further our knowledge of wireless sensor communication dynamics, we developed a measurement framework, which will be made available on the Internet, aimed at collecting and analysing a large database of packet-level traces. We not only collected a large quantity of sensor traces, but we also applied a rigorous methodology to gather useful insights from the observations: in other words, our aim has been to perform each measurement in an environment, which is as controlled as an experimental setup can be. In this section we discuss the most prominent issues of the experimental framework, describing the network setup, the sensor applications and the methodology followed in the measurement campaign.

11.3.1 Networking Setup

As already mentioned, we used two families of sensors, Mica2 and MicaZ, both manufactured by Crossbow Technologies Inc., and based on implementations of a CSMA-based MAC protocol; in particular MicaZ features a IEEE 802.15.4 compliant MAC. Being the aforementioned architectures among the most recent steps of the mote technology evolution started in (McLurkin 1999), (Hollar 2000), it is evident that the Mica2 and MicaZ families share a number of similarities: both the devices consist of a 7.3 MHz ATmega128L low-power and low-cost processor, with 128KB of code memory, 512KB EEPROM, 4KB of data memory. Their physical size, including two AA batteries, is about $5.7 \times 3.1 \times 1.8$ cm^3; MicaZ motes use 2.9 cm long $\lambda/2$-monopole antennas, whereas Mica2 use 8.2 cm long $\lambda/4$-monopole. However, we are more interested in their differences, which mainly concern the communication capabilities. Mica2 (Mica2 2002), (Hill2 2002), released in 2002, is still one of the most popular platforms in WSN research. The architecture uses ChipCon's CC1000 (ChipCon 2004) half-duplex radio, which exports a byte-level interface, operating in the Industrial, Scientific and Medical (ISM) band with a 868 MHz central frequency[1]. Besides, the radio is capable of a

[1] Although the CC1000 operating frequency is software-selectable, the external hardware is set to operate in one frequency band only, among 315,433 and 868/916 MHz. More precisely, in all the following experiments the Mica2 center frequency is set to be 867.9972 MHz, as preliminary experiments found this frequency to be the least subject to external interference in our environment.

noise-resilient FSK modulation and the maximum achievable data rate is 38.4 Kbaud Manchester-encoded, which yields a nominal bit rate of 19.2 Kbps. MicaZ (MicaZ 2004), released in 2004, continues the evolution by adopting a CC2420 (ChipCon 2005) RF transceiver that, operating in the 2.4 GHz unlicensed ISM band, achieves a significantly higher 250 Kbps data rate. The radio, which is capable of O-QPSK with half sine pulse shaping modulation, includes a digital Direct Sequence Spread Spectrum (DSSS) baseband modem, providing a spreading gain of 9 dBm. Moreover, a number of other features are fully implemented and automated in hardware, such as CRC computation and checking, Clear Channel Assessment (CCA), and several security-related operations. In both architectures, we used explicit link-layer acknowledgments and, to simplify the analysis, we considered a single transmission power setting P equal to 1 mW (0 dBm); in these conditions, data sheets report a typical current consumption of 17.4 mA for MicaZ and 16.5 mA for Mica2. Note that, while 0 dBm is the maximum possible RF power setting for the MicaZ architecture, the Mica2 maximum power is +5 dBm when operating at 868 MHz.

11.3.2 Praxis

In each of the following experiments, one or more motes transmit packets, while a single mote acts as a sink. Since we consider single-hop transmissions only, the *network* topology is always a star: therefore, in the remainder of the chapter the term *topology* will usually refer to the actual *physical placement* of the motes in the environment.

Though it is quite common, especially in indoor environments, to connect the motes to external *DC-powered* gateway-boards, our experience indicates that the external powering *does* influence experimental results. Therefore, in all our experiments all sensors are battery-operated (for every set of measurement runs, we use fully-charged batteries). The sink mote is attached to a serial programming board acting as a gateway toward a notebook PC running Linux, where a custom C application is used to gather the packet-level traces. Since application packets are small and the network utilisation very low, the *entire* packet content is stored. Each packet is time-stamped as soon as a start-packet delimiter byte is sent over the Universal Asynchronous Receiver/Transmitter (UART) interface; however, due to the unreliability of serial port communications for fine-grained time stamping, we also use the internal mote clock for this purpose.

Under any experiment, special care has been devoted to the *vertical* motes placement, i.e., to their height from the ground: indeed, if the motes are too close to the ground, a significant fraction of the radio-wave energy may be lost due to diffraction. This behavior, explained by the theoretical framework in (Green 2002) for wireless 802.11 devices (IEEE802.11 1999), has been confirmed in the WSN context by the experimental measurements in (Anastasi 2005), where the authors empirically found negligible power loss when sensors are over 1.50 m above the ground. In all our experiments, we fastened the motes to one end of 1.80 meter-high supports.

11.3.3 Measuring Application

The measurement system we developed is based on TinyOS (Hill 2000), an open-source operating system that, thanks to its popularity, has become the *de facto* standard for WSN applications development. Both the TinyOS operating system as well as its applications is written in nesC (Gay 2003), a programming language for networked embedded systems tailored to integrate reactivity to the environment, concurrency and communication.

We wanted the transmitter-receiver applications to be as simple as possible, in order to minimise their influence on the measured variables: on these premises, the *receiver* application constantly listens to the radio channel, time-stamping and forwarding over UART any received packet. The *transmitter* application is a simple Constant Bit Rate (CBR) source, where the packet inter-arrival is a discrete random variable with average T=200 ms, assuming values $T\pm4i$ in the support $[T-20,T+20]$ with uniform probability. The 4-ms granularity was chosen because the Mica2 stack is known to unreliably handle shorter timers, whereas the average inter-arrival is large enough not to limit the maximum number of transmission attempts at the MAC layer.

Furthermore, despite the many similarities shared by Mica2 and MicaZ motes, they have two different hardware architectures, supported by TinyOS through two distinct radio stacks. Thus, some of the quantities we want to derive (e.g., RSSI, time stamping, MAC-layer variables) are directly measured, and the measurement method is obviously different for either architecture. Other quantities, such as the packet loss rate and the loss burst length, are *inferred*, rather than actually measured, and are therefore unaffected by the underlaying hardware.

The correct packet reception or packet loss events are discriminated by using (i) a unique mote identifier and (ii) a per-mote incremental packet identifier. At the receiver, a packet loss is inferred when the difference L in the identifier space of two subsequently received packets originated from the same mote is greater than 1; in this case, the length of the loss burst is $L-1$.

Both mote and packet identifiers are 2-byte long integers, thus limiting the length of the longest loss burst to about 65k packets. At a rate of 5 packets per seconds, this corresponds to a blackout period of nearly four hours, which is more than enough for typical WSN operating conditions (e.g., the longest silence gap over all our experiments has been about 371 (222) packets long for the Mica2 (MicaZ) case). Indeed, though early measurement studies (Szewczyk 2004) reported that it is possible for motes deployed in outdoor environments to exhibit extended down-times (of the order of several hours) followed by self-recovery, it must be said that, in (Szewczyk 2004), the weather conditions to which the sensors were left unattended were rather demanding (i.e., heavy rain, strong wind, persistent humidity).

Both the Mica2 and MicaZ architectures support a ChipCon RF transceiver featuring a built-in Received Signal Strength Indicator (RSSI), with very similar precision characteristics. Indeed, although the RSSI dynamic range is different, the RSSI accuracy is about ±6 dBm in both cases. For the MicaZ, the RSSI is a digital value that can be read from an 8-bit signed 2-complement register of the CC2420.

The register value, which is always averaged over 8 symbol periods in accordance with the 802.15.4 standard (IEEE802.15.4 2003), can be directly referred to the power at the RF pin with a simple offset translation: $P_{MicaZ} = RSSI_{raw} - 45$ [dBm]. For the Mica2, the knowledge of the received RF power involves a more complicated process. The CC1000's RSSI output is an analog signal and two A/D conversions are needed to gather, for motes operating at 868 MHz, the RSSI as: $P_{Mica2} = -50.0 \times (1.223/V_{raw}) \times RSSI_{raw} - 45.5$ [dBm]. Thus, in the Mica2 case, the application must fetch the battery voltage besides reading the raw RSSI A/D channel; this implies that the packet payload has to carry some additional information (i.e., more bytes). To minimise the influence of this additional task, we sample the battery voltage sporadically, since the battery level is likely to stay constant over short time spans, such as one-minute intervals. Finally, although both devices estimate the RSSI with a ±6 dBm accuracy, some practical differences arise. Experimentally, the aforementioned formulas yield steps of 0.1 dBm (1 dBm) for the Mica2 (MicaZ) platform: as a result, Mica2 estimates are only apparently finer-grained, whereas MicaZ readings appear to be more stable, less oscillating and less noisy.

11.3.4 Application Packets

The application performance clearly depends on the length of the packets that have to be transmitted (e.g., in the multi-access case, the longer the packet the higher the collision probability for a given transmission rate). In the following, we will only consider 29-byte long application payloads, i.e., the maximum size allowed under TinyOS constraints: the application-layer packet, containing a 5-byte header and 2-byte CRC, is thus 36-byte long. At the MAC layer, a 18-byte preamble and a 2-byte synchronisation information are added, leading to a total of 56 bytes. This simplification follows from:

- The space requirement to transmit all the needed information: for example, time-stamping packets at both the transmitter and the receiver consumes *more than one fourth* of the full-size payload;
- The belief that many applications, aiming at reducing the overall power-consumption, will anyway aggregate as much information as possible in a single data packet (i.e., except for routing and short command packets), thereby reducing the wake-up periods by increasing the radio communication efficiency.

11.4 Methodology Description

To achieve meaningful results, experiments should be repeated a sufficient number of times, using as many different sensors as possible, deployed in as many different topologies as possible. However, a trade-off exists between the *controllabil-*

ity of an experiment and the number of parameters, which one can take into account. We therefore bounded each of the aforementioned requirements to the smallest possible amount that still allows significant statistics to be gathered. In other words, we

- Selected a subset of all the available sensors and ran some preliminary tests to identify the *N* most stable, reliable ones;
- Used a limited number of topologies;
- Repeated the experiments as many times as possible, under weather conditions as similar as possible.

11.4.1 Sensor Subset Selection

As observed in (Zhou 2004), sensor devices may have significantly different performance due to different hardware calibration. To identify an appropriate subset of nodes to be used, we ran some preliminary experiments and selected the sensor which yielded the most "stable" results, with the following criterion. We repeated a batch of simple tests looping over each of the available sensors, measuring (i) the RSSI and (ii) the packet loss probability at different spatial distances from the sink node. Then, we considered only the motes corresponding to the *central* portion of the loss probability and of the received signal strength distributions. Several runs of the tests were necessary to converge to a significant subset of motes – especially for the Mica2 architecture. Namely, in the following we will consider a subset of N=8 source sensors per technology, chosen among the initial 24-MicaZ and 24-Mica2 populations; indeed, this number is:

- Small enough to allow for repetition and topologies randomisation
- Small enough for the selection process to converge
- Big enough to smooth out physical differences among sensors

There are two important remarks that we want to highlight on this point. First, we point out that the buffer at the *receiver* side has been carefully dimensioned, in order to eliminate from our experiments any problems tied to buffer overflow. To this purpose, we performed some simple preliminary experiments suitable to quantify the buffer overflow probability, which is shown in Fig. 11.1. Considering a population size of N=8 motes transmitting at the same time within the radio range of the receiver, the figure depicts the probability that the receiver queue overflows as a function of the queue size itself, whereas the inset reports the same information using a logarithmic scale for the y-axis. Though in our experiments we used a single transmission rate corresponding to an *average* packet inter-arrival time of 200 ms, in order to conservatively set the buffer size, we also inspected higher transmission rates, corresponding to a packet inter-arrival time of 100 and 50 ms respectively. From the plot, it is easy to gather that the receiver buffer never overflows for packet inter-arrival times greater than or equal to 100 ms, provided that at least two packets can simultaneously be queued at the receiver. Conversely, at higher rates it may be necessary to significantly increase the buffer size in order to completely eliminate the occurrence of buffer overflow. In the following, we conservatively set the receiver queue size to 5 packets, which is more than twice the worst case observed in these preliminary tests.

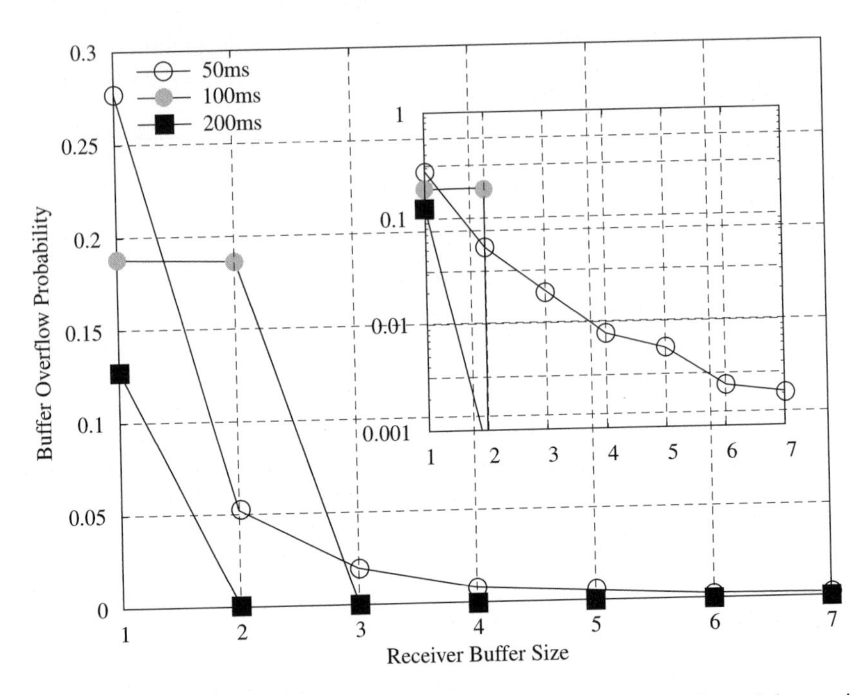

Fig. 11.1. Buffer overflow probability at the network sink as a function of the receiver buffer size, when N=8 Mica2 motes transmit at the same time and considering different transmission rates

Secondly, Fig. 11.2 depicts the sensor selection process in greater details. The picture reports, for the Mica2 architecture only, the per-mote packet loss probability measured at two different distances from the receiver, obtained aggregating the performance of individual motes over several runs. The performance is sorted from best to worst, and the *relative order* of the motes differs in the two cases: the motes ranking is subject to significant variations over different runs of the same test, i.e., it is very unlikely that the same mote always yields the same performance. However, we can identify three subsets of motes that exhibit similar behavior. Within a single subset, this is highlighted in the figure by the lines joining points corresponding to the performance of the same mote at different distance. In the following tests, we selected the motes whose performance lies around the median to be representative of the whole sensor population. It can be noted that, in the case of short-distance measurements, the median packet loss probability is very close to the average performance over all motes. Conversely, when transmitters are far from the receiver, the poor performance of few motes raises the mean packet loss probability well beyond the median of the distribution.

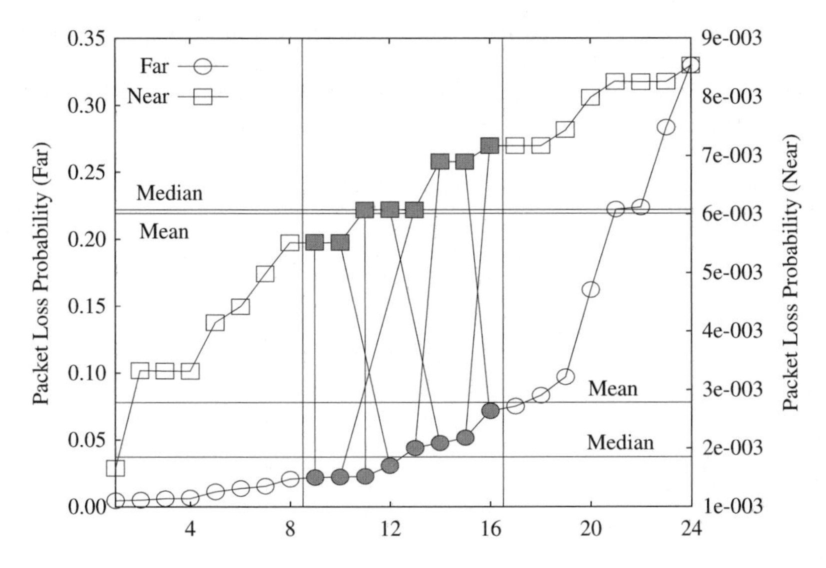

Fig. 11.2. The Mica2 motes selection process: Per-mote packet loss probability in the case of near and far communications, ordered from best to worst performance; the central portion of the population, whose performance is indicated by filled points, represents the selected subset used as representative of the whole population

11.4.2 Topologies Sampling

Topology sampling is necessary to reduce the bias possibly introduced by radio-connectivity anisotropy. Unless otherwise specified, the results presented in the following have been gathered with the following topological randomisation criterion. In the first series, we rotated each sensor over each topological point once, one for each experimental run; on subsequent runs we did the same, but we randomised the initial sensors position on each different series. In the case of single-sensor setup, where solely the distance D between transmitter and receiver describes the topology, we sampled several positions for the sink node, while we sampled different transmitter positions along the circle of radius D centred in the receiver mote. In the case of multi-sensor setup, we considered two simple topologies, namely a circular placement of transmitters centred around the sink mote, and an array of transmitters, where we varied the position of the receiver depending on the purpose of the test, as will be detailed below.

11.4.3 Weather and Day-time Sampling

We kept the experiment duration bounded in order to complete each experimental run under the same weather and daytime conditions. Also, we considered a whole run to be an outlier when the average values of the metric of interest were beyond a *rejection threshold*, given by the linear combination of the average and standard deviation of the considered metric. Namely, we discarded an experimental run when the difference of the average packet loss probability measured in the experiment exceeded the average obtained from the previous ones by at least twice the standard deviation. We point out that it is relatively simple for an experiment to fail but it may be surprisingly hard, whether possible at all, to identify the causes of its failure.

11.5 Experimental Results

This section presents an analysis of some of the most interesting scenarios that we explored. We considered both static and dynamic sensors, deployed in indoor and outdoor environments, either with single- or multi-sensor setup, collecting about 1.5 millions of packets. The aim of single-sensor experiments was to assess the importance of the radio channel conditions as well as the robustness of the two technologies; conversely, multiple sensors scenarios allowed us to investigate the impact of channel contention at the MAC layer.

11.5.1 Single-sensor Experiments

In these experiments only one sensor at the time transmits to the sink node (i.e., there is not channel contention). In each run, we recorded 250 correctly received packets per-mote; experiments have been repeated for 2 different topologies, in 2 different week-days, at 2 different day-times for a total of $250\times4\times2\times2\times2 = 8{,}000$ collected packets per mote. Note that, to achieve reliable results for the Mica2 architecture, we were forced to collect a double amount of data. For the sake of brevity, here we describe the results of outdoor experiments only.

11.5.2 Impact of Distance

It is well known that the transmission range of a wireless system is influenced by several factors – such as the transmission power, the antenna gain and efficiency, the operating frequency and the transmission data rate. The empirical motes transmission range has been the object of investigations, where the authors indirectly identify the transmission range as the distance beyond which the packet loss probability exceeds a given threshold. A similar approach is adopted, where authors take into account the impact of different transmission power settings as well. In our outdoor experiments, we analysed the performance of static sensors transmitting one at a time, placed at 4 different distances (namely, 4, 10, 20 and 40 m) *within* transmission range from the sink node: the aim of the experiments is to assess the fine-grained quality of the communication that can be set up, rather than just asserting whether it can be set up or not.

Table 11.1: Packet loss percentage and loss burst length for different communication radii

Motes	*Radius [m]*	*Loss Percentage*			*Loss Burst Length*		
		Best	Mean	Worst	Mean	95th	Maximum
Mica2	4	0.00	0.63	1.01	1.12	1	2
	10	0.16	0.88	1.85	1.38	1	6
	20	0.44	5.48	16.20	1.53	4	34
	40	1.15	6.23	22.34	1.87	6	47
MicaZ	<40	0.00	0.00	0.00	–	–	–
	40	0.00	8.65	47.26	2.83	9	222

The 150-meter transmission range for the Mica2 technology is about twice the length of the MicaZ range; therefore we were expecting the latter platform to be more sensitive to the communication distance. This is partly confirmed by the packet loss percentage and loss burst length performance, which are summarised in Table 11.1. Packet losses for the 2.4 GHz technologies begin to be more consistent (and the loss bursts longer) than those of 868 MHz motes at a distance of 40 m, which roughly corresponds to half the MicaZ transmission range. More striking are the facts that (i) no loss were recorded for distances below 40 m in the MicaZ case, whereas (ii) Mica2 sources experience faulty transmissions even when they are extremely near to the sink, and that (iii) some of the MicaZ motes experienced no losses even at the farthest distance. Let us explain this behavior with the help of Fig. 11.3, which presents some statistics of the RSSI as a function of the communication radius for both the architectures. The boxes delimit the 25th and the 75th percentile, the line inside the box gives the mean RSSI and the lines outside the box indicate the maximum and minimum values achieved. The plot clearly shows that, as expected, the mean RSSI decreases for both architectures as the communication distance increases; however, while the DSSS technique shows its efficacy for MicaZ motes, in the Mica2 case the minimum RSSI is (i) very low (about –95 dBm) even for short distances and (ii) almost unaffected by the explored distances. The latter effect, partly due to antenna directionality issues, is the main reason for the non-negligible amount of packet loss in nearby communications.

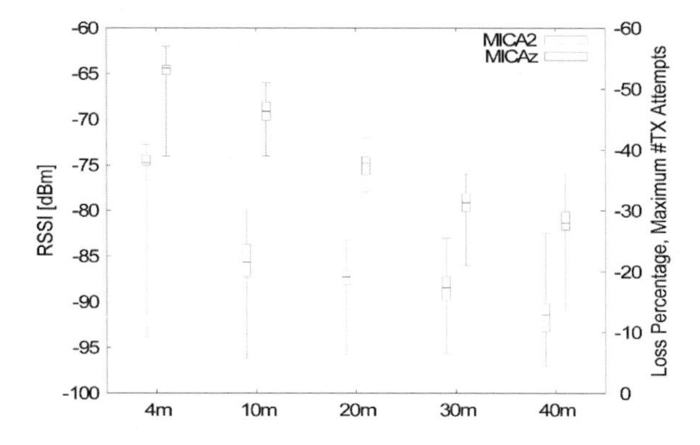

Fig. 11.3. Single-sensor static experiments: RSSI measurements

Finally, Fig. 11.4 reports, for the Mica2 architecture only, the average (left y-axis) and the maximum (right y-axis) number of transmission attempts as well as the packet loss percentage attempts, as functions of the communication distance. Rather than inferring the number of transmission attempts from delay measurements, we instructed the MAC-layer to add, on a reserved byte of each outgoing packet payload, the information concerning the previous[2] packet transmission. Since some of the packets were lost during the experiments, the information gathered at the sink is affected by losses due to the wireless channel: in case of multiple losses, there is no way of reconstructing the actual number of retransmission attempts of none but the last packet of the loss burst. The bias toward correctly received packets is clearly visible in that the average and maximum number of transmission attempts actually *decrease* as the distance grows beyond 20m, whereas sensors are experiencing consistent loss rates (about 20% at 50 m, coherently with (Anastasi 2005)) and the length of the loss burst is increasing as shown earlier in Table 11.1.

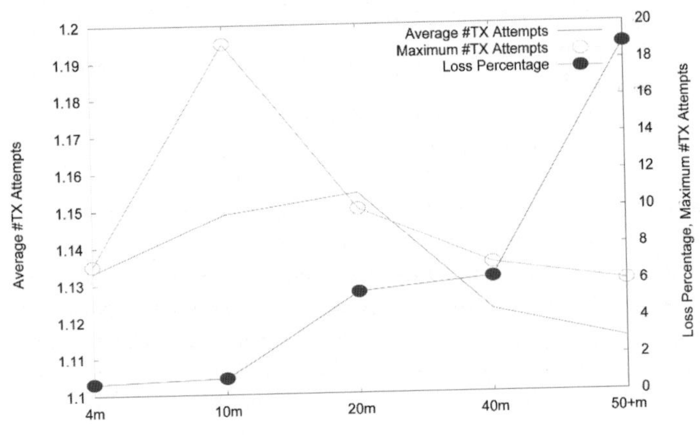

Fig. 11.4. Single-sensor static experiments: Average and maximum number of transmission attempts, and packet loss percentage for the Mica2 architecture as a function of the transmitter-receiver reciprocal distance

11.5.3 Impact of Mobility

In this second series of experiments, we moved the transmitting sensor back and forth along one quarter of a circular path having the sink as a centre; in this case we considered only two communication distances (4 and 20 m). The results reported in Table 11.2 show that mobility has a severe impact on packet loss and

---

[2] This architectural choice is dictated from (i) the need of a common cross-platform approach and (ii) the fact that outgoing MicaZ packets are transferred into a FIFO buffer prior of transmission, so that is difficult to write on-the-fly any information regarding the current transmission on the outgoing packet itself.

RSSI, especially in the case of Mica2 platforms. At short distance, mobility attenuates the received signal by about 10 dBm, increasing the average loss percentage by a factor of 4; at longer distance, communication is severely penalised by the source motion: the average received signal is few dBm above the radio sensitivity, causing the loss of nearly half of the transmitted packets. Conversely, the mobility impact on MicaZ motes, though non-marginal, is limited: for instance, it is still possible to have lossless communications when the moving sensors are 4 m apart. We also performed the same series of tests with static-transmitter and mobile-receivers, and we obtained similar results.

Table 11.2: Impact of mobility and communication radius on loss percentage and RSSI

Motes	*Radius [m]*	*Static* Loss percentage				*Mobile* Loss percentage			
		Best	Mean	Worst	RSSI [dBm]	Best	Mean	Worst	RSSI [dBm]
Mica2	4	0.00	0.63	1.01	-74.93	1.22	2.23	3.25	-83.57
	20	0.44	5.48	16.20	-87.25	28.51	42.09	51.51	-90.30
MicaZ	4	0.00	0.00	0.00	-64.37	0.00	0.02	0.05	-65.15
	20	0.00	0.00	0.00	-74.38	2.49	3.73	6.25	-78.29

11.5.4 Multi-sensor Experiments

The aim of this set of experiments was to assess the importance of the channel contention on the performance achieved by the different sensor technologies, highlighting which, among channel conditions and medium contention, has the greatest impact. To investigate the effect of several factors (such as contention, distance, topology, mobility, environment), we devised different scenarios and tested the impact of these factors either combined or in isolation.

11.5.5 Circular Topology

In this experiment we considered a simple circular topology, for both indoor and outdoor environments, the latter being a large $18 \times 18\ m^2$ open-space room with high ceiling and several sources of obstructions (such as desks, computers and other pieces of furniture). We considered N sensors contending the same radio channel, placed at the vertex of the N-sided regular polygon inscribed in a circle of radius r=4 m. Note that both Mica2 and *indoor* experiments required a double-sized database.

11.5.6 Variable Sensors

Initially, we varied the number of sensors to understand the importance of the medium access contention versus the physical surrounding environment. Fig. 11.5 shows, for both technologies, the packet loss probability in indoor and outdoor environments, where either 4 or 8 motes are contemporaneously deployed. As before, the boxes delimit the 1st and 3rd quantile, the inner thick line represents the average and the outer lines the maximum and minimum bounds. In this case, we can no longer directly compare the results *quantitatively*, since the channel contention has a smaller impact on the higher data-rate technology. However, the *qualitative* behavior is the same across platforms: increasing the number of motes increases the channel contention, resulting in higher packet loss rates. This is particularly evident in the indoor case where, due to the presence of obstructions and reflecting surfaces, Mica2 motes experience more than 40% packet loss and even MicaZ motes fail to transmit about one fourth of the packets. It is worth observing that the MicaZ results are much more stable than those for the Mica2: the 25th and 75th percentile are almost indistinguishable, and the difference among best and worst mote is always very small.

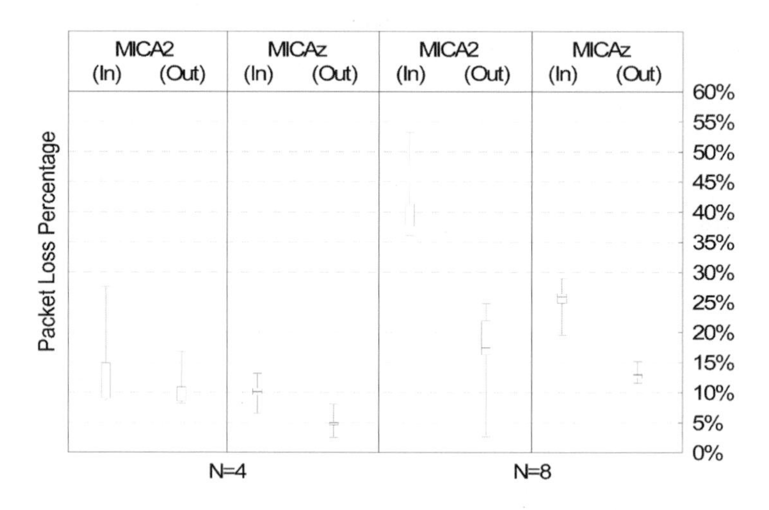

Fig. 11.5. Multi-sensor circular topology: Packet loss percentage distribution for different motes population size, hardware architectures and environments.

The length of the loss burst is the object of Fig. 11.6 where, as usual, the inner thick line represents the average and the outer line the maximum, whereas the box delimits the 95th percentile (since the 75th percentile is very close to the average). As expected, contention plays an important role especially for Mica2 motes, as the increase in the loss burst length testifies. Looking at the outdoor MicaZ experiments, we observe that the packet loss is actually more than doubled by the increase of the motes number from 4 to 8, while the loss burst length is practically unaffected: in either case, 95% of the loss bursts involve no more than 3 packets.

Finally, Fig. 11.7 presents the survival function of the number of transmission attempts, for both indoor and outdoor environments and both N=4 and N=8, in the MicaZ case only. In this case the length of the loss burst is very small: thus, when a single packet is lost, the next received packet carries the number of transmission attempts concerning the lost packet. Therefore, the bias toward correctly received packets observed in Fig. 11.4 is partly lessened. The transmission attempts behavior confirms that the surrounding environment plays as important a role as that played by contention for the MicaZ platform. Two transmission attempts are sufficient in the 95% of the cases when either N=8 motes are deployed outdoor or N=4 indoor; combining both the environment and contention effects (i.e., in the N=8 indoor case), one fourth of the transmissions require at least a second attempt.

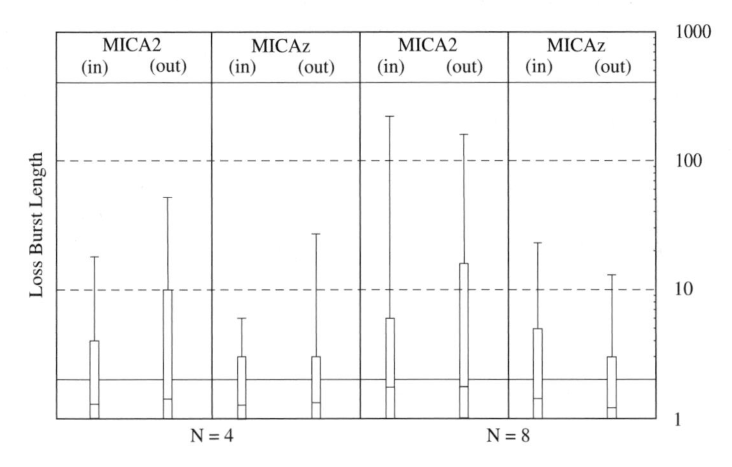

Fig. 11.6. Multi-sensor circular topology: Length of the loss burst distribution for different motes population size, hardware architectures, and environments

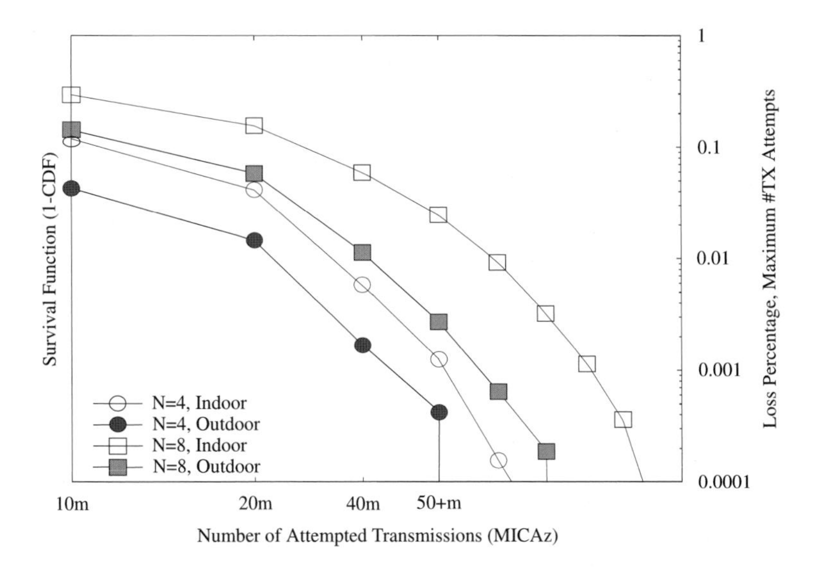

Fig. 11.7. Multi-sensor circular topology: Number of transmission attempts of the MICAz platform for the scenario with a variable number of sensors

11.5.7 Variable Communication Distance

Let now inspect the effects of the distance in the multi-sensor scenario; for the sake of brevity we present the results for outdoor environments only. We fixed the number of sensors to N=8; half of the sensors were at distance r=4 m from the center and the other half at distance 2r=8 m (between any two consecutive *near* sensors there was a *far* sensor). The performances obtained under this scenario are reported in Fig. 11.8 for both technologies and both far and near sensor groups. We averaged the performance of each group over 5 seconds temporal-windows and plotted the (packet loss percentage, loss burst length) couples recorded during the experiment. Note that, given transmission rate settings just outlined, each of the points approximately accounts for a 200-packets population. Additionally, the plot provides two reference lines for each group: the vertical lines are related to the packet loss percentage, the horizontal to the loss burst lengths. The segment position represents the average value, whereas its length represents the variance of the measured quantity: therefore, the point where the two segments cross each other indicates the packet loss percentage and loss burst length average values achieved by each group, whereas the extension of the cross arms is a measure of dispersion.

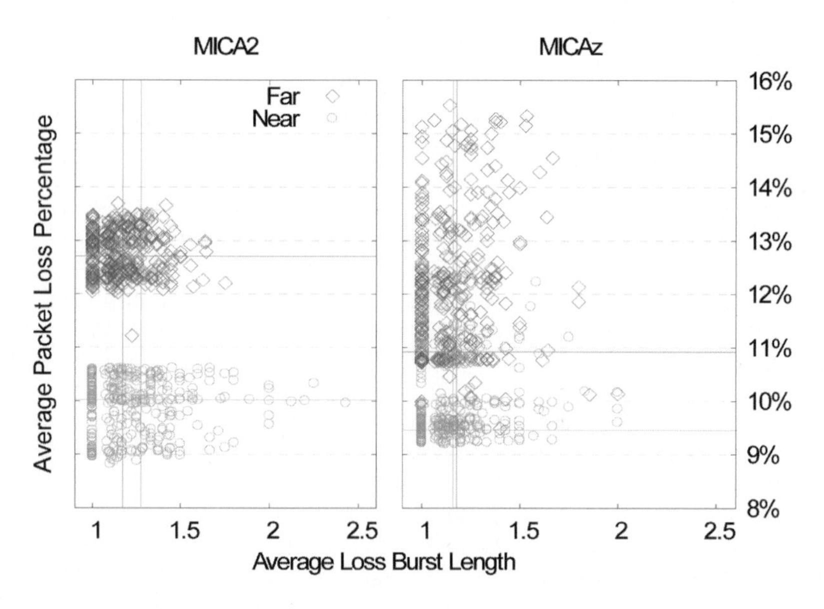

Fig. 11.8. Multi-sensor circular topology: Average packet loss versus average loss burst length for the variable communication distance scenario (five seconds averaging intervals)

Some remarks are in order. First of all, on the one hand, the performance of the two Mica2 groups are completely disjoint: i.e., motes at a lower distance always suffer lower losses with respect to farther motes; on the other hand, MicaZ performance fades more gracefully and the two groups partially overlap. Secondly, the far MicaZ motes performance seems to be more dispersed and even worse than Mica2 motes: however, looking at the average and variance, it is easy to see that *(a)* the MicaZ average performance of far and near motes are closer than for the Mica2 platform (thus, the MicaZ architecture is more fair) and *(b)* far motes performanceare better in the MicaZ case, to the slight detriment of near motes. A third interesting observation concerns the MicaZ performance cloud, whose dispersion in the upper part of the figure is essentially caused by two runs only. However, it can be noted that, due to the relatively high variance of the measurement dataset, for no averaging interval, neither the RSSI nor the packet loss percentage exceed the rejection threshold early introduced. Thus, although it might be desirable to discard these runs, they have to be considered. Indeed, the choice of more restrictive acceptance rules would, (i) unnecessarily eliminate many fluctuations that belong to the real world, yielding too artificial results; and (ii) imply a tremendous overhead for the measurement campaign, since many runs would have to be repeated indefinitely.

11.5.8 Array Topology

Next, we considered an array of N=8 transmitters, interleaved by a constant distance of D=5 meters; all sensors were active and competed to access the same channel. To simplify the analysis, we restrict out attention to the outdoor environment only. The sink node was placed along a line parallel to the transmitter array, and the two lines were D meters apart. We considered both a stationary and a mobile sink: for the former case, we present some results were we varied the sink *position* along the line, in the latter case we varied the sink *speed*.

Static-sink scenario: Let consider the static-sink scenario first, where clearly the physical channel conditions are no longer homogeneous for all sensors as they were in the circular topology. In this case, we used four different initial random placements of nodes on the array, but each mote has no longer been displaced on every other position. Rather, we decided to move the sink on several measurement points: specifically, for every different topology, the sink was placed in each different run exactly in correspondence of one of the transmitters. The per-mote packet loss, averaged over all the topologies and measurement points sampled in the static-sink array experiments, are reported in Fig. 11.9 ordered by ascending loss percentage. Although it is not possible to make a direct comparison with the multi-access circular topology, we expect performance to be negatively influenced by the coupled (i) channel contention and (ii) increased communication distance effects. With respect to the circular topology, both architectures face an average RSSI attenuation of 8 dBm. However, in the MicaZ case the performance degradation is limited to an additional 2% of lost packet, yielding an average loss of about 15%; in the Mica2 case, the average packet loss increases from 17.4% to 24.4% and, as usual, the variance is lower for MicaZ motes.

Let now ideally aggregate the sensors in two groups based on their position on the array: let the sensors placed at 1÷4 positions belong to the "lower" group and 5÷8 to the "upper" one. Let further denote with E_i the group of experimental runs where the sink was positioned in correspondence of the ith sensor. Let now average, using 5-seconds temporal windows, the RSSI values achieved within each sensor group over all the runs belonging to the same experiment group E_i: this methodology yields to the plot of Fig. 11.10. From the perspective of the upper MicaZ group, it can be seen that the RSSI performance improves as the sink is placed in correspondence of higher positions: when the sink occupies any position above the 4th sensor, then all the transmitters of the upper group are near enough to the sink, and the average RSSI value remains roughly constant afterward; the opposite happens for those sensors placed in low-labeled positions. The same behavior holds for Mica2 sensors, although the phenomenon is less intense because of the smaller RSSI values range.

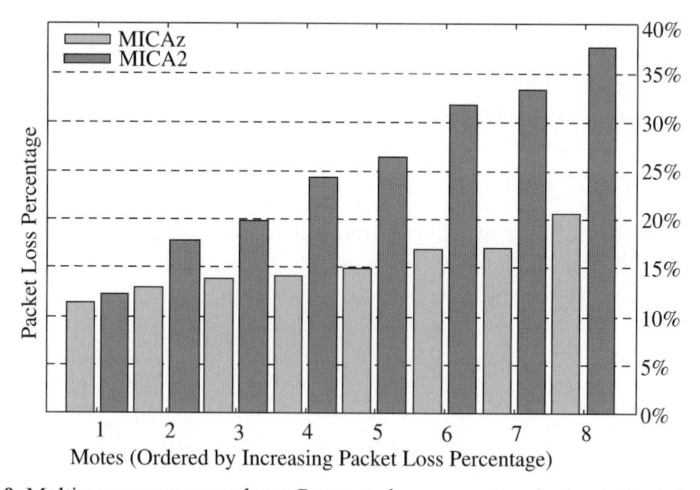

Fig. 11.9. Multi-sensor array topology: Per-mote loss percentage in the static-sink scenario

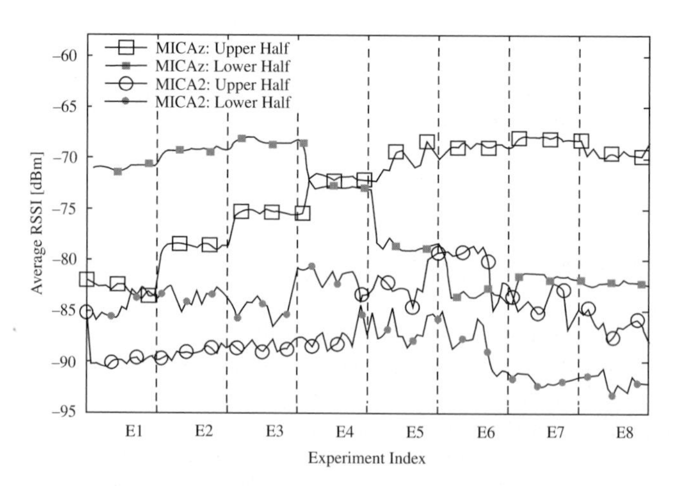

Fig. 11.10. Multi-sensor array topology: Per-group average RSSI in the static-sink scenario

11.5.9 Mobile-sink Scenario

Finally, let consider the impact of the mobility on the performance of the array scenario. While it is relatively difficult to perform variable-speed experiments while moving along a *circular* path with a 20 m radius, conversely an array topology is the ideal candidate to assess the impact of speed. In each run, we moved along a line 5 m apart from the array; adopting the same positional notation early introduced, the travelled path always begun in correspondence of the first sensor and ended in correspondence of the last one. At each run, we increased the sink speed, trying to keep the speed as constant as possible during each run; however, rather than actually controlling the speed *a priori*, we measured the duration of each experiment to gather the actual average sink speed. We repeated the experiments with the same four initial random nodes placement of the static-sink scenario; for each topology, we adopted 8 different sink speeds. On these premises, is not hard to imagine that in no case the sink travels at the same speed twice: therefore, it is not possible to average the results of the different topologies. Nevertheless, the qualitative behavior of the results is the same across the different sets: in the following we will consider the data gathered on a *single* topology for different sink speeds. Fig. 11.11 reports the average packet loss percentage (on the right y-axis) as well as the average length of the loss burst (on the left y-axis) for both Mica2 (on the top x-axis) and MicaZ (on the bottom x-axis) sink speeds – where it must be said that sink speeds are not in scale.

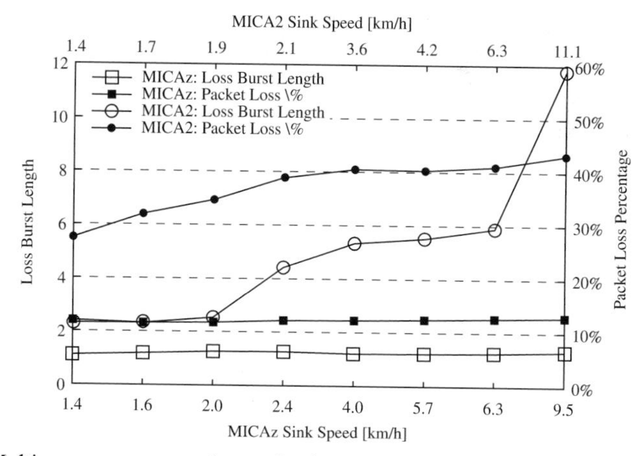

Fig. 11.11. Multi-sensor array topology: Packet loss percentage and loss burst length for the mobile-sink scenario, for different sink speed and different motes architectures

Results lend themselves to interesting considerations. The MicaZ performance is practically unaffected by the sink speed – at least when moving at such slow speeds. This is in agreement with the results obtained from the previous experiments: single-sensor results showed that the communication distance has a greater impact than the sensor mobility, while the static multi-sensor results highlighted the importance of the channel contention on packet loss. Conversely, the results show that for the Mica2 platform the speed at which sensors move monotonically influences the loss performance; indeed, as the sink speed increase, both the average packet loss probability as well as the average loss burst length increase.

11.6 Discussion and Conclusions

We presented an empirical analysis of the performance achievable by wireless sensor networks, using the Mica2 and MicaZ technologies. We developed a measurement framework, based on TinyOS that will be made available on the Internet, along with the database of packet-level traces collected in the experimental campaigns. Following an approach similar to (Reijers 2004), where EYES sensor performance are used, we aimed at assessing the link quality level, taking into account a broad range of factors influencing the communication. The link quality has been described focusing on performance metrics such as packet loss probability, length of the loss burst, number of transmission attempts at the MAC layer and received signal strength measurements. Among the several factors influencing the communication quality, we investigated the communication distance, the surrounding environment, the network size (i.e., the number of simultaneously deployed sensors), topology, sensor mobility as well as the motion speed. As a first observation, our experience is that, in accordance with (Zhou 2004), gathering reliable and repeatable results can be very difficult. Nevertheless, there are a number of major lessons that can be learned from the analysis of the experimental results, which we summarize here. First of all, Mica2 and MicaZ platforms are, despite their commonalities, two different technologies that achieve rather different performance. This suggests that, especially in wireless sensor networks, the application requirements may dictate lower-level choices.

A series of single-sensor tests were performed to investigate the impact of the radio channel conditions on the networking performance: in these tests, we considered stationary and mobile sensors, and varied the communication distance. We discovered that for the MicaZ platform the *distance* plays a greater role with respect to *mobility*: indeed, both static and nearby moving motes experience no loss at all, while motionless far motes do suffer remarkable packet losses. In the case of Mica2, the importance of distance and mobility is reversed, especially as the sensor speed increases: in mobile scenarios, the packet loss may increase by nearly one order of magnitude with respect to static Mica2 performance. Also, we observed that the impact of radio channel conditions on the Mica2 performance is much more significant than for the MicaZ technology. A partial explanation for this behavior lies in the RSSI behavior: the strength of the signal received by the Mica2 platform can be very close to the radio sensitivity even for very low communication distances; conversely, MicaZ performance benefits from the presence of a digital DSSS baseband modem.

Multi-sensor experiments assessed the importance of channel contention on the networking performance: in these tests, we considered static as well as mobile sensors, and we varied the network size and topology, the communication distance, the surrounding environment, the sensors speed. Our experimental findings confirm that medium access contention plays a decisive role in the performance of the sensor network. MicaZ motes, which are slightly affected by the radio channel conditions in nearby communication, suffer a non-negligible amount of losses when even a few motes compete for the same medium; contention effects are even more evident for Mica2 motes. However, it must be said that the role of the environment may still be critical for both architectures when considering indoor communications or harsh weather conditions (Szewczyk 2004) outdoors.

There are two future directions of this work. The first one involves further experimental tests, with special focus on mobile performance: on the basis of the former discussion, this will undoubtedly require a significant effort in order to gather meaningful results. The second one is to provide a wireless link model, based on our empirical results that can be sufficiently simple to be used for modelling and network-level simulations.

11.7 References

Anastasi G, Borgia E, Conti M, Gregori E, and Passarella A (2005) Understanding the real behavior of Mote and 802.11 ad hoc networks: an experimental approach. In: Pervasive and Mobile Computing (Elsevier), vol. 1, 237-256

Cerpa A, Busek N, and Estrin D (2003), SCALE: A tool for simple connectivity assessment in lossy environments. Technical Report, Center for Embedded Networked Sensing (CENS) at University of California, Los Angeles (UCLA)

Cerpa A, Wong JL, Kuang L, Potkonjak M, and Estrin D (2005) Statistical model of lossy links in wireless sensor networks. In: Proc. of the 4th ACM/IEEE Conf. on Information Processing in Sensor Networks (IPSN'05), Los Angeles, CA

ChipCon (2004) CC1000 Data Sheets, release 2.3 August 2004:
http://www.chipcon.com/files/CC1000_Data_Sheet_2_3.pdf

ChipCon (2005) CC2420 Data Sheets, release 1.3 Nov.
http://www.chipcon.com/files/CC2420_Data_Sheet_1_2.pdf

EYES Project, http://www.eyes.eu.org

Ganesan D, Krishnamachari B, Woo A, Culler D, Estrin D, and Wicker S (2002) Complex behavior at scale: an experimental study of low-power wireless sensor networks, Technical Report, University of California Los Angeles (UCLA), CSD-TR 02-0013

Gay D, Levis P, von Behren R, Welsh M, Brewer E, and Culler D (2003) The nesC language: a holistic approach to networked embedded systems. In: Proc. of Programming Language Design and Implementation (PLDI'03)

Green DB and Obaidat MS (2002) An accurate line of sight propagation performance model for ad-hoc 802.11 Wireless LAN (WLAN) devices. In: Proc. of IEEE Int. Conf. on Communications (ICC'02), New York, NY

Hill J, Szewczyk R, Woo A, Hollar S, Culler S, and Pister K (2000) System architecture directions for network sensors. In: Proc. of ACM Conf. on Architectural Support for Programming Languages and Operating Systems (ASPLOS'00), Cambridge, MA

Hill J, Culler D (2002) Mica: a wireless platform for deeply embedded networks. IEEE Micro, 6:12-24

Hollar S (2000), Cots dust. Master's thesis, University of California, Berkeley (UCB)
IEEE std. 802.11 (1999) Wireless LAN MAC and physical layer specifications
IEEE std. 802.15.4 (2003) Wireless MAC and physical layer specifications for low rate wireless personal area networks (LR-WPANs)
Lal L, Manjeshwar A, Herrmann F, Uysal-Biyikoglu E, and Keshavarzian A (2003) Measurement and characterisation of link quality metrics in energy constrained wireless sensor networks. In: Proc. of IEEE Global Communications Conf. (Globecom'03), San Francisco CA
Lymberopoulos D, Linsey Q, and Savvides A (2006) An empirical characterisation of radio signal strength variability in 3-D IEEE 802.15.4 networks using monopole antennas. In: Proc. of the 3rd Workshop on European Wireless Sensor Networks (EWSN'06), ETH Zurich, Switzerland
Malesci U and Madden S (2006) A measurement-based analysis of the interaction between network layers in TinyOS. In: Proc. of the 3rd Workshop on European Wireless Sensor Networks (EWSN'06), ETH Zurich, Switzerland
McLurkin J (1999) Algorithms for distributed sensor networks. Master's thesis, University of California, Berkeley, (UCB)
Mica2 (2002) Data Sheets,
http://www.xbow.com/Products/Product_pdf_files/Wireless_pdf/MICA2_Datasheet.pdf
MicaZ (2004) Data Sheets,
http://www.xbow.com/Products/Product_pdf_files/Wireless_pdf/MICAz_Datasheet.pdf
Reijers R, Halkes G, and Langendoen K (2004), Link layer measurements in sensor networks. In: Proc. of the 1st IEEE Conf. on Mobile Ad-hoc and Sensor Systems (MASS'04), Fort Lauderdale, FL
Szewczyk R, Polastre J, Mainwaring A, and Culler D (2004) Lessons from a sensor network expedition. In: Proc. of the 1st European Workshop on Wireless Sensor Networks (EWSN'04), Berlin, Germany
Whitehouse K, Woo A, Jiang F, Polastre J, and Culler D (2005) Exploiting the capture effect for collision detection and recovery. In: Proc. of the 2nd IEEE Workshop on Embedded Networked Sensors (EmNetS-II), Sydney
Woo A, Tong T, and Culler D (2003) Taming the Underlying Challenges of Reliable Multihop Routing in Sensor Networks. In: Proc. of the 1st ACM Conf. on Embedded Networked Sensor Systems, (SenSys'03), Los Angeles
Zhao J and Govindan R (2003) Understanding packet delivery performance in dense wireless sensor networks. In: Proc. of the 1st ACM Conf. on Embedded Networked Sensor Systems (SenSys'03), Los Angeles
Zhou G, He T, Krishnamurthy S, and Stankovic J (2004). Impact of radio irregularity on WSN. In: Proc. of the 2nd ACM Conf. on Mobile Systems, Applications, and Services, (MOBISYS'04), Boston

12 Energy Efficient Data Management

Haibo Hu[1], Jianliang Xu[2] and Xueyan Tang[3]

[1]Hong Kong University of Science and Technology, Hong Kong
[2]Hong Kong Baptist University, Kowloon Tong, Hong Kong
[3]Nanyang Technological University, Singapore

12.1 Introduction

Sensor networks have reached the stage of massive deployment for a wide range of applications such as ecosystem monitoring and traffic surveillance (Akyildiz et al. 2002; Szewczyk et al. 2004). A wireless sensor network typically consists of a base station (sometimes called *sink*, or *gateway*), and a set of sensor nodes (sometimes called *sources*, see Fig. 12.1). The sensor nodes are used to continuously capture environmental data such as temperature, humidity, lightness, etc. To send these data back to the base station and assist message routing, they are also equipped with radio transceivers that are capable of short-range transmission. The base station, as observed from the figure, serves as the gateway of the sensor network to exchange data and control messages with external applications to accomplish certain missions. As a typical example of the sensor network, suppose an external application wishes to know the average temperature of a certain area spanned by the sensor network. An explicit query (called *an aggregation query*) is then sent to the base station, which in turn forwards it to all the sensor nodes pertaining to this area. The temperature data (sometimes called *readings* or *sensor values*) are collected, averaged, and finally returned to the base station and the external application.

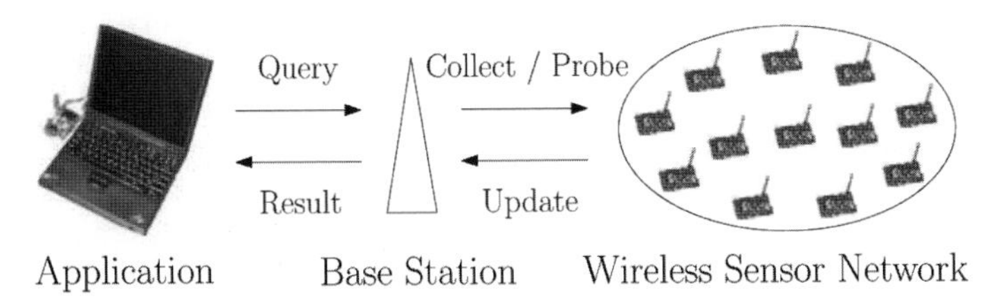

Fig. 12.1. Sensor network model

The aforementioned processing flow of queries in sensor networks is just a conceptual one. As sensors are equipped with certain computational and storage capacity, they can assist the processing procedure in many ways. In what follows, we briefly introduce some basic notions that are frequently used by sensor network systems, and gradually elaborate them as we explore this topic.

- *Snapshot query vs. continuous query.* If the query involves a temporal parameter, which spans a limited or unlimited period of time, it is a *continuous* query. Such queries are usually used to monitor the physical world and its evolution over time. In contrast to a continuous query, a snapshot query only asks for the sensor data at a specified time instant.
- *Query arrival rate vs. data capture (update) rate.* For a particular sensor node, query arrival rate denotes the rate of incoming queries whose querying area overlaps the node's coverage. This rate shows how popular this sensor node is, and it is often regarded as an external parameter (i.e., it cannot be controlled by the sensor network). The data capture (sometimes called "data update") rate, on the other hand, denotes how often the sensor *samples* (sometimes called *collects*) the data. Different from query arrival rate, data capture rate is often tunable by the base station or the sensor itself. It is noteworthy that, these two rates are the two competing factors that affect the performance of the sensor network, similar to the consumer-producer relations.
- *Push V.S. pull-based data dissemination.* The collected data from a single sensor can be either *pushed* to the central server, i.e., the base station, or still kept locally at the sensor until some query explicitly *pulls* the data.
- *Local V.S. centralized storage.* From the storage point of view, push-based dissemination advocates centralized data storage while pull-based dissemination advocates local storage. It is also understood that centralized storage favors high query arrival rate and low data capture rate while local storage favors low query arrival rate and high data capture rate. Some recent work suggests hybrid pull-and-push-based dissemination, or in the storage terms, distributed sensor storage (Zhang et al. 2003). The idea is neither to send an overwhelming volume of raw sensor readings directly to the base station, nor to keep them at sensing locations, but rather, to organize and distribute them to appropriate locations throughout the sensor network so that these data can be quickly retrieved and queried later. Among the distributed storage proposals, data-centric storage is the most popular one (Ratnasamy et al. 2003). It names (or more formally, indexes) the data being stored for future searches. Both the base station and sensor nodes can access the data by their names across the entire sensor network, which explains why this storage scheme is called "data-centric".

This chapter mainly concerns the energy consumption issue during query processing in sensor networks. While the base station is supplied with unlimited power, the sensor nodes are usually battery-powered. In most cases, these batteries are inconvenient or even impossible to replace once they are deployed in the wild field. On the other hand, most sensor applications require continuous monitoring of data, which further makes energy efficiency a critical factor in the design of a good sensor network.

Among all the sources that may consume sensor energy, wireless radio transmission is the most dominant source (Pottie et al. 2000). This is especially true in those networks where the sensor nodes fall far apart, since the power consumption for radio transmission is at least quadratic to the distance of transmission. The second major source that consumes significant energy is the sampling activities of the sensor nodes. Thus, a good design of energy-efficient sensor network should: (1) reduce unnecessary data transmission; (2) reduce the frequency of sensor reading. When a sensor node runs out of energy, its coverage is lost.

In most scenarios, the mission of an application would not be able to continue if the total coverage loss over the entire area is remarkable. Formally, the *network lifetime* of a sensor network is the time duration until it fails to carry out the mission due to insufficient number of "alive" sensor nodes (Akyildiz et al. 2002). Energy-efficient data management is crucial for the sensor networks to accomplish the mission and accomplish it economically. However, this is a challenging task due to the following reasons:

- The sensor nodes are inherently heterogeneous in energy consumption. Different types and brands of sensors are equipped with different transceiver chipsets, electrical parts, and etc. They vary a great deal in their energy consumption.
- Since external applications may have different areas of interest and different queries, the data capture rates of the sensor nodes throughout the entire area may vary and change all the time. This also implies that the data communication rates of these nodes are different and dynamically changing. As a result, the sensor nodes differ from one another in their energy consumption and residual energy.
- The energy consumption of radio communication heavily depends on the distance of transmission (Rappaport 1996). As the distribution of sensor nodes is affected by many factors, these sensor nodes are very likely to be located non-uniformly in their geographical positions so that the energy consumption of sending a message to the base station or neighboring nodes varies from node to node. In other words, even if all sensor nodes are of the same type and have the same data capture rate, their energy consumption could still be highly unbalanced.

In this chapter, we review recent studies on energy-efficient data management for sensor networks. We mainly focus on the following three topics of data management, namely, data storage, exact query processing, and approximate query processing.

- Data storage deals with the collection, compression, and allocation of the data from the sensor nodes for query processing.
- Exact query processing deals with executing various types of queries on the sensor data. Typical queries involve range queries, top-k queries, aggregation queries, etc.

- Approximate query processing also deals with executing various types of queries on the sensor data. However, unlike the strict query semantics in exact queries, approximate queries allow the results to deviate from the actual values with certain precision guarantee (Olston et al. 2003; Sharaf et al. 2003; Han et al. 2004; Deligiannakis et al. 2004). In many cases, the precision can be specified in the form of a quantitative error bound. For example, "retrieve the average temperature reading of all sensor nodes within an error bound of 1 °C" is an approximate query. By definition, if the actual average temperature is 20.5 °C, any query result between the range of 19.5 °C and 21.5 °C is valid. Approximate query alleviates the sensor nodes from reporting their readings to the base station all the time, since it is now possible for them to skip data sampling and transmission if the new value will not invalidate the previous result. In other words, only updates necessary to guarantee the desired level of precision need to be sent. As such, approximate query offers the flexibility for the sensor network to trade data quality for energy efficiency.

The next three sections of this chapter will explore these three topics in detail, followed by a summary. Before we start introducing various techniques and algorithms, two design principles in sensor networks should be mentioned:

- The design of an energy-efficient sensor network heavily depends on the applications of this network, i.e., what types of queries are going to be issued and how frequent/costly they are.
- A good design of an energy-efficient sensor network involves both the communication layer (MAC layer) and the data management layer. As we shall see later, even for the same application, if the underlying routing protocols are different, the query processing techniques may be different. Since a careful choice (or even adaptation) of an appropriate communication layer design is indispensable to energy saving, we will discuss the MAC layer if necessary, although it is not the focus of this chapter.

12.2 Data Storage

The collected sensor data can be stored either locally or in a centralized base station. In latter case all the readings from the sensors are pushed to the base station where the user queries are processed. Such client-server architecture eases the design logic and the deployment of the sensor nodes, but at the cost of high communication overhead. We start this section with the centralized storage scheme.

12.2.1 Centralized Storage Scheme

Obviously, the main research issue in centralized storage is to reduce the cost of communication when the sensor data is sent to the base station. Lazaridis and Mehrotra (Lazaridis et al. 2003) considered the data stream from a single sensor node as a time series. In order to reduce the number of transmission times between the sensor node and the base station, an online compression algorithm was proposed to capture the time series while guaranteeing an error bound. The algorithm is executed at the sensor side, and is based on the Poor Man's Compression (PMC). It always monitors and maintains the range of the input, i.e., the sensed data, and if the range exceeds twice of the allowed error, it outputs a new element to the compressed time series whose value is the midpoint of the range. The advantage of PMC is that it is easy to implement on a sensor node and requires little resources. The authors also proposed to predict the future time series so as to further reduce the transmission cost. The prediction can be performed at both the sensor side and the base station side. At the sensor side, the predicted value is generated from real values (as opposed to from the approximate values at the base station side due to PMC). The cost of selecting the prediction model and parameter estimation might be too overwhelming to the sensor. As such, an adaptive approach was introduced to strike the balance between the accuracy and the cost. The prediction is used when a query comes to the base station asking for the data value at some time t_q with bounded error ε. If t_q is old enough and ε is within the prediction precision, the query result can be returned immediately, otherwise the base station should either wait for the next update from the sensor or probe the sensor directly. At the end of this paper, a technique that combines both prediction and compression was also introduced. The idea is to first apply prediction at the sensor side, then compress the prediction error, and finally send it back to the base station. By this means, a precise prediction can further reduce the cost of communication since compressing the error of the prediction is more beneficial than compressing the raw sensor data streams.

12.2.2 Local Storage Scheme

The local storage scheme stores the data (or sometimes *events*) in the sensor node's local memory. In this scheme, a sensor node sends data to a base station (or *sink*, as in some literatures) only when the sink has sent a query for the data. A sink-sensor matching mechanism is required to help a sink find the sensor node that holds the requested data. The matching mechanism usually floods certain control messages to the entire or part of the network to look for the response. Obviously, this is the task of the underlying routing and data dissemination protocol. Directed diffusion (Intanagonwiwat et al. 2000) and two-tier data dissemination (TTDD) (Ye et al. 2002) are the two most commonly mentioned protocols.

In directed diffusion, the routing decisions are based on the name of the data rather than on the identities of the sending and receiving nodes. The dominant routing mechanism is flooding, however, in order to enhance the performance, additional mechanisms for the reinforcement of high-quality data delivery paths and for in-network aggregation[1] have also been proposed. In TTDD, the matching mechanism is reversed: instead of flooding the query request, TTDD makes the sensor nodes flood the advertisements of the data to the network, and the sinks that are interested in these data then send their queries directly to the source.

12.2.3 Data-Centric Storage

The aforementioned two storage schemes have their own advantages and disadvantages: local storage often overwhelms some sensor nodes and hits the limit of their storage capacity, while centralized storage incurs significant network traffic all the time. Therefore, distributed data storage is advocated in the research community. On one hand, the sensed data or detected sensor event must be forwarded to somewhere in case the storage limit of the detecting node might be reached; on the other hand, it may not be necessary to forward the data or event information directly to the base station if there is no query on such data or event. This more energy efficient way to handle the sensor data, which is frequently used by many existing sensor network systems (Zhang et al. 2003; Ratnasamy et al. 2003; Demers et al. 2003; Madden et al. 2003), is called *data-centric storage*. The basic idea is not to send an overwhelming volume of raw observations on the physical world directly to the base stations, but rather, to distribute the descriptions of certain events that are of interest, such as earthquakes or animal sightings, to the entire sensor network. The sensors can then forward or keep such event information for further querying. In order for the data or event information to be searched, they must be named, and this is why this storage scheme is called "data-centric". Once named, any user or node is allowed to access the data or event by its name across sensor network.

One storage scheme that falls into the data-centric category was proposed by Ratnasamy et al. (Ratnasamy et al. 2003). Different from directed diffusion where the data are stored locally but are routed by their names, they proposed to distribute data storage by the name of data. The analysis showed that as the sensor network grows larger and the detected events increase, such storage distribution is preferable to local storage. To distribute data storage, they proposed a geographic hash table (GHT), which is built on top of GPSR (Karp et al. 2000), a geographic routing system for multi-hop wireless networks. GPSR helps to route packets to the sensor node that is the closest to the destination among all the nodes, and GHT leverages this characteristic to route storage requests and user queries for the same data key to the same node.

Zhang et al. proposed an index-based distributed storage scheme (Zhang et al. 2003). In this scheme, the sensing data of an event are stored at the detecting nodes themselves or some nodes close to them (these nodes are called *storing*

[1]Rather than routing all the data to the query requester, it is less costly to aggregate them in advance at the intermediate nodes.

nodes). A storing node only sends data to a sink when it receives a query from the sink. Also, the location information (called index) of the storing nodes are pushed to and maintained at some nodes (called *index nodes*) based on the event type related to the stored data. Hence, queries for a particular event are first routed to the relevant index nodes before they are redirected to the storing nodes. The index-based scheme is more attractive than local storage schemes since it avoids both unnecessary transmission of the sensed data and the flooding of control messages.

The major challenge of the distributed storage scheme is how to maintain the indices in the network. They devised the Adaptive Ring-based Index (ARI) scheme, where index nodes for a particular event type are those surrounding one or more particular locations (called index centers of the event type). The index centers are determined by applying a predefined hash function, e.g., Geographic Hash Table (GHT), on the event type. The hash function maps an event type to one or more locations within the detecting region of the sensor network, and the index nodes for the same event type are connected through some forwarding nodes and thus form a ring. To achieve load balance, the number and locations of the index nodes in an index ring, as well as the shape of the ring, can be adaptively changed according to the current storage load of the nodes.

12.3 Exact Query Processing

In this section, we are to review recent literatures on sensor query processing. As we have already mentioned, a good solution to this problem involves an elaborate design from both the networking layer and the data managing layer. Therefore, we will first introduce some general frameworks and platforms for sensor query processing, followed by some discussions on processing several particular types of queries.

12.3.1 General Frameworks

Madden et al. have built a data management system called TinyDB on top of the Berkeley Mica mote hardware and the tiny operating system (TinyOS). Based on the supporting of TinyOS and TinyDB, they have further proposed a query-processing engine (ACQP) on this platform (Madden et al. 2003). ACQP supports three types of exact queries: fixed rate query, event-based query, and lifetime-based query. All of the three queries use SQL syntax, as sensor data is considered as a single table with one column per sensor type. The only difference between the three types lies in when (or more exactly, how often) the query is evaluated. For fixed rate query, the sample rate, i.e., how often the sensor value is sampled, is specified in the query. For example, a query "SELECT sensor-id, temperature FROM sensors SAMPLE INTERVAL 1s" means each sensor must read the temperature data once every second. For event-based query, the query is evaluated only when certain event (specified by the operating system or another query) occurs. For example, "ON EVENT fire-alarm SELECT sensor-id, temperature FROM sensors" means the query is evaluated on each sensor only when an event fire alarm occurs. The third query type, which is the most energy-related, lets the user specify for how long the sensor network system is expected to operate. For example, "SELECT sensor-id, temperature FROM sensors LIFETIME 180 days" means that the entire sensor network should last for 180 days before it runs out of battery. This query type is user-friendly to those who are particularly concerned about the lifetime of the sensor network; and ACQP automatically handles how frequent the query is evaluated in order to achieve the specified duration.

As motes are equipped with memory and processing power, the main task of query processing in ACQP is to optimize, disseminate, and finally execute the query. The optimization determines the evaluation order in the query plan. It treats data sampling as a special operator with selectivity equal to 1, and since the base station maintains some metadata about the cost of sampling and communication of each sensor, the optimization is quite similar to traditional RDBMS query optimization except that the cost to minimize is the energy consumption, rather than the I/O. As the communication in mote network is through broadcasting and routing tree (called the semantic routing tree, SRT), the second step, the query dissemination, goes all along from the root, i.e., the base station, down to the entire sensor network. During the propagation of the query, each node will decide whether it should execute the query and further forward it to the subtree rooted at this node. Since some metadata of the subtree (e.g., the value range) are stored in each node during the construction of SRT, query filtering is efficient to implement.

The last step is execution. Since the detailed query execution plan is forwarded to each of the corresponding sensor node by ACQP, the node has no difficulty in following the instruction, i.e., when to get awake, sample data values, and where to forward the results. In most cases, the result is forwarded directly to the parent, which in turn forwards the result upward along the SRT until it reaches the root, i.e., the base station. However, if the query is an event-based query, since the node that fires the particular event triggers the query, it is responsible for collecting the final query result and forwarding the result to the base station. During the result forwarding, ACQP applies *partial state record* at each intermediate node in the routing topology. This record represents the partially evaluated aggregation of the local sensor value and those sensor values received from child nodes as they flow up the routing tree. It reduces the amount of data transmitted compared to sending all sensors' readings to the root of the network for aggregation.

ACQP sets up a good paradigm for query processing on autonomous sensors, such as motes, which are equipped with processing power, wireless communication capability, and memory storage. Since each node is involved in the entire query execution process, the burden of the base station is significantly relieved. However, as TinyOS uses broadcasting and simple contention prevention policy, the data rate of each node cannot be high, in other words, it is not suitable for those heavy-loaded networks. Furthermore, the syntax of ACQP is for acquisitional query only: it cannot support complex queries such as "finding the region with highest temperature". This constraint limits the usage of ACQP in many sensor network applications.

Demers et al. developed the Cougar Sensor Database Project. In their sensor network development, energy-efficient data dissemination, and query processing are the major concerns (Demers et al. 2003). Different from the TinyDB project, they argued that the data management layer should not be considered in isolation from the communication layers. That is, cross-layer optimization can be designed to have the network communication protocols adapt to the particular needs of the data management layer. Bearing this principle in mind, they proposed three useful techniques to minimize the power consumption, namely wave scheduling, view selection and aggregation tree selection.

Wave scheduling is essentially a routing protocol that achieves scalability and energy-efficiency with modest delay penalties. The general routing problem determines if the direct communication between two nodes (normally known as an edge) is allowed. In Cougar project, the entire area is divided into a grid of square cells. Each cell is allowed to communicate only with its immediate neighbor in the grid. As such, the routing problem is much simplified. However, if nearby nodes transmit data at the same time, collision may occur, which degrades the efficiency. Therefore, the wave-scheduling scheme is designed in such a way that every edge in the network is activated exactly once per period, so it is free from collision. On the other hand, since every edge is activated within a period, any message between two nodes can always reach from source to destination. However, even though a path can be followed in principle, its latency may be unacceptably high if the path and the activation schedule for the edges do not "fit well". For example, suppose a path enters node n along edge e_1 and leaves it along e_2. Each message arriving at n along e_1 must be queued at n until the next time e_2 is scheduled. If e_2 is activated just before e_1 in the schedule, the message must wait for nearly a full period in the queue of n until it is forwarded in the next iteration of the schedule. If such phenomenon occurs at every node along the path, the resulting latency will be unacceptable for most applications. Wave scheduling, on the contrary, alleviates this problem by allowing a "wave" of messages to traverse the grid for a number of steps before it has to be queued. In other words, it activates a set of the neighboring edges simultaneously and then shifts the pattern by one cell, simulating a wave effect. Based on this idea, a family of scheduling schemes, tailored towards minimising different metrics, such as energy, latency or a combination of both, can be devised.

Aggregation tree and view selection are two inter-related techniques to speed up the query processing of aggregation query in wireless sensor networks. The basic idea is that, no matter how the underlying communication layer is organized, the sensor nodes in the network can be logically organized as a spanning tree, called the aggregation tree. A sensor node is responsible for maintaining the aggregation value of all the nodes in the subtree rooted at this node. The construction of an optimal aggregation tree is difficult because: (1) there are a huge number of possibilities for spanning tree construction; (2) the optimal solution depends on the query pattern, i.e., how often each node's value is concerned; and (3) even the simplest problem where each node is queried with probability 1 is NP-hard. As such, the authors showed some examples and heuristics on how to build sub-optimal aggregation trees. The second technique, view selection, goes one step further to decide if a node in the aggregation tree should leave the task of maintaining the aggregation value (i.e., maintaining the materialized view) at query time only, or at all the time. It is obvious that if the query arrival rate at this node is low and the data capture rate beneath this node is high, the cost of maintaining the aggregation value exceeds the cost saving. Similar to the tree construction problem, the optimal solution to view selection is difficult (NP-complete), however, the authors managed to devise some dynamic programming algorithms (with at least exponential worst-case complexity) that can return the optimal solution. It is noteworthy that, both aggregation tree construction and view selection are centralized operations that are only required at the root, i.e., the base station. Furthermore, since both queries and data capture rates are constant in most sensor networks, these two operations can be considered as one-time costs.

12.3.2 Advanced Processing Techniques for Special Query Types

The idea of aggregation tree and view selection is essentially the hybrid of pull and push-based data dissemination, which is the dominant trend in state-of-the-art wireless sensor data management. For special types of queries, the hybrid is shown to be superior to using pull or push-based dissemination alone.

Liu et al. proposed a *comb-needle* query processing technique for discovery queries (Liu et al. 2004). Discovery queries are those queries that are interested in the *existence*, for example, instead of asking "how many sensors are with temperatures higher than 90 degrees", such type of query asks "are there five sensors with temperatures higher than 90 degrees". Obviously, such queries can be answered, though not efficiently, by flooding the entire network. Liu et al., on the other hand, proposed an efficient approach to process them on the grid-based sensor network. In their approach, each sensor node pushes its data to a certain neighborhood and the query node disseminates its request to a subset of the network. More specifically, the query is first routed horizontally (or vertically) along the row (column) from the issuing node in the grid, and is then routed vertically (horizontally) along the columns (rows) of some nodes in this row (column), which resembles a comb. In order for the combing processing to find the needed data, each sensor node should push (i.e., duplicate) its data to certain neighboring nodes along the row (column). The problem is then how to strike a balance between the query efficiency (in terms of the density of the comb) and the maintenance cost. This is a typical problem in hybrid pull-push dissemination, in which both query arrival rate and data update rate should be considered and balanced. The paper presented a simplified analysis on how the optimal comb-needle can be derived.

Data-centric storage is essentially a push-based dissemination. Li et al. proposed a solution to multi-dimensional range queries using data-centric storage (Li et al. 2003). They assumed the sensor networks are deployed to collect events that are associated with a tuple of numerical data attribute values (readings). Multi-dimensional range queries are to retrieve those events whose corresponding attribute values fall into the specified range. For example, forest fire alarms should arise at those areas whose temperature exceeds 45 degrees and whose relative humidity is below 50%. Since it is not efficient (and even impossible if the dimension is large) to maintain a centralized index for these attributes, this paper proposed a distributed indexing scheme that relies on a locality-preserving geographic hashing algorithm and an underlying geographic routing scheme (namely, GPSR).

K-d trees inspire the geographic hashing algorithm. It looks like a space filling curve except that it maps a multi-dimensional cube to a two-dimensional square, by alternately splitting the space horizontally and vertically into two equal-sized zones until each zone contains at most one sensor. As a result, each sensor owns a zone. The sensor is responsible for storing the events whose value tuples fall into the corresponding multi-dimensional cube.

When a multi-dimensional range query is issued, it is routed to the nodes whose corresponding zone-mapping cubes overlap the range. However, if the range is large, it is not efficient to route the query individually to all these nodes, as many messages are redundant. The authors proposed an alternative approach to first route the query to the node whose corresponding cube covers the centroid of the query range. The query is then routed and split from this central node to all relevant nodes. In this way, the number of redundant messages is reduced.

12.4 Approximate Query Processing

In this section, we review our recent work on approximate query processing in wireless sensor networks. We start by looking at the approximate processing of snapshot queries. Then, we move on to the approximate processing of continuous queries including top-k monitoring and aggregate monitoring.

12.4.1 Processing Approximate Snapshot Queries

Taking advantage of users' error tolerances, we proposed a generic two-tier data storage strategy to support processing of various types of approximate queries (including ID-based, range, top-k, and aggregate queries) (Wu et al. 2006). The base station serves as the first tier (referred to as centric storage) that stores imprecise sensor data, while each sensor node serves as the second tier (referred to as local storage) that stores exact up-to-date data. Consider a sensor node i. The imprecision of the data stored at the base station is bounded by a certain error represented by an approximation range, i.e., a stored value v_i with an approximation range of e_i means that the actual value must lie in the approximate interval $[l_i, h_i]$, where $l_i = v_i - e_i/2$ and $h_i = v_i + e_i/2$. At each sampling instance, if the newly sensed value v_i' is within a difference of $e_i/2$ from the previously reported value v_i, the new value v_i' is kept at the sensor node locally, otherwise an update message is sent to the base station to replace v_i by v_i' in the centric storage. In this way, a lot of updating traffic can be saved. However, if the precision of stored data is insufficient to answer a query issued to the base station, we will have to refresh the data from the local storage, which incurs communication overhead.

The general query processing under the two-tier storage takes three steps. First, the base station computes a tentative result based on stored imprecise data. Second, if the tentative result is not sufficiently precise, the base station refreshes the readings from a (selected) subset of the sensor nodes. After refreshment, the approximation ranges of those refreshed nodes are shrunk to zero. Note that the refreshed values remain up-to-date till the next sampling instance, after which the approximation range of node i returns to e_i. Finally, the base station re-evaluates the query based on refreshed data. In the following, we detail the query processing techniques for different types of queries.

ID-based query. An approximate ID-based query is interested in the reading of a particular sensor node, with a precision constraint of R. It is acceptable as long as the returned value is within a deviation of R of the true reading.

Recall that the sensor reading kept at the base station is in the form of $[l_i, h_i]$. If $R \geq h_i - l_i$, meaning the stored data has a higher precision than the expected, it is immediately returned to the user. Otherwise, the stored data does not meet the precision requirement, thus we have to send a refresh message to the desired sensor node to probe its latest reading. By doing so, we shrink the approximation range to zero (till the next sampling instance), thereby satisfying the precision requirement of the query.

Range query. In this type of queries, we are interested in the sensor nodes whose readings are within a specified range $[L, H]$. With a precision constraint of R, we are required to find out all sensor nodes whose readings are in $[L + R, H - R]$ and to exclude those whose readings are not in $[L - R, H + R]$. The nodes whose readings are within $[L - R, L + R]$ or $[H - R, H + R]$ may or may not be returned.

By examining the approximate value $[l_i, h_i]$ of each node i, we divide the sensor nodes into three groups T^+, T^-, $T^?$, which respectively represent the nodes who can be returned, the nodes who are not returned, and the rest. A node i is categorized in T^+ if $l_i > L - R$ and $h_i < H + R$. It is categorized in T^- if $h_i < L + R$ or $l_i > H - R$. If none of these conditions is satisfied, the node is categorized in $T^?$. The nodes in $T^?$ must be refreshed because we are not sure whether they should be included in the query result.

Take the following query as an example: "finding the nodes whose readings are greater than 100 °C with a precision constraint of 5 °C." Thus, as illustrated in Fig. 8.2, for the nodes that hold $l_i > 95$, we throw them into T^+, and for the nodes who hold $h_i < 105$, we throw them into T^-. We will refresh all the other nodes and combine the qualified nodes with T^+ as the final result.

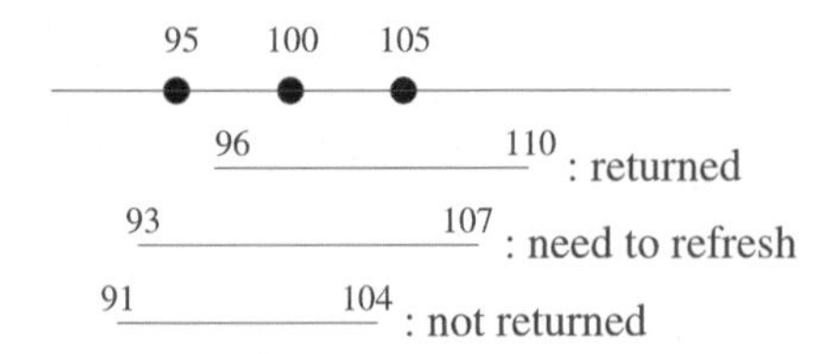

Fig. 8.2. Processing range query

Top-k query. In a top-k query, the user wants to get the k nodes with the highest (or lowest) readings. Recall that the reading of node i is approximated with an interval of $[l_i, h_i]$. Given a precision constraint of R, an approximate top-k query retrieves the (ordered) set of sensor nodes T with the highest readings:

$$T = < n_1, n_2, \ldots, n_k > \tag{12.1}$$

where,

$$\forall i > j, h_{n_i} \le l_{n_j} + R \tag{12.2}$$

and

$$\forall l \ne n_i (i = 1,\ 2, \cdots, k),\ \ h_l \le \min\{l_{n_1}, l_{n_2}, \cdots, l_{n_k}\} + R \tag{12.3}$$

Intuitively, if two sensor readings are within a difference of R, their order can be arbitrary. The evaluation of an approximate top-k query is much different from that of the previous two types of queries. Given an ID-based query or range query, the set of to-refresh nodes is uniquely determined. However, for a top-k query, whether a node needs to refresh depends on the relative order of its reading against the other sensor nodes. We divide the refreshing process into two steps: selecting to-refresh nodes and processing refreshment.

When the base station receives a top-k query with a precision constraint of R, it sorts the sensor nodes based on their current approximate readings. Without loss of generality, the nodes are sorted by the upper bounds of their approximate intervals. Suppose $n_1, n_2, \ldots, n_k$ is the tentative top-k list. We will return this list immediately if no node in the list has an overlap with any other node by greater than R in the approximate interval. Otherwise, we need to refresh some nodes to resolve the top-k order. To do so, we define *refreshing candidates* (RC_i) with respect to each node i in the tentative top-k list as follows:

$$RC_i = \begin{cases} \Phi & \text{if } \forall j,\ h_j - l_i \le R \\ \{i\} \cup \{j | h_j - l_i > R\} & \text{otherwise.} \end{cases} \tag{12.4}$$

Note that the refreshing candidate sets with respect to different nodes may overlap. Fig. 8.3 shows an example top-2 query among 4 nodes (with approximate intervals of [5, 8], [3, 7], [2, 6], and [1, 5], respectively). Assume the precision constraint $R = 1$. The RC_i sets for nodes 1 and 2 are {1, 2} and {2, 3, 4}, respectively. A straightforward refreshment strategy is to refresh all nodes in $RC = \bigcup_{i=1}^{k} RC_i$. We call it *full refreshment*. However, this might not be necessary because the refreshments of some nodes may eliminate the need to refresh other nodes. Consider the early example. RC = {1, 2, 3, 4}. Suppose we choose to refresh node 2 first, and assume that the current reading of node 2 is 5.5. After refreshment, the approximation interval [3, 7] of node 2 is replaced by its exact reading of 5.5. Hence, RC_1 and RC_2 are updated with empty sets. Thus, we can assert that the top-2 list is ⟨1, 2⟩ without refreshing nodes 1, 3, and 4 anymore.

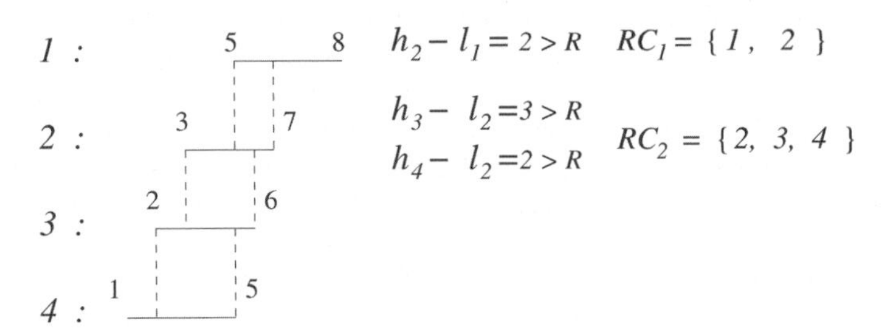

Fig. 8.3. Finding out RC_i in top-k query

This fact suggests that we can refresh in rounds. In each round, we choose a subset of RC to refresh. When we get the refreshed reading(s), we update the RC_i set for each node i in the tentative top-k list. This process is repeated until all the RC_i sets become empty. Two round-based refreshment strategies are proposed in (Wu et al. 2006):

- Batch: Starting from the top-1 node, the RC_i set of one top-k node is refreshed in each round.
- Sequential: One node is refreshed per round. In each round, the node that appears in most RC_i sets is selected to refresh. Refreshing such a node is expected to quickly resolve the order confusion.

Aggregate query. There are five types of standard aggregate queries: SUM, MAX, MIN, COUNT, and AVG. The MAX and MIN queries can be viewed as top-1 queries. While COUNT can be processed in a way similar to SUM, AVG and SUM queries differ by only a constant which is the number of sensor nodes. Therefore, we shall focus our discussion on the SUM query here.

If the reading of each sensor node i maintained at the base station has an approximation range e_i, the SUM aggregation can be computed with an approximation range:

$$E = \sum_{i=1}^{n} e_i \tag{12.5}$$

where, n is the number of sensor nodes in the network. If the query precision constraint R is greater than E, the result is returned by the base station without refreshing the reading of any sensor node. Otherwise, if R is smaller than E, some sensor readings have to be refreshed to refine the query result. Let T be the to-refresh node set. Since refreshing the reading of a sensor node reduces its approximation range to zero, to meet the precision requirement of the query, T must satisfy:

$$\sum_{i \in T} e_i \geq E - R \tag{12.6}$$

The refreshment can make use of in-network aggregation to improve energy efficiency. Specifically, on receiving up-to-date readings from more than one child, an intermediate node aggregates the readings before forwarding them upstream. For SUM aggregation, the partial aggregate result is simply the sum of the readings received. In-network aggregation cuts down the volume of data sent over the upper-level links in the routing tree. We define the subtree rooted at each child of the base station as a *region*. Since these children relay packets between the base station and the other nodes in their respective regions, they consume much more energy than the others. We therefore call these nodes the *hot-spot nodes*. In order to prolong the network lifetime, we should conserve the energy at these hot-spot nodes. This implies the following design philosophy of refreshing:

- We should distribute the to-refresh nodes in as few regions as possible. This is because due to in-network aggregation, the volume of data sent by a hot-spot node to the base station is independent of the number of sensor nodes refreshed in the corresponding region. To save the energy consumption at hot-spot nodes, it is desirable to reduce the number of regions involved in the refreshment.
- When selecting regions for refreshment, we favor those with more residual energy.
- When the number of to-refresh sensor nodes is smaller than that in one region, we should choose the nodes that lie closer to the base station. This helps reduce the number of sensor nodes involved in relaying the up-to-date readings and thus the network-wide total energy consumption.

The to-refresh node set is constructed as follows. Starting from an empty to-refresh node set, we continue to insert nodes into the set until the total approximation range of the nodes in the set adds up to $E - R$. In this process, the regions are sequentially examined. For each region, all nodes in the region are inserted to the to-refresh node set if the insertion does not make the total approximation range greater than $E - R$. Otherwise, only a subset of the nodes in the region are inserted to increase the total approximation range of the to-refresh node set to $E - R$. The subset of nodes is selected in increasing order of their distances to the base station. Two examination orders of the regions are proposed:

- Max-size: This strategy favors large regions to minimize the number of regions involved in the refreshment, i.e., the regions are examined in decreasing order of their sizes.
- Max-energy: The second strategy favors the regions with more residual energy to balance the energy consumption among regions. That is, the regions are examined in descending order of the residual energy of their hot-spot nodes. Note that since the hot-spot nodes are located near the base station, it

is practically easy to maintain the residual energy of these nodes (e.g., by piggybacking the energy information on the refresh messages).

12.4.2 Approximate Top-k Monitoring

Different from snapshot top-k queries, top-k monitoring continuously returns top-k results. An efficient filter-based monitoring approach called *FILA* was proposed for this type of queries (Wu et al. 2006). The basic idea is to install a filter at each sensor node to suppress unnecessary sensor updates. The base station also keeps a copy of the filter setting to maintain a view of each node's reading. A sensor node reports the reading update to the base station only when it passes the filter. The correctness of the top-k result is ensured if all sensor nodes perform updates according to their filters. Fig. 8.4 shows an example, where the base station has collected the initial sensor readings and installed three filters [20, 39), [39, 47), and [47, 80) at sensor nodes A, B, and C, respectively. At sampling instances 1 and 2, no updates are reported since the nodes' respective filters filter out all updates. At instance 3, the updated reading of node B (i.e., 48) passes its filter [39, 47). Hence, node B sends the reading 48 to the base station via node A (step 1). Since 48 lies in the filtering window of node C (i.e., [47, 80)), the top-1 result becomes undecided as either node B or C can have the highest reading. In this case, we probe node C for its current reading to resolve the ambiguity (steps 2 and 3). Thus, a total of four update messages and one probe message are incurred.

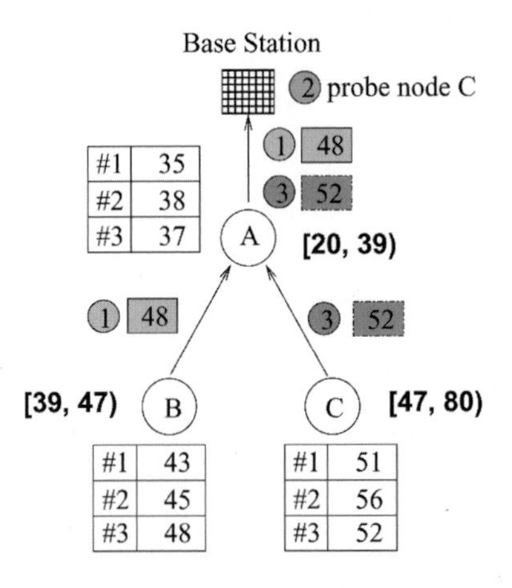

Fig. 8.4. An example of top-*k* monitoring

A crucial issue of this approach is how to set the filter for each sensor node in a coordinated manner such that the top-k result set is ensured to be correct. Represent the filter of each sensor node *i* by a window of $[l_i, u_i)$. Without loss of generality, we number the sensor nodes in decreasing order of their sensor readings, i.e., $v_1 > v_2 > \cdots > v_N$, where *N* is the number of sensor nodes under monitoring. Intuitively, to maintain the monitoring correctness, the filters assigned to the nodes in the top-k result set should cover their current readings. Moreover, given an error tolerance of R, an overlap of at most R between two neighboring filters is allowed. On the other hand, the nodes in the non-top-k set could share the same filter setting. Thus, we consider the filter settings only for the top-(k+1) nodes. A *feasible* filter setting scheme for approximate top-k monitoring, represented as $\{[l_i, u_i) \mid i = 1, 2, \ldots, k+1\}$, should satisfy the following conditions:

$$\begin{cases} u_1 > v_1; \\ v_{i+1} < u_{i+1}, u_{i+1} \le l_i + R, l_i \le v_i; \quad (1 \le i \le k); \\ l_{k+1} \le v_N. \end{cases} \tag{12.7}$$

Fig. 8.5 shows a feasible filter setting for top-3 monitoring, where sensor nodes 4 and 5 share a filter setting and u_{i+1} is set equal to $l_i + R$ for $1 \le i \le 3$ in order to maximize the filtering capability.

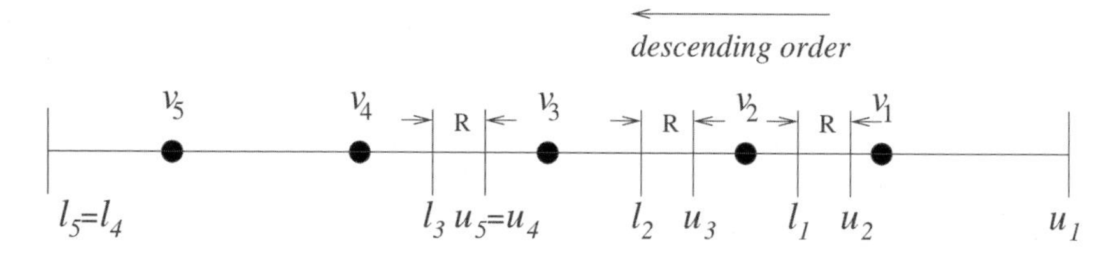

Fig. 8.5. Filter settings for top-3 monitoring

12.4.3 Approximate Aggregate Monitoring

In approximate data aggregation, not all sensor readings have to be sent to the base station. Data impreciseness is measured by the quantitative difference between an approximate value and the exact value. The application specifies the precision constraint of data aggregation by an upperbound E on data impreciseness (called the *error bound*). That is, on receiving an aggregate data value A' from the sensor network, the application would like to be assured that the exact aggregate value A lies in the interval $[A' - E, A' + E]$, i.e., $A' \in [A - E, A + E]$.

To reduce communication cost and energy consumption, the designated error bound on aggregate data can be partitioned and allocated to individual sensor nodes (called *precision allocation*). Each sensor node updates newly sensed data with the base station only when the new data significantly deviate from the last update and violate the allocated error bound. This is called a *BS-update*. Precision allocation must ensure that the designated precision of aggregate data is guaranteed as long as each sensor node updates data with the base station according to its allocated error bound. Therefore, the error bounds allocated to individual sensor nodes have to satisfy certain *feasibility constraints*. Different aggregations impose different constraints. Here, we consider two commonly used types of aggregations: SUM and AVG, which refer to the total and average values of the readings at all sensors respectively. For SUM aggregation, to guarantee a given error bound E on aggregate data, the total error bound allocated to the sensor nodes cannot exceed E, i.e.,

$$\sum_{i=1}^{n} e_i \leq E \tag{12.8}$$

where e_i is the error bound allocated to sensor node i. For AVG aggregation, the total error bound allocated to the sensor nodes cannot exceed $n \cdot E$, i.e.,

$$\sum_{i=1}^{n} e_i \leq n \cdot E \tag{12.9}$$

Here n is the number of sensor nodes. These two feasibility constraints share the common characteristic that the total error bound of the sensor nodes is capped by some given value. Without loss of generality, we focus on SUM aggregation in our discussion. The sensor nodes can be allocated the same or different error bounds under the feasibility constraint. For example, in a network of 10 temperature sensors, if the error bound on AVG aggregation is 1 °C, we can allocate an error bound of 1 °C to each sensor. Alternatively, we can also allocate an error bound of 5.5 °C to a selected sensor and an error bound 0.5 °C to each of the remaining sensors. This offers the flexibility to adjust the energy consumption rates of individual sensor nodes by careful precision allocation. In general, to achieve higher precision in data collection (i.e., smaller error bound), the sensor nodes need to send data updates to the base station more frequently, which translates to higher energy consumption. The quantitative relationship between BS-update rate and precision level depends on the *changing pattern* of sensor readings (e.g., the frequency and magnitude). Without loss of generality, we shall denote the BS-update rate of each sensor node i as a function $u_i(e)$ of the allocated error bound e. $u_i(e)$ is essentially the rate at which the sensed data captured by sensor node i change beyond e. Intuitively, $u_i(e)$ is a *non-increasing* function with respect to e. It is obvious that $u_i(\infty) = 0$ and $u_i(0) = R$, where R is the *sensing rate* at which the sensor node captures data. Consider a snapshot of the network. Let $e_1, e_2, \ldots, e_n$ be the error bounds currently allocated to sensor nodes 1, 2, … , n respectively. The energy consumption rate of sensor node i is given by:

$$u_i(e_i) \cdot c_i \tag{12.10}$$

where $u_i(e_i)$ is the rate at which sensor node i updates data with the base station and c_i is the energy consumed for each update. Suppose the residual energy of sensor i is p_i. Under static changing patterns of sensor readings, the expected lifetime of sensor node i is given by:

$$\frac{p_i}{u_i(e_i)\cdot c_i} \tag{12.11}$$

Therefore, the network lifetime is given by:

$$\min_{1\le i\le n}\frac{p_i}{u_i(e_i)\cdot c_i} \tag{12.12}$$

The objective of precision allocation is to find a set of error bounds $e_1, e_2, \dots,$ e_n that maximizes the network lifetime under the constraint $e_1 + e_2 + \cdots + e_n \le E$. An optimal precision allocation was developed in (Tang et al. 2006):

$$e_i^* = \begin{cases} u_i^{-1}(\frac{p_i}{l^*\cdot c_i}) & 1\le i\le j^*, \\ 0 & j^* < i \le n \end{cases} \tag{12.13}$$

where j^* satisfies,

$$\sum_{i=1}^{j^*} u_i^{-1}(\frac{p_i}{l_{j^*}\cdot c_i}) \le E < \sum_{i=1}^{j^*+1} u_i^{-1}(\frac{p_i}{l_{j^*+1}\cdot c_i}) \tag{12.14}$$

l_i is the minimum lifetime of sensor node I,

$$l_i = \frac{p_i}{u_i(0)\cdot c_i} \tag{12.15}$$

and l^* satisfies,

$$\sum_{i=1}^{j^*} u_i^{-1}(\frac{p_i}{l^*\cdot c_i}) = E \tag{12.16}$$

This implies that the sensor nodes with high residual energy or low communication costs may be assigned zero error bounds. The sensor nodes allocated non-zero error bounds in an optimal precision allocation must be equal in the energy consumption rate normalized by the residual energy:

$$r_i(e_i) = \frac{u_i(e_i)\cdot c_i}{p_i} \tag{12.17}$$

12.5 References

Akyildiz I, Su W, Sankarasubramaniam Y, and Cayirci E (2002) A survey on sensor networks. IEEE Communication Magazine 40: 102–114

Deligiannakis A, Kotidis Y, and Roussopoulos N (2004) Hierarchical in-network data aggregation with quality guarantees. Proc. of the 9th Int. Conf. on Extending Database Technology, March 14-18, Heraklion, Greece, pp 658–675

Demers A, Gehrke J, Rajaraman R, Trigoni J, and Yao Y (2003) The cougar project: A work-in-progress report. SIGMOD Record 32: 53–59

Han Q, Mehrotra S, and Venkatasubramanian N (2004) Energy efficient data collection in distributed sensor environments. Proc. of the 24th IEEE Int. Conf. on Distributed Computing Systems, 24-26 March, Hachioji, Tokyo, pp 590–597

Intanagonwiwat C, Govindan R, and Estrin D (2000) Directed diffusion: A scalable and robust communication paradigm for sensor networks. Proc. of the 6th ACM Annual Int. Conf. on Mobile Computing and Networking, Aug. 6-11, Boston, pp 56–67

Karp B and Kung HT (2000) GPSR: greedy perimeter stateless routing for wireless networks. Proc. of the 6th ACM Annual Int. Conf. on Mobile Computing and Networking, August 6-11, Boston, pp 243–254

Lazaridis I and Mehrotra S (2003) Capturing sensor-generated time series with quality guarantees. Proc. of the 19th Int. Conf. on Data Engineering, March 5-8, 2003, Bangalore, pp 429–440

Li X, Kim YJ, Govindan R, and Hong W (2003) Multi-dimensional range queries in sensor networks. Proc. of the 1st Int. Conf. on Embedded Networked Sensor Systems, November 5-7, Los Angeles, pp 63–75

Liu X, Huang Q, Zhang Y (2004) Combs, needles, haystacks: Balancing push and pull for discovery in large-scale sensor networks. Proc. of the 2nd Int. Conf. on Embedded Networked Sensor Systems, Nov 3-5, Baltimore, pp 122–133

Madden S, Franklin MJ, Hellerstein JM, and Hong W (2003) The design of an acquisitional query processor for sensor networks. Proc. of the 2003 ACM SIGMOD Int. Conf. on Management of Data, June 9-12, San Diego, pp 491–502

Olston C, Jiang J, and Widom J (2003) Adaptive filters for continuous queries over distributed data streams. Proc. of the 2003 ACM SIGMOD Int. Conf. on Management of Data, June 9-12, San Diego, pp 563–574

Pottie GJ and Kaiser WJ (2000) Wireless integrated network sensors. Communications of the ACM 43 : 51–58

Rappaport TS (1996) Wireless communications: principles and practice. Prentice Hall, New Jersey, USA

Ratnasamy S, Karp B, Shenker S, Estrin D, Govindan R, Yin L, and Yu F (2003) Data-centric storage in sensornets with GHT, a geographic hash table. ACM/Kluwer Mobile Networks and Applications 8:427–442

Sharaf MA, Beaver J, Labrinidis A, and Chrysanthis PK (2003) TiNA: A scheme for temporal coherency-aware in-network aggregation. Proc. of the 3rd ACM Int. Workshop on Data Engineering for Wireless and Mobile Access, Sept 19, San Diego, pp 69–76

Szewczyk R, Osterweil E, Polastre J, Hamilton M, Mainwaring A, and Estrin D (2004) Habitat monitoring with sensor networks. Communications of the ACM 47:34–40

Tang X and Xu J (2006) Extending network lifetime for precision-constrained data aggregation in wireless sensor networks. Proc. of the 25th IEEE INFOCOM Conf., April 23-29, Barcelona

Wu M, Xu J, and Tang X (2006) Processing precision-constrained queries in wireless sensor networks. Proc. of the 7th Int. Conf. on Mobile Data Management, May 10-12, Nara, Japan

Wu M, Xu J, Tang X, and Lee WC (2006) Monitoring top-k query in wireless sensor networks. Proc. of the 22nd IEEE Int. Conf. on Data Engineering, April 3-7, Atlanta

Ye F, Luo H, Cheng J, Lu S, and Zhang L (2002) A two-tier data dissemination model for large-scale wireless sensor networks. Proc. of the 8th Annual Int. Conf. on Mobile Computing and Networking, September 23-28, Atlanta, pp 148–159

Zhang W, Cao G, and Porta TL (2003) Data dissemination with ring-based index for wireless sensor networks. Proc. of the 11th IEEE Int. Conf. on Network Protocols, November 4-7, Atlanta, pp 305–314

13 Localisation

Guoqiang Mao, Barış Fidan and Brian DO Anderson

National ICT Australia, School of Electrical and Information Engineering, The University of Sydney, Australia

13.1 Introduction

Most wireless sensor network applications require knowing or measuring locations of thousands of sensors accurately. In environmental sensing applications such as bush fire surveillance, water quality monitoring and precision agriculture, for example, sensing data without knowing the sensor location is meaningless (Patwari et at. 2003). In addition, location estimation may enable applications such as inventory management, intrusion detection, road traffic monitoring, health monitoring, etc.

Sensor network localisation refers to the process of estimating the locations of sensors using measurements between neighbouring sensors such as distance measurements and bearing measurements. In sensor network localisation, it is typically assumed that small portions of sensors, called anchors, have *a priori* information about their coordinates. These anchor nodes serve to fix the location of the sensor network in the global coordinate system. In applications, which do not require a global coordinate system (e.g., monitoring in an office building or home environment), these anchor nodes define the reference coordinate system in which all other sensors are referred to. The coordinates of the anchor nodes may be obtained by using a global positioning system (GPS) or by installing the anchor nodes at fixed points with known coordinates. However due to constraints on cost and size of sensors, energy, implementation environment (e.g., GPS receivers cannot detect the satellites' transmission indoors) or the deployment of sensors (e.g., sensor nodes may be randomly deployed in the region), most sensors do not have *a priori* coordinate information. These sensor nodes without *a priori* coordinate information are referred to as the non-anchor nodes and their coordinates are to be estimated by the sensor network localisation algorithm. In this chapter, we shall provide an overview of sensor network localisation techniques as well as an introduction to the fundamental theory underpinning the sensor network localisation. While many techniques covered in this chapter can be applied in both 2-dimensions ($\Re^2$) and 3-dimensions ($\Re^3$), we choose to focus on 2-dimensional localisation problems. The rest of the chapter is organised as follows. In Section 2, we shall provide an overview of measurement techniques and the corresponding localisation algorithms. In Section 3 we shall focus on connectivity-based localisation algorithms. In Section 4 we shall focus on distance-based localisation techniques. Section 5 introduces the fundamental theory of the various distance-based localisation techniques and Section 6 summarises current research problems in distance-based localisation. Finally a summary is provided in Section 7.

13.2 Measurement Techniques

Most wireless sensor network localisation algorithms rely on measurements between neighbouring sensors for location estimation. Measurement techniques in sensor network localisation can be broadly classified into three categories: received signal strength (RSS) measurements, angle of arrival (AOA) measurements and propagation time based measurements. The propagation time based measurements can be further divided into three subclasses: one-way propagation time measurements, roundtrip propagation time measurements and time-difference-of-arrival (TDOA) measurements.

13.2.1 Received Signal Strength Measurements

Received signal strength indicator (RSSI) has become a standard feature in most wireless devices and the RSS based localisation techniques have attracted considerable attention in the literature for obvious reasons. The RSS based localisation techniques eliminate the need for additional hardware, and exhibit favourable properties with respect to power consumption, size and cost. As such, the research community has considered the use of RSS extensively (Patwari et al. 2003; Bergamo et al. 2002; Elnahrawy et al. 2004; Madigan et al. 2005; Bahl and Padmanabhan 2000; Prasithsangree et al. 2002). In general, the RSS based localisation techniques can be divided into two categories: the distance estimation based and the RSS profiling based techniques (Krishna et al. 2004)

13.2.2 Distance Estimation using the Received Signal Strength

In free space, the received signal strength at a receiver is given by the Friis equation (Rappaport 2001):

$$P_r(d) = \frac{P_t G_t G_r \lambda^2}{(4\pi)^2 d^2 L} \tag{13.1}$$

where, P_t is the transmitted power, $P_r(d)$ is the received power at a distance

d from the transmitter, G_t is the transmitter antenna gain, G_r is the receiver antenna gain, L is a system loss factor not related to propagation and λ is the wavelength of the transmitter signal. The gain of an antenna is related to its effective aperture, A_e, by,

$$G = \frac{4\pi A_e}{\lambda^2} \tag{13.2}$$

The effective aperture A_e is determined by the physical size and the aperture efficiency of the antenna. In a real environment, the propagation of an electromagnetic signal is affected by reflection, diffraction and scattering. Moreover, the environment will change depending on the particular application (e.g., indoor versus outdoor). It is very difficult, if possible, to obtain the received signal strength using analytical methods. However measurements have shown that at any value of d, the received signal strength $P_r(d)$ at a particular location is random and distributed log-normally about the mean distance-dependent value. That is,

$$P_r(d)[dBm] = P_0(d_0)[dBm] - 10 n_p \log_{10}\left(\frac{d}{d_0}\right) + X_\sigma \tag{13.3}$$

where, $P_0(d_0)[dBm]$ is a reference power in dB milliwatts at a close-in reference distance d_0 from the transmitter, n_p is the path loss exponent which indicates the rate at which the received signal strength decreases with distance and the value of n_p depends on the specific propagation environment, X_σ is a zero mean Gaussian distributed random variable with standard deviation σ and it accounts for the random effect of shadowing. In this chapter, we use the notation $[dBm]$ to denote that power is measured in dB milliwatts units. Otherwise, it is measured in watts. The log-normal model has been extensively used to obtain the received signal strength (Patwari et al 2003; Rappaport 2001). It is important to notice that the reference distance d_0 should always be in the far field of the transmitter antenna so that the near-field effect does not alter the reference power. In large coverage cellular systems, a 1 km reference distance is commonly used, whereas in microcellular systems, a much smaller distance such as 100 m or 1 m is used (Rappaport 200). The reference power $P_0(d_0)[dBm]$ is calculated using the free space Friis equation (Eq.13.1) or obtained through field measurements at distance d_0. Based on Eq.13.3, it is trivial to show that given the received signal strength measurement, p_{ij}, between a transmitter i and a receiver j, a maximum likelihood estimate of the distance, d_{ij}, between the transmitter and receiver is,

$$\hat{d}_{ij} = d_0 \left(\frac{P_{ij}}{P_0(d_0)} \right)^{-1/n_p} \quad (13.4)$$

It should be noted that P_{ij} and $P_0(d_0)$ in Eq.13.4 are measured in watts instead of dB milliwatts. Using Eq.13.3 and Eq.13.4, a simple calculation will show that the expected value of $\hat{d}_{ij}$ is related to the true distance d_{ij} by,

$$E\left(\hat{d}_{ij}\right) = d_{ij} e^{\frac{\sigma^2}{2\eta^2 n_p^2}} \quad (13.5)$$

where, $\eta = 10/\ln(10)$.

That is, the maximum likelihood estimate in Eq.13.4 is a biased estimate of the true distance and the bias can be removed when σ and n_p are known. Given the distances between neighbouring sensors, algorithms can be developed to estimate the coordinates of the non-anchor sensor nodes. A detailed discussion on these algorithms can be found in Section 4.

13.2.3 RSS Profiling based Localisation Techniques

The RSS profiling based localisation techniques have been mainly used in location estimation in wireless local area networks (WLANs). In the RSS profiling based localisation techniques, a large number of sample points are distributed throughout the coverage area of the sensor network. One or more signal strength measurements are taken from all visible anchor nodes (e.g., access points in WLAN, sniffing devices) at each sample point. Based on the collected information, an RSS model is established where each sample point is mapped to either a signal strength vector or a signal strength probability distribution. The RSS model constructed is essentially a map of the signal strength behaviour in the coverage area. The RSS model generated is unique with respect to the anchor locations and the sensor network environment. The RSS model is then stored in a central location.

A non-anchor node that is not aware of its location collects one or more signal strength measurements from all the visible anchor nodes and creates its own RSS fingerprint which is sent to the central station. The central station matches the presented signal strength vector to the RSS model using either nearest neighbour based methods or probabilistic techniques, from which an estimate of the non-anchor node's location can be obtained. The estimated location is then sent to the non-anchor node by the central station. A non-anchor node may request the RSS model from the central station and perform the location estimation by itself.

The accuracy of the RSS profiling based localisation techniques depends on the particular technique used to build the model and the algorithm used to match the measured signal strength to the model. Generally speaking, the RSS profiling based localisation techniques produce relatively small location estimation errors in comparison with the distance estimation based techniques. Elnahrawy *et al.* (Elnahrawy et al. 2004) proposed several area-based localisation algorithms based on the RSS profiling. Different from the earlier point-based algorithms, which estimate the exact location of the non-anchor node, the area-based algorithms estimate the possible area that might contain the non-anchor node. The performance of the area based algorithms is measured by two parameters: accuracy and precision, where accuracy is the likelihood the object is within the area and precision is the size of the area. Three techniques were proposed for the area-based algorithms, i.e., single point matching, area based probability and Bayesian networks. By comparing the performance of these area based algorithms with the point based algorithm in (Bahl and Padmanabhan 2000), they observed that the performance of these algorithms is quite similar and there is a fundamental limit in the RSS profiling based localisation algorithms. Elnahrawy *et al.* also found a general rule of thumb is that using 802.11 technologies, with dense sampling and a good algorithm, one can expect a median localisation error of about 3 m and a 97th percentile error of about 9 m. With relatively sparse sampling, every 6 m, or 37 $m^2 / sample$, one can still get a median error of 4.5 m and 95th percentile error of 12 m.

The major obstacle in the RSS profiling based localisation a technique is that in order to build an accurate RSS model an extensive amount of profiling data is required. This involves substantial initial investment and effort for deployment and also adds significantly to the complexity of maintaining the model. Furthermore, even in normal office environments, changing environmental, building and occupancy conditions can affect the RSS model. Therefore a static RSS model may not be suitable in some applications. Recently, online profiling has been proposed as a way to reduce or eliminate the amount of profiling required before deployment at the expense of deploying a large number of additional devices called “sniffers”. These “sniffers” are deployed at known locations. Together with a large number of stationary emitters also deployed at known locations, these “sniffers” can be used to construct the RSS model online.

13.2.4 Angle of Arrival Measurements

It is convenient to divide the angle of arrival measurement techniques into two categories: those making use of the receiver antenna’s amplitude response and those making use of the receiver antenna’s phase response. The accuracy of the AOA measurements is limited by the directivity of the antenna, by shadowing and by multipath reflections. The first category of AOA measurement techniques is based on the principle that anisotropy in an antenna’s reception pattern can be used to find a transmitter’s bearing with regards to the receiver. The measurement unit can be of small size (in comparison with the wavelength of the transmitter signal) when using this technique. Fig. 13.1 shows the antenna pattern of a typical anisotropic antenna. When rotating the beam of the receiver antenna, the direction of maximum signal strength is determined as the direction of the transmitter. Using a rotating beam has the simplifying advantage that the only detailed knowledge of the antenna pattern required is the beam width. But this simplicity has the potential problem that signals of varying strength can completely distort the measurements. This problem can be overcome by employing an extra non-rotating omnidirectional antenna at the receiver that also receives the transmitter signal (Koks 2005). By normalising the signal strength received by the rotating anisotropic antenna with regards to the signal strength received by the non-rotating omnidirectional antenna, the impact of varying signal strength can be removed.

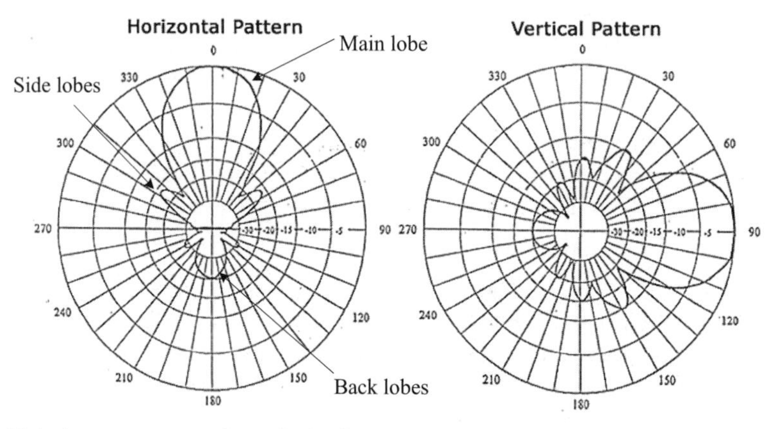

Fig. 13.1. Antenna pattern of a typical anisotropic antenna

Alternatively, the problem caused by varying signal strength can be solved by using at least two stationary antennas with known, anisotropic antenna patterns. Overlapping these patterns and comparing the signal strength from each then gives the transmitter direction, even when the signal strength is changing. This method of amplitude comparison has been used widely for AOA measurements (Koks 2000). Typically four or more antennas are used, with coarse tuning being done by simply measuring which antenna has the strongest signal, followed by fine-tuning by comparing the amplitude responses.

The second category of measurement techniques is based on the principle that a transmitter's bearing with regards to the receiver can be obtained from measurements of the phase differences in the arrival of a wave front. The measurements typically require a large receiver antenna (in comparison with the wavelength of the transmitter signal) or an antenna array. Fig. 13.2 shows an antenna array of N antenna elements. The adjacent antenna elements are separated by a uniform distance d. For a transmitter far away from the antenna array, the distance between the transmitter and the ith antenna element is approximately given by

$R_i \simeq R_0 - id\cos\theta$, where R_0 is the distance between the transmitter and the 0th antenna element and θ is the bearing of the transmitter with regards to the antenna array. Therefore the transmitter signal received by adjacent antenna elements will have a phase difference of $2\pi \times (d\cos\theta / \lambda)$. By measuring the phase difference, the bearing of the transmitter can be obtained.

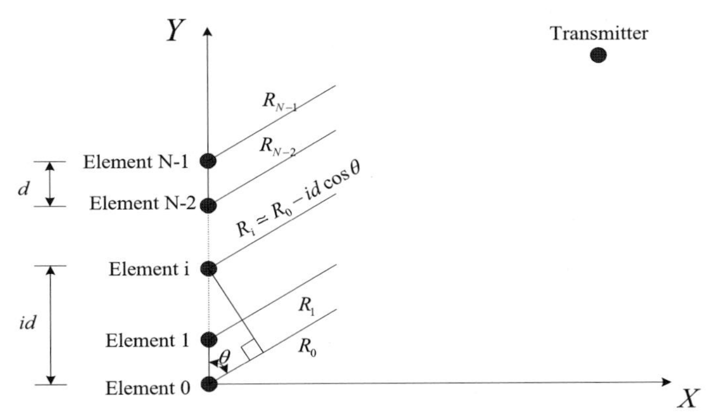

Fig. 13.2. Antenna array of N antenna elements

In the absence of noise and interference, bearing lines from two or more receivers will intersect to determine a unique location, which becomes the location estimate of the transmitter. In the presence of noise, more than two bearing lines will not intersect at a single point. The reader may refer to (Torrieri 1984) for a detailed discussion on location estimation techniques with noisy measurements.

13.2.5 Propagation Time based Measurements

Propagation time based measurements can be further divided into three subclasses: one-way propagation time, roundtrip propagation time and time-difference-of-arrival (TDOA) measurements.

13.2.6 One-way Propagation Time and Roundtrip Propagation Time

Both one-way propagation time and roundtrip propagation time measurements have been widely used for estimating distance between neighbouring sensors. The principle of one-way propagation time measurements is straightforward. By measuring the difference between the sending time and the receiving time of a signal, the distance between the transmitter and the receiver can be obtained. However, one-way propagation time measurements require very accurate synchronisation between the transmitter and the receiver. This requirement significantly adds to the cost of the sensor. Moreover, due to the high propagation speed of wireless signals, a small error in propagation time measurements may cause a large error in the distance estimate. These disadvantages make one-way propagation time measurements a less attractive option. In (Priyantha et al. 2000), Priyantha *et al.* obtained the one-way propagation time measurements by using a combination of RF and ultrasound hardware. On each transmission, a transmitter concurrently sends an RF signal, together with an ultrasonic pulse. When the receiver hears the RF signal, it turns on its ultrasonic receiver and listens for the ultrasonic pulse, which will usually arrive a short time later. Utilising the fact that the speed of sound in the air is much smaller than the speed of light (RF) in the air, the receiver uses the time difference between the receipt of the RF signal and the ultrasonic signal as an estimate of the one-way propagation time. Their method gave fairly accurate distance estimate at the expense of additional hardware and complexity of the system because ultrasonic reception suffers from severe multipath effects caused by reflections from walls and other objects.

In comparison with the one-way propagation time measurements, roundtrip propagation time measurements are more popular as there is no synchronisation problem. When measuring the roundtrip propagation time, one sensor transmits a signal to another sensor, which immediately replies with its own transmission. At the first transmitter, the measured delay between its transmission and its reception of the reply is the roundtrip propagation time, which is equal to twice the one-way propagation time plus a delay internal in the second sensor in handling the reply. This internal delay is either known via *a priori* calibration, or measured and sent to the first sensor to be subtracted. Recently the use of ultra wide band (UWB) signals for accurate distance estimation has received significant research interest (Lee and Scholtz 2002). An UWB signal is defined to be a signal whose bandwidth to centre frequency ratio is larger than 0.2 or a signal with a total bandwidth of more than 500 MHz. Fundamentally, UWB can achieve higher accuracy because its bandwidth is very large and its pulse has a very short duration, which allows easy separation of multipath signals.

13.2.7 Time-difference-of-arrival Measurements

Location estimation using TDOA is a technique for estimating the location of a transmitter from measurements of TDOA of the transmitter's signal at a number of receivers. Fig. 13.3 shows an TDOA localisation scenario with a group of four receivers at locations $\boldsymbol{r}_1, \boldsymbol{r}_2, \boldsymbol{r}_3, \boldsymbol{r}_4$ and a transmitter at $\boldsymbol{r}_t$. The time of arrival t_i for a receiver located at $\boldsymbol{r}_i$ of a signal emitted from the transmitter at time t_t is:

$$t_i = t_t + \frac{\| \boldsymbol{r}_i - \boldsymbol{r}_t \|}{c}, \qquad i = 1, \ldots, N \tag{13.6}$$

where, N is the number of receivers, $\|\bullet\|$ denotes the Euclidean norm and c is the propagation speed of the transmitter signal. The TDOA between a pair of receivers i and j is given by:

$$\Delta t_{ij} \triangleq t_i - t_j = \frac{1}{c}\left(\| \boldsymbol{r}_i - \boldsymbol{r}_t \| - \| \boldsymbol{r}_j - \boldsymbol{r}_t \|\right), \qquad i \neq j \qquad (13.7)$$

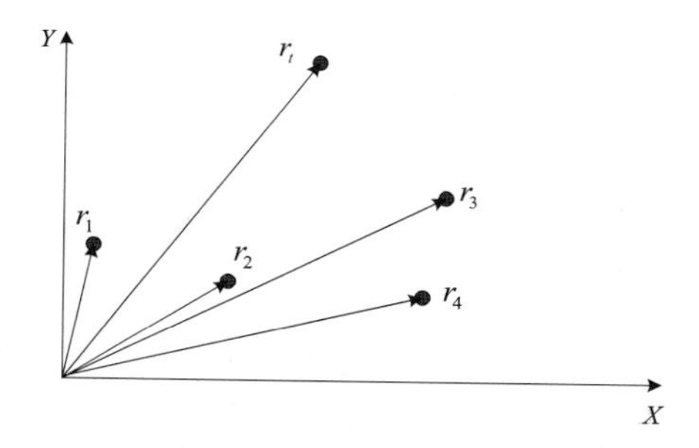

Fig. 13.3. Localisation using TDOA measurements

The location of the transmitter located at $\boldsymbol{r}_t$ is estimated from TDOA measurements from four receivers located at known locations r_1, r_2, r_3, r_4 Given the TDOA measurement Δt_{ij} and the coordinates of receivers i and j, Eq.13.7 defines one branch of a hyperbola whose foci are at the locations of receivers i and j and on which $\boldsymbol{r}_t$ must lie. In $\Re^2$, measurements from a minimum of three receivers are required to uniquely determine the location of the transmitter. This is illustrated in Fig. 13.4. In practice, Δt_{ij} is not available; instead we have the noisy TDOA measurement $\Delta \tilde{t}_{ij}$ defined by,

$$\Delta \tilde{t}_{ij} = \Delta t_{ij} + n_{ij} = \frac{1}{c}\left(\| \boldsymbol{r}_i - \boldsymbol{r}_t \| - \| \boldsymbol{r}_j - \boldsymbol{r}_t \|\right) + n_{ij} \qquad (13.8)$$

where, n_{ij} denotes an additive noise, which is usually assumed to be a Gaussian distributed random variable with zero mean. In a system consisting of N receivers, there are $N-1$ linearly independent TDOA Eq.13.8. A maximum likelihood solution to the transmitter location can be obtained numerically by using an iterative gradient descent algorithm (Torrieri 1984). Recently, Doğançay developed a closed-form transmitter location estimator using TDOA measurements based on triangulation of hyperbolic asymptotes (Dognacay 2005). In a WSN, which measurement technique to use for location estimation will depend on the specific application. Typically, both AOA and propagation time based measurements are able to achieve better accuracy than RSS based measurements. However, that accuracy is achieved at the expense of higher equipment cost. Patwati *et al.* gave a Cramér-Rao lower bound for location estimation using some measurement techniques in (Patwari et al. 2005).

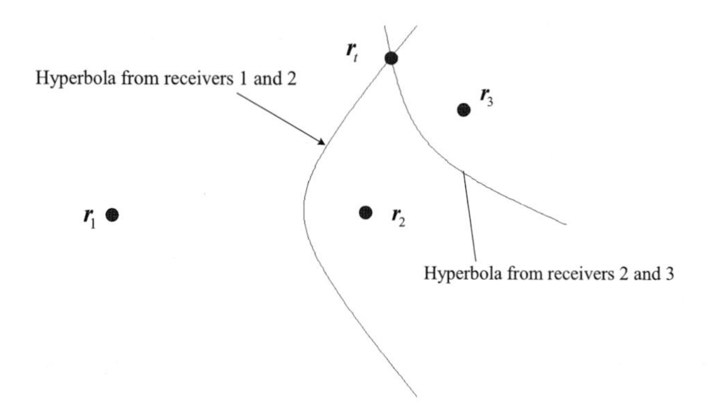

Fig. 13.4. Intersecting hyperbolas from three receivers. The transmitter is located at the intersection of hyperbola obtained from TDOA measurement of receivers 1 and 2 and hyperbola obtained from TDOA measurement of receivers 2 and 3

13.3 Connectivity-based Localisation Algorithms

It is worth noting that there is a class of localisation algorithms that do not rely on any of the measurement techniques in the last section. They are the so-called connectivity-based or "range free" localisation algorithms. They use the connectivity information, i.e., "who is within the communications range of whom" (Shang et al. 2005) to derive the locations of the non-anchor nodes. In (Doherty et al. 2001), Doherty *et al.* formulated the connectivity-based localisation problem as a convex optimisation problem and locations of the non-anchor nodes are obtained using polynomial-time numerical algorithms based on interior point methods for solving linear programs or semidefinite programs (SDP). A linear program is of the form,

$$\begin{aligned} &\text{Minimise} \quad \boldsymbol{c}^T \boldsymbol{x} \\ &\text{Subject to} \quad \mathrm{A}\boldsymbol{x} < \boldsymbol{b} \end{aligned} \tag{13.9}$$

where, $\boldsymbol{x} = [\boldsymbol{x}_1, \boldsymbol{x}_2, \ldots, \boldsymbol{x}_n]^T$ and $\boldsymbol{x}_i$ represents the coordinate vector of node i, i.e. $\boldsymbol{x}_i = [x_i, y_i]$. A connection between node i and j can be represented by a "radial constraint" on the node locations: $\| \boldsymbol{x}_i - \boldsymbol{x}_j \| \leq R$, where R is the transmission range. By leaving the objective function $\boldsymbol{c}^T \boldsymbol{x}$ blank and solving the problem, a solution to the coordinates of the non-anchor nodes satisfying the radial constraints can be obtained. The solution may not be unique. By slightly changing the cost function, i.e., setting the element of $\mathbf{c}$ corresponding to x_i (or y_i) to be 1 or -1 and all other elements of $\mathbf{c}$ to be 0, a lower bound or an upper bound on x_i (or y_i) satisfying the radial constraints can be obtained. Therefore, a rectangular box bounding the location estimates of the non-anchor nodes can be obtained. Simulation was performed using a total of 200 randomly placed nodes in a square of size $10R \times 10R$ and the average node degree (i.e., the average number of neighbours per node) is 5.7 (Doherty et al. 200). It was found that the mean error, which is defined as the average of the Euclidean distance between the estimated coordinates and the true coordinates, decreases monotonically with the number of anchor nodes. When the number of anchors is small, the estimation performs as poorly as a random guess of the node's coordinates. The mean error reduces to R when the number of anchors is increased to 18; and reduces further to $0.5R$ when the number of anchors is increased to 50.

In (Bulusu et al. 2000), Bulusu *et al.* proposed an algorithm, which locates each non-anchor node at the centroid of its neighbours. For a specific transmitter-receiver pair, a connectivity metric is defined as the ratio of the number of transmitter signals successfully received to the total number of signals from that transmitter. A receiver uses those transmitters whose connectivity metric exceeds a certain threshold (e.g., 90%) as its reference points. The location of the receiver is estimated to be the centroid of its reference points. An experiment was conducted by placing four reference points at the four corners of a $10m \times 10m$ square in an outdoor parking lot. The $10m \times 10m$ square was further divided into 100 smaller $1m \times 1m$ grids and the receivers were placed at the grid points. Experimental results showed that for over 90 percent of the data points the localisation error falls within 30 percent separation distance between two adjacent reference points.

In the "DV (distance vector) -hop" approach developed by Niculescu et al. the anchors first flood their locations to all nodes in the network. Each node counts the least number of hops that it is away from an anchor. When an anchor receives a message from another anchor, it estimates the average distance of one hop, which is sent back to the network as a correction factor. When receiving the correction factor, a non-anchor node is able to estimate its distance to anchors. Trilateration is then performed to estimate the location of the non-anchor node. Simulation using a total of 100 nodes uniformly distributed in a circular region of diameter 10 showed that the algorithm has a mean error of 45% transmission range with10% anchors. This error reduces to about 30% transmission range when the percentage of anchors increases above 20%. The average node degree is 7.6.

Shang *et al.* solved the problem by using multi-dimensional scaling (MDS). MDS has its origin in psychometrics and psychophysics. It can be seen as a set of data analysis techniques that displays the structure of distance-like data as a geometric picture. The algorithm consists of three steps: firstly, compute the shortest paths between all pairs of nodes; secondly, apply MDS to the distance matrix and find an approximate solution to the relative coordinates of all nodes; finally, transform the relative coordinates to the absolute coordinates by aligning the estimated coordinates of anchors with their true coordinates. Using the location estimates obtained earlier as an initial solution, a least-squares minimisation can be used to refine the location estimates. Simulation using 100 nodes uniformly distributed in a square of size 10×10 and four randomly placed anchors showed a localisation error of 0.35. The average node degree is 10. The MDS algorithm can be easily extended to incorporate distance measurements for location estimation. Shang *et al.* further improved the algorithm in (Shang and Ruml 2004). The main idea is to compute a local map using MDS for each node consisting only of nearby nodes, and then to merge these local maps together to form a global map. The improved algorithm avoids using the shortest path between far away nodes and can be implemented in a distributed fashion. It performs better than the original method on irregularly shaped networks.

Generally, the connectivity-based localisation algorithms are able to obtain a coarse grained estimate of each node's location. The localisation error decreases with increasing density of the network and the number of anchors. The localisation error is smaller in a regular network topology and becomes much larger in an irregular network topology. Although not able to achieve an accurate estimate of location, the simplicity of the connectivity-based localisation algorithms makes them attractive for applications requiring an approximate location estimate only.

13.4 Distance-Based Localisation Algorithms

In the remaining sections, we shall focus on distance-based localisation algorithms. Distance-based localisation algorithms can be divided into centralised algorithms and distributed algorithms. In centralised algorithms, all measurements are sent to a central processor where location estimation is performed. In distributed algorithms, there is no central processor. Each non-anchor node estimates its location using measurements between neighbouring nodes and the location estimates of its neighbours.

Centralised algorithms can be used in applications where a centralised architecture already exists. For example, in applications such as health monitoring, road traffic monitoring and control, environmental monitoring and precision agriculture, a central system already exists, which gathers information from all nodes in the network. In that case, it is convenient to piggyback the distance measurements onto the monitoring information sent to the central processor. Centralised algorithms are likely to provide more accurate location estimates than distributed algorithms. However centralised algorithms suffer from the scalability problem and generally are not suitable for a large-scale sensor network. Centralised algorithms also suffer from higher computational complexity, and lower reliability, which is caused by multi-hop transmission over a wireless network. In comparison, distributed algorithms are more scalable and have lower computational complexity. However distributed algorithms are difficult to design because of the potentially complicated relationship between local behaviour and global behaviour. Algorithms that are locally optimal may not perform well in a global sense. How to optimally distribute the computation of a centralised algorithm in a distributed implementation continues to be a research problem. Moreover, distributed algorithms generally require multiple iterations to converge to a stable solution. The relatively slow convergence speed of distributed algorithms may be a concern in some applications. Error propagation is also a potential problem in distributed algorithms.

Another important factor affecting the choice between centralised algorithms and distributed algorithms is communication cost. Sensors are energy constrained. Depending on the hardware and the transmission range, the energy required for transmitting a single bit could be used to execute 1,000 to 2,000 instructions (Chen et al. 2002). Centralised algorithms in large networks require each sensor's measurements to be sent over many hops to a central processor. Distributed algorithms require only local information exchange between neighboring nodes but possibly many such exchanges are required, depending on the number of iterations needed to arrive at a stable solution. The energy efficiency of centralised and distributed algorithms was compared in (Rabat and Nnowak 2004). In general, when the average number of hops to the central processor exceeds the necessary number of iterations, distributed algorithms will likely save communication energy costs.

13.4.1 Centralised Algorithms

There are three major approaches to distance-based location estimation in a centralised architecture, i.e., multi-dimensional scaling, linear (and/or semi-definite) programming, and stochastic optimisation (e.g., simulated annealing).

The connectivity-based localisation algorithm in (Shang et al. 2004) using MDS can be easily extended to incorporate distance measurements into the optimisation problem. Ji *et al.* presented an algorithm similar to that in (Shang et al. 2004). They first divided the sensors into smaller groups where adjacent groups may share common sensors. Each group contains at least three anchors or sensors whose locations have already been estimated. MDS is used to estimate the relative positions of sensors in each group and build local maps. Local maps are then stitched together to form a larger global map by utilising common sensors between adjacent local maps. An initial estimate of nodes' locations is then obtained. By aligning the estimated locations of anchors with the true locations of anchors, the estimated locations of sensors in the stitched maps will approach their true locations iteratively. The algorithm can be implemented distributedly or in a centralised architecture. As a large number of iterations are required for the algorithm to converge, their algorithm is more appropriately executed in a centralised architecture. Simulation was conducted using a total of 400 nodes uniformly distributed in a square of 100 by 100 and a transmission range of 10. The distance measurement error is uniformly distributed in the range $[0,\eta]$. A normalised error is used to measure the performance of the algorithm and it is defined as the mean location estimation error divided by the transmission range, R. With 10% anchors, when

η is 0, $0.05R$, $0.25R$ and $0.5R$, the normalised error is 0.1, 0.15, 0.3 and 0.45 respectively. The connectivity-based localisation algorithm proposed by Doherty *et al.* can also be extended to incorporate distance measurements and the location estimation problem can be solved using numerical algorithms for linear programming or semidefinite programming. Biswas *et al.* presented quadratic formulations of the location estimation problem and used SDP to solve the locations of non-anchor nodes. Liang *et al.* improved the result in (Biswas and Ye 2004) by using gradient search to fine tune the initial solution obtained using SDP. Yet another approach to solve the distance-based location estimation problem is via stochastic optimisation, particularly simulated annealing (SA) (Kanan et al. 2005). SA is a generalisation of the Monte Carlo method, a well-known technique in combinatorial optimisation. The concept is based on the manner in which liquids freeze or metals recrystalise in the process of annealing. Considering a sensor network of m anchor nodes numbered from 1 to m and n non-anchor nodes numbered from $m+1$ to $m+n$, the location estimation problem can be formulated as an optimisation problem:

$$\min_{\hat{x}_i,\ m+1\le i\le m+n} \sum_{i=m+1}^{m+n} \sum_{j\in N_i} \left(\hat{d}_{ij} - \tilde{d}_{ij}\right)^2 \tag{13.10}$$

$$\hat{d}_{ij} = \| \hat{\boldsymbol{x}}_i - \hat{\boldsymbol{x}}_j \| \tag{13.11}$$

N_i is the set of neighbors of node i, $\tilde{d}_{ij}$ is the measured distance between node i and node j and $\hat{\boldsymbol{x}}_i$ is the estimated coordinate vector of node i. An SA approach to the optimisation problem Eq.13.10 is presented in (Kannan et al 2005). It is also observed in (Kannan et al 2006) that there is some information about the sensors' locations hidden in the knowledge of whether given pair of sensors are neighbours or not. This observation is utilised to further improve the performance of the SA based localisation algorithms by mitigating a certain kind of localisation error caused by flip ambiguity, a concept which is described in detail in Section 5. In (Kannan et al 2006), a simulation was performed in a sensor network of 200 nodes uniformly distributed in a square of size 10 by 10. Fig. 13.5 shows the performance of the SA algorithm with flip ambiguity error mitigation (Kannan et al 2006) and the performance of the SDP with gradient search improvement (Liang et al. 2004). The location estimation error is normalised by the transmission range. When the transmission range is 1.1, 1.2, 1.4, 1.6, 1.8 and 2.0, the corresponding average node degree is 6.86, 8.19, 10.87, 13.96, 17.87 and 21.36 respectively. The following noise model is used in the simulation:

$$\tilde{d}_{ij} = d_{ij}\left(1+\rho\times n\right) \tag{13.12}$$

where, n is a random noise with normal distribution and $\rho = 0.1$.

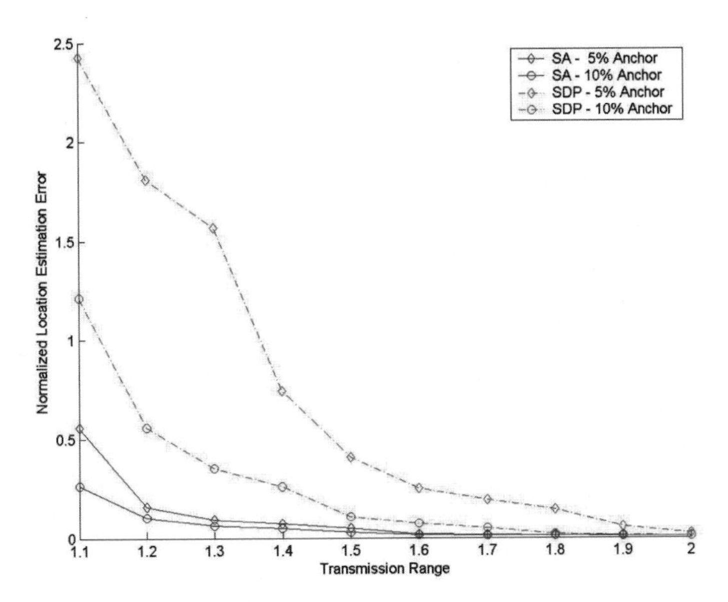

Fig. 13.5. Performance of SA algorithm with flip ambiguity mitigation and SDP algorithm with gradient search improvement

As shown in the figure, the SA algorithm has better accuracy than the SDP algorithm with gradient search however this improvement is achieved at the expense of higher computation cost. A possible reason for the better performance is the SA algorithm is robust against being trapped into a local minimum.

13.4.2 Distributed Algorithms

The "DV-hop" algorithm developed by Niculescu *et al.* can be readily modified to include distance measurements into the location estimation. This is done by letting neighbouring nodes propagate measured distance instead of hop count. The corresponding algorithm is referred to as the "DV-distance" algorithm (Niculescu and Nath 2001). Savarese *et al.* developed a two-stage localisation algorithm in (Savarese Rabaey 2002). In the first stage, a "Hop-terrain" algorithm, which is similar as the "DV-hop" algorithm, is used to obtain an initial estimate of the nodes' locations. In the second stage, the measured distances between neighbouring nodes are used to refine the initial location estimate. To mitigate location estimation errors caused by error propagation and unfavourable network topologies, a confidence value is assigned to each node's position. An anchor has a higher confidence value (close to 1) and a non-anchor node with few neighbours and poor constellation has lower confidence value (close to 0). Simulation using 400 nodes uniformly placed in a 100 by 100 square showed that the algorithm is able to achieve an average location estimation error of less than 33% of the transmission range in the presence of 5% distance measurement error (normalised by the transmission range) and 5% anchors. The average node degree is greater than 7.

The localisation algorithm developed by Savvides *et al* is divided into four stages. In the first stage, those nodes whose locations are "tentatively unique" are selected based on the assumption that solution to a node location is "tentatively unique" if it has at least three neighbours that are either non-collinear anchors or their solutions are "tentatively unique". The locations of these tentatively uniquely localisable nodes are estimated in stage two and three. In the second stage, each non-anchor node obtains the estimated distances to at least three anchors using a "DV-distance" like algorithm. An estimated distance to an anchor node allows the location of the non-anchor node to be constrained inside a square centred at that anchor node. The estimated distances to more than three anchors allow the location of the non-anchor node to be confined inside a rectangular box, which is the intersection of the squares. The location of the non-anchor node is estimated to be at the centre of the rectangular box. The initial location estimates obtained in the second stage are refined in the third stage by a least-squares trilateration using the location estimates of the neighbouring nodes and the measured distances. Finally, the location of each node deemed not "tentatively unique" in stage one is estimated using the location estimates of its tentatively uniquely localisable neighbours.

Langendoen *et al.* identified the common features of the aforementioned three algorithms as well as some of their variants. By comparing the performance of the three algorithms and some variants of them, they found that these algorithms have comparable performance and which algorithm has better accuracy depends on conditions such as distance measurement error, vertex degree and percentage of anchors. Another category of distributed localisation algorithms first constructs local maps using distance measurements between neighbouring nodes; then uses common nodes between local maps to stitch them together to form a global map. The localisation algorithm by Ji *et al.* is a typical example. In the algorithm of Čapkun *et al.*, each node builds its local coordinate system and the locations of its neighbours are calculated in the local coordinate system. Then the directions of the local coordinate systems are aligned to be the same using common nodes between adjacent local coordinate systems. Finally, the local coordinate systems are reconciled into a global coordinate system using linear translation. The major problems in these algorithms are error propagation and the large number of iterations required for the algorithm to converge. These problems have impeded the implementation of these algorithms in a distributed system. They are also the reasons why we categorise the algorithm of Ji *et al.* as a centralised algorithm.

Recently the use of particle filters (Kwok et al. 2004) for localisation has attracted significant research interest and a number of papers were published on using particle filters for robot localisation and sensor network localisation. Ihler *et al.* formulated the localisation problem as an inference problem on a graphical model and applied nonparametric belief propagation (NBP) to obtain an approximate solution to sensor locations. NBP is a variant of the popular belief propagation (BP) algorithm. Ihler *et al.* formulated NBP as an iterative, local message-passing algorithm, in which each node computes its "belief" about its location estimate, communicates this belief to and receives messages from its neighbors, then updates its belief and repeats. This process is repeated until some convergence criterion is met, after which each sensor is left with an estimate of its location and the uncertainty of the estimate. NBP has the advantage that it is easily implemented in a distributed fashion and only requires a small number of iterations to converge. Simulations showed that NBP's performance is comparable to the central MAP (maximum *a posteriori*) estimate. Given its promising performance, capability to provide uncertainty of location estimation and accommodate non-Gaussian distance measurement error and many possible avenues of improvement, NBP appears to provide a useful tool for estimating unknown sensor locations.

13.5 Distance-Based Localisation Using Graph Theory

The problem of distance-based sensor network localisation discussed in Section 4 can be restated in a compact form as follows: Consider a planar or possibly three-dimensional array of sensors. Assume that a collection of inter-sensor distances as well as the Euclidean coordinates of a small number of sensors (i.e., ***anchor*** sensors) is known. The network localisation problem is then one of determining the Euclidean coordinates of all the sensors in the network (Anderson et al. 2005; Eren et al. 2004; Aspnes et al. 2005; Goldenberg et al. 2005).

13.5.1 Problem 1 (Distance-Based Localisation Problem)

Consider a sensor network $N(S,D,S_a)$ with a set S of sensor nodes, a set D of known distances d_{ij} between certain pairs of nodes $s_i, s_j \in S$, and a set $S_a \subset S$ of ***anchor*** nodes s_{a_i} ($a_i \in \{1,\ldots,|S|\}, i \in \{1,\ldots,|S_a|\}$)[1], whose coordinates $\boldsymbol{r}_i$ are known. Find a mapping $p : S \to \Re^d$ ($d \in \{2,3\}$) that assigns a d-dimensional coordinate to each node $s_i \in S$ such that $\|p(s_i) - p(s_j)\| = d_{ij}$ holds for all node pairs $s_i, s_j \in S$ for which d_{ij} is given (i.e., $d_{ij} \in D$), and the assignment is consistent with any (***anchor***) node assignments provided, i.e., $p(s_{a_i}) = \boldsymbol{r}_i, \forall s_{a_i} \in S_a$.

The sensor network localisation problem, which is formulated in Problem 1 and illustrated in Fig. 13.6, can be split up into an analytic existence/solvability problem and an algorithmic problem. The analytic problem is to determine the proper-

[1] In this chapter, $|\overline{S}|$ for a given set $\overline{S}$ denotes the number of elements in $\overline{S}$.

ties of a sensor network that ensure unique solvability of the localisation problem. The algorithmic problem is to find a method of solving the localisation problem and determine the computational complexity of this method. A specific pair of questions to be answered related to the algorithmic problem are how to deal with the presence of errors in the inter-sensor measurements and how such errors translate into errors in the algorithm's output of sensor coordinates.

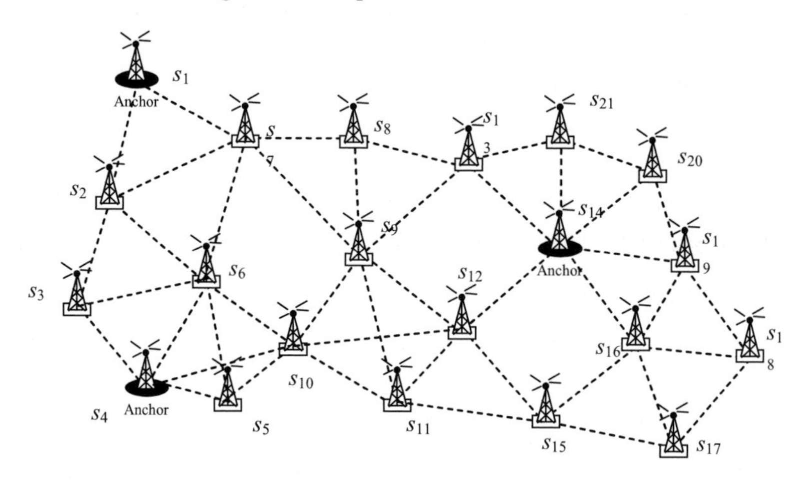

Fig. 13. 6. A two-dimensional sensor network localisation problem: Absolute positions $\boldsymbol{r}_i$ of the three anchor sensors $s_{a_i}, a_1 = 1, a_2 = 4, a_3 = 14$ are known. The distance d_{ij} between each sensor pair s_i, s_j connected with a (dashed) line segment is measurable. The task is to estimate the absolute position of every sensor s_i in the network

13.5.2 A Graph Theoretical Framework

In the sequel, both the analytic and the algorithmic problems above will be studied in the framework of graph theory, particularly using the notions of ***rigid*** graph theory. A sensor network $N(S, D, S_a)$, as described in Problem 1, can be represented by a graph $G = (V, E)$ with a vertex set V and an edge set E, where each vertex $i \in V$ is associated with a sensor node s_i in the network, and each edge $(i, j) \in E$ corresponds to a sensor pair s_i, s_j for which the inter-sensor distance d_{ij} is known. In this case, we call $G = (V, E)$ the *underlying graph* of the sensor network $N(S, D, S_a)$. A d-dimensional ($d \in \{2,3\}$) *representation* of a graph $G = (V, E)$ is a mapping $\overline{p} : V \rightarrow \Re^d$. Given a graph $G = (V, E)$

and a d-dimensional ($d \in \{2,3\}$) representation of it, the pair $(G, \overline{p})$ is called a d-dimensional *framework.* A *distance set* $\overline{D}$ for G is a set of distances $\overline{d}_{ij} > 0$, defined for all edges $(i, j) \in E$. Given a distance set $\overline{D}$ for the graph G, a d-dimensional ($d \in \{2,3\}$) representation $\overline{p}$ of G is a (d-dimensional) *realisation* if it results in $\|\overline{p}(i) - \overline{p}(j)\| = \overline{d}_{ij}$ for all pairs $i, j \in V$ where $(i, j) \in E$. In this case, we call $\overline{p}$ a d-dimensional *realisation* of $(G, \overline{D})$ and $(G, \overline{p})$ a d-dimensional *realisation framework.*

Application of the above notions in modelling sensor networks is illustrated in Fig. 13.7. In this figure, the sensor network $N(S, D, S_a)$ in Fig 6. is modelled by a 2-dimensional realisation framework $(G, \overline{p})$, where $G = (V, E)$ is the underlying graph of $N(S, D, S_a)$, the distance set $\overline{D}$ is selected to be the union of the known inter-sensor distance set D and the set D_a of distances between pairs of anchor sensors (which can be calculated using the known positions of the anchor nodes), and $\overline{p}$ is a 2-dimensional realisation of (G, D). In such a setting, the problem of localising a d-dimensional ($d \in \{2,3\}$) sensor network $N(S, D, S_a)$ with an underlying graph G is equivalent to finding the d-dimensional realisation $\overline{p}$ of (G, D) that satisfies $\overline{p}(a_i) = \boldsymbol{r}_i, \forall s_{ai} \in S_a$.

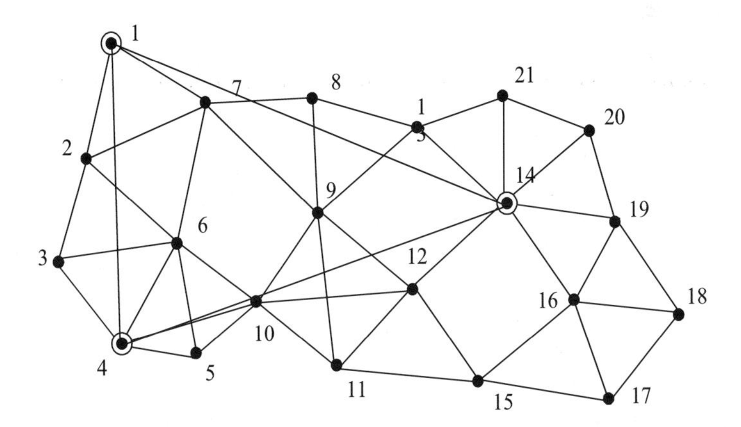

Fig. 13.7. A (graph) realisation corresponding to the sensor network in Fig. 6. The realisation can be thought as a physical structure of rigid bars and joints, where the bar lengths are equal to the known/measured distances. Note that the vertices 1, 4, and 14 correspond to the three anchor nodes. These vertices are connected by edges between each other since the corresponding distances, although not measured via sensors, can be calculated using the known positions of the anchor nodes

13.6 Rigid and Globally Rigid Graphs in Localisation

Use of graph rigidity notions in network localisation is well described and their importance is well demonstrated in the recent literature (Anderson et al. 2005; Eren et al. 2004; Aspnes et al. 2005; Goldenberg et al. 2005). Particularly, it is established in (Eren et al. 2004; Aspnes et al. 2005) that a necessary and sufficient condition for unique localisation of a sensor network is ***generic global rigidity*** of its underlying graph, which will be defined later in this section. This result forms the basis for application of rigid graphs in localisation of sensor networks, from both the algorithmic and the analytic aspects. Next we define the fundamental notions used in rigid graph theory.

Two frameworks $(G, \overline{p})$ and $(G, \overline{q})$, where $G = (V, E)$, are ***equivalent*** if $\|\overline{p}(i) - \overline{p}(j)\| = \|\overline{q}(i) - \overline{q}(j)\|$ for any vertex pair $i, j \in V$ for which $(i, j) \in E$. The two frameworks $(G, \overline{p})$ and $(G, \overline{q})$ are ***congruent*** if $\|\overline{p}(i) - \overline{p}(j)\| = \|\overline{q}(i) - \overline{q}(j)\|$ for any pair $i, j \in V$, whether or not $(i, j) \in E$. This means that $(G, \overline{q})$ can be obtained from $(G, \overline{p})$ applying a combination of translations, rotations and reflection only. A framework $(G, \overline{p})$ is called ***rigid*** if there exists a sufficiently small positive constant ε such that if $(G, \overline{q})$ is equivalent to $(G, \overline{p})$ and $\|\overline{p}(i) - \overline{q}(i)\| < \varepsilon$ for all $i \in V$ then $(G, \overline{q})$ is congruent to $(G, \overline{p})$. Intuitively, a rigid framework cannot flex, and if a framework $(G, \overline{p})$ (with a distance set $\overline{D}$) is non-rigid then continuous deformations can be applied to produce an infinite number of different realisations of G (for $\overline{D}$).

Note that there exist rigid frameworks $(G, \overline{p})$ and $(G, \overline{q})$ which are equivalent but not congruent, as demonstrated in Fig. 13.8. A framework $(G, \overline{p})$ is *globally rigid* if every framework, which is equivalent to, $(G, \overline{p})$ is congruent to $(G, \overline{p})$. It is easy to see that if G is a complete graph then the framework $(G, \overline{p})$ is necessarily globally rigid.

If a framework $(G, \overline{p})$ (with a distance set $\overline{D}$) is rigid but not globally rigid, like the ones in Fig. 13.8, although a continuous deformation does not exist as in the case of non-rigid frameworks, there are two types of discontinuous deformations that can prevent a realisation of G (for the distance set $\overline{D}$) from being unique (in the sense that it differs from the other possible realisations of the same graph at most by translation, rotation or reflection) (Hendrickson 1992; Moore et al. 2004): *flip* and *discontinuous flex ambiguities*. In *flip ambiguities* in $\Re^d$ ($d \in \{2,3\}$), a vertex (sensor node) v has a set of neighbours which span a

$(d-1)$-dimensional subspace, e.g., v has only d neighbours; which leads to the possibility of the neighbours forming a mirror through which v can be reflected. Fig. 13.8(a) depicts an example of flip ambiguity. In *discontinuous flex ambiguities* in $\Re^d$ $(d \in \{2,3\})$, the removal of an edge allows the remaining part of the graph to be flexed to a different realisation (which cannot be obtained from the original realisation by translation, rotation or reflection) and the removed edge reinserted with the same length. Fig. 13.8(b) depicts an example.

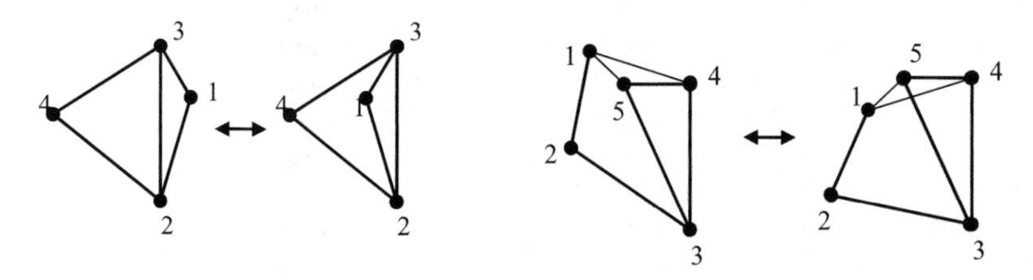

Fig. 13.8. Two pairs of equivalent rigid frameworks in $\Re^2$ which are not congruent: **(a)** The frameworks suffer from flip ambiguity: Vertex 1 can be reflected across the edge (2,3) to a new position without violating the distance constraints. **(b)** The frameworks suffer from discontinuous flex ambiguity: Removing the edge (1,4), flexing the edge triple (1,5),(1,2),(2,3), and reinserting the edge (1,4) so that the distance constraints are not violated in the end, we obtain a new realisation

Knowing the graph and distance set of a globally rigid framework is not sufficient to position the framework absolutely in $\Re^d$ $(d \in \{2,3\})$. In order to do this, we need the absolute positions of at least three vertices in $\Re^2$ or four vertices in $\Re^3$ as well, and in fact they must be *generically positioned*, i.e., these vertices should not lie on the same line in $\Re^2$ or the same plane in $\Re^3$. Fig. 13.9 illustrates a case in $\Re^2$, where the three vertices with known absolute positions are collinear. For this example, there are two possible absolute positions for the whole framework, which are symmetric with respect to the line passing through the three collinear vertices with known absolute positions.

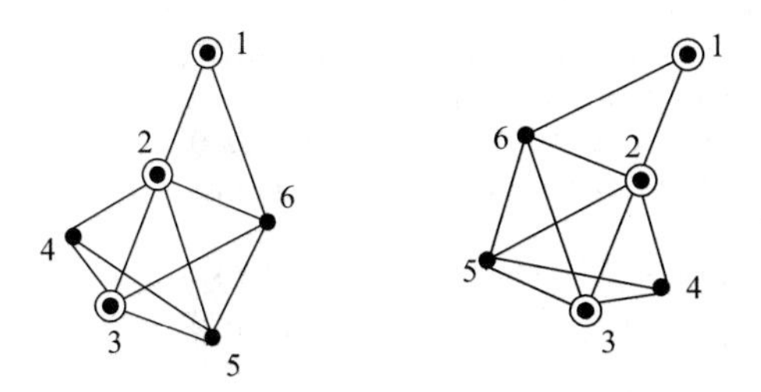

Fig. 13.9. Absolute localisation of a framework in $\Re^2$: The vertices 1,2,3 have known positions but are collinear. The length of each edge shown in the figure is known. Using this information, there are two candidates for the absolute location of the whole framework

The analytic existence/solvability problem for sensor network localisation in Section 5.1 can be thought of as follows: Suppose a realisation framework underlying the sensor network is constructed, i.e., the edge lengths in the framework correspond to the known inter-sensor distances. The framework may or may not be rigid; and even if it is rigid, there may be a second and differently shaped framework, which is a realisation (constructible with the same vertex, edge and length assignments), similar to the case depicted in Fig. 13.8. If the framework is globally rigid, then the sensor network can be thought of as like a rigid entity of known structure, and one then only needs to know the Euclidean position of several sensors in it to locate the whole framework in $\Re^2$ or $\Re^3$. The above discussions on the relation between the localisability of the sensor network $N(S,D,S_a)$ in Problem 1 and the global rigidity of a realisation framework underlying $N(S,D,S_a)$ is summarised in the following theorem:

13.6.1 Theorem 1 (Eren et al. 2004; Aspnes et al. 2005)

Consider a d-dimensional ($d \in \{2,3\}$) sensor network $N(S,D,S_a)$ as specified in Problem 1. Assume that at least $d+1$ of the $|S_a|$ anchor nodes are generically positioned in $\Re^d$. Let $(G,\overline{p})$ be a d-dimensional realisation framework underlying $N(S,D,S_a)$, i.e., a d-dimensional realisation of the underlying graph of $N(S,D,S_a)$ with the edge lengths corresponding to the known inter-sensor distances in D. Then Problem 1 is solvable if and only if $(G,\overline{p})$ is globally rigid.

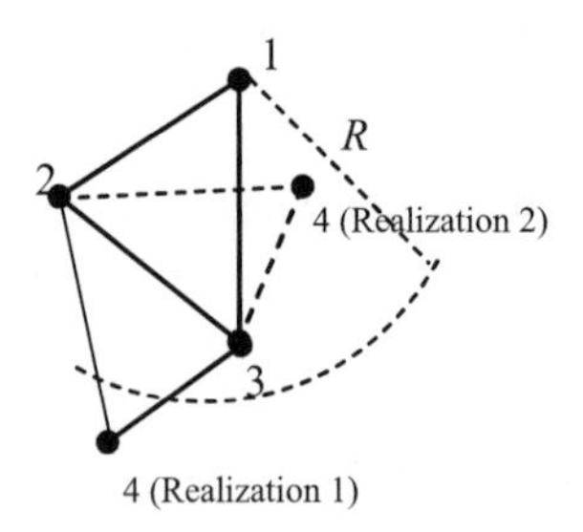

Fig. 13.10. Absolute localisation of a (non-globally) rigid unit-disk framework in $\Re^2$: The positions of vertices 1,2,3 and the lengths of all the edges are known. In general, there are two possible realisations of such a framework. However, Realisation 2 is ruled out using the unit-disk property (if it was the correct realisation, there would exist an edge between the vertices 1 and 4, requiring $d_{14} < R$), and Realisation 1 is fixed as the correct unique realisation

Remark 1: The necessity of global rigidity for unique solution of Problem 1, as stated in Theorem 1, is valid for general situations where other *a priori* information is not helpful. Rigidity is needed, in any case, to have a finite number of solutions. However, in some cases where $(G, \overline{p})$ is rigid but not globally rigid, some additional *a priori* information may compensate the need for global rigidity. For example, assume that a sensor network can be represented by a unit disk graph, where there is an edge between two nodes (the nodes can sense each other) if and only if the distance between them is less than a certain threshold $R > 0$. Then the ambiguities due to the non-globally rigid nature of the underlying graph may sometimes be eliminated using the unit disk graph properties as demonstrated in Fig. 13.10. Given a rigid framework (G, p) in $\Re^d$ $(d \in \{2,3\})$, if p is *generic*, for example, if no combination of $d+1$ vertices in G lies on a $d-1$ dimensional hyperplane (which is a line in $\Re^2$ and a plane in $\Re^3$), then the graph G alone determines the rigidity of the framework, and so we can define *generic rigidity of a graph*. There is a linear algebraic test for rigidity based on the rank of a matrix, whose entries are formed from the coordinates of the vertices. In $\Re^2$, there exists the following alternative combinatorial (essentially graph theoretic) necessary and sufficient condition for rigidity, termed Laman's Theorem (Laman 1970; Whitely 1996).

13.6.2 Theorem 2 (Laman's Theorem)

A graph $G = (V, E)$ in $\Re^2$, where $|V| > 1$, is generically rigid if and only if there exists a subset $E' \subseteq E$ satisfying the following conditions, (i) $|E'| = 2|V| - 3$, and (ii) For any non-empty subset $E'' \subseteq E'$, $|E''| \leq 2|V(E'')| - 3$; $V(E'')$ denotes the set of all end-vertices of the edges in E''. No such combinatorial necessary and sufficient condition is currently available in $\Re^3$, where the generalisations of the two dimensional results are only necessary, but not sufficient. We have the following necessary counterpart in $\Re^3$:

13.6.3 Theorem 3 (Whitely 1996)

If a graph $G = (V, E)$ in $\Re^3$, where $|V| > 2$, is generically rigid, then there exists a subset $E' \subseteq E$ satisfying the following conditions, (i) $|E'| = 3|V| - 6$, and (ii) For any non-empty subset $E'' \subseteq E'$, $|E''| \leq 3|V(E'')| - 6$, where $V(E'')$ denotes the set of all end-vertices of the edges in E''.

We can define *generic global rigidity of a graph* in $\Re^2$ in a similar way (Connelly 2005). In order to state a combinatorial test for generic global rigidity in $\Re^2$, we need to define two other terms: For any positive integer k, a graph is called *k-connected* if between any two vertices of the graph, there exist at least k paths which have no edge or vertex in common, or equivalently, it is not possible to find $k-1$ vertices whose removal (together with the removal of the edges incident on them) would render the graph unconnected (Godsil and Royel 2001). A graph is termed *generically redundantly rigid* if with the removal of any edge, it remains generically rigid (Aspens et al 2005; Jackson and Jordan 2005). In $\Re^2$, there is a variant of Laman's Theorem for checking generic redundant rigidity; the reader may refer to (Hendrickson 1992; Jackson and Jordan 2005) for details. Based on these definitions, the following theorem states an elegant necessary and sufficient condition for *generic global rigidity* of a framework (i.e., global rigidity of a generic framework) in $\Re^2$.

13.6.4 Theorem 4 (Jackson and Jordan 2005)

A graph $G=(V,E)$ in $\Re^2$ with $|V|\geq 4$ vertices is generically globally rigid if and only if it is 3-connected and redundantly rigid. In $\Re^3$, it is necessary that a graph be 4-connected and generically redundantly rigid for the graph to be generically globally rigid. However, these conditions are known to be insufficient (Connelly 2005). No necessary and sufficient condition for generic global rigidity has been established in $\Re^3$ yet. In fact, in contrast to rigidity, it is by no means obvious that global rigidity in $\Re^3$ is a generic property, i.e., if a particular network is globally rigid, it is not obvious that for any other network with the same underlying graph, that network will also be globally rigid, unless its nodes satisfy a certain set of algebraic constraints.

So far we have seen the relation between the solvability of the distance-based sensor network localisation and global rigidity, and necessary and sufficient conditions for rigidity and global rigidity, which are all related to the analytic aspects of the localisation problem. Next, we move our focus to the algorithmic aspects. Based on Theorem 1, the algorithmic sensor localisation problem can be seen as a graph realisation problem, i.e., realisation of the underlying graph of a given sensor network $N(S,D,S_a)$ using the information about the positions of the anchor sensors in S_a and the known inter-sensor distances in D, as also described in Section 5.2. Fundamentally, the task is to solve a number of simultaneous quadratic equations, in which the unknowns are the coordinates of the sensors, and there is one equation for each known distance. For various specific sensor network settings, different ad hoc approaches to the above problem can be developed based on Theorem 1. More general methodologies have been studied in the literature for certain classes of sensor networks as well. Examples of certain classes of sensor networks as well. Examples of such methodologies can be found in the studies mentioned in Section 4.

For certain classes of underlying graphs including the ones described in Section 5.4, iterative[2] methods can be used to achieve localisation (sometime also with reduction of complexity). Such a class is the class of *bilateration graphs*, viz. graphs $G=(V,E)$ with an ordering of vertices $v_1,v_2,\ldots,v_{|V|}$ (called *bilaterative ordering*) such that (i) the edges $(v_1,v_2),(v_1,v_3),(v_2,v_3)$ are all in E, (ii) each vertex v_i for $i=4,5,\ldots,|V|-1$ is connected to (at least) two of the vertices $v_1,v_2,\ldots,v_{i-1}$, and (iii) the vertex $v_{|V|}$ is connected to (at least) three of the vertices $v_1,v_2,\ldots,v_{|V|-1}$. An iterative localisation algorithm for sensor networks with bilateration underlying graphs is given in (Fang et al 2006).

The computational complexity of an arbitrary localisation procedure is concluded in the literature to be, in general, exponential in the number of sensor nodes, unless an iterative procedure is applied to sensor networks with certain classes of underlying graphs such as the *trilateration* and *quadrilateration graphs* described in the next subsection (Eren et al. 2004; Aspnes et al. 2005; Saxe 1979) (but not bilateration graphs, for which the complexity can be exponential, despite the existence of an iterative algorithm). These results are particularly relevant for sensor networks whose underlying graphs are unit disk graphs, i.e., underlying graphs in which any pair of vertices have an edge joining them if and only the sensor nodes they represent are closer than a pre-specified constant distance R, which is called the transmission range or *sensing radius* (Aspens et al 2005). In the next section, we focus on the trilateration and quadrilateration graphs, leaving further discussions on computational complexity of the rigid graph theory based localisation algorithms.

13.7 Trilaterations and Quadrilaterations

The notions *trilateration* and *quadrilateration* are used both as globally rigid graph growing operations and for labelling certain classes of globally rigid graphs. The *trilateration operation* on a globally rigid graph $G=(V,E)$ in $\Re^2$ is addition of a new vertex v to G together with (at least) three new edges connecting v to the vertices in V. A *trilateration graph* $G=(V,E)$ in $\Re^2$ is one whose vertices can be ordered as $v_1,v_2,\ldots,v_{|V|}$ such that (i) the edges $(v_1,v_2),(v_1,v_3),(v_2,v_3)$ are all in E and (ii) each vertex v_i for

[2] Here, the notion "iterative" is used in a different way from Section 4. The "iteration" here sequentially passes through sensor nodes.

$i = 4,5,\ldots,|V|$ is connected to (at least) three of the vertices $v_1, v_2, \ldots, v_{i-1}$, i.e., the graph $G = (V, E)$ can be obtained by applying $|V| - 3$ consecutive trilateration operations starting with the graph formed by the vertices v_1, v_2, v_3 and the edges between them and adding vertex v_{i+3} at each step $i = 1,2,\ldots,|V| - 3$. The ordering $v_1, v_2, \ldots, v_{|V|}$ of the vertices above is called a *trilaterative ordering*.

Similarly, the *quadrilateration operation* on a globally rigid graph $G = (V, E)$ in $\Re^3$ is addition of a new vertex v to G together with (at least) four new edges connecting v to the vertices in V. A *quadrilateration graph* $G = (V, E)$ in $\Re^3$ is one whose vertices can be ordered as $v_1, v_2, \ldots, v_{|V|}$ such that (i) $(v_i, v_j) \in E$ for any $1 \le i < j \le 4$ and (ii) each vertex v_i for $i = 5,6,\ldots,|V|$ is connected to (at least) four of the vertices $v_1, v_2, \ldots, v_{i-1}$, i.e., the graph $G = (V, E)$ can obtained by applying $|V| - 4$ consecutive quadrilateration operations in a similar way as the construction of trilateration graphs. Similarly, the ordering $v_1, v_2, \ldots, v_{|V|}$ of the vertices is called a *quadrilaterative ordering*. Note that trilateration and quadrilateration graphs are necessarily bilateration graphs as well.

For a proof of generic global rigidity of 2-dimensional trilateration graphs and 3-dimensional quadrilateration graphs, the reader may refer to (Eren et al. 2004; Aspens et al. 2005). Note that the ordering $v_1, v_2, \ldots, v_{|V|}$ of the vertices for a given trilateration graph $G = (V, E)$ (based on application of trilateration operations) may not be unique, which needs to be considered in localising sensor networks with trilateration underlying graphs. The same argument applies to quadrilateration graphs as well. Nevertheless, it is established in (Eren et al. 2004; Aspens et al. 2005) that a 2-dimensional sensor network with a trilateration underlying graph or a 3-dimensional one with a quadrilateration underlying graph can be localised with an iterative algorithm that is polynomial (and in some cases even linear) in the number of sensor nodes. The computational complexity of localising such networks is further discussed in Section 5.7.

Since trilateration and quadrilateration underlying graphs provide proven reduced computational complexity in localisation and actually there are systematic methods to locate networks with such underlying graphs, it is of interest to develop mechanisms to make these methods applicable for certain other classes of underlying graphs as well. One way of doing that is development of techniques to construct trilateration (quadrilateration) graphs from non-trilateration (non-quadrilateration) graphs, which is further discussed in the next subsection.

13.7.1 Acquiring Globally Rigid, Trilateration, Quadrilateration Graphs

Systematic construction of generically globally rigid trilateration and quadrilateration graphs from graphs without these properties via addition of some extra edges is studied in (Anderson et al. 2005). In sensor network localisation, addition of extra edges to the underlying (representative) graph corresponds to increasing the sensing radius for each sensor, e.g., by adjusting the transmit powers. This will let some sensors in the network determine their distances not just to their immediate neighbours (in the original setting), but also to their two-hop, three-hop, etc. distant neighbours (in the original setting).

For example, doubling the sensing radius in a network will let all the two-hop neighbour[3] pairs in the original setting become immediate neighbours. From the perspective of underlying graphs, this means construction, from an original graph $G=(V,E)$, of a new graph $G^2=(V,E\cup E^2)$, where for any $v_i \neq v_j \in V$, $(v_i,v_j)\in E^2$ if and only if there exists a $v_k \in V$ such that $(v_i,v_k)\in E$ and $(v_k,v_j)\in E$. For such a construction we have the following theorem drawn from (Anderson et al. 2005), which applies to a special superset of 2-connected graphs. Before stating the theorem we need to make the following definition: For any positive integer k, a graph is called *edge k-connected* if between any two vertices of the graph, there exist at least k paths which have no edge in common[4].

13.7.2 Theorem 5 (Anderson et al. 2005)

Let $G=(V,E)$ be an edge 2-connected graph in $\Re^2$. Then $G^2=(V,E\cup E^2)$ is generically globally rigid. Theorem 5 indicates that, given a 2-dimensional sensor network with an underlying graph that is not generically globally rigid but edge 2-connected, one can make the underlying graph generically globally rigid and hence the sensor network localisable (provided certain other conditions in Theorem 1 hold) by doubling the sensing radius of each sensor.

Note that bilateration graphs are 2-connected, and hence edge 2-connected although not globally rigid. Another example of non-globally rigid edge 2-connected graphs is the class of cycle graphs $C=(V_C,E_C)$, where the vertices can be ordered as $v_1,v_2,\ldots,v_{|V_C|}$ such that $E_C=\{(v_1,v_2),(v_2,v_3)\ldots,(v_{|V_C|-1},v_{|V_C|}),(v_{|V_C|},v_1)\}$. This example is elabo-

---

[3] Here for any sensor s in the network, a neighbour sensor is any sensor that can be sensed by s. A 2-hop neighbour is neighbour of a neighbour, a 3-hop neighbour is a neighbour of a 2-hop neighbour, etc

[4] Note that any k-connected graph is necessarily edge k-connected.

rated and a method to realise the globally rigid graph $C^2 = (V_C, E_C \cup E_C^2)$ for a given cycle graph $C = (V_C, E_C)$ is given (and hence localise the corresponding sensor network) in (Anderson et al. 2005), which can be easily generalised for $G^2 = (V, E \cup E^2)$ corresponding to an arbitrary 2-dimensional edge 2-connected graph $G = (V, E)$. In $\Re^3$, we have the following counterpart of Theorem 5.

13.7.3 Theorem 6 (Anderson et al. 2005)

Let $G = (V, E)$ be an edge 2-connected graph in $\Re^3$. Then $G^3 = (V, E \cup E^2 \cup E^3)$ is generically globally rigid, where for any $v_i \neq v_j \in V$, $(v_i, v_j) \in E^3$ if and only if there exist $v_k, v_l \in V$ such that $(v_i, v_k) \in E$ and $(v_k, v_l) \in E$ and $(v_l, v_j) \in E$. Based on Theorem 6, for a given 3-dimensional sensor network with an underlying graph that is not generically globally rigid but edge 2-connected, one can make the underlying graph generically globally rigid and hence the sensor network localisable (provided certain other conditions in Theorem 1 hold) by tripling the sensing radius of each sensor, as opposed to doubling in $\Re^2$. Next, we review counterparts of the global rigidity acquisition results summarised in Theorems 5 and 6 for trilateration and quadrilateration properties discussed in Section 5.5. The following result establishes that given any sensor network with a connected underlying graph, one can make the underlying graph a trilateration (quadrilateration) graph and apply the results in Section 5.5 to localise the sensor network by tripling (quadrupling) the sensing radius of each sensor:

13.7.4 Theorem 7 (Anderson et al. 2005)

Let $G = (V, E)$ be a connected graph in $\Re^d$ $(d \in \{2,3\})$ and let $v_1, v_2, \ldots, v_{|V|}$ be an ordering of the vertices in V such that for any integer $1 \leq k \leq |V|$, the subgraph of G induced by the vertex set $V_k = \{v_1, v_2, \ldots, v_k\}$ (i.e., the largest subgraph of G having V_k as its vertex set) is connected. Then: (i) $G^3 = (V, E \cup E^2 \cup E^3)$ is a trilateration graph. (ii) $G^4 = (V, E \cup E^2 \cup E^3 \cup E^4)$ is a quadrilateration graph, where E^4 is defined to E^2 and E^3. (iii) $v_1, v_2, \ldots, v_{|V|}$ is a trilaterative ordering for G^3 and quadrilaterative ordering for G^4.

Remark 2: The process of doubling, tripling, or quadrupling the sensing radius, in practice, is not an easy task. In order to double the radius it is needed to amplify the associated power by a scale of 4-8, in order to triple by a scale of about 9-27, depending on the particular path loss exponent. Nevertheless, the idea can be used in the cases where the sensors are preferred to operate at power levels much less than the maximum (full power) value and are permitted to increase their powers occasionally (when needed). As an alternative method for obtaining $G^2 = (V, E \cup E^2)$ from an underlying graph $G = (V, E)$, use of the cosine law is proposed in (Anderson et al. 2005) for the cases where each sensor can determine the angle between two of its neighbours in addition to the distances to those neighbours. In such a case, noting that the cosine law allows determination of the distance between any two neighbours of each sensor and that any pair of neighbours of a given sensor are either neighbours of each other or at a two hop distance from each other, all two-hop distances can be determined.

13.8 Localisation of Large-Scale Sensor Networks

Modelling sensor networks with a large number of sensor nodes using the deterministic graph settings of the previous subsections is not feasible in general, due to various sources of uncertainties. An often-appropriate way of abstracting such large-scale networks is found to be via *random geometric graph* notions (Eren et al. 2004; Asnes et al. 2005).

A 2-dimensional *random geometric graph* $G_n(R)$ parameterised by a pair of positive real numbers n, R is a random graph framework $((V, E), \overline{p})$ where V is a random vertex set and $\overline{p}$ is a random realisation such that each vertex $i \in V$ with realisation $\overline{p}(i)$ is a random point in $[0,1]^2$ generated by a 2-dimensional Poisson process of intensity n, and E is a random edge set, for any pair $i \neq j \in V$, satisfying $(i, j) \in E$ if and only if $|\overline{p}(i) - \overline{p}(j)| < R$. The parameters n and R are called the *sensor density* and the *sensing radius*, respectively.

Modelling of a large-scale sensor network using a random geometric graph allows analysis of the probability (of the underlying graph) of the network being connected, k-connected, and globally rigid, etc. It further allows one to analyse the effects of changing the sensing radius on such probabilities. The results of a series of such analyses can be found in (Eren et al. 2004; Asnes et al. 2005; Penrose 1999). Next, we present some of these results together with the implications of the earlier theorems in the section for random geometric graphs.

Remark 3: Theorem 5 implies that (in $\Re^2$), for a given sensor density – sensing radius pair n, R, the probability of $G_n(2R)$ to be globally rigid is greater than or equal to the probability of $G_n(R)$ being 2-connected.

13.8.1 Theorem 8 (Eren et al. 2004; Asnes et al. 2005)

Consider a set of random graphs $G_n(R)$ with n varying, and with R dependent on n and such that for all n, there holds $\{nR^2/(8\log n)\} \geq 1+\varepsilon$ for some positive ε. Then the probability that $G_n(R)$ is not a trilateration graph tends to zero as $n \to \infty$.

13.8.2 Theorem 9 (Eren et al. 2004; Asnes et al. 2005)

For any positive integer (sensor density) n, there exists some sensing radius $R \in O\left(\sqrt{\log n / n}\right)$ such that a realisation of the 2-dimensional random geometric graph $G_n(R)$ is computable in linear time if the positions of three vertices in $G_n(R)$ connected with edges to each other are known. Further discussions about sensor network localisation protocols based on the above results as well as the analysis of 3-dimensional large-scale sensor networks, as opposed to 2-dimensional ones, can be found in (Eren et al. 2004; Asnes et al. 2005). As a relevant issue an analysis of the localisation of so-called *partially localisable sensor networks*, where some of the sensors in the network cannot be uniquely localised, can be seen in (Goldenberg et al. 2005).

13.9 Complexity: Rigid Graph Theory based Localisation

The computational complexity of distance-based network localisation algorithms has been investigated in the literature (see, Eren et al. 2004; Asnes et al. 2005; Saxe 1979). In general, the computational complexity of localisation is exponential in the number of sensor nodes (Saxe 1979). As mentioned in Section 5.3, this conclusion is particularly valid for sensor networks whose underlying graphs are unit disk graphs. Nevertheless, there are exceptional classes of sensor network underlying graphs, for which this conclusion does not apply. The trilateration and quadrilateration graphs examined earlier constitute one of these classes.

Localisation of sensor networks modelled by trilateration and quadrilateration graphs is shown to have polynomial (and on occasions linear) computational complexity in the number of sensor nodes (Eren et al. 2004; Asnes et al. 2005). For the trilateration graphs, the computational complexity is linear if a so-called ***seed*** of the trilateration graph is known. If such a seed is not known, it takes in general (non-linear) polynomial time to locate the whole sensor network. While many graph theory results available for $\Re^2$ do not generalise straightforwardly, if not at all, to $\Re^3$, quadrilateration, as a three-dimensional generalisation of the trilateration, also allows a corresponding sensor network localisation process to be polynomial (or linear, if a seed is available) in the number of sensor nodes.

Sensor networks having certain classes of globally rigid underlying graphs that are not trilateration graphs may have reduced complexity of localisation as well. For example, if a generically globally rigid graph is a union of a small number of trilateration subgraphs (*trilateration islands*) connected with a number of edges, one may expect to have a corresponding localisation algorithm composed of a subalgorithm whose computational complexity is exponential in the number of trilateration islands and another one polynomial in the number of the maximum number of vertices in a trilateration island.

13.10 Current Problems In Distance-Based Localisation

Despite the recent developments in the literature, there exist quite many unsolved problems in distance-based sensor network localisation. The challenges to be addressed are both in analytical characterisation of the sensor networks (from the aspect of localisation) and development of (efficient) localisation algorithms for various classes of sensor networks under a variety of conditions. In this section, we present some of these current problems together with the preliminary approaches to them. An important issue not well addressed in the existing literature is the effects of imprecise distance measurement on sensor network localisation. No matter which algorithm is used in localisation, one must expect to have some difference between the actual inter-sensor distances and the corresponding distance measurements/estimates because of measurement noise and estimation errors. Nevertheless, it is of great interest to reduce these inaccuracies (and hence their affects on the overall network localisation task) as much as possible.

As a general approach to the above problem of reducing inter-sensor distance calculation inaccuracies, joint and full manipulation of the geometric and algebraic relations among the inter-sensor distances is proposed in (Cao et al. 2005). In the same paper, based on this idea, a concrete method to describe the geometric relations among the ordinary sensor – anchor sensor distances as an array of quadratic equalities is presented as well. The key tool in the method is use of the so called *Cayley-Menger matrix*, a well-known matrix in Distance Geometry Theory (see Havel et al. 1983) used to describe the distance relations between pairs of point sequences, and the classical results about this matrix. Although the Cayley-Menger matrix is demonstrated in (Cao et al. 2005) to be a very useful tool in reducing certain inter-sensor distance calculation inaccuracies, integration of this tool with the existing localisation algorithms is an open problem. Note that the problem of integration is not a trivial one, especially for large-scale networks where linear estimation methods would be preferred, because of the nonlinearities introduced by the Cayley-Menger matrix.

A dual inaccuracy issue is the errors in sensor position estimates using which other sensors are localised. This issue emerges especially in estimation of the location of non-immediate neighbours of anchor sensors, i.e., k -hop neighbours of anchor nodes with large k. As k increases the magnitude of the error in the corresponding sensor's position estimate is expected to increase due to accumulation of such errors in the position estimates of all the sensors between this sensor and the nearest anchor sensor. Although a sensor network with a globally rigid underlying graph is theoretically free of non-unique localisation problems as indicated in Theorem 1, the above inaccuracy sources may cause a localisation algorithm to suffer from certain ambiguities described in Section 5.3, which are typical in non-globally rigid (but rigid) settings. Examples of such situations are given in (Moore et al. 2004). A sample situation where measurement errors may cause flip ambiguity problems is depicted in Fig. 13.11.

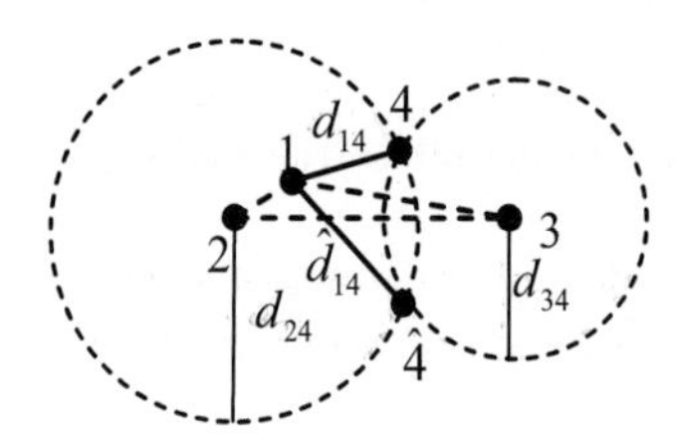

Fig. 13.11. Measurement noises causing a flip ambiguity in a globally rigid setting: Sensor node 4 is desired to be localised employing the known positions of nodes 1,2,3 and measurements of the distances between node pairs (4,1), (4,2), and (4,3). The distance measurements d_{24} and d_{34} constrain the position of node 4 to lie on one of the two intersection points of the circles shown in the figure. An erroneous measurement of the distance d_{14} as $\hat{d}_{14}$ will cause node 4 to be located as $\hat{4}$

A recent work focusing on robust distributed localisation of sensor networks with certain measurement/estimation errors of the above type and ambiguities caused by these errors is presented in (Moore et al. 2004). In this paper, certain criteria are provided in selection of the subgraphs of the underlying graph of a network to be used in a localisation algorithm robust against such errors. The analysis in (Moore et al. 2004), however, is not complete and there may be other criteria that may better characterise robustness of a given sub-network against distance measurement and location estimation errors in localisation, which both show future research directions in the area. A closely related current problem is to understand how graph properties characterise sensor networks with robust localisation to (complete) sensor failures rather than measurement and estimation inaccuracies.

Another area of active research is understanding and utilising the data and error propagation characteristics in a sensor network. Some of the available tools related to this area can be seen in (Ihler et al 2005). Using the tools here together with some other methods, approaches can be developed to localisation based on analysis of accumulation of gross (ambiguity-type) errors of low probability and small errors of high probability. It is worth to recall the idea of using trilateration islands stated in Section 5.7 to reduce the computational complexity of localisation of networks having certain classes of globally rigid underlying graphs, which are not trilateration graphs. Given such an underlying graph, the problem of partitioning the graph to trilateration islands in an optimal way is open. Once such a partitioning is known, development of an optimal algorithm employing trilateration ideas is another future research direction.

13.11 References

Anderson BDO, Belhumeur PN, Eren T, Goldenberg D, Morse AS, Whiteley W, and Yang RY (2005) Graphical properties of easily localisable sensor networks, submitted for publication in Wireless Networks

Aspnes J, Eren T, Goldenberg DK, Morse AS, Whiteley W, Yang YR, Anderson BDO, and Belhumeur PN (2005) A theory of network localisation, submitted to IEEE Trans. on Mobile Computing

Bergamo P and Mazzini G (2002) Localisation in sensor networks with fading and mobility. The 13th IEEE Int. Symp.on Personal, Indoor and Mobile Radio Communications, pp 750-754

Bahl P and Padmanabhan VN (2000) RADAR: an in-building RF-based user location and tracking system. IEEE INFOCOM, pp 775-784

Biswas P and Ye Y (2004) Semidefinite programming for ad hoc wireless sensor network localisation. 3rd Int. Symp. Information Processing in Sensor Networks, pp 46-54

Bulusu N, Heidemann J, and Estrin D (2000) GPS-less low-cost outdoor localisation for very small devices, IEEE Personal Communications, vol. 7, pp 28-34

Capkun S, Hamdi M, and Hubaux J (2001) GPS-free positioning in mobile ad-hoc networks. 34th Hawaii Int. Conf. on System Sciences, pp 3481-3490

Cao M, Anderson BDO, and Morse AS (2005) Localisation with imprecise distance information in sensor networks. Proc. Joint IEEE Conf on Decision and Control and European Control Conf., pp 2829-2834

Chen JC, Yao K, and Hudson RE (2002) Source localisation and beam forming, IEEE Signal Processing Magazine, vol. 19, pp 30-39

Connelly R (2005) Generic global rigidity, Discrete and Computational Geometry, vol. 33, pp 549-563

Dogancay K (2005) Emitter localisation using clustering-based bearing association, IEEE Trans. on Aerospace and Electronic Systems, vol. 41, pp 525-536

Doherty L, pister KSJ, and El Ghaoui L (2001) Convex position estimation in wireless sensor networks. IEEE INFOCOM, pp 1655-1663

Elnahrawy E, Li X, and Martin RP (2004) The limits of localisation using signal strength: a comparative study. 1st Annual IEEE Communication Society Conf. on Sensor and Ad hoc communications and Networks, pp 406-414

Eren T, Goldenberg D, Whiteley W, Yang RY, Morse AS, Anderson BDO, and Belhumeur PN (2004) Rigidity and randomness in network localisation. IEEE INFOCOM, pp 2673-2684

Fang J, Cao M, Morse AS, and Anderson BDO (2006) Sequential localisation of networks. submitted to the 17th Int. Symp. on Mathematical Theory of Networks and Systems-MTNS 2006

Goldenberg DK, Krishnamurthy A, MAness WC, Yang RY, Young A, Morse AS, Savvides A, and Anderson BDO (2005) Network localisation in partially localisable networks. IEEE INFOCOM, pp 313 - 326

Godsil C and Royle G (2001) Algebraic Graph Theory. New York: Springer-Verlag

Havel TF, Kuntz ID, and Crippen GM (1983) The Theory and Practice of Distance Geometry, Bulletin of Mathematical Biology, vol. 45, pp 665-720

Hendrickson B (1992) Conditions for unique graph realisations, SIAM J. Comput., vol. 21, pp 65-84

Ihler AT, Fisher JW, III, Moses RL, and Willsky AS (2005) Nonparametric belief propagation for self-localisation of sensor networks, IEEE Journal on Selected Areas in Communications, vol. 23, pp 809-819

Jackson B and Jordan T (2005) Connected rigidity matroids and unique realisations of graphs, J. Combinatorial Theory Series B, vol. 94, pp 1-29

Ji X and Zha H (2004) Sensor positioning in wireless ad-hoc sensor networks using multidimensional scaling. IEEE INFOCOM, pp 2652-2661

Kannan AA, Mao G, and Vucetic B (2005) Simulated Annealing based Localisation in Wireless Sensor Network. The 30th IEEE Conf. on Local Computer Networks, pp 513-514

Kannan AA, Mao G, and Vucetic B (2006) Simulated Annealing based Wireless Sensor Network Localisation with Flip Ambiguity Mitigation. to appear in IEEE Vehicular Technology Conf.

Koks D (2005) Numerical calculations for passive geolocation scenarios, Edinburgh, SA, Australia, Report No. DSTO-RR-0000

Krishnan P, Krishnakumar AS, Ju W-H, Mallows C, and Gamt SN (2004) A system for LEASE: location estimation assisted by stationary emitters for indoor RF wireless networks. IEEE INFOCOM, pp 1001-1011

KWOK C, FOX D, and MEILA M (2004) Real-time particle filters, Proc. of the IEEE, vol. 92, pp 469-484

Langendoen K and Reijers N (2003) Distributed localisation in wireless sensor networks: a quantative comparison, Computer Networks, vol. 43, pp 499-518

Liang T-C, Wang T-C, and Ye Y (2004) A gradient search method to round the semidefinite programming relaxation for ad hoc wireless sensor network localisation Standford University, Report no. Aug. 26

Laman G (1970) On graphs and rigidity of plane skeletal structures, J. of Engineering Mathematics, vol. 4, pp 331-340

Lee J-Y and Scholtz RA (2002) Ranging in a dense multipath environment using an UWB radio link, IEEE Journal on Selected Areas in Communications, vol. 20, pp 1677-1683

Madigan D, Einahrawy E, Martin RP, Ju W-H, Krishnan P, and Krishnakumar AS (2005) Bayesian indoor positioning systems. IEEE INFOCOM 2005, pp 1217-1227

Moore D, Leonard J, Rus D, and Teller S (2004) Robust distributed network localisation with noisy range measurements. the 2nd ACM Conf. on Embedded Networked Sensor Systems (SenSys'04), pp 50-61

Niculescu D and Nath B (2001) Ad hoc positioning system (APS). IEEE GLOBECOM, pp 2926-2931

Patwari N, Ash JN, Kyperountas S, Hero AO, III, Moses RL, and Correal NS (2005) Locating the nodes: cooperative localisation in wireless sensor networks, IEEE Signal Processing Magazine, vol. 22, pp 54-69

Patwari N, Hero AO, Perkins M, Correal NS, and O'Dea RJ (2003) Relative location estimation in wireless sensor networks, IEEE Trans. on Signal Processing, vol. 51, pp 2137-2148

Penrose MD (1999) On k-connectivity for a geometric random graph, Random Structures and Algorithms, vol. 15, pp 145-164

Prasithsangaree P, Krishnamurthy P, and Chrysanthis P (2002) On indoor position location with wireless LANs. The 13th IEEE Int. Symp. Personal, Indoor and Mobile Radio Communications, pp 720-724

Priyantha NB, Chakraborty A, and Balakrishnan H (2000) The Cricket Location-Support System. Proc. of the Sixth Annual ACM Int. Conf. on Mobile Computing and Networking

Rabbat M and Nowak R (2004) Distributed optimisation in sensor networks.3rd Int. Symp. on Information Processing in Sensor Networks, pp 20-27

Rappaport TS (2001) Wireless Communications: Principles and Practice. 2nd ed, Prentice Hall PTR

Saxe J (1979) Embeddability of weighted graphs in k-space is strongly NP-hard. 17th llerton Conf. in Communications, Control and Computing, pp 480-489

Shang Y and Ruml W (2004) Improved MDS-based localisation. IEEE INFOCOM 2004, pp 2640-2651

Shang Y, Ruml W, Zhang Y, and Fromherz M (2004) Localisation from connectivity in Sensor Networks, IEEE Trans. on Parallel and Distributed Systems, vol 15, pp 961-974

Savarese C and Rabaey J (2002) Robust positioning algorithms for distributed ad-hoc wireless sensor networks. Proceedings of the General Track: 2002 USENIX Annual Technical Conference, pp 317-327

Savvides A, Park H, and Srivastava MB (2002) The bits and flops of the n-hop multilateration primitive for node localisation problems. International Workshopon Sensor Networks Application, pp 112-121

Torrieri DJ (1984) Statistical theory of passive location systems, IEEE Transactions on Aerospace and Electronic Systems, vol. AES-20, pp 183-198

Whiteley W (1996) Some matroids from discrete applied geometry in Contemporary Mathematics. vol. 197, J. E. Bonin, J. G. Oxley, and B. Servatius, eds.: American Mathematical Society

14 Location Estimation

SH Gary Chan and Victor Cheung

Department of Computer Science and Engineering, The Hong Kong University of Science and Technology, Hong Kong

14.1 Introduction

In recent years, there has been increasing interest in WSNs in academic, industrial, and commercial sectors. In wireless sensor networks, wireless nodes are equipped with measuring devices in addition to wireless communication capabilities. With such networks one can monitor habitats of interest, detect intruders within private premises, or explore unknown environment (Mainwaring et al. 2002; Wang et al. 2003; Akyildiz et al. 2002). It is often useful to know the relative or absolute nodal locations in order to improve the quality of service delivered. Nodal location information is important in the following two cases:

- *Location-based routing:* As the transmission range of each node is limited, intermediate nodes are often required to forward data packets. When a packet is forwarded, its route may be chosen according to the relative positions of the nodes. Consequently, estimating the relative positions of the nodes is important for route correctness and efficiency. Location-based routing has been widely studied, most of which determine the route by making use of the network topology as obtained in the intermediate nodes and the location of the destined node (Karp and Kung 2000; Bose and Morin 2000).
- *Location-based services:* This refers to offering services to users depending on their locations. These services include identifying the location of a patient in a hospital, locating the parking position of one's car, monitoring a variety of ambient conditions, etc. The main concern in such a system is hence the estimation error in terms of both absolute and angular distance between nodes, as well as the distance error between the estimated and real locations.

There has been much work on location estimation. Here we survey various techniques of estimating nodal locations applicable in wireless sensor networks. These techniques form the basics for location estimation. We consider a wireless sensor network as a network consisting of many stationary sensor nodes (in the order of hundreds to thousands of nodes) closely placed together with wireless communication capability. Depending on the specific technique a node may also be equipped with some extra capabilities such as angle measurement, GPS, etc. We also allow a small proportion of entities, which are more powerful than the other sensor nodes in the network, whose role is to provide some special features so as to facilitate the location estimation process. We discuss the state-of-the-art beyond what was presented in (Hightower and Borriello 2001), and evaluate the pros and cons of each of them. We first go over a number of location estimation techniques where a node either gathers information from neighbouring entities to estimate its own location or shares information with each other to estimate locations cooperatively. We classify the estimation approaches into four categories, namely, distance-based, angle-based, pattern-based and connectivity-based, and provide examples in each category. We then provide a comparison between these techniques in terms of estimation steps required, message complexity, power consumption, density dependency and special hardware requirement.

14.2 Location Techniques

Due to limited processing power and energy in sensor nodes, as well as the requirement of low-cost and large number of nodes in a sensor network, equipping every node with Global Positioning System (GPS) (Wellenhop et al. 1997) capability is not feasible. However, GPS is necessary when a user wants to know the exact locations of the nodes (absolute positioning), as opposed to the nodal locations relative to each other (relative positioning). For absolute positioning, usually only a few (powerful) nodes have GPS capability, based on which the remainder nodes estimate their own locations. In this section we discuss some of the recent location estimation techniques. We classify them into four categories, namely, distance-based, angle-based, pattern-based and connectivity-based. We first explain its basic working mechanism, followed by its general strengths and weaknesses.

- *Nodes:* Sensor nodes, which possess communication capabilities. Depending on the estimation technique, they may also be equipped with some special measuring devices such as signal strength measurer, directional antenna, etc.
- *Landmarks:* Landmarks are nodes, which know their locations, either by GPS or via some other means (by, for example, placing them at known locations). A landmark can be part of a fixed infrastructure, or any node that has its location estimated independently. In location estimation, either the landmarks estimate the locations of other nodes in a centralised manner, or the nodes estimate their own locations in a distributed manner given landmark locations.

14.2.1 Distance-based

In this category, nodes first measure distances among themselves. Given the distances, tools such as *trilateration*, *least square method* and *multidimensional scaling (MDS)* may be used to estimate the locations of the nodes. In trilateration, a node estimates its own location using its distances from a number of non-collinear landmarks. For a two dimensional plane, at least three landmarks are needed. In case there are more than three landmarks available, a node can perform trilateration on taking average on some combinations of any three of them, or simply use any three of them. There is also a generalised version of trilateration, which allows more than three distances and uses linear equations to estimate the position iteratively. We only highlight the basic principles here.

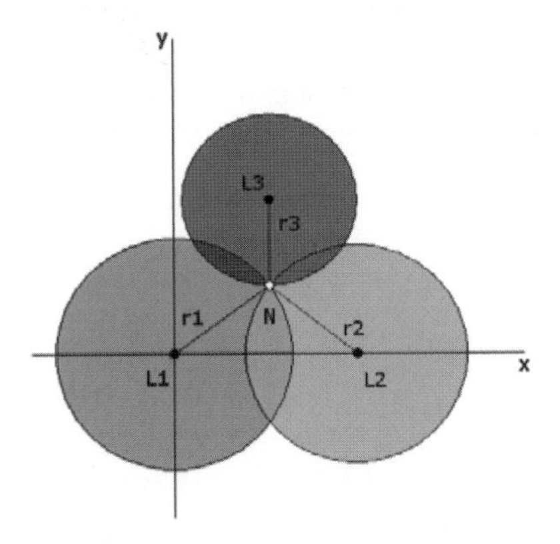

Fig. 14.1. Example of trilateration: L_1, L_2 and L_3 are landmarks. With the knowledge of their locations, together with r_1, r_2 and r_3, location of N can be obtained. For simplicity the coordinates are transformed such that L_1 is at the origin and L_2 lies on the x-axis

Fig. 14.1 illustrates how the location of node N is estimated, given the locations of and distances from three landmarks labelled as L_1, L_2 and L_3. For simplicity in discussion, let's assume for now that L_1 is at the origin with coordinate (0, 0), L_2 is on x-axis with coordinate $(x_{L2}, 0)$, and L_3 is with coordinate (x_{L3}, y_{L3}). (This assumption can be realised by proper transformation of coordinates.) Node N measures its distances to L_1, L_2 and L_3 as r_1, r_2 and r_3, respectively. To estimate the coordinate (x_N, y_N) of N, we first write down the following system of equations consistent with the measurements:

$$\begin{aligned} x'^2_N + y'^2_N &= r_1^2 \\ (x'_N - x'_{L_2})^2 + y'^2_N &= r_2^2 \\ (x'_N - x'_{L_3})^2 + (y'_N - y'_{L_3})^2 &= r_3^2 \end{aligned} \qquad (4.1(a\text{-}c))$$

From the first two equations we can calculate x'_N by,

$$x'_N = \frac{r_1^2 - r_2^2 + x'^2_{L_2}}{2x'_{L_2}} \qquad (14.2)$$

Substituting x'_N into the first equation and together with the third equation, we obtain,

$$y'_N = \frac{r_1^2 - r_3^2 + (x'_N - x'_{L_3})^2}{2y'_{L_3}} + \frac{y'_{L_3}}{2} - \frac{(r_1^2 - r_2^2 + x'^2_{L_2})^2}{8x'^2_{L_2} y'_{L_3}} \tag{14.3}$$

In reality there may be measurement noise such that r_1, r_2 and r_3 have random components. In this case, we may obtain a Region of Confidence (RoC) within which the node N lies with some high probability (Gwon et al. 2004).

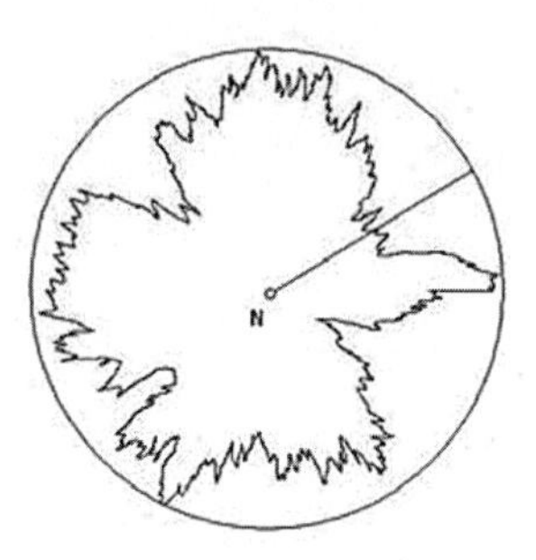

Fig. 14.2. An illustration of the irregularity of signal pattern: The inscribing circle is the ideal signal pattern, with radius equal to the maximum radio range of node N

The advantage of this tool is that, given the locations of the landmarks and their distances to the node, the location of the node can be easily calculated. This is particularly useful when the computational power of the node is low. The accuracy of this tool depends on how accurate the distances are obtained. Usually one obtains the distance by means of received signal strength, as it decreases with distance. However, due to unpredictable surrounding environment, the radio pattern is often not circular (Zhou et al. 2004). Fig. 14.2 illustrates a possible irregular signal pattern of a sensor node. This is even worse if the environment is dynamic, for example in a shopping mall with many customers. Another possible source of error is due to multi-path, where a receiver receives multiple reflected signals as well as the original signal. Techniques such as GPS and Cricket address the signal pattern irregularity by using constant propagation speed of the radio signal (Time of Arrival and Time Difference of Arrival) (Priyantha et al. 2000). This comes with the cost of synchronisation with the landmarks (satellites in the case of GPS and ultrasonic emitters in Cricket).

In order to address the irregularity problem in signal pattern, some other trilateration techniques estimate the distance by using *hop counts* together with some accurate distance information. One example is the DV-HOP system as illustrated in Fig. 14.3 (Niculescu and Nath 2003). In this system, all nodes know their distances in terms of hops to the landmarks via distance vector exchange. Once a landmark gets the actual coordinates of other landmarks, it estimates the average one-hop distance by dividing the total distances to the landmarks by the total hop counts, and broadcasts this to the entire network. For example, when L_2 receives the coordinates of L_1 and L_3, it knows that it is 70 m from L_1 (and being 5 hops away) and 100 m from L_3 (and being 6 hops away). It hence estimates the one-hop distance as (70+100)/(5+6) = 15.45 m. Nodes receiving the one-hop distance may then estimate distances to landmarks and perform trilateration. The system does not work so well when the distribution of the nodes is non-uniform with clusters. Furthermore, estimation errors may propagate with increasing network size.

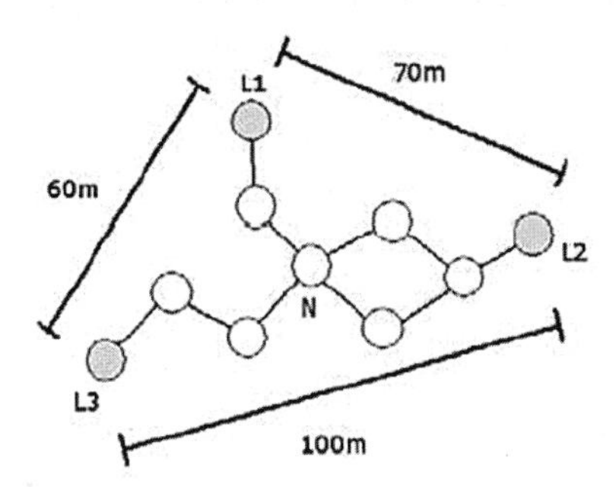

Fig. 14.3. A DV-HOP system: Distance between landmarks L_1, L_2 and L_3, together with hop counts from node N to each of them is used for the input of trilateration

Recently, there has been work addressing the impact of distance error on location estimation using some mathematical models. One popular candidate is called Multidimensional Scaling (MDS), which takes distances among nodes as constraints and generates a map preserving them (more details on MDS will be provided in Section II-D). Example of techniques using MDS together with distance information can be found in (Ji and Zha 2004). In this example, pair-wise distances among neighbouring nodes are measured using the signal strength received, which are then passed to MDS instead of trilateration to obtain a local map containing the neighbouring nodes. After aligning the local maps generated by the nodes, a global map is generated with locations found.

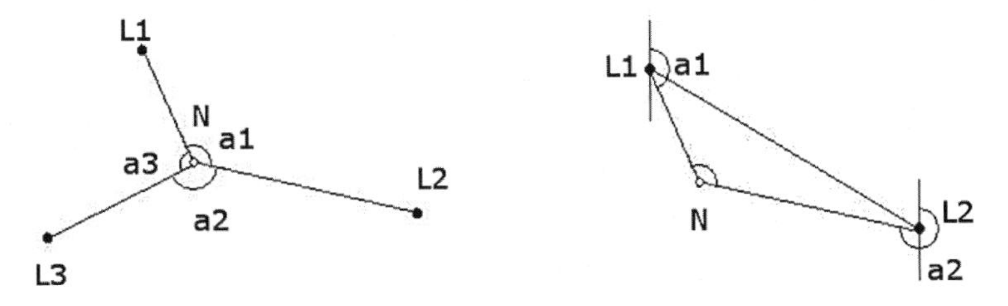

Fig. 14.4. Two cases of triangulation: In the first case, with the knowledge of a_1, a_2 and a_3 in addition to the location of the landmarks, location of node N can be obtained (left Fig); in the second case, with the knowledge of a_1 and a_2 in addition to the locations of the landmarks, location of node N can be obtained (right Fig)

14.2.2 Angle-based

The techniques in this category make use of measured angles among nodes to estimate nodal locations. *Triangulation* is usually used to calculate locations using angles. It requires special antenna for angle measurements, usually in the form of rotating directional receiver, which determines the transmitter direction by sampling the strongest signal (Niculescu and Nath 2004). This involves some mechanical parts and consumes more power than simple signal strength measurements. Another way to measure angle is to use a receiver array and the time difference between signal arrivals (Priyantha et al.).

Triangulation is a tool that a node estimates its location using the angles it bears with a number of landmarks taking two at a time (in case of a two-dimensional plane, a total of at least three landmarks are needed). With the location information of the landmarks, the node can then deduce its own location. Note that it is also possible that only two angles are needed, but in this case the angles have to be measured at the landmarks (instead of at the node), and an additional distance value between the landmarks is required.

Fig. 14.4 illustrates the two cases of trilateration using three angles and using two angles. For triangulation using three angles, consider landmarks L_1 and L_2 by referring to Fig. 14.5. With the knowledge of angle a_1 and locations of L_1 and L_2, one can obtain a circle with centre C_1 and radius r_1 whose circumference is the locus of all the possible positions of node N by using the circle property of *angles in the same segment*. Similarly, one can obtain two other circles by considering other angle-landmark pairs (a_2, L_2, L_3), and (a_3, L_3, L_1), respectively. With these three circles, we then essentially transform the triangulation problem into a trilateration problem with distances and centre coordinates r_1 and c_1, r_2 and c_2, and r_3 and c_3, respectively. Therefore the location of node N can be calculated using the technique described in 14.2.1.

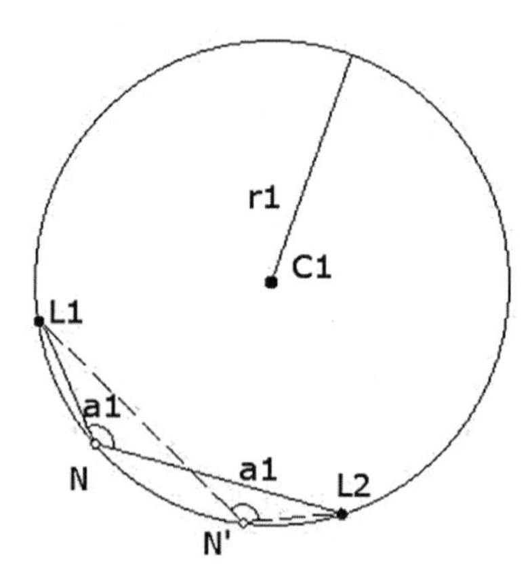

Fig. 14.5. Triangulation using three angles. Each time one angle (a_1) and the corresponding two landmarks (L_1 and L_2) are used. With these, a circle is formed, with all the possible positions of node N on its circumference

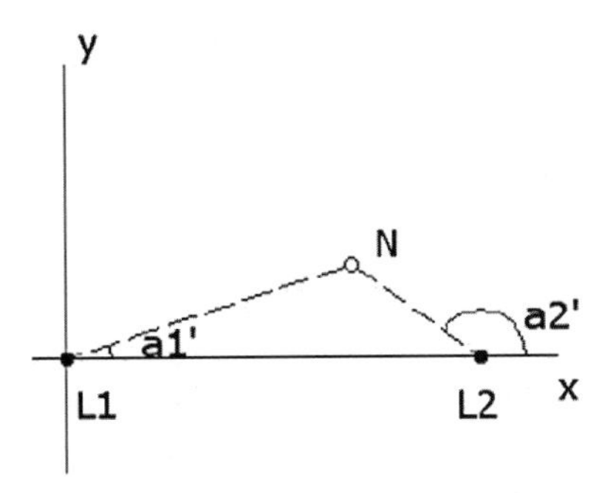

Fig. 14.6. Triangulation using two angles. L_1 and L_2 are the landmarks; their positions are transformed for ease of calculation

As for triangulation using two angles, we need only two landmarks. Let the two landmarks be L_1 and L_2, and the angles to node N are known. To simplify the discussion, we transform the coordinates such that L_1 is at the origin and L_2 lies on the x-axis, as shown in Fig. 14.6. (The actual coordinates can be obtained by transforming the coordinates back at the end) After the transformation we have two angles a'_1 and a'_2, which can be derived from angles a_1 and a_2. Let the coordinates of N, L_1 and L_2 in the transformed coordinate system be (x'_N, y'_N), (0, 0) and $(x'_{L_2}, 0)$, respectively. Clearly, the following system of two linear equations holds:

$$x'_N \tan a'_1 - y'_N = 0 \tag{14.4a}$$

$$-(x'_{L_2} - x'_N) \tan a'_2 - y'_N = 0 \tag{14.4b}$$

Equating the above two equations we get

$$x'_N = \frac{x'_{L_2} \tan a'_2}{\tan a'_2 - \tan a'_1} \tag{14.5}$$

Substituting the above into any of the two equations, we get

$$y'_N = \frac{x'_{L_2} \tan a'_2 \tan a'_1}{\tan a'_2 - \tan a'_1} \tag{14.6}$$

An example of angle-based technique is an Ad Hoc Positioning System using AOA (Angle of Arrival) (Niculescu and Nath 2003). Another example is the VOR (VHF Omni-directional Ranging) base stations in which each landmark is equipped with a custom-made revolving antenna and uses Maximum Likelihood Estimates (MLEs), a statistical tool inferring distributional parameter, to estimate location (Niculescu and Nath 2004).

14.2.3 Pattern-based

In pattern-based category, a certain specific pattern (e.g. in terms of signal strength) is associated with a location. When the pattern is observed later on, the location can be deduced. There is a wide range of pattern from visual pattern (such as a particular scene) to measurement pattern (such as a particular signal strength value). We only focus on signal strength measurement pattern here. This category of techniques usually employs a mapping of received signal strength to some pre-measured pattern to estimate nodal location. Because of that it is often referred to as fingerprinting. In general techniques in this category consist of two phases:

- The *Training Phase:* The system gathers signal patterns from various pre-defined locations.
- The *Online Phase*: A certain signal pattern is collected and is compared with the training results. The location is then estimated by weighted-averaging a couple of closest pre-defined locations.

We illustrate this category using RADAR as an example, which have landmarks as the base stations (Bahl and Padmaabham 2000):

- *Training Phase:* First the nodes and the base stations are synchronised. The node then starts broadcasting UDP packets to all the nearby base stations periodically. Each base station then records the measured signal strength of the packet together with the time when the packet is received. In addition, the nodal coordinates are also recorded. Therefore, each base station has a list of tuples: the location of the node, the signal strength of its packet and the time of measurement. All the tuples from the base stations are merged into a single unified table for lookup during the Online Phase.
- *Online Phase:*A node estimates its location by first broadcasting beacons to the base stations. Upon receiving the beacons, the base stations compare the signal strength of the beacons with the tuples in the unified table to find the closest match. They estimate the nodal location based on the received beacon and a number of previously measured tuples with similar reading as the received beacon (this is called *Multiple Nearest Neighbours Approach*).

The strength of pattern-based techniques is their accuracy if the environment is rather stable over time (2 to 3 meters accuracy in RADAR). It is also fast since the estimation is essentially just a mapping. However, if the environment is dynamic, the fluctuating measurement may affect the accuracy of the system. One way to deal with that is to perform the Training Phase regularly to keep the training data updated. However, this incurs inconvenience as no location can be estimated during the training phase. Another problem with pattern matching is that it usually requires multiple pattern measurements from different places for a single point in space. For example, since radio signals propagate in all directions, when signal strength pattern is used, it is not sufficient to have the pattern measured only at one place. To correctly relate with a location, the pattern has to be measured at multiple places simultaneously. This implies that the area of interest has to be within the coverage of multiple signal transmitters or receivers.

To reduce the measurement time and complexity of the Training Phase over large-scale 802.11 wireless networks, a Gaussian distribution is assumed in estimating the patterns of the signal (Haeberlen et al. 2004). Pattern matching techniques have been widely employed in robotics area, and recently these are applied in location estimation as well. For example, there is a work using a tool called Monte Carlo Localisation (MCL), which was originally developed in robotics localisation (Hu and Evans 2004). Its main idea is to represent the distribution of possible locations with a set of weighted samples, which are maintained by periodic re-sampling. The estimation is an ongoing process, which involves a prediction step to compute the possible locations, and a filtering step to remove impossible locations.

14.2.4 Connectivity-based

This category refers to using connectivity information to estimate nodal locations. It is different from distance-based and angle-based techniques where precise measurement is important, and from the pattern-based techniques where a training phase is required. The tool often used in connectivity-based technique is Multi-Dimensional Scaling (MDS). MDS has long been used in psychology and social science as well as machine learning for data visualisation and extraction. In brief, it is a tool capturing inter-correlation of high dimensional data and maps them to a lower dimension. In location estimation, MDS is used to obtain nodal coordinates so that their inter-nodal distances are consistent with the measured values (usually in the form of a connectivity matrix) (Shang et al. 2003; Shang and Ruml 2004; Wong et al. 2005).

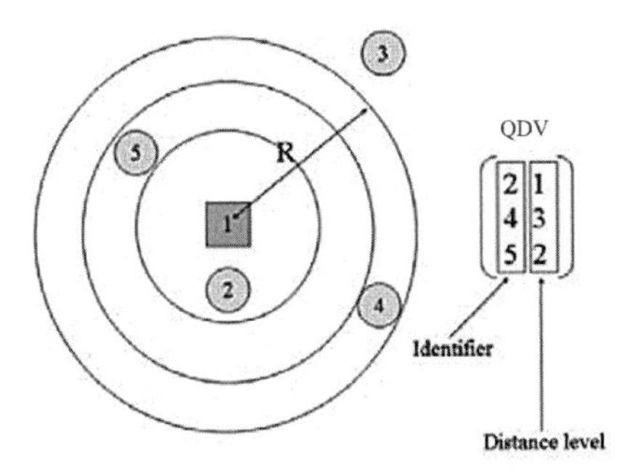

Fig. 14.7. Quantised distance and quantised distance vector (QDV)

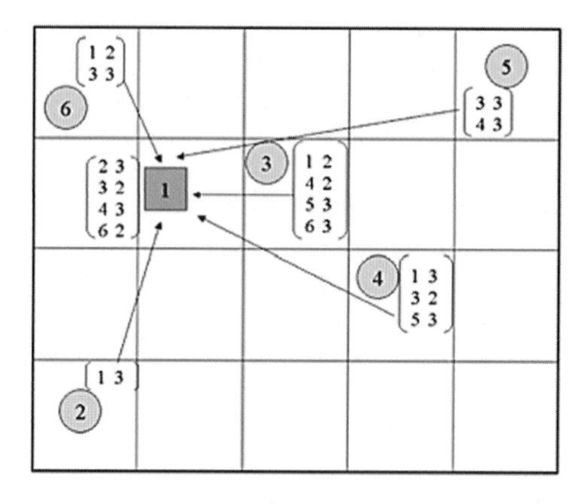

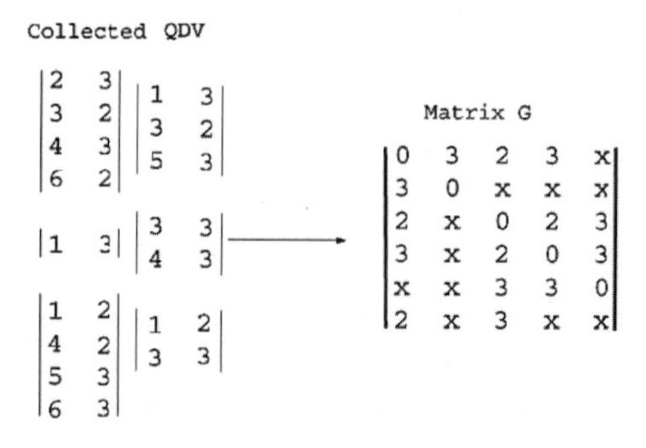

Fig. 14.8. Illustration of the bootstrap node requesting QDVs from its neighbours. Left part is the sending of QDVs to the bootstrap node (Node 1); Right part is the compilation of the distance matrix G with collected QDVs

As an illustration of how MDS is used, consider a node, say node 1, in a sensor network (Wong et al. 2005). It first finds all the nodes in its neighbourhood by adjusting its power level and put them into a Quantised Distance Vector (QDV). In Fig. 14.7, node 1 finds that nodes with identifiers 2, 4 and 5 are at distance level 1, 3 and 2, respectively, and hence puts them into its QDV accordingly. The levels are discrete and equidistant from each other with a maximum range of R. A bootstrap node in the network requests the QDVs from a number of its neighbours, and performs MDS on those QDVs to obtain their coordinates. This is done as shown in the left of Fig. 14.8. The bootstrap node 1 contacts all its neighbouring nodes 2, 3, 4, 5 and 6 and obtains their QDVs as indicated. Given these QDVs and its own, node 1 compiles a distance matrix G as shown towards the right of Fig. 14.8. Note that some of the entries in G may not be known, which may be estimated by some shortest path algorithm. Given the completed G, MDS is then used to obtain the coordinates of each of the nodes.

The computed coordinates are sent to the neighbouring nodes, each of which update its own coordinates, re-computes its neighbours' and send its updated locations to them. This process continues like ripples and after a few iterations the coordinate of each node converges to some stable value. Below is a description of MDS modified for location estimation, which involves solving a system of equations. We go through the steps only briefly here. Interested readers may refer to the details in (Coxand Cox 1994).

Suppose the n nodes in the system of m dimension have positions represented as $\{x_1 \dots x_n\}$ in R^m. The Euclidean distance between any two points x_i and x_j can be computed as:

$$d_{ij} = \left\| x_i - x_j \right\|^2 = x_i^T x_i + x_j^T x_j - 2x_i^T x_j \tag{14.7}$$

With the assumption that the mean $\bar{x} = \sum_{i=1}^{n} x_i$ is zero, a matrix $M = \lfloor M_{ij} \rfloor_{n \times n}$ can be written as:

$$M = -\frac{1}{2}\left(D - \frac{1}{n} 11^T D - \frac{1}{n} D11^T + \frac{1}{n^2} 11^T D11^T \right) = -\frac{1}{2} HDH \tag{14.8}$$

where $M_{ij} = x_i^T x_j$, $D = [d_{ij}^2]_{n \times n}$, $H = I - 11^T$ (the so-called centring matrix), I is the $n \times n$ identity matrix and 1 is the n-dimensional column vector with all ones.

Consider the eigen-decomposition of matrix M, $Mv_k = \lambda_k v_k$, with $m \leq n$ non-negative eigenvalues, then we have $M = V\Lambda V^T = (V\Lambda^{1/2})(V\Lambda^{1/2})^T$, where $\Lambda = \mathrm{diag}(\lambda_1, \dots, \lambda_m)$ and $V = [v_1, \dots, v_m]$. The coordinates of each of the n nodes x_i can thus be obtained as $(\sqrt{\lambda_1} v_1, \dots, \sqrt{\lambda_m} v_m)^T$. A nice feature of MDS is that the system does not require any landmark to estimate the relative location of the nodes. Without landmarks, the coordinates obtained differ from the geographical ones by an arbitrary rotation, reflection, and/or translation. This is sufficient in cases like location-based routing, which requires only relative locations. Fig. 14.9 illustrates the performance of location estimation using MDS. Each cross in the figure represents a wireless node. The left figure shows the real locations, while the right figure shows the estimated locations based on MDS. Clearly the estimated location differs by a rotation (about 20° clockwise) and a translation of the origin. To obtain the geometrical coordinates, GPS-enabled landmarks can be used as reference points.

Tools like MDS may incur some computational overhead, as it needs to gather and process enough measurement data in order to have an accurate estimation. This overhead increases with the number of nodes, as the node calculating the coordinates would have long a list of measurement data from other nodes. One possible solution is a distributed mechanism where each node performs part of the location estimation, which may then be combined (Shang and Ruml 2004; Wong et al. 2005). Clearly, this increases information exchanges among nodes.

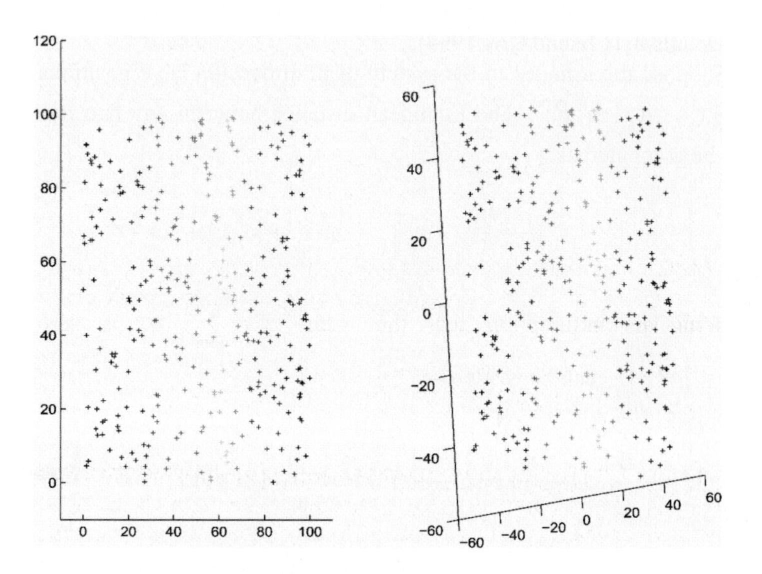

Fig. 14.9. Relative position estimation for a 100 X 100 grid with density 40 using MDS: Left – Real locations of the nodes; Right - estimated locations of the nodes

14.3 Comparisons

In evaluating location estimation techniques, the following metrics are often used:

14.3.1 Accuracy

Accuracy refers to how precise the estimated location of a node is. Usually Euclidean Error, defined as the Euclidean distance from the estimated location to the real location, represents accuracy. Sometimes *Angular Error* is also used, defined as the angle between the line connecting the estimated locations of two nodes and the line connecting the real locations of the same nodes. This is useful when the technique may lead to rotational errors. Fig. 14.10 illustrates both Euclidean Error and Angular Error. Another useful metric in case of rotated and/or translated map is the pair-wise distance error between nodes, which evaluates how good the technique preserves inter-nodal distances. Note that for some applications, close to 100% accuracy is not essential. For example, one may be interested in just which region a node is in. This may allow some error (e.g., several meters), which in turn simplifies the technique.

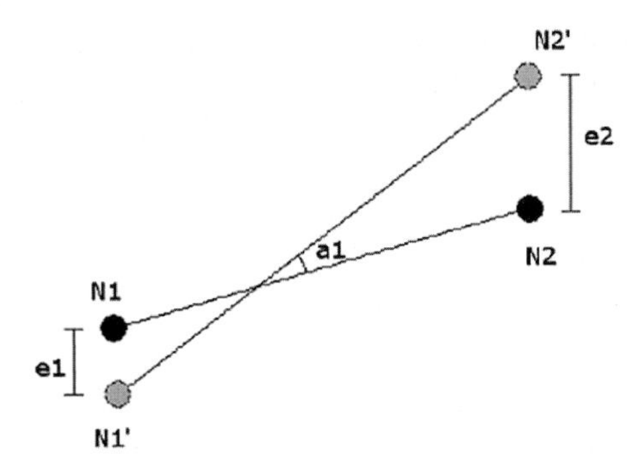

Fig. 14.10. Illustration of the accuracy metric. N'_1 and N'_2 are estimated locations of N_1 and N_2, respectively. e_1 is the Euclidean error of N_1's estimation while e_2 is that of N_2's. a_1 is the angular error

14.3.2 Estimation Steps

When a location is queried, an immediate reply is usually preferred. However, this is not always possible because for acceptable accuracy it may take a couple rounds of message exchanges for the location to converge. While measuring the convergence time required is probably the most direct way to evaluate speed, it highly depends on the hardware used. We therefore define and compare speed as how many steps are required for a node to estimate its location. The steps may include synchronisation, broadcasting, network flood, map formation and calculations. Synchronisation refers to having a consistent value among nodes; for example, a GPS device has to synchronise its clock with the satellites'. Broadcast refers to transmitting measured values to nearby nodes, while network flood refers to transmissions to the entire network. For some techniques, a node may form a map containing the locations of its neighbours and itself. This is referred to as map formation. Calculation refers to the manipulation of the received or measured values for location estimation.

14.3.3 Message Complexity

An important factor in evaluating the system performance in wireless communication is the amount of messages exchanged. Very often a node requires message exchanges with other entities before it can estimate its location. Message complexity is defined as the number of message exchanges for a node to collect enough information in estimating its own location.

- *Power consumption:* As sensor nodes often run on batteries, low power consumption is desirable. Power may be consumed in many ways, such as intensive computation, long-distance communication, algorithmic complexity, etc. Among these, communication distance is often the dominant factor. Power consumption may be indicated by with which nodes a node communicates. For example, less power would be consumed if the communication is limited to nearby neighbours, while more power would be consumed if the node's signal has to reach distant base stations. In general, if a technique requires a base station type of landmarks, the landmarks would consume relatively high power.

- *Density dependency:* When estimating its location, a node may either gather information from a number of neighbouring nodes, or communicate with a few designated nodes. For the former type, decrease in number (density) of neighbouring nodes would result in a greater degree of accuracy degradation. Furthermore, error tends to propagate along the path as information is exchanged. Generally, techniques that are less dependent on node density tend to be more robust. The trade-off, however, is the careful positioning of the more powerful designated nodes at places within which nodes can communicate. We may compare techniques in terms of accuracy dependency on the density. Note that if there is an infrastructure and a node only needs to interact with it, then the technique would be less sensitive to nodal density and can maintain a higher accuracy even when node density becomes low (or some nodes go out of battery).
- *Special Hardware Requirement:* Due to the differences in the underlying mechanisms, some techniques require specific hardware equipment. For example, an angle-based technique requires special devices that can measure angles. A node requires a radio device in order to send and receive RF signals to and from other entities.

We present in Table 14.1 a comparison among some schemes based on the location estimation techniques discussed (they all achieve high accuracy). As we can see in the table, even techniques within the same category may markedly differ in certain metrics, depending on how the information is processed. This difference in performance provides users with trade-off on employing particular technique best suits their own requirements.

14.4 Conclusions

Wireless location estimation for sensor networks is an emerging field with growing importance. With the location information of the nodes, a sensor network may be designed with better performance (e.g. in terms of routing efficiency) and enhanced services (e.g. location-based services). In this article we have presented four categories of location estimation techniques depending on the way location is estimated, namely, 1) distance-based category, where location is estimated based on the distances between nodes; 2) angle-based category, where location is estimated based on the angles between nodes; 3) pattern-based category, where location is estimated by matching the observed pattern with pre-measured patterns; and 4) connectivity-based category, where location is estimated using connectivity information. We illustrate with examples how each category works. We also compare a number of recent techniques in these categories, in terms of metrics such as their estimation steps required, message complexity, power consumption, density dependency and special hardware requirement.

Table 14.1: A comparison between a number of typical location estimation techniques

TN	CT[1]	EST	MC[2]	PC	DDp	SHR
Cricket [11]	DB	A	*O(L)*	AA	Low	URD
DV-HOP [12]	DB	B	*O(K)*	BB	High	RD
MDS in [13]	DB	C	*O(1)*	CC	Medium	RD
APS AoA [16]	AB	D	*O(k)*[3]	DD	Medium	AMRD
VOR [14]	AB	E	*O(1)*	EE	Low	RSRD
RADAR [17]	PB	F	*O(1)*	FF	Low	RSRD
Gaussian [18]	PB	G	*O(1)*	GG	Low	RSRD
MCL [19]	PB	H	*O(1)*[4]	HH	Medium	RD
MDS-MAP [20]	CB	I	*O(K)*[3,5]	II	Medium	RD
MDS-MAP(P) [21]	CB	J	*O(K)*[3,5]	JJ	Medium	RD
MDS in [22]	CB	X	*O(K)*[5]	KK	Medium	RD

A: Clock synchronisation & calculation; B: Broadcast & calculation; C: Network flood & propagating map information; D: Angle measurement, broadcast & calculation; E: Angle measurement & calculation; F: Training & online phases; G: Training & Online phases; H: Location prediction & filtering & filtering; I: Path & map formation; J: Individual map formation & map merging; X: Individual map formation & location updates; AA: Low (Receive localised beaconing); AB: Angel-Based; BB: Low (Neighbour communication); CC: Low (Neighbour communication); CB: Connectivity-Based; CT: Category; DD: Low (Neighbour communication); DDp: Density Dependency; EE: High (Signal to base stations); FF: High (Signal to base stations); GG: High (Signal to base stations); HH: Low (Neighbour communication); II: Low (Neighbour communication); JJ: Low (Neighbour communication); KK: Low (Neighbour communication); MC: Message Complexity; PB: Pattern-Based; PC: Power Consumption; RD: Radio Devices; TN: Technique Name; EST: Estimation Steps Required; MCL: Monte Carlo Localisation; SHR: Special hardware requirement; URD: Ultrasonic and radio devices; AMRD: Angle measuring and radio devices; RSRD: Revolving base stations and radio devices; [1]DB: Distance-Based; [2]L: number of landmarks; K: number of rounds in distance vector exchange; [3]For distance vector setup, we use wireless broadcasting so one message is needed to be transmitted to notify all the neighbours; [4]This technique requires an ongoing sampling process; [5]Assuming connectivity information has already been known (via, e.g. standard distance vector algorithm).

14.5 References

Akyildiz IF, Su W, Sankarasubramaniam Y, and Cayirci E (2002) Wireless sensor networks: a survey. Computer Networks 38-4:393–422

Cox T and Cox M (1994) Multidimensional scaling. Chapman & Hall, London

Bose P, Morin P, Stojmenovic I, and J Urrutia (2001) Routing with guaranteed delivery in ad hoc wireless networks. Wireless Networks 7-6:609–616

Bahl P and Padmanabhan VN (2000) Radar: An in-building RF-based user location and tracking system. In: IEEE Int. Conf. on computer communication, pp 775–784

Gwon Y, Jain R, and Kawahara T (2004) Robust indoor location estimation of stationary and mobile users. In: IEEE Int. Conf. on computer communication. Hong Kong, pp 1032–1043

Hightower J and Borriello G (2001) Location sensing techniques. University of Washington, Department of Computer Science and Engineering, Seattle

Hightower J and Borriello G (2001) A survey and taxonomy of location sensing systems for ubiquitous computing. University of Washington, Department of Computer Science and Engineering, Seattle

Hofmann-Wellenhof G, Lichtenegger H, and Collins J (1997) Global positioning system: Theory and practice, Fourth Edition. Springer Verlag

Haeberlen A, Flannery E, Ladd Am, Rudys A, Wallach DS, and Kavraki LE (2004) Practical robust localisation over large-scale 802.11 wireless networks. In: the 10th Annual Int. Conf. on mobile computing and networking. ACM Press, New York, pp 70–84

Hu L and Evans D (2004) Localisation for mobile sensor networks. In: the 10th Annual Int. Conf. on mobile computing and networking. ACM Press, New York, pp 45–57

Ji X and Zha H (2004) Sensor positioning in wireless ad-hoc sensor networks using multidimensional scaling. In: IEEE Int. Conf. on computer communication. Hong Kong, pp 2652–2661

Karp B and Kung HT (2000) GPSR: Greedy perimeter stateless routing for wireless networks. In: the 6th Annual Int. Conf. on mobile computing and networking. : ACM Press, New York, pp 243–254

Mainwaring A, Culler D, Polastre J, Szewczyk R, and Anderson J (2002) Wireless sensor networks for habitat monitoring. In: the 1st ACM Int. workshop on wireless sensor networks and applications. ACM Press, New York, pp 88–97

Niculescu D and Nath B (2003) DV based positioning in ad hoc networks. Telecommunication Systems, 1-4:267–280

Niculescu D and Nath B (2004) VOR base stations for indoor 802.11 positioning. In: the 10th Annual Int. Conf. on mobile computing and networking. ACM Press, New York, pp 58–69

Niculescu D and Nath B (2003) Ad hoc positioning system (APS) using AoA. In: IEEE Int. Conf. on computer communication. San Francisco, pp 1734–1743

Priyantha NB, Chakraborty A, and Balakrishnan H (2000) The cricket location-support system. In: the 7th Annual Int. Conf. on mobile computing and networking. ACM Press, New York, pp 32–43

Priyantha NB, Miu AK, Balakrishnan H, and Teller S (2001) The cricket compass for context-aware mobile applications. In: the 7th Annual Int. Conf. on mobile computing and networking. ACM Press, New York, pp 1–14

Shang Y, Ruml W, Zhang Y, and Fromherz MPJ (2003) Localisation from mere connectivity. In: the 4th ACM Int. Symp. on mobile ad hoc networking & computing. ACM Press, New York, pp 201–212

Shang Y and Ruml W (2004) Improved MDS-based localisation. In: IEEE Int. Conf. on computer communication. Hong Kong, pp 2640–2651

Wong KS, Tsang IW, Cheung V, Chan SHG, and Kwok JT (2005) Position estimation for wireless sensor networks. In: IEEE Global communications conference. St. Louis, pp 2772–2776

Wang H, Elson J, Girod L, Estrin D, and Yao K (2003) Target classification and localisation in habitat monitoring. In: IEEE Int. Conf. on speech and signal processing. Hong Kong, pp IV844–847

Zhou G, He T, Krishnamurthy S, and Stankovic JA (2004) Impact of radio irregularity on wireless sensor networks. In: the 2nd Int. Conf. on mobile systems, applications, and services. ACM Press, New York, pp. 125–138

15 Application-driven Design

Mark Halpern, Wanzhi Qiu, Khusro Saleem and Stan Skafidas

National ICT Australia, Department of Electrical and Electronic Engineering, University of Melbourne, Australia

15.1 Introduction

It is widely predicted that wireless sensor networks (WSN's) will proliferate in a vast range of potential applications including academic, industrial, agricultural, domestic and military. The purpose of this chapter is to indicate the ways in which the application affects the design of a wireless sensor network. We make use of the framework illustrated in Fig. 15.1. For any attribute of a wireless sensor network, it is the application, which determines the importance of that attribute in the design specification. The available technology then determines the manner in which that desired attribute might be achieved. The final design solution is then made by selecting from the allowable choices, tempered by other constraints such as cost. For example, considering the attribute of water-resistance, it may be that a particular outdoor application dictates the need for a water-resistant enclosure. Available technology then allows selection of an enclosure made from ABS plastic, polycarbonate, diecast aluminium or stainless steel. The actual selection might then be made on the basis of cost or weight. In this work we consider the manner in which the application dictates the design specification and also the possible design solutions allowed by current technology.

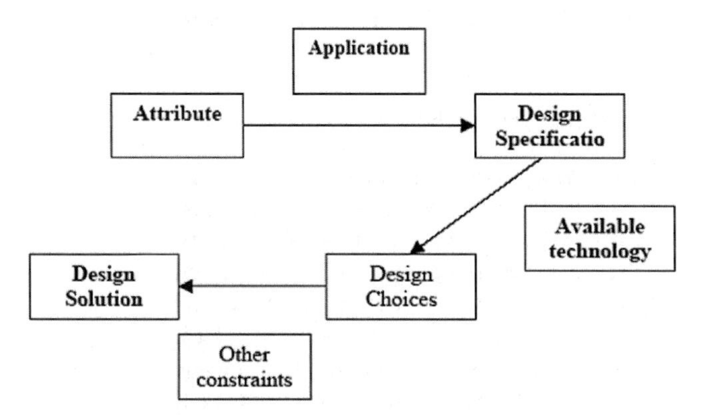

Fig. 15.1. Framework by which application affects design for each attribute

The next section contains a list of the specific attributes that will be addressed. In the remainder of this chapter, each attribute is discussed in the context of application requirements and consequences in the context of hardware and software design are discussed.

15.1.1 Attributes

The following list captures a large subset of attributes relevant to wireless sensor networks. Whilst this list is not exhaustive, it represents some of the common attributes that arise in the current wireless sensor network research and practical deployments.

- Localisation
- Power supply constraints
- Physical operating environment
- EMC compatibility
- Sensing requirements
- Control requirements
- Cost constraints
- Security requirements
- Field programmability
- Presentation layer (GUI, web enabled data queries)
- Reliability of nodes and the network
- Deployment of networks and nodes

15.2 Localisation

Applications such as target tracking (Zhao and Guibas 2004 Ch. 2) and wildlife monitoring (Zhang et al 2004) require nodes have the capability to determine their geographical positions. This self-location is referred to as localisation. Some applications require global absolute position information while others only require positioning relative to certain reference points (Hightower and Borriello 2001). The type of position information as well as the accuracy of localisation dictate the design specification for localisation.

When absolute position information is required for each node, a GPS receiver can be incorporated into each node which allows nodes to localise themselves within a few meters accuracy by listening to signals emitted by a number of satellites and assistive transmitters (Zhao and Guibas 2004 p. 118). In order to reduce power consumption and cost, if accuracy requirements can be met, only some of the nodes, called landmarks, are equipped with a GPS receiver. Nodes without a GPS receiver can calculate their absolute locations by reference to a certain number of landmarks, using various ranging or direction-of-arrival techniques that allow them to determine the distance or direction of landmarks. If localisation by GPS is not feasible (such as in an indoor environment) or too expensive, a relative reference framework has to be used where each node works out its position relative to certain reference points. As in the case of absolute localisation using landmarks, ranging or direction-of-arrival (DOA) techniques need to be used to calculate the distance or direction of reference points.

In cases where relative localisation or absolute localisation using landmarks is used, dedicated resources can be incorporated into the node to perform ranging or direction-of-arrival and corresponding positioning algorithms to determine the position.

As an example, Chipcon's IEEE 802.15.4 radio transceiver CC2431 has a dedicated location engine for this purpose (Chipcon 2006). Software for processing and reporting position information is required. Position computing software using DOA techniques (Moses et al 2003) or various ranging techniques exploiting received signal strength (RSS), time of arrival (TOA) or time difference of arrival (TDOA) may also be required (Zhao and Guibas 2004 p. 120).

15.3 Power Supply Constraints

15.3.1 General

WSN's are electronic devices and consume electrical power. There are two main aspects to powering WSN's, both determined by the application. The first aspect is the available power sources. The second is the amount of power required. Available power sources include non-rechargeable batteries, rechargeable batteries using a solar panel, and mains power. In most low-cost applications, it is envisaged that non-rechargeable batteries will power WSN's. The power required includes power for the sensors, the node electronics, and any equipment such as pumps or alarms, controlled by the node.

15.3.2 Hardware Design to Save Power

The usual design approach for low-powered electronic systems is to minimise the *average* current drawn, by spending as much time as possible in a low power consumption mode, sometimes called 'sleep'. When response is needed the device wakes into an active state where it can carry out actions. During this time while the device is capable of carrying out actions it draws current. After the device has carried out its work it returns to sleep. The duty cycle of the device in the system is the fraction of time for which the device is active. For those WSN applications where long life is required from dry cells, we might aim for a duty cycle from 0.1 % to 1 %.

15.3.3 Effect of Supply Voltage and Clock Frequency

Most digital IC's, and indeed many modern analogue (Razavi 2001) and RF (Lee 2004) IC's are based on CMOS (Complementary Metal Oxide Semiconductor) technology. CMOS technology is based around a particular type of transistor called a MOSFET (Metal Oxide Semiconductor Field Effect Transistor). Devices are characterised by the Process technology, which refers to the channel length, a physical dimension, of the transistors. Much of the following is based on (Benini et al 2001, Landernäs 2005). The main sources of power dissipation, P, in a CMOS circuit are,

$$P = P_{\text{Switching}} + P_{\text{Short-circuit}} + P_{\text{Leakage}}$$

The switching power $P_{Switching}$ of a complementary transistor pair is due to the charging and discharging of capacitors driven by the switching transistors and is given by,

$$P=1/2\, f\, C\, V^2\, K$$

where f is the clock frequency, C is the capacitance being driven, V is the voltage to which the capacitor is charged, K is the fraction of clock cycles which involve the pair switching.

Formerly, when transistors were larger (technologies of 0.8 μm and greater) due to manufacturing limitations, $P_{Switching}$ was the most significant source of power dissipation. The short circuit power, $P_{Short\text{-}circuit}$, is due to the currents which flow while both transistors in a complementary pair are on briefly during switching. The leakage power, $P_{Leakage}$, is due to currents which flow when both transistors are off. As transistor sizes are decreasing, the leakage power is becoming more significant. This new issue is attracting research interest (Elgharbawy and Bayoumi 2005). The supply voltages required for digital systems take on discrete values, typically 5.0 Volts, 3.3 Volts or 1.8 Volts. Formerly, most general purpose digital electronics used a 5 V supply. Presently the most common value is 3.3 V with 1.8 V reserved for more specialised systems. As device sizes fall, one could expect that 1.8 V devices will come into more widespread use.

15.3.4 Microcontroller and Radio Power Saving Modes

More highly integrated IC's such as microcontrollers and radios possess features allowing them to be operated in modes where they consume only a small amount of power. Switching between these modes occurs without losing information stored in internal registers. Using this mechanism is preferable to actually removing the power supply to the devices since that approach would cause internal data to be lost which would make it difficult to resume previous states and would prevent seamless operation. Most microcontrollers have at least two low-power modes sometimes called "idle" and "sleep". These are entered from the active state by software instructions. In the idle mode, the supply current to the microcontroller is typically of the order of 40% of the active current. In the idle mode, it is not possible to carry out computations or control actions, but internal timers keep running and can wake the microcontroller back into the active mode through internally generated interrupts. In sleep mode, the MCU supply current is typically of the order of 10 μA. Internal RAM and register contents are maintained during sleep, but the timers do not run, and consequently an external hardware signal source is required to wake the microcontroller. Some microcontrollers allow further control of power consumption through controlling the power to internal subsystems. Usually the microcontroller current drain in the active state is approximately proportional to the clock frequency and takes values from 0.5-1.5 mA/MHz for the small microcontrollers found in WSN's. For such microcontrollers, the maximum clock frequency is of the order of 10 MHz. Radio chips also have low power consumption modes. These are usually selected by signals from the microcontroller. Other peripheral devices are handled in the same way.

15.4 Software Design to Save Power

- *Sleep pattern schedules and coordination:* This is carried out under the control of the microcontroller
- Switching off peripheral devices: The de-selection of peripheral components is carried out under the control of the microcontroller software.
- *Minimising cycle count:* Algorithms which reduce the time for which the microcontroller is active will reduce the average current drawn. In certain systems a worthwhile reduction in cycle count can be achieved by replacing floating point calculations by integer calculations. Small microcontrollers tend to execute floating point routines relatively slowly: times of the order of 250 μS might be expected, while integer calculations might require execution times of the order of 1 μS. It is clearly desirable to avoid using floating point routines if possible.
- *Use data aggregation to reduce transmission times:* Compression and merging of data can be carried out in order to reduce the amount of information transmitted over a radio link.
- *Power control in the transmitter:* Software algorithms may be used to control transmitted power based on the ambient signal level.

15.5 Physical Operating Environment

The physical operating environment is characterised by temperature, humidity, destructive chemicals, water ingress, dust ingress, solar radiation, animals, potential for tampering, obstacles to radio propagation (mountains, water bodies, vegetation, buildings, vehicles). The first line of defence against these unwanted environmental factors is the equipment housing or enclosure.

15.5.1 IP, NEMA Ratings of Equipment Enclosures

A rating system which categorises the ability of components to exclude solid bodies ranging in size from large objects down to dust and moisture is the Ingress Protection (IP) rating. This comprises two digits. The first describes the ability to exclude solid objects and is a digit from zero to six, where zero denotes no special protection and six denotes dust tight. The second digit from zero to eight describes the ability to exclude moisture. The IP rating system is well established (The Engineering Tool Box 2005). In addition to enclosures, connectors for conveying electrical signals between the outside and the inside of an enclosure can have an IP rating. A rating system describing the degree of protection against certain environmental conditions often used in the U.S. is from the National Electric Manufacturers' Association (NEMA). This is called the NEMA rating system (Reliance Electrical 2000).

15.5.2 Materials for Equipment Enclosures

Following materials are used for external enclosures: plastics and metals.

- *Plastic:* ABS, polycarbonate: ABS (Acrylonitrile-Butadiene-Styrene) and polycarbonate are probably the most common plastics used in electronic equipment enclosures. Polycarbonates (Polymer Technology and Services LLC 2004) have many useful properties. There are many variants of these materials with different characteristics; in general polycarbonate is physically stronger than ABS, but has less resistance to chemicals and is more expensive.
- *Metals:* Aluminium, Zincalume Steel: Most aluminium enclosures are cast. Zincalume steel is alloy coated steel which has good forming properties. Aluminium has good corrosion properties, is easily drilled and has density about one third of that of steel.

15.5.3 Electromagnetic Compatibility (EMC)

EMC is an important part of electronic system design. It is necessary to ensure that (i) Electronic systems do not generate emissions which interfere with other equipment, a) Transmitters must keep out-of-band radiation below allowable level and b) Unintentional radiators should not radiate above allowable level; (ii) Electronic systems operate properly in the presence of external interference. This property is called immunity; and (iii) Electronic systems do not cause interference to themselves which could cause malfunctions. The mechanism behind radiated electromagnetic interference (EMI) is that a conductor carrying an electrical signal will radiate the signal if the signal wavelength is comparable with the conductor length. (In other words the conductor behaves like a transmitting antenna). Similarly a conductor can behave as a receiving antenna and can introduce an external RF signal into a system. Digital signals with fast edges contain high frequency components which can radiate. The same effect occurs with short pulses. On a PCB, fast switching of currents can generate voltage spikes that interfere with other parts of the circuit. All electronic circuitry is required to satisfy regulatory EMC requirements. (There are some situations where these requirements are reduced eg for education, police, military.)

15.6 Hardware Design for EMC

Shielding of cables and equipment enclosures can sometimes be used to prevent interference being radiated or being received. This aspect of electronic system design (Atmel 2003) requires the system clock frequency be kept as low as possible, whilst maintaining required system throughput. Another aspect of this design approach is the avoidance of digital component technologies with unnecessarily short signal rise times. The physical layout of the electronic components in a system has a great influence on EMC performance. In many electronic systems, this is determined largely by the printed circuit board (PCB). The most basic design guidelines for PCBs to reduce EMC are,

- Use short interconnecting leads (traces on a PCB). This is easier if components are small.
- Use multilayer boards with separate power and ground planes.

More detailed guidelines on designing PCB's to reduce EMC are available (Henry Ott Consultants 2000).

15.7 Selection of RF

The application will determine whether to use Industrial Scientific and Medical (ISM) bands or a dedicated frequency band which would require payment of a licence fee. If a WSN is used for triggering emergency services, if other users are present in an ISM band then one might select a dedicated frequency band.

15.7.1 Use of Algorithms Allowing Low System Clock Frequency

The considerations here are similar to those of minimising cycle count to reduce power consumption discussed.

15.7.2 Mitigation of Link Errors

Using a communication protocol with error detection/correction and repeat-request transmission can mitigate interference to the radio link. Performance requirement is based on a desired bit error rate (BER) or packet error rate (PER).

15.8 Sensing Requirements

15.8.1 Hardware Aspects of Sensors

Interfacing with sensors may affect the hardware in a number of ways. The interface circuitry must ensure proper signal timing and levels for either analogue or digital signals going in/out sensors. Electrical isolation may be needed, and avoiding use of common ground may be important if multiple on-board sensors require different ground levels. Also, ubiquitous protocols for received digital signals such as the UART functions may be implemented in hardware.

- Signal levels: The signal levels used by the sensor must be compatible with those acceptable to the WSN node input circuitry. Sensors with either analogue or digital outputs are available for many physical quantities of interest. A commonly used digital signal level is 5 Volt signaling, but larger voltages are used to obtain improved noise rejection. RS-232 and RS-485 signaling levels are used to convey digital signals between the devices. Analogue signals are in the 0-5 Volt DC or 0-10 Volt DC range. A common approach is to use a 4-20 mA DC current signal to indicate an analogue signal level.
- Signal Condition: Sensed signals in industrial environments may be contaminated with additive noise. This is of particular concern with analogue signals since the levels are affected. Digital signals have more noise immunity. The noise may be reduced by measures such the use of low-pass analogue filtering, shielded leads and the use of differential common mode chokes.
- Electrical isolation: In some applications, the WSN node electronics will be connected to sensors which have their own power supplies. Particularly if there is more than one such sensor to a single WSN node, it may be very undesirable to in effect connect these sensors together through the node. In such cases, it is necessary to provide electrical isolation between the inputs to the WSN node and the "downstream electronics" in the node. For digital signals, this isolation can be achieved through the use of opto-isolators. For analogue signals, one approach is to use differential amplifiers chosen to allow large common mode voltages to be rejected.
- Providing and controlling power to sensors: In some applications, the WSN will be required to control and in some situations provide power to the sen-

sors. It may be that the third party supplied sensors consume more power than the carefully designed electronics in the WSN node.

- Microcontroller performance: The MCU processing power expressed in mega instructions per second (MIPs) may be important, for example if a significant amount of digital signal processing is required to filter a number of noisy analogue inputs.

15.8.2 Software

Software is required on each node to control sensing process and ensure sensing speed, frequency and accuracy meet application's requirements. Raw data collected must be transmitted and/or processed to perform the given task. Sensor driver software will be needed. Protocols for received digital signals can be implemented in software. It may be possible to arrange processing of the sensed signals to use spare MCU capacity, or to provide a distributed processing facility, or to reduce data transmitted over the wireless link, thereby reducing transmission times, radio power consumption, reducing clashes. Some applications such as target tracking (Zhao and Guibas 2004 p. 23 - 25) require nodes to process data locally in order to achieve a desired global objective. Applications can only be realised in this way when there is no super-node with adequate resources to perform the centralised computing. Since nodes themselves are distributed to collect data and they all have computing capacity, it is natural to adopt a distributed computing strategy. Typically, distributed computing algorithms only utilise local data or that of neighbours hence leading to reduced traffic and better scalability with increase in network size. However, a distributed algorithm may be non-optimal or sacrifice immunity to cascading failure and security attacks. Based on the application, decision has to be made on whether a centralised or distributed computing approach should be taken. A distributed computing algorithm has to be carefully designed to achieve the global goal. Issues need to be considered include the degree of information sharing between nodes, and how many and which nodes to participate in the computing. The actual data fusing algorithm and the mechanism to control nodes to participate in the computing are difficult tasks (Zhao et al 2003) and may require extensive theoretical study, simulation and experiment.

15.9 Control Requirements

Here the WSN provides an output signal to control an actuator. There may also be input signals to confirm that that the actuation has been carried out. These are similar to sensor requirements, although there is additional "responsibility" since there is a direct physical outcome arising from the MCU executing its code.

- *Hardware:* The hardware will need to be capable of driving the control actuator. Depending on the type of actuator, some kind of high power driver may be needed. Care should be taken that high power drive circuitry does not interfere with the operation of the digital circuitry and that it does not add interference to low level analogue signals being sensed.

- *Software:* Actuator driver software will be needed. Arranging to operate the actuator at times when low-level analogue signals are not being read may reduce interference from the actuator driver. A control algorithm may also be required. This will depend in a specific manner on the application.

15.10 Cost Constraints

Cost is a critical consideration in sensor network systems. The success and utilisation of wireless sensor network applications depends critically on the overall cost of the system. There are many systems, both wired and wireless, in existence today that can be used to implement sensor network systems for many applications. However in many cases the cost of the system can be prohibitive for the requisite application. Today when the end user performs a return on investment analysis for a given application they may find the benefit they derive does not warrant the cost in deploying and maintaining such a system. Future wireless systems will have an overall low cost of ownership. The cost in deploying a sensor network system includes everything from the cost of (i) Sensor network hardware, (ii) Powering sensor network system, (iii) Software and applications, (iv) Network deployment, (v) Licensing and utilising spectrum and wide area communication access, (vi) Maintaining the sensor network, and (vii) Physical sensors. An extremely important component to the overall cost of a sensor network is the cost of the transducers that convert physical signals and processes to digital data

- *Cost of hardware:* This is mainly due to cost of design, cost of components, cost of manufacturing and packaging. In many cases the cost of the wireless sensor network system may be a small component of the overall system cost.
- Cost of supplying power: In many cases the benefits of having a wireless sensor network can be lost if a wired connecting to a power source is required. In many cases solar cells can be used to recharge batteries. However good solar cells are expensive, require access to sunlight and may be expensive to install.
- *Availability of software and applications:* Although sensor networks can provide an efficient means of providing a lost cost communication infrastructure an important cost consideration is the availability of configuration software and applications that can run on and control the sensor network system. Software will need to be available that can be used to control, configure, maintain and process the data such that useful business application decisions is derived from the data that is collected and distributed over the network.
- *Setting up the network is an important cost consideration:* Any sensor network system must require minimal configuration. The system should not require a network engineer to deploy. In order to derive maximum benefit from sensor network systems a unsophisticated user with no prior networking ex-

perience should be able to deploy such a system. The network should automatically be able to configure network parameters and be able to seamlessly deal with nodes being added to the network and with nodes being subtracted from the network with end user intervention.

- Cost of licensing or using spectrum and wide area communications: Sensor network systems can utilised both licensed and unlicensed spectrum. In the case of licensed spectrum the end user buys or leases spectrum from a regulatory body. This cost can in some circumstance be substantial. An alternative is to utilise unlicensed spectrum such as spectrum in the 2.4 GHz ISM band. This is spectrum that regulatory bodies, such as the Federal Communications Committee, permits people to use free of any fee provided that adhere to a set of rules. The cost or disadvantage of using unlicensed spectrum is that other users can also use this spectrum. In doing so that can cause interference that in most cases reduces system performance. However intelligent interference avoidance techniques can be instituted that make unlicensed spectrum as reliable. An important component in a wireless sensor network is access to wide area communications. For many systems the data processing and command centre can be many hundreds of kilometers away. In many cases the wide area network is controlled by a telecommunication carrier, which will charge for the transmission of data.
- Cost of maintaining sensor network system: Another important consideration in sensor networks is the cost of maintaining a sensor network. This includes everything from detecting and replacing failed nodes, adding new nodes to the network and upgrading software, application and sensors in the field.

15.11 Security Requirements

Effective network security is critical for some applications to ensure proper safeguards for information. If an application requires access control capability, that is, to prevent unauthorised access to the network, certain form of authentication procedure has to be in place to force nodes accessing the network (for example, nodes joining the network) to prove their identities. A simple authentication procedure (Barkin 2004) may just check to see whether the requesting node has a valid password. A more sophisticated authentication procedure (Edney and Arbaugh 2004) usually involves multiple encrypted message exchanges between the nodes accessing the network and an access controller. An access controller can either be the node receiving the access request or a node dedicated to authentication. During the authentication procedure, the requesting node is challenged to produce certain responses by which its knowledge of the required credentials (keys) can be validated.

If an application requires detection of tampered data, a mechanism for data integrity has to be implemented. This normally involves inserting a check code (IEEE 2003) into each data packet at the sender and verifying the check code at the receiver. At the receiver, data packets detected to have been altered are discarded and countermeasures such as retransmissions are triggered. If an application wishes to keep confidentiality on the information transmitted over the air, that is, to prevent eavesdropping by entities outside the network, an encryption mechanism has to be used which scrambles data to be transmitted over the air into unreadable forms for all except the part who has the proper keys. Encryption algorithms can be symmetric using a shared secret key to encrypt, decrypt or be asymmetric using a public key (Tanenbaum 1996) to encrypt data so that only authorised party with the proper private key can correctly decrypt. The Advanced Encryption Algorithm (AES) (Daemen and Rijmen 2002) is considered to be secure and has been adopted in a number of standards such as IEEE 802.11i and ZigBee (ZigBee Alliance 2005). Depending on the level of security required, implementation of security mechanisms can put high pressure on resources such as memory and CPU time in order to manage credentials, perform encryption and decryption and run the authentication protocol. The powerful IEEE 802.11i security suite used for WLANs is not applicable for most sensor network applications due to limitations on resources available. For example, unlike WLANs, a wireless sensor network normally does not have such a node with adequate energy and computing resources to act as an authentication server. The ZigBee security mechanisms are less stringent and require less computing resources, and, therefore, more suitable for wireless sensor network applications. When implementing security mechanisms, dedicated balance and judgement on performance and cost is required when deciding what security mechanisms are to be used. Physical memory is required to store the keys for security purposes. One-way to mitigate computing pressure is to use a dedicated chip for encryption and decryption. Some radio chips (for example, Chipcon CC2420) used in wireless sensor networks implement in hardware the AES algorithm (Chipcon 2006). Software handling security issues such as authentication, data integrity, encryption and decryption is required to implement desired security measures.

15.12 Field Programmability

This refers to replacing or modifying part or all of the executable microcontroller code image on one or more of the nodes comprising the WSN. It is necessary when WSN's require firmware upgrades, for example sensor drivers to accommodate new sensors or patches to fix existing bugs. The ability to download code at runtime (Dunkels et al 2004) is a generally accepted system requirement. For networks comprising a small number of readily accessible nodes this field programming could be done by connecting a laptop PC to the node, but more generally, "over the air" (OTA) programming is required. The microcontroller must provide support for reprogramming the executable image. Most small MCU's of the kind used in WSN's use internal Flash memory for storing their code image. Reprogramming the code image involves the MCU under program control writing its own program memory and the details of the approach used depend on the microcontroller architecture. Small MCU's generally have one of two processor architectures. The first is the so called Princeton or Von Neumann architecture with code and data sharing the same address space on the microcontroller. The second is the so called Harvard architecture in which the code and data spaces are separate. An example of the Harvard architecture is the Atmel ATmega series of microcontrollers, while an example of the Von Neumann architecture is the Texas Instruments MSP430 series of microcontrollers. More powerful processors tend to have larger numbers of address spaces. With MCU's using the Harvard architecture, the code for reprogramming the Flash memory can be executed from RAM. With MCU's having the Princeton architecture, a segment of flash memory is reserved for a boot loader program which has access to special instructions for reprogramming the flash. In either case, it seems prudent to use external non-volatile data storage to store safe as well as new code images in case of a mishap such as receipt of a corrupted code image. After the boot loader successfully reprogrammed the node, it restarts the system and the new code image begins executing. The software needs to control the receipt over the air of new parts of the code image, as well as its storage in external memory and subsequent retransmission of the image to nodes downstream. To ensure low probability of using a corrupted image, error detection in conjunction with either error correction or retransmission of images containing detected errors would be required.

15.13 Presentation Layer

WSN is generally be deployed at remote locations. However, a user of the information gathered by the network may be located far from the network deployment site. For example, the user of a wildlife monitoring sensor network may need to access and analyse the collected information from an office many miles away (Zhang et al 2004). It is often too costly or unrealistic for the sensor network to extend its reach to provide this kind of connectivity. A more realistic option is to utilise the existing communication infrastructure to transport the sensing information beyond the scope of the sensor network. Dedicated radio transceivers can be incorporated into each node or some of the node (for example, the gateway node or cluster-heads) to use available telecommunication infrastructure such as GSM and WCDMA (Martinez et all 2004). Internet access capability can also be built into the gateway node to send information out or allow users to access sensor network information and query the network via a Web browser (Nath and Ke 2003).

15.14 Reliability of Nodes and Network

Limited resources (power, bandwidth and computing capability) and harsh working environment impose challenges in achieving high reliability in wireless sensor networks. Reliability is a system-wide issue and must be addressed in a system level, and achieved collectively though all hardware and software modules. In some applications, redundancy in components or/and nodes can be created to improve reliability. Redundancy contributes to robust networks that can overcome signal propagation degradation or the loss of a node. In mission critical applications, it is often essential for a sensor network to autonomously detect and report faults to achieve the required reliability. The fault can be either a malfunction of a local node or, more importantly, a problem of the network as a whole. The specific application dictates whether and what diagnostic capability is to be built in to the system. Some applications may require only simple diagnostic mechanisms. These include the ability of a node to generate and send an alarm to neighbours when its software detects a fault condition so that the neighbours can route data around the suspect node, or to a central location where the diagnostic information can be examined and proper actions taken. More powerful diagnostic strategies require sophisticated software algorithms running on each node which constantly monitors the health of each node and the whole network. This diagnostic software checks the correctness of each executing process and monitors the states of each software and hardware module. It may also sends probing signals to neighbouring nodes to examine the states of its neighbours and exchange health information. At each node, the diagnostic algorithm makes a decision based on gathered information from local and other nodes. The time, location, identity and severity of any detected problem are logged locally and reported to a pre-specified location.

It is required in some applications that users of the sensor network can add, remove or relocate nodes as required. In addition to this requirement, dynamics exist as the resources (energy, bandwidth and processing power) of a network and the environment in which the network is deployed are ever changing. This requires the system operate autonomously, changing its configuration (number of nodes, topology, data packet sizes, routes, etc.) as needed from the algorithms implemented in software. The networking software needs to have network formation, discovery and joining capabilities. When a node moves from one location to another, or is removed either by itself (e.g. as a result of a diagnosed fatal fault) or by the user, the software needs to adjust network topology and routing tables adaptively without operator intervention. Traditional routing algorithms such as Internet Protocol (IP) are not likely to be viable candidates in this context, since they need to maintain routing tables for the global topology and routing table updates incur heavy overhead in terms of time, memory and energy (Chong and Kumar 2003). Alternative routing algorithms such as Ad hoc On-Demand Distance Vector Routing (AODV) (Perkins and Royer 1999) need to be used which build up routes on an as-needed basis and, therefore, better suit the dynamic requirements. Networking software, which supports dynamic topology, routing, and network maintenance needs to be implemented.

15.15 Deployment

Deployment of wireless sensor networks is often overlooked in the design process. Techniques and methods that make the deployment of wireless sensor networks a simple and straightforward task as essential if such networks are to be widely used and deployed. Often the deployment and maintenance costs in existing wireless networks often outweigh the initial cost of the system; consider for example established GSM and 3G networks. Various problems arise during the deployment phase, these include: (i) How many nodes are required?, (ii) Where shall a node be located?, and (iii) If a node is to be relocated or removed, how will it affect the network?

The number of nodes required in a network is a function of the application requirements, for example, the number of sites that require monitoring. But it is also a function of the required radio coverage. For example, a site that requires monitoring may be located far from the network coordinator, or there may be a radio obstruction between the network coordinator and the site. In this case, if one is using a mesh network, it is necessary to add one or more router nodes between the network coordinator and the site. However, unless an RF site survey is carried out, it is difficult to automate such a task. Where a node is to be located is also a function of the radio coverage. Depending on the surrounding environment, RSSI measurements may change dramatically from one location to another. Such variations will affect a node's ability to join a network and exchange data with other nodes. Again, without an RF site survey, it is difficult to determine if a node can be located at a particular site. Often, each time a node is to be installed at a location, one has to experimentally determine, given the current network configuration, if the node can join the network. Despite the redundancy inherent in mesh networks, relocating, or removing nodes can have catastrophic consequences on a network's performance. For example, many nodes often serve as routers connecting one or more disjointed sub-networks. If such a router is removed or relocated it can effectively isolate one of the sub-networks. In such cases, without a-priori knowledge of the network topology, it is not possible to relocate nodes.

15.16 Conclusions

Wireless sensor networks will proliferate in a vast range of potential applications including academic, industrial, agricultural, domestic, military, not-for-profit. In this chapter we described how the application affects the design of a wireless sensor network. For any attribute of a wireless sensor network, it is the application, which determines the importance of that attribute in the design specification. The available technology then determines the manner in which that desired attribute might be achieved.

15.17 References

Atmel (2003) 8051 Microcontrollers application note, EMC Improvement Guidelines, URL: http://www.atmel.com/dyn/resources/prod_documents/doc4279.pdf

Barken L (2004) How secure is your wireless network? Pearson Education Inc, p. 70

Benini L, De Micheli G, and Macii E (2001) Designing low-power circuits: Practical recipes, IEEE Circuits and Systems magazine, 1:1, pp. 6 - 25

Chipcon (2006) http://www.chipcon.com

Chong C and Kumar SP (2003) Sensor networks: Evolution, opportunities, and challenges, Proc. IEEE, 91:8, pp. 1247 - 1256

Daemen J and Rijmen V (2002) The design of Rijndael AES: The advanced encryption standard, Springer Verlag

Dunkels A, Grönvall B, and Voigt T (2004) Contiki: A lightweight and flexible operating system for tiny networked sensors, Proc. of the First IEEE Workshop on Embedded Networked Sensors, Tampa, FL

Edney J and Arbaugh W (2004) Real 802.11 security: Wi-Fi protected access and 802.11i, Addison-Wesley, pp. 120-122

Elgharbawy WM and Bayoumi MA (2005) Leakage sources and possible solutions in nanometer CMOS technologies, IEEE Circuits and Systems magazine, 5:4, pp. 6-17

Henry OTT Consultants (2000) EMC design guide: PCB design guidelines, URL: http://www.hottconsultants.com/pdf_files/pcb_guide.pdf

Hightower J and Borriello G (2001) Location system for ubiquitous computing, Computer, 34:8, pp. 57-66

IEEE (2003) Wireless medium access control (MAC) and physical layer (PHY) specifications for low rate wireless personal area networks (WPANs), New York: IEEE Press

Landernäs K (2005) Low-power digital CMOS design: A survey MRTC report ISSN 1404-3041 ISRN MDH-MRTC-179/2005-1-SE, Mälardalen Real-Time Research Centre, Mälardalen University

Lee TH (2004) The design of CMOS RF integrated circuits, Cambridge University Press

Martinez K, Ong R, and Hart J (2004) Glacsweb: A sensor network for hostile environments, Proc. of the 1st IEEE Int. Conf. Sensors and ad-hoc Networks, London, pp. 81-7

Moses RL, Krishnamurthy D, and Patterson R (2003) A self-localisation method for wireless sensor networks, EURASIP Journal on Applied Sig. Processing (4): pp. 348 - 358

Nath S and Ke Y (2003) IrisNet: An architecture for enabling sensor-enriched internet service, Ntel Research Pittsburgh Technical Report IRP-TR-03-04

Perkins CE and Royer EM (1999) Ad hoc on-demand distance vector routing, Proc. of the 2nd IEEE Workshop on Mobile Computing Systems and Appl., New Orleans, pp. 90 - 100

Polymer Technology and Services LLC (2004) A guide to polycarbonate in general, URL: http://www.ptsllc.com/polcarb_intro.htm

Razavi B (2001) Design of analog CMOS integrated circuits, McGraw Hill

Reliance Electrical (2000) Does this NEMA rating match your required IP rating? URL: http://www.reliance.com/pdf/drives/whitepapers/D7745.pdf

Tanenbaum A (1996) Computer networks, Prentice Hall PTR, pp. 602 - 613

The Engineering Tool Box (2005) IP: Ingress protection ratings, URL: www.engineeringtoolbox.com/ip-ingress-protection-d_452.html

Zhang P, Sadler CM, Lyon SA, and Martonosi M (2004) Hardware design experiences in ZebraNet, Proc. of the 2nd Int. Conf. on Embedded Networked Sensor Syst, pp. 227 - 238

Zhao F and Guibas LJ (2004) Wireless sensor networks: An information processing approach, Morgan Kaufmann

Zhao F, Liu J, Liu JJ, Guibas L, and Reich J. (2003) Collaborative signal and information processing: an information direct approach, Proc. IEEE, 91:8, pp. 1199 - 1209

16 ZigBee and Their Applications

Meng-Shiuan Pan and Yu-Chee Tseng

Department of Computer Science, National Chiao Tung University, Hsin-Chu, 30010, Taiwan

16.1 Introduction

The rapid progress of wireless communication and embedded micro-sensing microelectromechanical systems (MEMS) technologies has made wireless sensor networks (WSN) possible. A WSN consists of many inexpensive wireless sensors, which are capable of collecting, storing, processing environmental information, and communicating with neighboring nodes. In the past, sensors are connected by wirelines. With the development of *ad hoc* networking technologies, tiny sensors can communicate through wireless links in a more convenient manner (Pottie and Kaiser, 2000; Sohrabi et al. 2000). A lot of applications of WSN have been proposed. For example, wildlife-monitoring applications are discussed in (FireBug 2004; GreatDuckIsland 2004) and mobile object tracking issues are addressed in (Lin and Tseng, 2004; Tseng et al., 2003). How to ensure network coverage/connectivity is discussed in (Huang et al., 2005; Yan et al., 2003). Guiding applications based on wireless sensor networks are presented in (Li et al, 2003; Tseng et al., 2006). Applications of mobile sensors are presented in (Tseng et al., 2005). Many WSN platforms have been developed, such as MICA2, MICAz, TelosB MOTE (Xbow, 2005), and Dust Network (DustNetworks, 2005). To allow different systems to work together, standards are needed. ZigBee/IEEE 802.15.4 protocols are developed for this purpose. ZigBee/IEEE 802.15.4 is a global hardware and software standard designed for WSN requiring high reliability, low cost, low power, scalability, and low data rate. Table 16.1 compares ZigBee/IEEE 802.15.4 against several other wireless technologies. The ZigBee alliance (ZigBee, 2004) is to work on the interoperability issues of ZigBee/IEEE 802.15.4 protocol stacks. The IEEE 802.15 WPAN Task Group 4 (IEEE Std 802.15.4, 2003) specifies physical and data link layer protocols for ZigBee/IEEE 802.15.4. The relationship of ZigBee and IEEE 802.15.4 is shown in Fig. 16.1. In the current development, IEEE 802.15 WPAN working group creates two task groups 15.4a and 15.4b. The former is to specify an alternate physical layer, the ultra wide band (UWB) technologies. The latter is to enhance the IEEE 802.15.4 MAC protocol so that it can tightly couple with the network layer functionalities specified by ZigBee. ZigBee alliance published the version 1.0 standard in Dec. 2004.

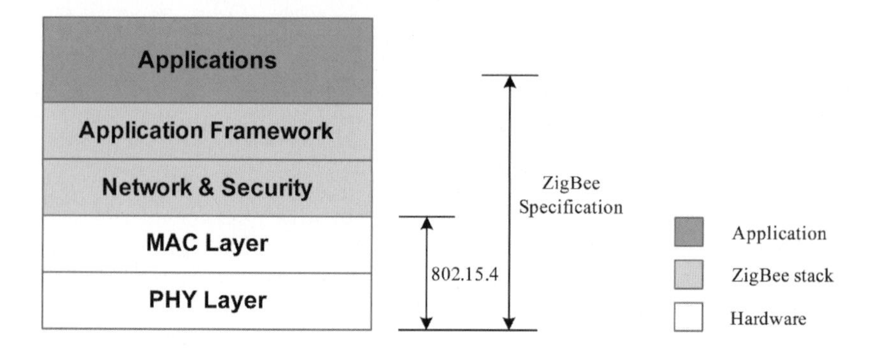

Fig. 16.1. The ZigBee/IEEE 802.15.4 protocol stack

Table 16.1: Comparison of different wireless technologies (ZigBee 2004)

Standard	ZigBee/IEEE 802.15.4	Bluetooth	UWB	IEEE 802.11 b/g
Working frequency	868/915 MHz, 2.4GHz	2.4 GHz	3.1 - 10.6 GHz	2.4 GHz
Range (m)	30 – 75+	10 – 30	~10	30 – 100 +
Data rate	20/40/250 kbps	1 Mbps	100+ Mbps	2 – 54 Mbps
Devices	255 – 65k	8		50 – 200
Power consumption	~1 mW	~40 – 100 mW	~80 – 300 mW	~160 mW – 600W
Cost ($US)	~2 – 5	~4 – 5	~5 – 10	~20 – 50

Companies such as Chipcon (Chipcon, 2005), Ember (Ember 2005), and Freescale (Freescale 2005) provide system-on-chip solutions of ZigBee/IEEE 802.15.4. For home networking, ZigBee/IEEE 802.15.4 can be used for light control, heating ventilation air conditioning (HVAC), security monitoring, and emergency event detection. For health case, ZigBee/IEEE 802.15.4 can integrate with sphygmomanometers or electronic thermometers to monitor patients' statuses. For industrial control, ZigBee/IEEE 802.15.4 devices can be used to improve the current manufacturing control systems, detect unstable situations, control production pipelines, and so on. In the rest of this chapter, we will review IEEE 802.15.4 and ZigBee network layer protocols in Section 2 and Section 3, respectively. Section 4 discusses the beacon scheduling issue in a ZigBee tree network. Section 5 introduces the broadcast procedures in ZigBee. Some application examples of WSN are introduced in Section 6. Finally, we conclude this chapter in Section 7.

16.2 IEEE 802.15.4 Basics

IEEE 802.15.4 specifies the physical layer and data link layer protocols for low-rate wireless personal area networks (LR-WPAN), which emphasise on simple, low-cost applications. Devices in such networks normally have less communication capabilities and limited power, but are expected to operate for a longer period of time. As a result, energy saving is a critical design issue. In IEEE 802.15.4, there are two basic types of network topologies, the star topology, and the peer-to-peer topology. Devices in a LR-WPAN and can be classified as *full function devices (FFDs)* and *reduced function devices (RFDs)*. One device is designated as the *PAN coordinator*, which is responsible for maintaining the network and managing other devices. A FFD has the capability of becoming a PAN coordinator or associating with an existing PAN coordinator. A RFD can only send or receive data from a PAN coordinator that it associates with. Each device in IEEE 802.15.4 has a unique 64-bit *long address*. After associating to a coordinator, a device will be assigned a 16-bit *short address*. Then packet exchanges between the coordinator and devices will use short addresses. In the following, the IEEE 802.15.4 physical layer and data link layer protocols are introduced.

16.2.1 Physical Layer (PHY)

In IEEE 802.15.4 PHY, there are three operating frequency bands with 27 radio channels. These bands are 868 MHz, 915 MHz, and 2.4 GHz. The channel arrangement is shown in Fig. 16.2. Channel 0 is in the frequency 868.0~868.6 MHz, which provides a data rate of 20 kbps. Channels 1 to 10 work in frequency 902.0~928.0 MHz and each channel provides a data rate of 40 kbps. Channels 11~26 are located in frequency 2.4~2.4835 GHz and each channel provides a data rate of 250 kbps.

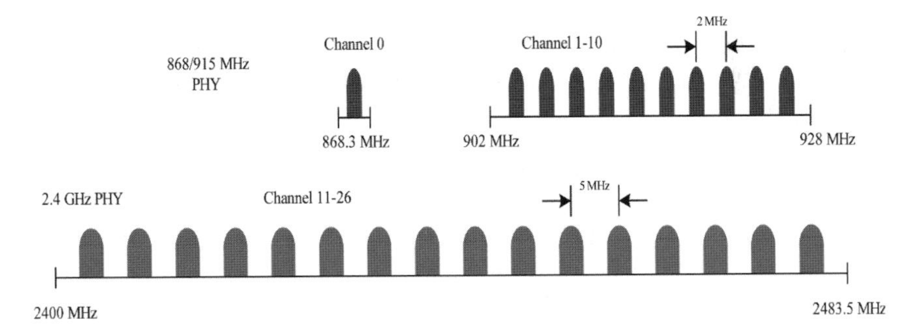

Fig. 16.2. Arrangement of channels in IEEE 802.15.4

Channels 0 to 10 use the binary phase shift keying (BPSK) as their modulation scheme, and channels 11 to 26 use the offset quadrature phase shift keying (O-QPSK) as their modulation scheme. The required receiver sensitivity should be larger than -92 dBm for channels 0 to 10, and larger than -85 dBm for channels 11 to 26. The transmit power should be at least -3 dBm (0.5 mW). The transmission radius may range from 10 meters to 75 meters. Targeting at low-rate communication systems, in IEEE 802.15.4, the payload length of a PHY packet is limited to 127 bytes.

16.2.2 Data Link Layer

In all IEEE 802 specifications, the data link layer is divided into two sublayers: *logical link control (LLC)* sublayer and *medium access control (MAC)* sublayer. The LLC sublayer in IEEE 802.15.4 follows the IEEE 802.2 standard. The MAC sublayer manages superframes, controls channel access, validates frames, and sends acknowledgements. The IEEE 802.15.4 MAC sublayer also supports low power operations and security mechanisms. In the following subsections, we introduce the MAC layer protocols in IEEE 802.15.4.

16.2.3 Superframe Structure

In IEEE 802.15.4, its coordinator defines the superframe structure of a network. The length of a superframe is equal to the time interval of two adjacent beacons sent by a coordinator. A superframe can be divided into an active portion and an inactive portion. An active portion consists of 16 equal-length slots and can be further partitioned into a *contention access period (CAP)* and a *contention free period (CFP)*. The CAP may contain i slots, $i = 1, 2, \ldots, 16$, and the CFP, which follows the CAP, may contain 16-i slots. The coordinator and network devices can exchange packets during the active portion and go to sleep during the inactive portion. The superframe structure is shown in Fig. 16.3.

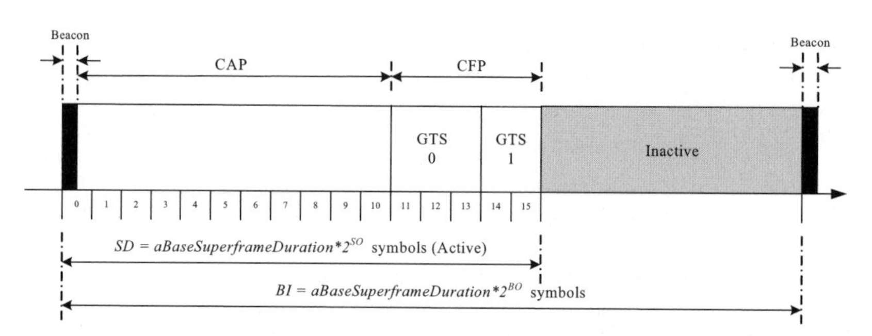

Fig. 16.3. The superframe structure in IEEE 802.15.4

Beacons are used for starting superframes, synchronizing with other devices, announcing the existence of a PAN, and informing pending data in coordinators. In a beacon-enabled network, devices use the slotted CAMA/CA mechanism to contend for the usage of channels. FFDs, which require fixed rates of transmissions, can ask for *guarantee time slots (GTS)* from the coordinator. A CFP can include multiple GTSs, and each GTS may contain multiple slots. For example, in Fig. 16.3, GTS 0 uses three slots and GTS 1 uses two slots. A coordinator can allocate at most seven GTSs for network devices. In IEEE 802.15.4, the structure of superframes is controlled by two parameters: *beacon order (BO)* and *superframe order (SO)*, which decide the length of a superframe and its active potion, respectively. For a beacon-enabled network, the setting of BO and SO should satisfy the relationship $0 \leqq SO \leqq BO \leqq 14$. For channels 11 to 26, the length of a superframe can range from 15.36 *ms* to 215.7 *s*, so can an active potion. Specifically, the length of a superframe is:

$$BI = aBaseSuperframeDuration \times 2^{BO} \text{ symbols}$$

where each symbol is 1/62.5 *ms* and *aBaseSuperframDuration* = 960 symbols. Note that the length of a symbol is different for channels 0 to 10. The length of each active portion is:

$$SD = aBaseSuperframeDuration \times 2^{SO} \text{ symbols}$$

Therefore, each device will be active for $2^{-(BO-SO)}$ portion of the time, and sleep for $1-2^{-(BO-SO)}$ portion of the time. Change the value of (BO-SO) allows us to adjust the on-duty time of devices, and thus estimate the network lifetime. Table 16.2 shows the relationship between (BO-SO) and duty cycle of network devices. When (BO-SO) becomes larger, the duty cycle of network devices will reduce. For a non-beacon-enabled network, the values of both BO and SO will be set to 15 to indicate that superframes do not exist. Devices in a non-beacon-enabled network always use the unslotted CSMA/CA to access the channels. Because of no superframe structure, devices cannot go to sleep to save energy.

Table 16.2: The relationship between (BO-SO) and the DC (duty cycle) of network devices

BO-SO	0	1	2	3	4	5	6	7	8	9	>10
DC (%)	100	50	25	12	6.25	3.125	1.56	0.78	0.39	0.195	< 0.1

16.2.4 Data Transfer Models

IEEE 802.15.4 defines three data transfer models: 1) data transmission to a coordinator, 2) data transmission from a coordinator, and 3) data transmission between peers. The first two models are for star networks and the last one is for peer-to-peer networks. In the following, we introduce these three data transfer models.

- *Data transmission to a coordinator*: In a beacon-enabled network, devices that have data to send use the slotted CSMA/CA mechanism to contend for channels after receiving beacons. In a non-beacon-enabled network, devices contend for channels using the unslotted CSMA/CA mechanism. In both kinds of networks, after successfully obtaining a channel, a device can send data to its coordinator directly. A coordinator that receives a data frame from a device may reply an acknowledgement (optional). Fig. 16.4 shows the procedures of data transfer to a coordinator.
- *Data transmission from a coordinator*: Data transmission from a coordinator is based on requests from devices. In a beacon-enabled network, a coordinator

should notify devices that it has buffered packets by its beacons, instead of directly sending data frames to devices. A device that receives a beacon first checks whether its ID appears in the *pending data fields* in the beacon. If so, this device sends a data request command to the coordinator. The coordinator, after receiving the data request, will reply an acknowledgement and forward the data frame to that device. On the other hand, in a non-beacon-enabled network, a device should periodically send data request frames to query the coordinator if there are buffered packets for itself. The coordinator, on receipt of a data request frame, should check if there are frames for the sender. If so, the coordinator will reply an acknowledgement and then send a data frame to the corresponding device. The procedures of data transmission from a coordinator are shown in Fig. 16.5.

- *Data transmission between peers*: In a beacon-enabled network, peers cannot send data to each other directly. However, peers can directly transmit data to each other in a non-beacon-enabled network. The unslotted CSMA/CA mechanism is used to contend for channels.

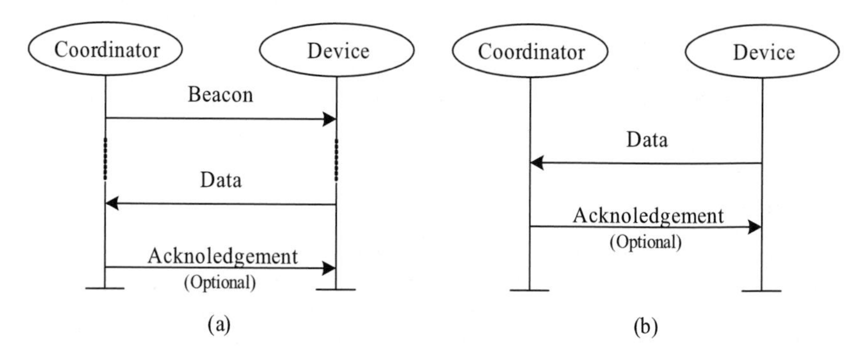

Fig. 16.4. (a) Data transmission to a coordinator in a beacon-enabled network. (b) Data transmission to a coordinator in a non-beacon-enable network

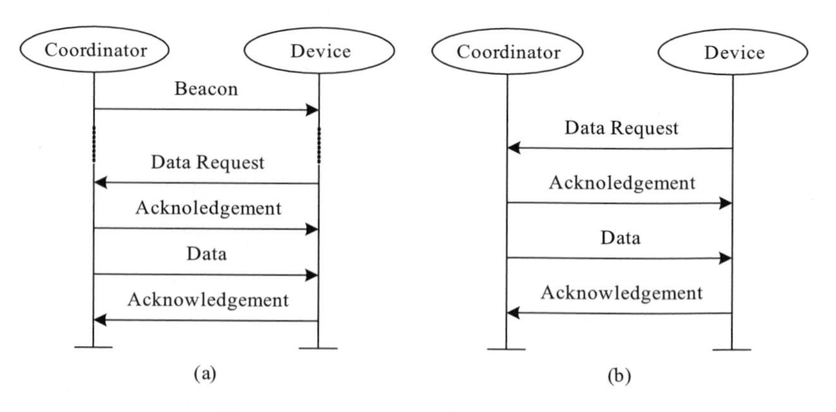

Fig. 16.5. (a) Data transmission from a coordinator in a beacon-enabled network. (b) Data transmission from a coordinator in a non-beacon-enable network

16.2.5 CSMA/CA Mechanisms

There are two channel access mechanisms in IEEE 802.15.4. One is unslotted CSMA/CA and the other is slotted CSMA/CA. The operations of unslotted CSMA/CA are similar to the ones in IEEE 802.11 CSMA/CA. A device that has a data or command frame to send will randomly backoff a period of time. If the medium is idle when the backoff expires, this device can transmit its frame. On the other hand, if the medium is busy, this device will increase its backoff window and waits for another period of time. The slotted CSMA/CA works differently from unslotted CSMA/CA. In the slotted CSMA/CA mechanism, the superframe structure is needed. A superframe can be further divided into smaller slots called backoff periods, each of length 20 symbols[1]. The start of the first backoff period in a superframe is aligned to the start of beacon transmission. Before transmission, a device first calculates a random number of backoff periods. After timeout, the device should perform *clear channel assessment (CCA)* twice in the upcoming two backoff periods. If the channel is found to be clear in two CCAs, the device can start to transmit a frame to the coordinator. If the channel is found to be busy in any of the two CCAs, the device should double its contention window and perform another random backoff. Fig. 16.6 shows the procedures of the slotted CSMA/CA mechanism in IEEE 802.15.4.

16.2.6 Association and Disassociation Procedures

A device becomes a member of a PAN by associating with its coordinator. At the beginning, a device should scan channels to find potential coordinators. After choosing a coordinator, the device should locate the coordinator's beacons and transmit an association request command to the coordinator. In a beacon-enabled network, the association request is sent in the CAP of a superframe. In a non-beacon-enabled network, the request is sent by the unslotted CSMA/CA mechanism. On receipt of the association request, the coordinator will reply an ACK. Note that correctly receiving an ACK does not mean that device has successfully associated to the coordinator; the device still has to wait for an association decision from the coordinator. The coordinator will check its resource to determine whether to accept this association request or not. In IEEE 802.15.4, association results are announced in an indirect fashion. A coordinator responds to association requests by appending devices' long addresses in beacon frames to indicate that the association results are available. If a device finds that its address is appended in a beacon, it will send a data request to the coordinator to acquire the association result. Then the coordinator can transmit the association result to the device. The association procedure is summarised in Fig. 16.7.

[1] The time required to transmit a symbol varies according to working bands of PHY. For example, in the 2.4 GHz band, the length of a symbol is 16 us; hence, in the 2.4 GHz band, a unit backoff period is 320 us.

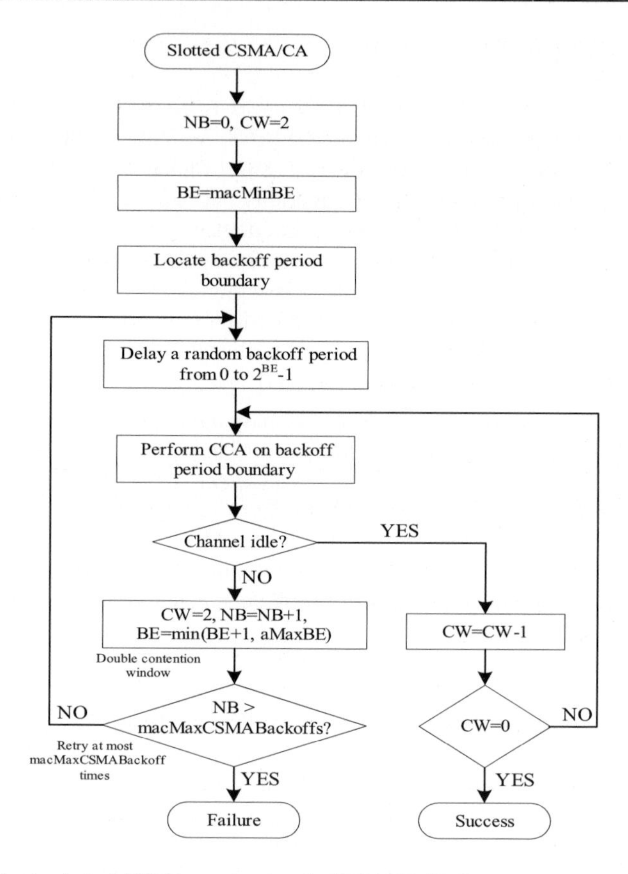

Fig. 16.6. The basic slotted CSMA mechanism in IEEE 802.15.4

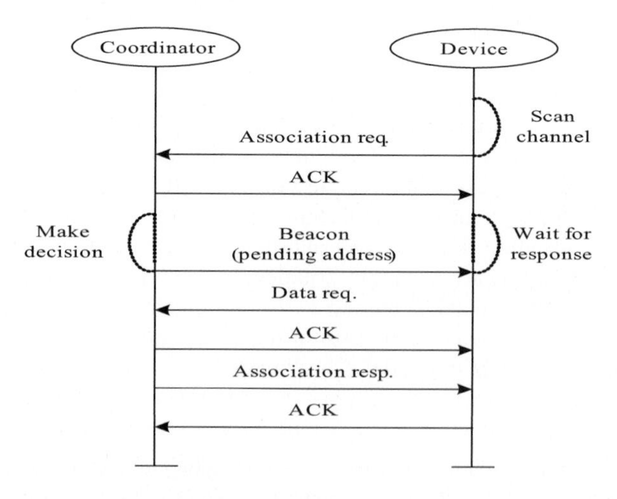

Fig. 16.7. The association procedure in IEEE 802.15.4

When a coordinator would like an associated device to leave its PAN, it can send a disassociation notification command to the device. After receiving this command, the device will reply an ACK. If the ACK is not correctly received, the coordinator will still consider that the device has been disassociated. When an associated device wants to leave a PAN, it also sends a disassociation notification command to the coordinator. On receipt of the command, the coordinator will reply an ACK and remove the records of the correspond device. Similar to the above case, the device considers itself disassociated even if it does not receive an ACK from the coordinator.

16.2.7 Summary of IEEE 802.15.4

IEEE 802.15.4 specifies the physical layer and data link layer protocol for low-rate wireless personal area networks. However, this specification only concerns communications between devices that is within each other's transmission range. For larger sensor networks, the support of network layer protocols is needed. In the next section, we will introduce a developing standard, ZigBee, which supports protocols above the data link layer for connecting IEEE 802.15.4 devices together.

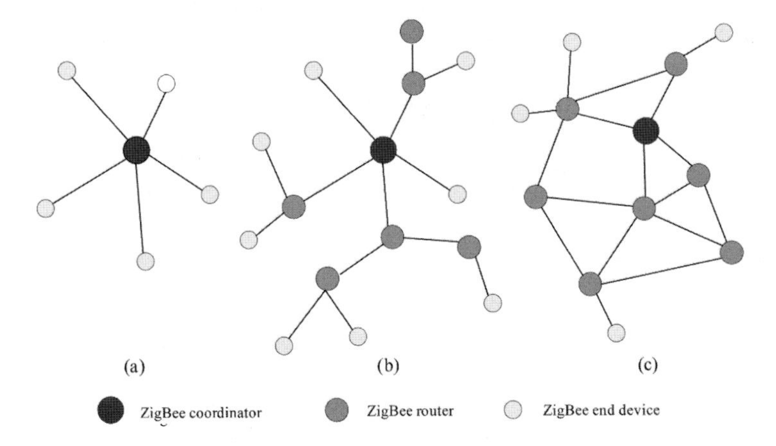

Fig. 16.8. Zigbee network topologies: (a) star, (b) tree, and (c) mesh

16.3 ZigBee Network Layer

In ZigBee, the network layer provides reliable and secure transmissions among devices. Three kinds of networks are supported, namely star, tree, and mesh networks. A ZigBee coordinator is responsible for initializing, maintaining, and controlling the network. A star network has a coordinator with devices directly connecting to the coordinator. For tree and mesh networks, devices can communicate with each other in a multihop fashion. The network backbone is formed by one

ZigBee coordinator and multiple ZigBee routers. RFDs can join the network as end devices by associating with the ZigBee coordinator or ZigBee routers. In a tree network, the coordinator and routers can announce beacons. However, in a mesh network, regular beacons are not allowed. Devices in a mesh network can only communicate with each other by peer-to-peer transmissions specified in IEEE 802.15.4. Examples of ZigBee network topologies are shown in Fig. 16.8.

16.3.1 Network Formation

Devices that are coordinator-capable and do not currently join a network can be candidates of ZigBee coordinators. A device that desires to be a coordinator will scan all channels to find a suitable one. After selecting a channel, this device broadcasts a beacon containing a PAN identifier to initialise a PAN. A device that hears beacons of an existing network can join this network by performing the association procedures and specifying its role, as a ZigBee router or as an end device. The beacon sender will determine whether to accept this device or not by considering its current capacity and its permitted association duration. Then its beacons can carry the association response. If a device is successfully associated, the association response will contain a short 16-bit address for the request sender. This short address will be the network address for that device.

16.3.2 Address Assignment in a ZigBee Network

In a ZigBee network, network addresses are assigned to devices by a distributed address assignment scheme. After forming a network, the ZigBee coordinator determines the maximum number of children (*Cm*) of a ZigBee router, the maximum number of child routers (*Rm*) of a parent node, and the depth of the network (*Lm*). Note that $Cm \geq Rm$ and a parent can have (*Cm-Rm*) end devices as its children. In this algorithm, their parents assign addresses of devices. For the coordinator, the whole address space is logically partitioned into *Rm*+1 blocks. The first *Rm* blocks are to be assigned to the coordinator's child routers and the last block is reversed for the coordinator's own child end devices. In this scheme, a parent device utilises *Cm*, *Rm*, and *Lm* to compute a parameter called *Cskip*, which is used to compute the starting addresses of its children's address pools. The *Cskip* for the ZigBee coordinator or a router in depth *d* is defined as:

$$Cskip(d) = \begin{cases} 1 + Cm \cdot (Lm - d - 1), & \text{if } Rm = 1 \\ \dfrac{1 + Cm - Rm - Cm \cdot Rm^{Lm-d-1}}{1 - Rm}, & \text{Otherwise} \end{cases} \tag{16.1}$$

The coordinator is said to be at depth 0; a node which is a child of another node at depth *d* is said to be at depth *d*+1. Consider any node *x* at depth *d*, and any node *y* which is a child of *x*. The value of *Cskip(d)* indicates the maximum number of nodes in the subtree rooted at y (including y itself). For example, in Fig. 16.9, since the *Cskip* value of B is 1, the subtree of C will contain no more than 1 node; since the *Cskip* value A is 7, the subtree of B will contain no more than 7 nodes. To understand the formulation, consider again the nodes x and y mentioned above. Node y itself counts for one node. There are at most Cm children of y. Among all children of y, there are at most Rm routers. So there are at most $CmRm$ grandchildren of y. It is not hard to see that there are at most $CmRm^2$ great grandchildren of y. So the size of the subtree rooted at y is bounded by,

$$Cskip(d) = 1 + Cm + CmRm + CmRm^2 + \ldots\ldots + CmRm^{Lm-d-2} \tag{16.2}$$

Since the depth of the subtree is at most $Lm\text{-}d\text{-}1$. We can derive that,

$$\text{Eq. 16.2} = 1+Cm(1+Rm+Rm^2+\ldots+Rm^{Lm-d-2}) = 1+Cm(1-Rm^{Lm-d-1})/(1-Rm) \tag{16.3}$$

Address assignment begins from the ZigBee coordinator by assigning address 0 to itself and depth d=0. If a parent node at depth d has an address A_{parent}, the nth child router is assigned to address $A_{parent}+(n-1)\times Cskip(d)+1$ and nth child end device is assigned to address $A_{parent}+Rm\times Cskip(d)+n$. An example of the ZigBee address assignment is shown in Fig. 16.8. The *Cskip* of the ZigBee coordinator is obtained from Eq. 1 by setting d=0, Cm=6, Rm=4, and Lm=3. Then the 1st, 2nd, and 3rd child routers of the coordinator will be assigned to addresses 0+(1-1)×31+1=1, 0+(2-1)×31+1=32, and 0+(3-1)×31+1=63, respectively. And the two child end devices' addresses are 0+4×31+1=125 and 0+4×31+2=126.

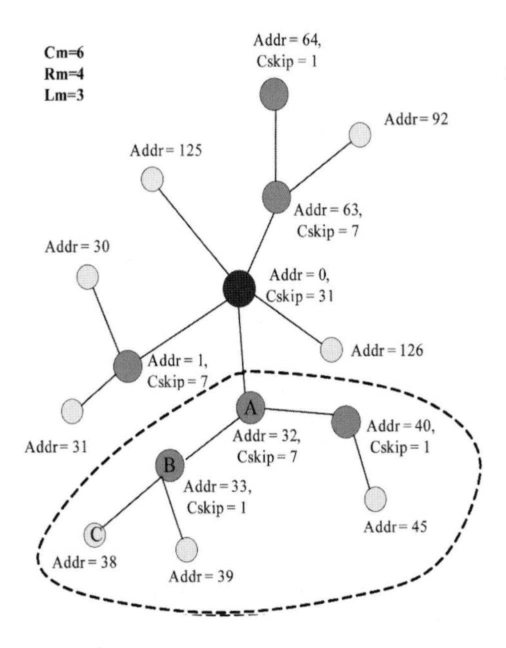

Fig. 16.9. An address assignment example in a ZigBee network

16.3.3 Routing Protocols

In a tree network, the ZigBee coordinator and routers transmit packets along the tree. When a device receives a packet, it first checks if it is the destination or one of its child end devices is the destination. If so, this device will accept the packet or forward this packet to the designated child. Otherwise, it will relay packet along the tree. Assume that the depth of this device is d and its address is A. This packet is for one of its descendants if the destination address A_{dest} satisfies $A < A_{dest} < A + Cskip(d\text{-}1)$, and this packet will be relayed to the child router with address,

$$A_r = A + 1 + \left\lfloor \frac{A_{dest} - (A+1)}{Cskip(d)} \right\rfloor \times Cskip(d) \tag{16.4}$$

If the destination is not a descendant of this device, this packet will be forwarded to its parent. In a mesh network, ZigBee coordinators and routers are said to have *routing capacity* if they have routing table capacities and route discovery table capacities. Devices that are routing-capable can initiate routing discovery procedures and directly transmit packets to relay nodes. Otherwise, they can only transmit packets through tree links. In the latter case, when receiving a packet, a device will perform the same routing operations as described in tree networks. When a node needs to relay a received packet, it will first check whether it is routing-capable. If it is routing-capable, the packet will be unicast to the next hop. Otherwise, the packet will be relayed along the tree. A device that has routing capacity will initiate route discovery if there is no proper route entry to the requested destination in its routing table. The route discovery in a ZigBee network is similar to the AODV routing protocol (Perkins et al., 2003). Links with lower cost will be chosen into the routing path. The cost of link l is defined based on the packet delivery probability on link l. However, how to calculate the packet delivery probability is not explicitly stated in the ZigBee specification. At the beginning of a route discovery, the source broadcasts a route request packet. A ZigBee router that receives a route request packet first computes the link cost. If this device has routing capacity, it will rebroadcast this request if it does not receive this request before or the link cost recorded in route request plus the cost it just computed is lower than the former received request. Otherwise, it will discard this request. For the case that a ZigBee router that is not routing capable receives a route request, it also determines whether to resend this request based on the same comparison. If this device determines to resend this route request, it will check the destination address and unicast this route request to its parent or to one of its children (in the tree network). An example is shown in Fig. 16.10. In Fig. 16.10, device S broadcasts a route request for destination T and devices A and D receive this packet. Since device A has no routing capacity, it will check the address of destination T and unicast this request to device C. Since device D has routing capacity, it will rebroadcast this request. A device that has resent a route request packet will record the request sender in its route discovery table. This information will be discarded if this device does not receive a route reply within a time interval.

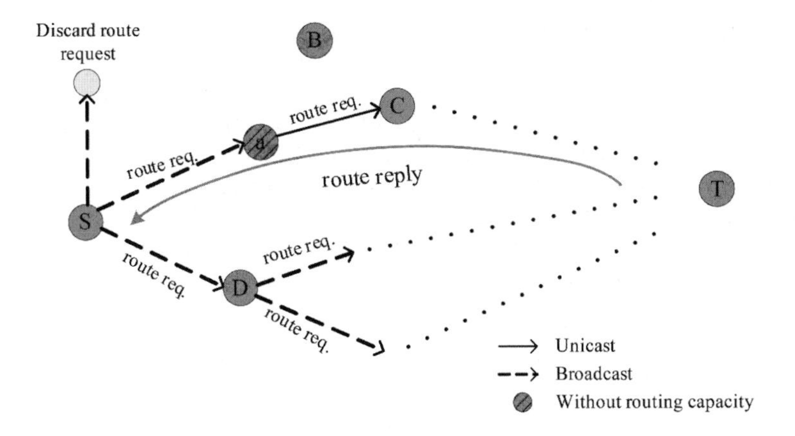

Fig. 16.10. An example of route request dissemination in a ZigBee network

When the destination receives route request packets from multiple paths, it will choose the routing path with the lowest cost and send a route reply packet to the source. The route reply packet will be sent by unicast. An intermediate node that receives the route reply packet checks its route discovery table and sends the route reply to the request sender. After the source node successfully receives the route reply, it can send data packets to the destination node along the discovered route. The ZigBee network layer specifies route maintenance mechanisms for mesh and tree networks. In a mesh network, a failure counter detects route failure. If the counter of a ZigBee router exceeds a threshold, the router can start the route maintenance procedure. For those routers that have routing capacity, they can flood route request packets to find destinations else, they will unicast route request packets to their parents or children according to the destination addresses. However, in a tree network, a router does not broadcast route request packets when it loses its parent. Instead, it disassociates with its parent and tries to re-associate with a new parent. After re-association, it will receive a new short 16-bit network address and can transmit packets to its new parent. Note that a device that re-associates to a new parent will disconnect all its children. Those children that lose their parents will also try to find new parents. On the other hand, when a router cannot send packets to a child, it will directly drop this packet and send a route error message to the packet originator. The router sends a disassociation notification command to the child. The disassociated child may reconnect to the same parent or find a new parent depending on its new scan result.

16.3.4 Summary of the ZigBee Network Layer

ZigBee is designed to support low-cost network layer. It supports three kinds of network topologies, which are star, tree, and mesh networks. Network developers can choose a suitable network topology for their applications. The pros and cons of these three topologies are summarised in Table 16.3.

Table 16.3: Pros and cons of different kinds of ZigBee network topologies

	Pros	Cons
Star	1. Easy to synchronise 2. Support low power operation 3. Low latency	1. Small scale
Tree	1. Low routing cost 2. Form superframes to support sleep mode 3. Allow multihop communication	1. Route reconstruction is costly 2. Latency may be quite long
Mesh	1. Robust multihop communication 2. Network is more flexible 3. Lower latency	1. Cannot form superframes (cannot support sleep mode) 2. Route discovery is costly 3. Storage for routing table

16.4 Beacon Scheduling in ZigBee Tree Networks

In a tree network, the ZigBee coordinator and routers can transmit beacons. Sending beacons facilitates devices to synchronise with their parents and thus can support devices to go to sleep and save energy. Recall that after forming a network, the network coordinator will determine the beacon order (BO) and superframe order (SO). When BO is larger than SO, devices can go to sleep during the inactive portions of superframes. In the ZigBee network specification version 1.0, a superframe can be divided into 2^{BO-SO} non-overlapping time slots. A router can choose a slot to announce its beacon. The start time of its beacons is also the start time of superframes of that router. Therefore, routers' superframes will be shifted away from those of the coordinator's by multiples of SD. To avoid collisions, a device should not arbitrarily choose a slot to transmit its beacons. A device should avoid using the same beacon transmit slots as its neighbors' and its parent's; otherwise, its children may lose beacons due to collisions. Beacon collisions may occur in two ways: *direct beacon conflict* between two neighbors (refer to Fig. 16.11(a)) and *indirect beacon conflict* between non-neighbors (refer to Fig. 16.11(b)). In Fig. 16.11(b), since A and B are not neighbors, the conflict is more difficult to detect.

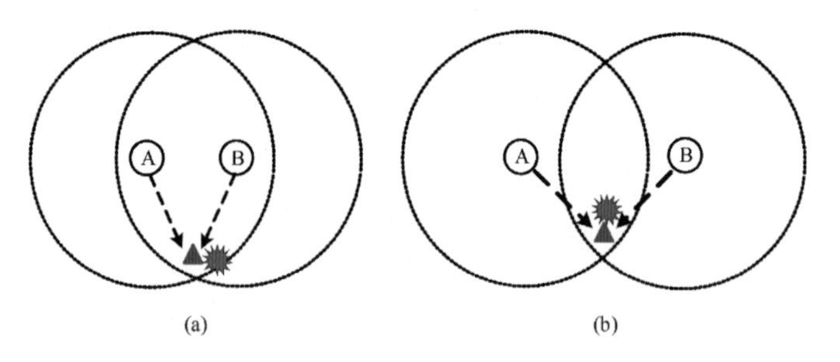

Fig. 16.11. Beacon conflicts in a ZigBee tree network: (a) direct beacon conflict and (b) indirect beacon conflict.

The ZigBee network specification version 1.0 does not provide an explicit solution to this problem. In the current specification, a device should keep the beacon transmission schedules of its neighbors and its neighbor's parents. In other words, beacon transmission schedules of nodes within two hops should be maintained. The same slots should be avoided. When sending beacons, a device will add the time offset between its beacon transmission time and its parent's in the beacon payload. This will help a device to choose a conflict-free slot. In a tree network, a device decides its beacon transmission time when joining the network. During the joining procedure, a device listens to the beacons from its parent and its neighbors for a period of time. Then the device calculates an empty slot as its beacon transmission slot. If there is no available slot, this device will join this network as an end device. After deciding beacon transmission time, the network layer will inform the MAC layer the time difference between its beacon transmission time and its associated parent's beacon transmission time.

16.5 Broadcasting in ZigBee Networks

The network layer informs the MAC layer to broadcast network-layer packets. In ZigBee, the broadcast initiator can specify the scope of this broadcast. A device that receives a broadcast packet will check whether the radius field in the broadcast packet is larger than zero. If so, the device will rebroadcast the packet; otherwise, this packet will not be further broadcast. ZigBee defines a passive acknowledgement mechanism to ensure the reliability of broadcasting. After broadcasting, the ZigBee device records the sent broadcast packet in its *broadcast transaction table (BTT)*. The BTT will be combined with its neighbor table. This allows devices to track whether their broadcast packets have been properly rebroadcast or not. If a device finds that a neighbor does not rebroadcast, it will rebroadcast to guarantee reliability. In ZigBee, devices use different strategies to broadcast packets according to the parameter *maxRxOnWhenIdle* in the MAC layer. *maxRxOnWhenIdle* controls whether a device can receive data when idle. By the nature of wireless communication, devices can detect radio signals when idle. However, they will refuse to process the received signals if *maxRxOnWhenIdle* is False. When broadcasting is needed, a device with *maxRxOnWhenIdle* = True will do so immediately. This device will also unicast the broadcast packet to those neighbors with *macRxOnWhenIdle* set to False. On the other hand, a device with *macRxOnWhenIdle* set to False can only unicast the broadcast packet to its neighbors. This is because that the device may miss passive acknowledgements from neighbors. Unicasting can ensure reliability. Fig. 16. 12 shows an example that router A sets *macRxOnWhenIdle* to False. After receiving the broadcast packet from S, A will relay the packet to B and C by unicasting. However, broadcasting in ZigBee network may cause redundant transmissions. Reference (Ding et al., 2006) introduces a tree-based broadcast scheme to relieve this problem. The authors utilise the properties of ZigBee address assignment to find a set of forwarding nodes in the network. The proposed algorithm incurs low computation cost.

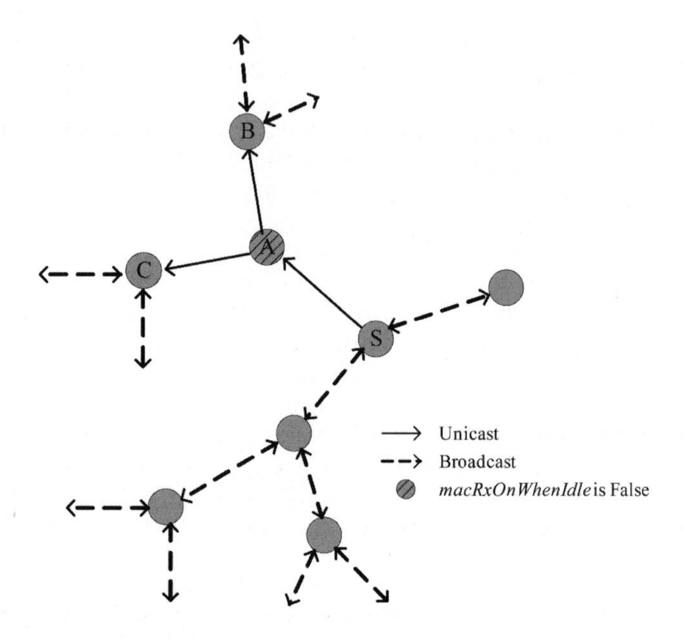

Fig. 16.12. A broadcast example in a ZigBee Network

16.6 Applications of Wireless Sensor Networks

Many research institutes have developed various applications of wireless sensor networks. In this section, we introduce two such applications: medical care and emergency guiding. The underlying communication protocols in these systems are all based on ZigBee/IEEE 802.15.4 or similar protocols.

16.6.1 Medical Care Applications

Harvard University and Boston University have developed a medical care application called CodeBlue based on wireless sensor networks (Lorincz et al, 2004). CodeBlue is a remote medical care system. Patients can be well monitored no matter they are in hospital or at home. Two wearable sensors, wireless pulse oximeter sensor and 2-lead electrocardiogram sensor, have been developed on MICA2 MOTE and Telos MOTE platforms (Xbow, 2005) to detect patients' physiological statuses. These sensors can be used to collect heart rates and oxygen saturation information. Sensors use an ad hoc routing protocol to send information to a sink. The sink keeps track of patients' statuses so that doctors can make remote diagnoses. Moreover, wireless sensors are integrated with PDAs. Patients and rescuers can watch for vital signs on PDAs in a real-time fashion. When detecting abnormal events from patients, sensors will automatically alert nearby hospitals or doctors. Rescuers can quickly locate patients by location tracking services implemented in CodeBlue.

Reference (Ho et al., 2005) integrates wireless sensor networks with the RFID technology to develop an elder healthcare application. This system utilises RFID to identify medications and uses wireless sensors to monitor if elders take right medications. Each medication has a unique ID. A patient wears a RFID reader connected to a MOTE. When the patient moves a medication bottle, a weight sensor on the bottom of this bottle can detect this event. This system can monitor whether the patient doses the right medication or the correct amount of drugs and generate alarms when necessary. Moreover, this system tracks elders by equipping RFID tags on them. When elders do not take medications on time, this system will generate alarms to them.

16.6.2 Fire Emergency Applications

The University of California at Berkeley developed a project called Fire Information and Rescue Equipment (FIRE, 2005) to improve fire-fighting equipments. It can protect fire fighters and help to relieve victims. Fire fighters are equipped with a Telos mote, a wearable computer, and a fireproof mask called FireEye. Wireless smoke and temperature sensors are deployed in a building to monitor the environment. When a fire emergency occurs, wireless sensor can provide real-time emergency-related information, which can be shown on the FireEye. FireEye mask can also show other fire fighters' locations. The Telos sensor monitors fire fighter's heart rate and air tank level and propagates the information to other fire fighters.

In some situations, fire fighters may not be able to go into the fire scenes. The integrated mobile surveillance and wireless sensor (iMouse) system (Tseng et al., 2005) supports a mobile sensor car to help monitoring a fire emergency. The iMouse system consists of some wireless sensors deployed in a building and a mobile sensor car. This mobile sensor car is equipped with a WebCam, a wireless LAN card, and a Stargate (Xbow, 2005). At normal time, sensors are responsible for monitoring the environment. When a fire emergency is detected, the mobile sensor car can utilise the information reported from sensors and find a shortest path to visit all emergency sites. After moving to each emergency site, the mobile sensor car will take snapshots around this site and send back these pictures through its WLAN interface. The architecture of the iMouse system is shown in

Fig. **16.13.**.

In a fire emergency, another important issue is to safely guide people inside the emergency scene to exits. Reference (Tseng et al., 2006) designs a distributed emergency navigation algorithm to achieve this goal. As in the previous systems, sensors are responsible for monitoring the environment at normal time. When emergencies occur, sensors will identify hazardous regions. After locating hazardous regions, each sensor will compute a safe guidance direction to one of the exits in a distributed manner. In this system, sensors will avoid leading people to go through hazardous regions. When passing hazardous regions is inevitable, sensors can also guide people as farther away from emergency locations as possible. Fig. 16.14 shows the guidance interface developed in (Tseng et al., 2006). The guiding direction will be shown on the LED panel depending on the judgement of the sensor attached to the LED panel.

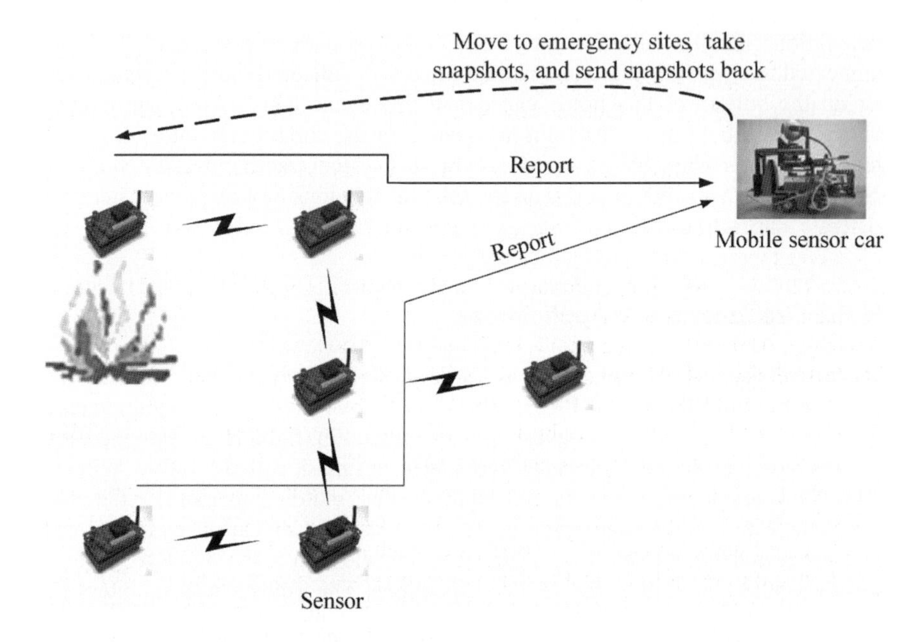

Fig. 16.13. The iMouse system architecture

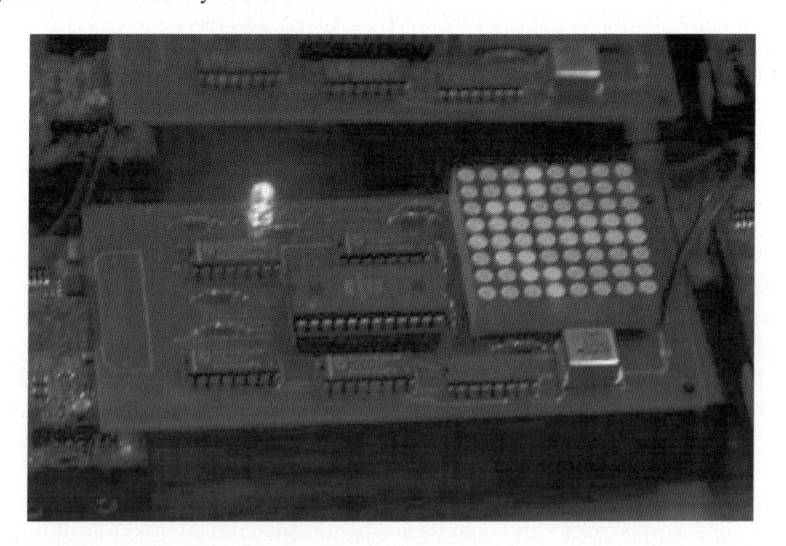

Fig. 16.14. The guiding interface developed in (Tseng et al., 2006) for emergency guidance applications

16.7 Reference

Chipcon (2005) Chipcon corporation. http://www.chipcon.com/

Dam T and Langendoen K (2003) An adaptive energy-efficient MAC protocol for wireless sensor networks. In Proc. of ACM Int'l Conf. on Embedded Networked Sensor Systems (SenSys)

Ding G, Sahinoglu Z, Orlik P, Zhang J, and Bharhava B (2006) Tree-based data broadcast in IEEE 802.15.4 and ZigBee Networks, IEEE Trans. on Mobile Computing (to appear)

DustNetworks (2005) Dust network Inc. http://dust-inc.com/flash-index.shtml

Ember (2005) Ember – Wireless semiconductor. http://www.ember.com/

FireBug (2004) Design and construction of a wildfire instrumentation system using networked sensors. http://firebug.sourceforge.net/

FIRE (2005) The fire information and rescue equipment (FIRE) project. http://kingkong.me.berkeley.edu/fire/

Freescale (2005) Freescale semiconductor. http://www.freescale.com

Gandham S, Dawande M, and Prakash R (2005) Link scheduling in sensor networks: Distributed edge coloring revisited. In: Proc. of IEEE INFOCOM

GreatDuckIsland (2004) Habitat monitoring on great duck island. http://www.greatduckisland.net/technology.php

IEEE Std 802.15.4 (2003) IEEE standard for information technology - telecommunications and information exchange between systems - local and metropolitan area networks specific requirements part 15.4: wireless medium access control (MAC) and physical layer (PHY) specifications for low-rate wireless personal area networks (LR-WPANs)

Ho L, Moh M, Walker Z, Hamada T, and Su C (2005) A prototype on RFID and sensor networks for elder healthcare: progress report, In: Proc. of SIGCOMM workshop on Experimental approaches to wireless network design and analysis

Huang C, Lo L, Tseng Y, and Chen W (2005) Decentralised energy-conserving and coverage-preserving protocols for wireless sensor networks, In: Proc. Int'l Symp. on Circuits and Systems (ISCAS)

Intanagonwiwat C, Govindan R, Estrin D, Heidemann J, and Silva F (2003) Directed diffusion for wireless sensor networking. IEEE/ACM Trans. Networking, 11:1, pp 2–16

Li Q, DeRosa M, and Rus D (2003) Distributed algorithm for guiding navigation across a sensor network. In: Proc. of ACM Int. Symp. on Mobile Ad Hoc Networking and Computing (MobiHOC)

Lin C and Tseng Y (2004) Structures for in-network moving object tracking in wireless sensor networks. In: Proc. of Broadband Wireless Networking Symp (BroadNet)

Lorincz K, Malan D, Fulford-Jonse T, Nawoj A, Clavel A, Shnayder V, Mainland G, Moulton S, and Welsh M (2004) Sensor networks for emergency response: challenges and opportunities, IEEE pervasive computing, special issue on pervasive computing for first response, Oct-Dec

Perkins C, Belding-Royer E, and Das S (2003) Ad hoc on-demand distance vector (AODV) routing. IETF RFC (3561)

Pottie G and Kaiser W (2000) Wireless integrated network sensors. Commun. ACM, 43:5, pp 51–58

Schurgers C and Srivastava M (2001) Energy efficient routing in wireless sensor networks. In Proc. of Military Communications Conf. (MILCOM)

Sohrabi K, Gao J, Ailawadhi V, and Pottie G (2000) Protocols for self-organisation of a wireless sensor network. IEEE Personal Commun., 7:5, pp 16–27

Tseng Y, Kuo S, Lee H, and Huang C (2003) Location tracking in a wireless sensor network by mobile agents and its data fusion strategies. In: Proc. of Int'l Symp. on Information Processing in Sensor Networks (IPSN)

Tseng Y, Wang Y, and Cheng K (2005) An integrated mobile surveillance and wireless sensor (iMouse) system and its detection delay analysis. In: ACM/IEEE International Symposium on Modeling, Analysis and Simulation of Wireless and Mobile Systems (MSWiM)

Tseng Y, Pan M, and Tsai Y (2006) A distributed emergency navigation algorithm for wireless sensor networks. IEEE Computer (to appear)

Xbow (2005) Crossbow technology inc. http://www.xbow.com/

Yan T, He T, and Stankovic J (2003) Differentiated surveillance for sensor networks. In: ACM Int'l Conf. on Embedded Networked Sensor Systems (SenSys)

Ye W, Heidemann J, and Estrin D (2002) An energy-efficient MAC protocol for wireless sensor networks. In Proc. of IEEE INFOCOM

ZigBee (2004) ZigBee Alliance. http://www.zigbee.org

17 MANET versus WSN

JA Garcia-Macias[1] and Javier Gomez[2]

[1]CICESE, Código Postal 22860, Ensenada, B.C. Mexico
[2]Electrical Engineering Department, National University of Mexico, Ciudad Uni versitaria, Coyoacan, C.P. 04510, D. F. Mexico

17.1 Introduction

A mobile ad-hoc network (MANET) is a self-configuring network where nodes, connected by wireless links, can move freely and thus the topology of the network changes constantly. A great amount of resources has been devoted to research in the MANET field in the past three decades; many conferences have been held, many projects have been funded, many articles have been written; however very few MANET-type applications have emerged from all this hard work. A great body of knowledge about MANETs has been produced and many researchers in the field are now trying to apply this knowledge to the field of wireless sensor networks (WSN). The reasoning is that both MANETs and WSNs are auto-configurable networks of nodes connected by wireless links, where resources are scarce, and where traditional protocols and networking algorithms are inadequate. However, as we discuss in this chapter, great care should be taken before applying algorithms, protocols, and techniques to WSNs, if they were originally developed for MANETs. Although, both types of networks indeed have many similarities, the differences are also such that WSN can arguably be considered a whole different research field.

17.2 Similarities

Probably the main reason why WSNs immediately resemble an ad hoc network is because both are distributed wireless networks (i.e., there is not a significant network infrastructure in place) and the fact that routing between two nodes may involve the use of intermediate relay nodes (also known as multihop routing). Besides, there is also the fact that both ad hoc and sensor nodes are usually battery-powered and therefore there is a big concern on minimizing power consumption. Both networks use a wireless channel placed in an unlicensed spectrum that is prone to interference by other radio technologies operating in the same frequency.

Finally, self-management is necessary because of the distributed nature of both networks. Wireless ad hoc networks were developed in the early 70's with the US military as the main customer. Three decades later when commercial applications based on ad hoc technology are finally emerging, one wonder if there is any more work to do in this field or it is enough to simply leverage all these previous research. The answer to this question is that these commercial applications are quite different from traditional military applications and therefore they require a new fresh look (Gerla 2005). Assumptions such as single purpose-application, cost unaware, large scale, and unique hardware/radio commonly given in military ad hoc networks can not be exported to emerging ad hoc nets such as disaster recovery, long-lived applications, peer-to-peer, WSNs, human context interaction, P2P etc.

Recently there is a re-emergence of ad hoc networks pushed by two confluent forces, on one hand there is a technology push resulting in smaller more powerful mobile devices, and on the other hand new types of ad hoc applications are emerging. Higher chip integration and hardware architectures optimized for low power operation tighten with new UWB and MIMO radios taking advance of wider unlicensed spectrum are creating new type of mobile devices with unseen capabilities. The key Internet paradigm that the network core should be kept simple (i.e., only care for the delivery of data packets) while the intelligence is at the edges, does not fit well in commercial ad hoc networks, which several people argue are quite different from traditional Internet or WLAN. New applications are changing the face of traditional ad hoc networks (i.e., pure routing) to networks where there is need for networking, processing, and storage everywhere in the network.

17.3 What Makes WSN Different

Although there are important similarities between WSNs and MANETs, there are also fundamental differences. Some of these differences derive from the nature of both types of networks: MANETs are usually "close" to humans, in the sense that most nodes in the network are devices that are meant to be used by human beings (e.g., laptop computers, PDAs, mobile radio terminals, etc.); conversely, sensor networks do not focus on human interaction but instead focus on interaction with the environment. Indeed, nodes in a sensor network are usually embedded in the environment to sense some phenomenon and possibly actuate upon it; this is why some people say that WSNs can be considered as a "macroscope". As a consequence, the number of nodes in sensor networks, as well as the density of deployment, can be orders of magnitude higher than in ad hoc networks; this of course involves thinking about scalability issues. The vision of seminal projects such as SmartDust (Kahn et al. 1999) contemplates networks with thousands or millions of nodes, although the largest deployment up to now had about 800 nodes.

If a network is going to be deployed in the outdoors, on the top of an active volcano, in the middle of the ocean, or in some other environment where sensor network applications typically take place, some nodes will eventually get damaged and fail. This means that the topology of the network may change dynamically, not due to node mobility like in ad hoc networks, but because some nodes will fail. In this case reconfiguration mechanisms will have to be used, so the network design should consider that nodes are prone to failure. It is worth saying that there are some applications where nodes are attached to animals, cars, or moving objects, but in the majority of applications nodes remain static, so some issues that are important in mobile networks may not be of great importance in wireless sensor networks. Besides failure, topology may also change due to the sleep-awake cycle observed in some protocols designed with sensor networks in mind. These protocols go through these cycles in order to achieve energy savings, which is one of the biggest concerns and design requirements in resource-scarce sensor networks. This scarcity of resources, again, constitutes a differentiating feature in sensor networks: nodes are typically left unattended for extended periods of time (i.e., months, years) and they are expected to operate on batteries; the range of communications is typically within a few meters and at low rates (some kilobits per second); there is typically a few kilobytes of memory and the processor may operate at speeds of only some megahertz.

It should also be pointed out that the service offered by wireless sensor networks is not simply to move bits from one place to another, but to provide answers instead. These answers should respond to questions such as: what are the regions of the network where the temperature is above the specified threshold? What is the path followed by the herd? Thus, responding to these types of questions implies taking into account geographic scopes, which is a requirement that is not needed in most other networks. Indeed, in some applications the ID (e.g., the address) of individual nodes is irrelevant and location becomes a more important attribute. In general, communication paradigms are affected by the application-specific nature of sensor networks, and we will discuss this point in more detail ahead.

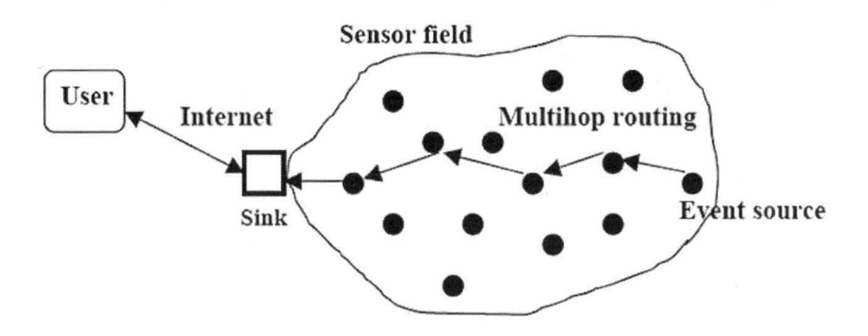

Fig. 17.1. Typical functional architecture for a WSN

17.4 Protocol Stack

A protocol stack for WSN is shown in Fig. 17.1 for illustration purposes only. This stack is similar to the one used in MANET networks (which is also the same stack used in TCP/IP networks) except for the addition of a power, mobility and task management planes that operate across all layers. While this layered approach has remained accepted and mostly untouched in MANET networks for quite a long time, most researchers find serious difficulties to adhere to it in WSN. The main arguments for this opposition are that WSN is very application-specific and resource-constrained, so a layered architecture may not be the best way to approach the wide range of applications and optimize the limited resources. In fact, a cross-layer view of WSN is becoming more and more accepted in the research community.

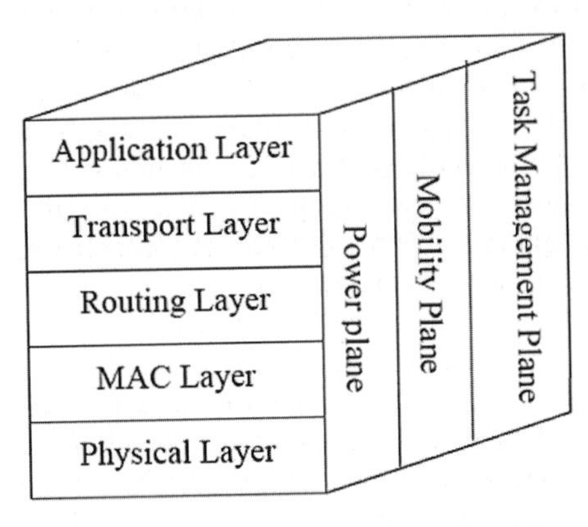

Fig. 17.2. Protocol stack for WSN

The protocol stack shown in Fig. 17.2 consists of the *physical layer, MAC layer, routing layer, transport layer and application layer*. The physical layer addresses the lower-level operations of the radio interface for a robust transmission and reception of packets in a harsh wireless environment. These operations include the frequency selection; transmit power, modulation, signal detection and coding. The MAC layer is responsible for proper channel access among competing transmitters. The MAC should avoid collisions as much as possible and turn-off the radio whenever a sensor is not actively transmitting or receiving packets in order to save energy. The routing layer is responsible for node addressing and routing in a network that is commonly multihop. Terms such as unicast and multicast common in MANETs are hardly applicable in WSN where we find other forms of routing such as one-to-many, many-to-one, many-to-many, etc. The transport layer addresses proper delivery of packets. A provision for congestion control either within the transport layer or as a separate module should be included in order to reduce the probability of network overflow. Finally, the application layer creates an abstraction of the main classes of applications found in WSN. General-purpose software associated with a given class can be reused in various applications, thus reducing prototyping time.

Across all layers in Fig. 17.2 we find the power, mobility and task management planes. The power plane emphasizes the power-awareness that should be included in each layer and across all layers in WSN. For example, a sensor may keep its radio on after sensing some activity in the channel, or it may turn it off if it is not generating any data or it does not belong to any active route. A sensor that is running low in energy may turn-off its radio and save its energy for sensing activities only. The mobility plane is responsible for maintaining the full operation of the sensor network even in the event of sensor mobility. While most sensing applications we can think of are static, we cannot discard that sooner or later mobile sensing applications will emerge. This could be the case when sensors are mounted on mobile platforms such as robots, persons, animals, cars, etc. Routes used to carry information across the network have a limited lifetime and need to be periodically repaired because of node mobility. Even without mobility, routes may change due to the fact that nodes run out of power or follow an awake/sleep duty cycle, so a route that is valid at some point in time may no longer be valid a little later (Niculescu 2005). In both cases, the routing layer is mainly responsible for route maintenance. The task management plane should be capable of coordinating all nodes toward a common objective in a power-aware manner. Some sensors in a given region, for example, may be temporarily turn-off if there is enough sensing-redundancy from other sensors in that region.

Below we present a detailed description of each layer. It should be noted that this layered-protocol approach is just a reference model commonly used in the literature, so we are using it for presentation purposes in this chapter only. However, very active attention has been paid in the research community to cross-layer approaches where layers and their functions are not as strictly defined.

In a traditional layering approach, layers in a protocol stack provide services and interact only with contiguous layers through well-defined interfaces. Thus, there is a clear separation of functions and strict boundaries are imposed between layers. Diverging from this traditional layering approach, the cross-layer approach is more flexible and allows a more intensive feedback between layers (Shakkottai et al. 2003). For instance, using a cross-layer design, adaptive modulation and coding at the physical layer can be designed considering the radio link level error control technique (e.g., ARQ) to maximize network capacity under constrained QoS requirements. Cross-layer techniques can also be developed at the application layer for wireless multimedia services which can exploit physical and radio link layer information, thus performing adaptations according to varying conditions in the lower layers. These tight interactions between different layers are very beneficial in wireless networks, but the benefits are exacerbated in MANETs and in WSNs; indeed, in these types of wireless networks resources are scarce (power, bandwidth, etc.) and should be managed very efficiently.

17.5 Physical Layer

The physical layer, as seen in traditional layered-network architectures, is responsible for the lower-level operations of the radio interface including frequency selection, transmit power, modulation, signal detection and coding. Signal detection and coding are strongly related to hardware capabilities like processor speed or memory size. For this reason, we focus on the selection of frequency, transmission power, and modulation issues.

Frequencies used in today's sensor networks include the 915 MHz and recently the higher 2.4 GHz of the industrial, scientific and medical (ISM) band, although the 310 MHz and 433 MHz bands can also be found. Lower frequencies should be preferred because of the higher signal attenuation experienced by higher frequencies. Unfortunately, the limited bandwidth spectrum available in the lower frequency bands is pushing sensor networks to higher frequencies where more bandwidth is available, allowing for higher transmission rates. Signal attenuation in wireless channels is also affected by terrain conditions. The attenuation experienced by a transmitted signal over a distance d is proportional to d^n, with $2<n<4$. Ground-lying sensor networks are likely to observe attenuations with the exponent n closer to 4 (n is equal to 2 in free-space conditions only). Higher attenuation means a higher transmit power is required in order to guarantee a proper packet reception. This results in higher energy consumption in a power-scarce sensor network. All techniques available to reduce transmission power over the wireless channel should be used in order to save energy in sensor networks, this include spatial, frequency and time diversities.

The choice of a good modulation scheme is a key factor for the correct delivery of information among sensor nodes. Different modulation schemes may differ in various aspects including the number of bits per symbol, bit error rate (BER), power efficiency, and spectrum efficiency among others. Complex modulation schemes such as M-ary are capable of transmitting several bits per symbol, but this is at the expense of a higher transmitted power and increased BER. Simpler binary modulation schemes like PSK or QPSK transmit fewer bits per symbol but require less power and are more robust against channel errors. In (Shih et al. 2001) it is shown that, considering the transmit power as the dominant factor in a sensor node, binary modulation schemes are more energy-efficient. In near future it is expected that sensors could implement some sort of adaptive modulation; allowing the radio to dynamically change the modulation that better match current channel conditions. Ultra wideband (UWB) has also been proposed for future sensor networks requiring high transmission rates. Only the baseband signal is transmitted in UWB (i.e., no carrier frequency is used), which makes it simpler to built and more resilient to attenuation and multipath effects.

Opposite to ad hoc networks where the IEEE 802.11 radio interface has become a de-facto standard for communications (and its underlying physical-layer settings), the choice of a physical layer in sensor networks can vary significantly among the different radio-hardware choices available in the market today. The new IEEE 802.15.4 standard is an effort to set a radio standard for general-purpose sensor deployments. This radio is aimed at low-power low–range communications devices that may allow for years of battery-life without replacement. This standard provides support for one-hop reliability and some basic QoS support. There are several good reasons to built sensor applications above a common radio-interface, the most important one being the possibility to recycle functionality (e.g., code, algorithms, etc.) among different applications, thus reducing deployment time and costs. A sensor radio interface produced in large quantities is also likely to be cheaper and more robust than a prototype radio. Recycling functionality is one of the main reasons behind the layered design of IP networks. Having a single common radio interface in sensor networks, although desirable, is not feasible in practice. Different sensing applications may simply need a different type of radio: consider the extreme case of passive RFID tags operating without batteries. It is likely that in the future, as it is case today, there will be several radio interfaces available to fit the wide-range of sensing applications with the underlying differences found in the physical layer of each radio.

17.6 MAC

The MAC layer is responsible for ordered channel access among competing transmitters. As previously mentioned, the IEEE 802.11 has become a de-facto standard in most MANET network deployments. The IEEE 802.11 uses Channel Sense Multiple Access (CSMA) with collision avoidance (CA). In IEEE 802.11, the CSMA/CA protocol is also known as the Distributed Coordination Function (DCF). The main goals of DCF are achieving strong connectivity among nodes and transmission fairness.

The main components of CSMA/CA are listening, backoff and collision avoidance. The listening component let potential transmitters know if the channel is occupied by an ongoing transmission in order to avoid unwanted collisions. Upon detecting the channel occupied or after a collision, a node triggers an exponential backoff algorithm to re-schedule its transmission. This mechanism has the effect of time-spreading competing transmitters, thus reducing the probability of future collisions. The collision avoidance component reduces the impact of collisions created by hidden terminals.

There are several issues why it will be inappropriate to use DCF in WSNs. First DCF follows the always-ready paradigm of the Internet. To achieve this goal MANET nodes remain in a continuous awake mode in order to be ready for either transmission or reception of packets. This always-ready operation pays a high price on power consumption, inappropriate for a power-scarce sensor node. Second, the DCF function operates better when packet births are stochastically distributed in time. This assumption is opposite to the high data correlation found in WSN where periodic streams of sensed data may be common.

Given the good knowledge (and why not popularity also) of CSMA/CA in multihop (MANET) networks, a good deal of research has focused on modifying this protocol to suit WSN requirements, in particular the power consumption issue. In this line of thinking we find the different versions of SMAC (Ye et al. 2002). The Energy Efficient MAC protocol for WSNs (SMAC) is based on a listen/sleep duty cycle specifically designed for WSN. In SMAC, a sensor node transmits SYNC packets carrying the node's listen/sleep schedule so that other nodes know exactly when they can communicate with it. SMAC schedules communications without the need for a local or global synchronization entity. Because nodes in SMAC operate with a low duty cycle (i.e., sleeping periods are much longer than listening periods), energy consumption is reduced significantly.

periods), energy consumption is reduced significantly. A node wanting to transmit a packet but knowing the intended receiver is currently in sleep mode must queue its data and wait for the next receiver's listening period, resulting in a delay. This delay is particularly onerous in multihop networks such as WSN where a packet may travel through several intermediate sensors before reaching the intended receiver or sink node.

The adaptive MAC (Ye et al. 2004) allows a node to briefly wake up in the middle of a sleep period if future activity in the MAC is predicted to occur (e.g., after a NAV timeout). Cleary adaptive S-MAC reduces the delay compared with SMAC at the expense of a slight increase in energy consumption. DSMAC (Lin et al. 2004) doubles the duty cycle for faster data transmission based on the presence of queued data and the average one-hop latency. DMAC (Lu et al. 2004) adjusts listening periods according to the traffic load and performs optimization based on a data-gathering tree-structure. DMAC reduces the long delays observed in multi-hop routes compared with fixed listen/sleep based protocols. The authors in (Sichitiu 2004) use routing information to predict future activity in the channel in order to turn on/off the radio.

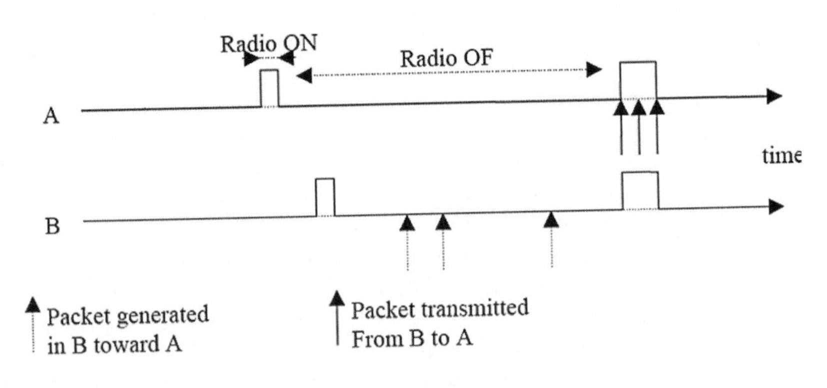

Fig. 17.3. Example of the operation of SMAC: Node B generates three packets that are destined to node A. Node B, however, must wail until the next active interval of node A before transmitting those packet

Although the previously discussed MAC protocols successfully address the power consumption issue of CSMA, they do not properly address the problem of the highly correlated data found in WSN. A hybrid TDMA/FDMA protocol for WSN proposed in (Shih et al. 2001) is an attempt to address both issues simultaneously. This MAC assumes power-constrained sensor nodes can communicate directly with a nearby-located high-powered base station. TDMA is used to accommodate a single sensor in order to minimize delays whereas FDMA is used to guarantee a minimum bandwidth to each sensor. TDMA is not always preferred in this scheme because of the costs associated with time synchronization.

The very-low-power limited-range IEEE 802.15.4 MAC deserves a special note in this category. This standard was designed for several applications including home networking, automotive networks, industrial networks, interactive toys and remote sensing. Network topologies include star and peer-to-peer using the well-known CSMA/CA channel access protocol described before and operating in the 2.4 GHz ISM band. The standard allows for two types of devices. Full function devices (FFD) can become network coordinators and can talk to any other device. Reduced function devices (RFD) are limited to star topologies and cannot become a network coordinator. RFD devices have a very simple implementation and therefore can become extremely low cost. An optional super-frame structure with contention-access and contention-free periods allows for nodes requiring guaranteed bandwidth.

17.7 Routing

Any textbook on computer networking will tell you that the address of a node is a fundamental concept to understand routing. After all, since the beginning of computer networking history, routing mechanisms have relied on knowledge of the addresses of nodes in order to establish routes between them. For instance, the Internet routing mechanism relies on IP addresses and a hierarchical structure to establish routes and to route data between nodes. Under this view, the same name is used to identify individual nodes and also to identify communication endpoints; this coupling has resulted in problems when trying to achieve mobility of nodes in IP networks (Bhagwat et al. 1996).

Mobile ad-hoc networks follow a traditional node-centric approach for routing, *i.e.*, routing relies on individual nodes and their corresponding addresses. Routing in ad-hoc networks has been classified as proactive, reactive, and hybrid, based on how the network reacts to route invalidation. With proactive routing, the network is under constant survey in order to know all possible routes between nodes at any given time; this means that routes are constantly being discovered, even if routes have not been invalidated. In contrast, reactive routing attempts to establish routes between nodes only when they are needed or when routes are no longer valid. The hybrid approach, as the name suggests, uses a mix of both proactive and reactive routing.

Due to their data-centric and application-specific nature, node-centric approaches do not constitute the best communication paradigm for sensor networks. Instead, data-centric communications are preferred (Niculescu 2005), since it is more adequate for applications where the data read from sensors is important, and not the address of specific nodes. Indeed, a typical application may be interested in knowing the regions of a field where temperature is beyond a certain threshold; here what is important are the values of temperatures read by the sensor nodes, and communications are established according to this criteria, without communicating with specific nodes by their addresses. This way, a network-wide request is issued and only those nodes whose read values satisfy the criteria respond; then a data aggregation process takes place at various points along the path from the data sources to a data-gathering node commonly referred to as a sink. Along this process, the identity (*e.g.*, address) of the involved nodes is not important, as data is forwarded and aggregated from node to node according to its value. This type of many-to-one communications, where data is sent from different sources to a sink, is sometimes called gathercast. Directed diffusion (Intanagonwiwat et al. 2000) is a typical data-centric routing scheme, where following a data request a reverse tree rooted at the sink and with leaves at the data sources is set up. The tree is called a gradient tree, where the routing entries are the gradients and data matching these gradients is forwarded from sources. A gradient reinforcement process is then performed, where the best paths are kept while others simply time out and are removed.

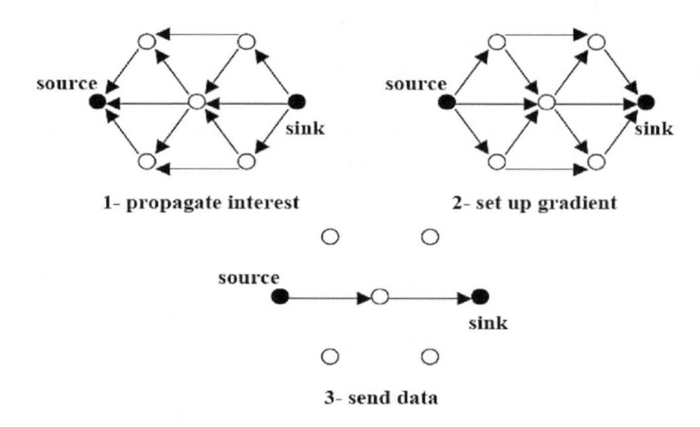

Fig. 17.4. Basic operation of the direct diffusion data-centric routing protocol

As previously stated, node mobility presents serious challenges in node-centric networks because communication endpoints and paths are tightly coupled to the names or identifiers of nodes. Being node-centric and also presenting high mobility of nodes, MANETs inherit these mobility problems. For instance, when one or more nodes move, paths involving these nodes are affected and a new path should be rediscovered by the routing mechanism (either proactively or reactively) generating important overhead. Mobility is not as important in WSNs, as typical applications do not involve moving nodes[1]. However, routes may change due to the fact that nodes follow an awake/sleep duty cycle, so a route that is valid at some point in time may no longer be valid a little later. The same consequence will be true due to the high rate of node failure that is expected in sensor networks, where not only hardware is cheap, but also is exposed to adverse conditions. Data-centric routing in sensor networks is not seriously affected by the on-off nature of individual nodes, as data is forwarded from the sources to the sink through any available nodes that match specific criteria.

[1] Although it is true that there are applications where sensor nodes change their position over time, currently the most usual application involves sensing (and probably actuating) with fixed networks.

It is also worth noting that, in both MANETs and WSNs, new routing schemes that take advantage of knowledge of the physical location of nodes are being used. Although many proposals have been made for position-based routing in MANETs (Giordano et al. 2003) (Mauve et al. 2001) not all these are adequate for sensor networks, mainly due to restrictions in power and size of sensor nodes. Ganesan (Ganesan et al. 2003)) identify at least a couple of benefits for position-based routing in wireless sensor networks:

- Sensor data is likely to be geographically correlated. Data reduction or aggregation schemes would need to route geographically to exploit such correlations.
- Queries that are geographically scoped are likely in many applications where users would prefer to query a small geographical region rather than the entire network. For instance, in a tracking application, the query is efficiently answered by querying only nodes on the trajectory of the target rather than all nodes in the network. Similarly, weather monitoring that is targeted at understanding local characteristics of data rather than global ones can be handled efficiently using geographically scoped queries.

Of course, as in the case of data-centric routing, the mobility of nodes is not a problem when using position-based routing; here it is only necessary to know the position of the endpoints and of any intermediate nodes, without the need to construct routing tables or perform routing updates.

17.8 Transport and Congestion Control

Transport protocols are yet another important area where MANET and WSN diverge significantly. MANET traditionally implement the full TCP/IP protocol stack, meaning MANET nodes will have IP addresses or something similar, support broadcast, unicast and multicast routing, and more important, be fully compatible with UDP/TCP transport protocols. Some researchers argue there is a need for native TCP support in WSN also. This way the WSN can be directly connected to an outside network without the need for special proxy servers or protocol converters. Bringing TCP/IP to wireless sensor networks is a difficult task, however. First, because of their limited physical size and low cost, sensors are severely constrained in terms of memory and processing power. Traditionally, these constraints have been considered too limiting for a sensor to be able to use the TCP/IP protocols. Second, the harsh communication conditions make TCP/IP perform poorly in terms of both throughput and energy efficiency. Sensor networks may exhibit higher packets losses (2% to 30%) compared to ad hoc networks. A good deal of research has been devoted to improving TCP performance in MANET in the past decade. Although the main perceived trend in the research community is not to consider the use of TCP in WSN due to the issues presented before, there are few researchers who argue there should be some TCP support in WSN.

The common trend in the research community is that transport protocols in WSN should refrain from copying TCP ideas. Most sensor applications are event-driven and therefore do not need a reliable transport protocol such as TCP. They will be likely optimized for a particular task/operation and may tune their transport protocol to suit specific requirements.

Some researchers argue, however, that even if not related to TCP, there is some need for reliable transport in WSNs in the near term. They argue future WSN may become general-purpose sensor platforms to some extend, requiring the ability to reprogram the functionality of the sensor network periodically. Reprogramming the sensor network necessarily requires a reliable transport protocol. An example of this way of thinking is the PSFQ protocol (Wan et al. 2002). PFSQ (Pump Slowly and Fetch Quickly) recovers from losses locally and avoids using end-to-end ACK messages. This results in minimum signaling involved for loss detection and recovery. When a packet is lost in PSFQ, the packet is retransmitted locally while copies of received packets with higher sequence numbers are buffered and transmitted only until successful retransmission of the lost packet occurs. Another protocol in this category is the Reliable Data Transport in Sensor Networks (RSMT) (Stann et al. 2003). This protocol operates as a filter within the directed diffusion stack (Intanagonwiwat et al. 2000). Reliability in RSMT refers to the delivery of all fragments of a large packet (called entity) to all the subscribing sinks in a WSN.

The ability of controlling the rate of transmitted packets (i.e., congestion control) to match the available bandwidth in the network has always been a primary source of concern in packet networks. Congestion control can be implemented end-to-end as a part of the transport protocol (e.g., TCP) or as a separate protocol. TCP implements congestion control by means of a sliding window that grows slowly when no packet losses are detected, and decreases fast when packet losses do occur. In this way TCP attempts to transmit information between end points as soon as possible (e.g., files, web pages, etc.) without overloading the network. Opposite to TCP, a UDP connection does not have any congestion control provisioning. Without congestion control, UDP packets can easily overload the MANET network, possibly disrupting other connections including TCP sessions. An example of a congestion control mechanism for MANET with both TCP and UDP traffic is presented in SWAN (Ahn et al. 2002). SWAN uses rate control for UDP and TCP packets and sender-based admission control for UDP real-time traffic. SWAN uses explicit congestion notification (ECN) to dynamically regulate admitted real-time traffic in the face of network congestion.

Event-driven WSN suffer from a different source of congestion, here an idle or lightly loaded sensor network may suddenly become active in response to a detected or monitored event. Transport of these events to the sink points may result in sudden congestion in the network depending on the sensing application. It is during this period of activity in the network that the probability of congestion is greatest and the importance of the monitored information most significant. An example of a congestion control algorithm for WSN is CODA (Wan et al. 2003). This protocol uses two complementary congestion control techniques. First there is an open loop hop-by-hop backpressure mechanism that signals nodes upstream (from the congested node toward the source) to reduce their pace of forwarded packets (e.g., drop packets). Second, there is a closed-loop multi source regulation to specifically tell sources to slow down their transmit rate. The ESRT (Sankarasubraniam et al. 2003) is an event-to-sink reliable transport protocol, which also implements congestion control. In ESRT, any forwarding node experiencing buffer overflow sets the congestion flag on in each forwarded packet. Upon reception of packets with the congestion flag set, the sink node signals all sources to slow down the transmission rate by using a high power transmission.

17.9 QoS Issues

In the beginning of wireless packet networks in the 70s nobody cared much about QoS as long as nodes in the network were connected to each other some how. Transmission of data packets was then the dominant type of traffic, and a best-effort delivery by the network was considered good enough. In the multimedia world we are immersed now, connectivity alone can not guarantee that the different types of media will be properly delivered by the network. Applications such as VoIP, real-time video, etc. require tight bandwidth, jitter and delay guarantees to work properly.

Providing QoS in ad hoc networks is quite complex and has become one big obstacle in the deployment of commercial ad hoc networks. This situation is mainly due to the poor end-to-end channel utilization found in current ad hoc networks based on IEEE 802.11 technology. For example, considering mobility and hidden terminals only, measured end-to-end channel utilization can get below 18% even for routes with few hops (Garcia-Luna-Aceves 2005). Forwarding over a common channel and packet header overhead can bring this utilization down to 1% even with RTS-CTS in place. The picture is even worse if we consider that data and control packets share this 1%, indistinctly. For routes with several hops and high node mobility the end-to-end channel utilization approaches zero. The poor QoS performance shown by current ad hoc networks is making several researchers to rethink how ad hoc networks should be built. J. J Garcia-Luna-Aceves (Garcia-Luna-Aceves 2005) identified some key factors that may need a fresh look from the community in order to improve QoS:

- *Traditional packet switching.* Current ad hoc networks do not make any distinction about how different types of packets are handled by the network. In order to guarantee bandwidth and delay constrains for real-time applications, it is necessary to distinguish the way packets are queued and forwarded by each node. Soft-state approaches and switching flows of packets rather than packets in isolation are promising approaches to be tested.
- *End to end connectivity.* The famous end-to-end Internet paradigm assumes that the network can connect any pair of nodes transparently. The presence of obstacles and network partitions make impossible to guarantee end-to-end connectivity in ad hoc networks. New directions in this area call for use of

storage, processing, and communication resources opportunistically in order to live with network disruption.

- *Resource allocation.* Current ad hoc networks use a common channel that is shared by all nodes. This approach results in high levels of interference and low channel utilization. New trends to improve performance in this area consider the use of several channels. This strategy has several advantages over the common-channel approach beyond having an increased network capacity. Separating which applications are allowed to use a given channel reduces interference and can provide some coarse QoS control. Similarly, a communication that failed in one channel can be attempted in a different channel providing a richer connectivity.
- *One-to-one competitive communications.* Probably the main culprit of the poor QoS performance shown by ad hoc networks is the use of WLAN technology. New directions in this area call for new radios and new communications models. New types of radios that could exploit multi-user detection and equipped with directional antennas could improve performance in case only one channel is available (Bao et al. 2002). Current communication models are based on competition-driven approaches that try to fight interference. Because in most cases a common channel is used, any transmission creates interference almost everywhere in the network, leading to scaling problem (Gupta et al. 2000). A better communication model could be, for example, the one proposed by Grossglauger (Glossglauger et al. 2001) where information is delivered taking advantage of node mobility.

Although there has been some research on different aspects of QoS (mainly QoS routing) for MANETs, little has been done in the field of WSN. This may be due to the fact that sensor networks are very resource-constrained, thus providing not only any kind of service, but service with quality guarantees, poses an extremely complex problem.

But, the question arises: is there a need for QoS in WSNs? Clearly, traditional monitoring and control applications (*e.g.*, greenhouse temperature control) do not require strict observation of common QoS parameters such as bandwidth, delay and jitter. Real-time applications may not require bandwidth guarantees but certainly will need temporal guarantees, *i.e.*, delay and jitter. Some recent applications involve audio and/or video traffic, so bandwidth along with temporal guarantees may be needed; take for instance a recently presented application (Rahimi et al. 2005) where sensor nodes (called cyclops) equipped with a tiny camera provide a network for image sensing and interpretation. Younis et al. (Younis et al. 2004) identify important design considerations for handling QoS traffic in wireless sensor networks:

- *Bandwidth limitation.* Applications may generate both real-time and non real-time traffic, so using the limited available bandwidth to accommodate both may result difficult to say the least. The traditional approach of reserving bandwidth for QoS traffic is simply unacceptable in WSNs.

- *Removal of redundancy.* Data fusion, data aggregation, and many other data-handling techniques are common in sensor networks; these take advantage of the fact that many applications generate considerable amounts of redundant data. However, these techniques can not be readily applied to QoS traffic (*e.g.*, audio, video) that require more complex manipulations, which in turn generate more processing overhead that would deplete energy supplies.
- *Energy and delay trade-off.* Multihop transmission is one aspect that helps WSNs reduce energy consumption, at the cost of delaying the delivery of packets. An important element in this cumulative delay will be the time for queuing and classifying packets that handling QoS traffic will require. Thus, the energy and delay trade-off commonly present in WSNs is only exacerbated when QoS traffic is introduced.
- *Buffer size limitation.* The buffers required for routing QoS traffic may suffer the same fate of other resources: scarcity. Not having adequate buffer sizes would complicate classification, introduces delays, and generally would reduce the possibility of granting QoS guarantees. The introduced delays would also have a negative impact on medium access scheduling.
- *Support of multiple traffic types.* Currently emerging sensor network applications are increasingly complex as they involve not only monitoring temperature, light, and other similar parameters, but also transmitting audio/video, tracking objects, etc. Consequently, managing such diversity of traffic implies handling different data rates, different QoS constraints, and multiple data delivery models. This heterogeneity raises the challenges for providing QoS, as routing becomes more complex and more exhaustive processing is needed.

Although some proposals for QoS routing (Akkaya et al. 2004) (He et al. 2003) and providing adequate MAC support for QoS (Lu et al. 2002) have been made, the issue of providing QoS in WSNs remains largely an open issue.

17.10 Application Issues

An oft-cited distinctive feature of wireless sensor networks is that they are very application-specific. The applications of these networks are becoming increasingly sophisticated, but some common classes of applications can be identified. The most common class involves communicating sensed data from the sensor field to a sink node; some other applications do some level of in-network processing, and more complex ones involve multiple kinds of distributed interaction and communication. Of course, all this diversity of applications poses a diverse set of requirements (*e.g.*, sensor field to sink vs. sensor to sensor communications, long-lived vs. ephemeral data streams, etc.) that have a great impact on the architecture, algorithms, and protocols of the network. For instance, a routing algorithm (or MAC or transport protocol, for that matter) that is suited for an environmental-monitoring application, where data is read at specific time intervals, may be almost useless for an intelligent road application where automobiles should be constantly informed of road conditions, presence of other vehicles, etc.

The address of a node is a fundamental element for communications in "traditional" networks. Indeed, at the core of IP-based network is the concept of the network address used to identify nodes and endpoint communication entities (MANETs are usually IP-based). Contrary to these node-centric networks, sensor networks are data-centric (Niculescu 2005) since the identifier of individual nodes is not really as important as the data gathered by sets of nodes is. As nodes in the network will frequently and randomly turn on and off, due to reduced duty cycles, it would be inefficient to base communications on constantly changing endpoint identifiers; instead, communications should be oriented toward the actual data gathered in certain regions of the network. The ultimate goal of most sensor networks is to answer to requests of the type "obtain the data that satisfies this (or these) condition(s)". In order to answer these queries, the identities of the nodes satisfying the given conditions is not known, and it does not really matter, so network wide discovery should be used in order to find the nodes that have the needed data. As an analogy, responses to queries in a database do not need to include the addresses of the records satisfying the queries, as only the actual data therein will suffice. In fact, the similarity with databases has generated much research efforts in the sensor network community.

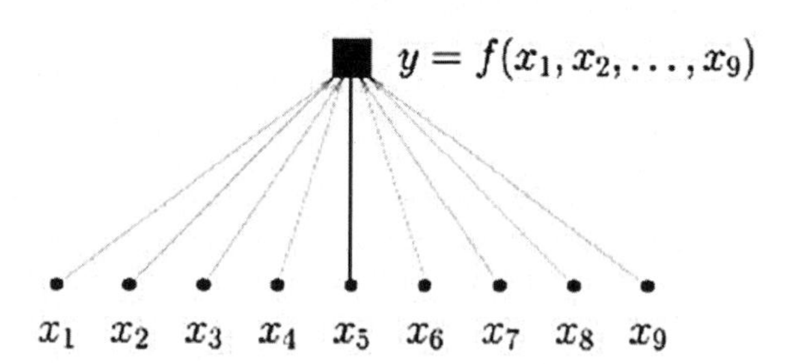

Fig. 17.5. With data aggregation, the result (y) is determined by an aggregator node and a function (f) taking as inputs data sensed by a set of nodes ($x_1,\ldots,x_9$)

Given that sensor networks are best designed in a data-centric manner, several research groups have explored a novel view of the sensor network as a database (Govindan et al. 2002). In most projects adopting this view, an SQL-like language is used for querying the network and sensor data is considered as a single table with one column per sensor type. In the TinyDB project (Madden et al. 2003), each sensor node has its own query processor, while other projects such as Cougar (Bonnet et al. 2001) perform query processing in a database front-end, leaving only some basic functions to the sensor nodes. Another point of concern is how to efficiently handle the flow of data from sources to sinks, taking into account that communication activities take a heavy toll on available energy. Observing that when a given phenomena occurs several sensors in a region will likely have similar or redundant data, techniques involving in-network filtering and processing have been proposed; data aggregation is one of the most widely used techniques, and the main idea is to combine the data coming from different sources enroute, thus eliminating redundancy, minimizing the number of transmissions and saving energy (Krishnamachari et al. 2002).

As sensor networks grow, it becomes increasingly important to raise the level of abstraction for programmers. The sensor network as a database paradigm, although providing a good abstraction for some applications, has already shown important limitations; as some have pointed out, real-world data issues such as probabilistic availability of data, various levels of confidence in data, and missing or late data, can make this paradigm insufficient. Another approach in providing appropriate abstractions is to use a middleware, which is the software that resides between the applications and the underlying operating systems and networks. Middleware systems should provide reusable services that can be composed, configured, and deployed for the rapid creation of networked robust applications. Although distributed middleware (e.g., CORBA, DCOM, etc.) have been in use for a long time, they are not suitable for WSNs, due to the fact that they demand a lot of memory, computational power, and other resources. Middlewares for sensor networks should be simple, easy to implement, and lightweight, and they also have to take into account the unique operating modes that make WSN different from traditional networks, including ad-hoc deployments, flexible operation and dynamic operating environments (Krishnamachari et al. 2004) (Gerla 2005).

Besides providing higher-level abstractions, it is important to also provide programming mechanisms that scale to the foreseen size of future sensor networks. Current sensor networks are programmed node by node ("manually"), using low-level programming languages, interfacing directly to the network and the hardware via primitive operating system constructs; this is of course a cumbersome and error-prone method that can not scale to networks of hundreds, thousands, or even million of nodes. Over-the-air programming techniques, in which programs are sent to the nodes, have been proposed in order to solve the problem of programming large networks. Some of these techniques involve novel operating systems (Han et al. 2005) where modules can be inserted or extracted dynamically, while others take an approach of having a virtual machine in every node to interpret code sent to them (Levis et al. 2002) (Levis et al. 2005).

17.11 Network Design

Traditional network design requires careful engineering to determine the right topology, conduct appropriate network dimensioning, test typical network performance, etc. When mobile ad-hoc networks appeared, network design requirements had to be revisited, as fixed topologies could not be assumed, network dimensioning could not be precisely performed due to the dynamic nature of the network itself, and additionally, new requirements had to be introduced including energy consumption considerations. This situation was exacerbated with the introduction of wireless sensor networks. Indeed, as Römer and Mattern point out (Römer et al.

2004), there are several dimensions through the design space of WSNs that should be closely examined; some of them include:

- *Deployment*. Typical applications for WSN mentioned in the literature include dropping sensor nodes off an airplane for military purposes and installing sensor nodes in fields for agricultural monitoring. Thus, the diversity of applications implies that some networks will have a pre-designed topology while others will have nodes randomly placed. Also, some networks will remain fixed once their nodes are in place, while others will change as nodes are added, removed, or replaced. All these factors have implications on the density of the network, the degree of network dynamics, the available links and routing hops, etc.
- *Mobility*. Once sensor nodes are dropped off a plane, or deliberately placed in selected locations, the most common situation is that they remain in their place for the rest of their lifetime. However, some applications require placing these nodes on buoys at the sea, inside automobiles, or attached to some other moving entities; nodes may even have their own mobility means. Either way, mobility is a factor that should be considered in these cases, as it will affect the design of communication protocols and distributed algorithms.
- *Node features*. In order for sensor networks to be practical, they have to be economical, operate unattended for long periods of time, and be sufficiently powerful. Achieving these goals involves sometimes-conflicting requirements, as making a node more powerful normally has an impact on the size, energy consumption, and cost. There are currently a great variety of wireless sensor nodes, ranging from the millimeter-scale ones of the Smart Dust project to brick-sized nodes found in some environmental monitoring applications. Although the traditional approach is to have very homogeneous sensor networks, with the increasing diversity of node types these networks are also becoming increasingly more heterogeneous; as a result the networks are becoming more complex, including the software executed on them and the management of whole systems.
- *Network size and coverage*. As sensor networks grow from current prototypes of tens or hundreds of nodes, to the envisioned ubiquitous networks of millions of nodes, so grows the scalability requirements of their algorithms and protocols. The geographic coverage, combined with the number of nodes determine the density of the network. High-density networks will obviously be more expensive than sparse ones, but may result in more accurate sensing and involve more sophisticated data-processing algorithms.

With these and other designs considerations (Römer et al. 2004), it should be clear that WSNs have very distinctive characteristics that imply particular considerations not commonly present in other types of networks. MANETs, although requiring similar design considerations (e.g., energy-saving, mobility, etc), normally present less stringent requirements, as they are by definition spontaneous, short-lived networks with more powerful nodes.

17.12 References

Ahn GS, Campbell AT, Veres A, and Sun LH (2002) SWAN: Service differentiation in stateless wireless ad hoc networks. Proc. of IEEE INFOCOM, New York

Akkaya K and Younis M (2004) Energy-aware routing of delay-constrained data in wireless sensor networks. Journal of communication systems, vol. 17, pp 663-687

Bao L and Garcia-Luna-Aceves JJ (2002) Transmission scheduling in ad hoc networks with directional antennas. Proc. ACM MobiCom

Bhagwat P, Perkins C, Tripathi SK (1996) Network layer mobility: an architecture and survey. IEEE personal communication, vol 3(3)

Bonnet P, Gehrke J, and Seshadri P (2001) Towards sensor database systems. Proc. of 2nd int. conf. on mobile data management, Hong Kong, pp 3-14

Ganesan D, Cerpa A, Yu Y, Ye W, Zhao J, and Estrin D (2003) Networking issues in sensor networks. Journal of parallel and distributed computing Elsevier Publishers

Garcia-Luna-Aceves JJ (2005) Another look at ad hoc wireless networks. Keynote presentation, 4th International conference on ad hoc networks and wireless (AdhocNOW), Cancun, Mexico

Gerla M (2005) From battlefields to urban grids: new research challenges in ad hoc wireless networks. Pervasive and mobile computing 1, pp 77-93

Giordano S, Stojmenovic I, and Blazevic L (2003) Position based routing algorithms for ad hoc networks: a taxonomy. Ad hoc wireless networking, Kluwer

Glossglauger M and Tse D (2001) Mobility increases the capacity of ad hoc wireless networks. Proc. of INFOCOM

Govindan R, Hellerstein JM, Hong W, Madden S, Franklin M, and Shenker S (2002) The sensor network as a database. USC technical report No. 02-771

Gupta P and Kumar R (2000) The capacity of wireless networks. IEEE trans. info. theory, vol. 46, no. 2, pp 388-404

Han C, Kumar R, Shea R, Kohler E, and Srivastava M (2005) A dynamic operating system for sensor nodes. Proc. of 3rd int. conf. on mobile systems, applications, and services. Seattle

He TJ, Stankovic A, Lu C, and Abdelzaher T (2003) SPEED: a stateless protocol for real-time communication in sensor networks. Proc. of intl. conf. distributed computing systems, Providence

Intanagonwiwat C, Govindan R, and Estrin D (2000) Directed diffusion: a scalable and robust communication paradigm for sensor networks. Proc. of MOBICOM, Boston

Kahn JM, Katz RH, Pister KJ (1999) Next century challenges: mobile support for smart dust. Proc. ACM MOBICOM, Seattle

Krishnamachari B, Estrin D, and Wicker SB (2002) The impact of data aggregation in wireless sensor networks. Proc. of 22nd int. conf. on distributed computing systems, Washington, pp 575-578

Krishnamachari Y, Yu B, and Prasanna VK (2004) Issues in designing middleware for wireless sensor networks. IEEE Network Magazine

Levis P and Culler D (2002) Maté: a tiny virtual machine for sensor networks. Proc. of tenth international conference on architectural support for programming languages and operating systems. San Jose, pp 85– 95

Levis P, Gay D, and Culler D (2005) Active sensor networks. Proc. of second symposium on networked systems design and implementation, Boston

Lin P, Qiao C, and Wang X (2004) Medium access control with a dynamic duty cycle for sensor networks. Proc of IEEE WCNC

Lu C, Blum BM, Abdelzaher TF, Stankovic JA, and He T (2002) RAP: a real-time communication architecture for large-scale wireless sensor networks. IEEE real-time and embedded technology and applications symposium, San Jose

Lu G, Krishnamachari B, and Raghavendra CS (2004) An adaptive energy-efficient and low-latency MAC for data gathering in wireless sensor networks. IEEE IPDPS

Madden S, Franklin MJ, Hellerstein JM, Hong W (2003) The design of an acquisitional query processor for sensor networks. Proc. of ACM SIGMOD, pp 491-502

Mauve M, Widmer J, and Hartenstein H (2001) A survey on position-based routing in mobile ad hoc networks. IEEE network magazine, 15(6)

Niculescu D (2005) Communications paradigms for sensor networks. IEEE communications, pp 116-122

Rahimi M, Baer R, Iroezi O, Garcia J, Warrior J, Estrin D, and Srivastava M (2005) Cyclops: in situ image sensing and interpretation in wireless sensor networks. Proc. of ACM conference on embedded networked sensor systems, San Diego

Römer K and Mattern F (2004) The design space of wireless sensor networks. IEEE wireless communications, vol. 11 No. 6, pp 54-61

Sankarasubraniam Y, Akan O, and Akyildiz I (2003) Event-to-sink reliable transport in wireless sensor networks. Proc. of the 4th ACM symposium on mobile ad hoc netowrking & computing (Mobihoc), Maryland

Shakkottai S, Rappaport TS, ad Karlsson PC (2003) Cross-layer design for wireless networks. IEEE commun. mag., vol. 41, no. 10, pp 74-80

Shih E *et al* (2001) Physical layer driven protocol and algorithm design for energy-efficient wireless sensor networks. Proc. of ACM MobiCom, Rome

Sichitiu ML (2004) Cross-layer scheduling for power efficiency in wireless sensor networks. Proc. of IEEE INFOCOM, Hong Kong

Stann F and Heidemann J (2003) RMST: reliable transport in sensor networks. Proc. of international workshop on sensor net protocols and applications, Anchorage

Wan CY, Campbell AT, and Krishnamurthy L (2002) PSFQ: a reliable transport protocol for wireless sensor networks. Proc. of 1st ACM international workshop on wireless sensor networks and applications, Atlanta

Wan CY, Eisenman SB, and Campbell AT (2003) CODA: congestion detection and avoidance in sensor networks. Proc. of first ACM conference on embedded networked sensor systems, Los Angeles

Ye W, Heidemann J, and Estrin D (2002) An energy-efficient MAC protocol for wireless sensor networks. Proc. of IEEE INFOCOM, New York

Ye W, Heidemann J, and Estrin D (2004) Medium access control with coordinated adaptive sleeping for wireless sensor networks. Proc. of IEEE/ACM transactions on networking

Younis M, Akayya K, Eltowiessy M, and Wadaa A (2004) On handling QoS traffic in wireless sensor networks. Proc. of intl. conf. on system sciences, Hawaii

18 Industrial Sensor and Actuator Busses

N. P. Mahalik

Department of Mechatronics, Gwangju Institute of Science and Technology, 1 Oryong dong, Buk gu, Gwangju, 500 712, Republic of South Korea

18.1 Introduction

In recent years, industrial automation and control systems have been preferred in the implementation of Distributed Control Systems (DCS) instead of centralised systems, because of their advantage of great flexibility over the whole operating range. Other benefits include low implementation cost, easy maintenance, configurability, scalability, modularity, and extendibility. Conventional centralised control is characterised by a central processing unit that communicates with the field devices (sensors, actuators, switches, valves, drives etc) with separate (parallel link) individual point-to-point links. On the other hand, DCS interconnects devices with a single serial link, as illustrated in Fig. 18.1.

Since I/O (Input/Output) points within the industrial control systems are distributed, and the number of auxiliary components and equipments are progressively increasing, DCS architecture is seen to be appropriate. Each I/O point can be defined in terms of smart device. Potential operational benefits of adopting DCS schemes can be summarised. These include:

- Sharing of the processing load to avoid the bottleneck of a single centralised controller
- Replacement of complex point-to-point wiring harnesses with control networks to reduce weight and assembly costs
- Freedom to vary the number and type of control nodes on a particular application in order to easily modify its functionality
- Ability to individually configure and test segments (components) of the system before they are combined
- Build and test each intelligent sub-unit separately
- Provisions for interfacing for data exchange between the run-time control system and other factory/business systems (e.g., management information, remote monitoring and control, etc.)

DCS can be leveled into four layers of automation services. The bottom layer called the component level, including the physical components such as intelligent devices (e.g., PC, industrial PC, PLC, microprocessors, micro-controllers, etc.), and non-intelligent devices such as sensors, actuators, switches, A/D, D/A, ports, transceivers, communication media, etc. The interface layer is similar to a MAC sub-layer of the link layer protocol. The process layer includes application layer features. Since control systems do not transfer raw data through a physical media an application layer has to exist. The application layer defines the variables, which are responsible for transferring data from one place to other when they are logically connected. Additionally, the management layer generates object code from the source code by the use of resources and methods such as compilers and OLE/DDE (Object Linking and Embedding/Dynamic Data Exchange), respectively. The management layer manages the control design. There is a set of generic functions within the management layers, which are accountable for providing services for all aspects of control management. Typical management functions are installation, configuration, setting, resetting, monitoring, operator interfacing, testing, and so on.

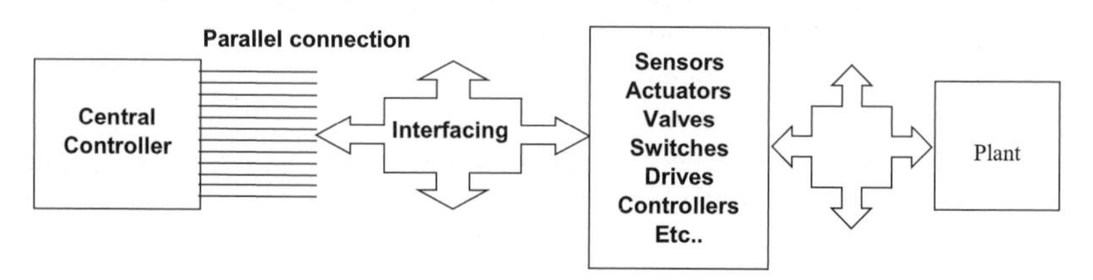

A schematic diagram of Centralised Control

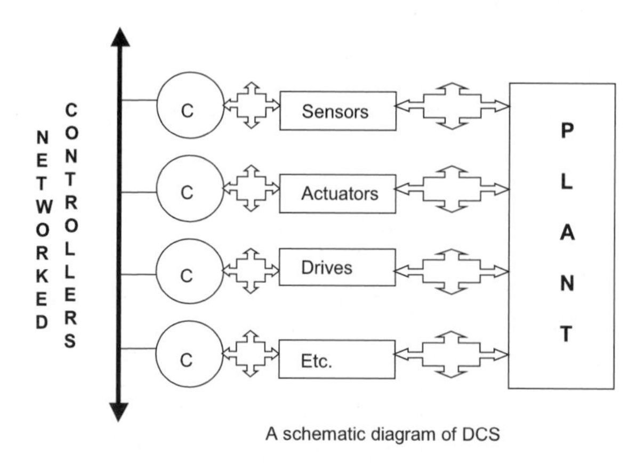

A schematic diagram of DCS

Fig. 18.1. Centralised versus DCS

18.2 Fieldbuses and their Applications

The industrial control systems have a generic set of essential requirements in terms of co-ordination, synchronisation, acknowledgement, timing etc. In addition to achieving the required sequence of operations it is vital that operator information requirements are also satisfied. The control system should also satisfy the features such as, interlock checking time out checking status display (At each step in the sequence, an application relevant message can be displaced to the operator), and error messaging (Each step in the sequence has an associated list of messages to display when any error has been detected) (Harrison 1996)

Fieldbus is a technology for the realisation of DCS. Fieldbus can be classified based on topology, processing power, type, and speed (Fig. 18.2 and Table 18.1). This technology is being developed for most of the industrial automation and control applications. The leading fieldbuses with their characteristics can be seen in (Mahalik and Lee 2001). The technology includes a protocol, which can transceive digital plant data through multiple media and channels. Recently, many fieldbuses are available in the technology market place and their application areas vary. For industrial control applications, the type of bus used is described as a *sensor-bus* or *device-bus*, usually deals with complex devices such as motors, drives, valves, etc. providing process variable information. In summary, fieldbus provides a good basis for distributed real-time control systems. It has been successfully tried and tested in many applications including, wide range of,

- Industrial applications (food processing, conveyor system automation, packaging plant automation etc.)
- Ship automation
- FMS
- Monitoring and control in hazardous area
- Flight simulator
- Building automation
- SCADA systems (automotive power distribution applications)
- Deterministic applications (robotics, air craft, space)
- Rehabilitation, and so on.

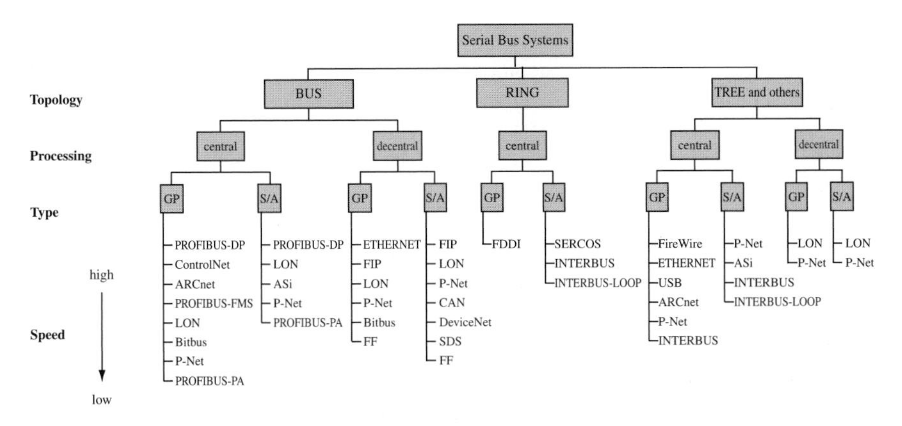

Fig. 18.2. Fieldbus classification (Source and Copyright (1992-2006): STEINHOFF Automation & Fieldbus-Systems; Courtesy: STEINHOFF A)

Table 18.1: Abstract-level fieldbus comparison (Source and Copyright (1992-2006): STEINHOFF Automation & Fieldbus-Systems)(http://www.steinhoff.de/fb_comp.htm)

Busses	Master	Topology	Max. Segment Length	Max. Speed	Wire	Max data/PDU	Standard	URL
ASI	single	bus tree	100m	167kb/s	2	4 bits	EN50295	1
BITBUS	multi	bus	300m @ 75kb/s	375kb/s	2	248-bytes	IEEE1118	2
CAN	multi	bus	500m @125kb/s 40m @ 1Mb/s	1Mb/s	2	8 bytes	ISO11898 ISO11519	3
ControlNet	multi	bus star tree	5km 250m/48 nodes	5Mb/s	coax	510 bytes	open specified	4
DeviceNet	multi	bus	500m @ 125kb/s 100m @ 500kb/s	500kb/s	4	8 bytes	open specified	5
Foundation Fieldbus	multi	bus	2000m 9.5km total	31.25kb/s	2	246 bytes	open specified	6
FIP	multi	bus	2000m @ 1Mb/s	2.5Mb/s	2	32 bytes	EN50170	7
INTERBUS	single	ring	12.8km	500kb/s	2/8	64 bytes	EN50253	8
LON	multi	bus tree	6.1km @ 5kb/s	1.2Mb/s	2	228 bytes	ANSI	9
Modbus plus	multi	bus	1.8km	1Mb/s	2	32 bytes	proprietary	10
P-Net	multi	bus tree	1200m	76.8kb/s	2	56 bytes	EN50170	11
PROFIBUS FMS	multi	bus	19.2km @ 9.6kb/s 200m @ 500kb/s	500kb/s	2	246 bytes	EN50170	12
PROFIBUS DP	multi	bus	1km @12Mb/s (4 repeater)	12Mb/s	2	246 bytes	EN50170	13
PROFIBUS PA	single	bus	1.9km	93.75kb/s	2	246 bytes	EN50170	14
SERCOS	single	ring	250m	16Mb/s	2/fiber	16 bytes	IEC61491	15
Seriplex	single	bus	1000 feet	~250kb/s	4	32 bytes	proprietary	16
SwiftNet	multi	bus	360m	5Mb/s	2	896 bytes	proprietary	17

[1]http://www.as-interface.com; [2]http://www.bitbus.org, [3]http://www.can-cia.de; [4]http://www.controlnet.org; [5]http://www.odva.org; [6]http://www.fieldbus.org; [7]http://www.worldfip.org; [8]http://www.interbusclub.com; [9]http://www.echelon.com; [10]http://www.modicon.com; [11]http://www.p-net.dk; [12]http://www.profibus.com; [13]http://www.profibus.com; [14]http://www.profibus.com; [15]http://www.sercos.org; [17]http://www.seriplex.org; [18]http://www.shipstar.com

18.3 OSI RM for Fieldbus Protocol (Murugesan 2003)

The Open Systems Interconnect Reference Model (OSI RM), for modular networking architecture, is based on the International Organisation for Standardisation (ISO) model for computer networking. OSI RM has seven layers, one layer responsible for each discrete aspect of the communication process. The model describes the flow of data in a network, from the bottom-most layer (# 1 layer for physical connections) up to the top-most layer (# 7 layer containing the user's applications). A set of data stream travelling on a network is passed from one layer to another. Each layer is able to communicate with the layer immediately above it and also the layer immediately below it. When a layer receives a packet of information, it checks the destination address, and when it is not meant for it, the layer passes the packet to the next layer. The layered diagram of the OSI RM is depicted in Fig. 18.3.

- Physical Layer: This layer transmits bit streams from one device to another and regulates the transmission of a stream of bits over a physical medium. It defines how the medium is attached to the network adapter and the transmission technique used to send the data over it. The Physical Layer defines the physical medium itself. All media are functionally equivalent. The main difference is in convenience and cost of installation and of course maintenance. Converters from one media to another operate at this level. Data are represented as electronic bits, 0's and 1's. Data travel using specific transmission devices and media. At this layer, the concern is on *how* the message travels.
- Data Link Layer: This layer packages raw bits of data from the Physical layer into frames (logical, structured packets for data). After sending a frame, it waits for acknowledgment from the receiving device. The Data Link Layer defines the format of data on the network. A network data frame includes data packet checksum, source and destination address. The largest packet that can be sent through a data link layer defines the Maximum Transmission Unit (MTU). The Data Link Layer handles the physical and logical connections to the packet's destination, using a network interface. A host with an Ethernet interface to handle connections with the outside world would have a loop back interface to send packets to itself. Ethernet addresses a host using a unique, 48-bit address called its Ethernet address or Media Access Control (MAC) address. MAC addresses are usually represented as six colon-separated pairs of hex digits, e.g., 6:0:28:13:ac:82. This number is unique and is associated with a particular Ethernet device. Hosts with multiple network interfaces should use the same MAC address on each. The data link layer's protocol-specific header specifies the MAC address of the packet's source and destination. When a packet is sent to all hosts (broadcast), then a special MAC address (ff:ff:ff:ff:ff:ff) is used. At this layer, the concern is on how the message looks. As the first point for logical organisation, this layer formats electronic bits into logical chunks of data, as frames. A frame is a contiguous series of data with a common purpose. Data link framing is accomplished using standard field definitions. Working together, these fields establish a standard frame format called the synchronous data link control (SDLC).
- Network Layer: This layer addresses messages and translates logical addresses and names into physical addresses. This layer determines the route from the source to the destination device and manages traffic problems, such

as switching and routing, to control traffic congestion. An Internet Protocol (IP) is used as the network layer interface. The IP is responsible for routing and directing datagrams from one network to another. The network layer may break large datagrams, larger than MTU, into smaller packets. The host receiving such packets will have to reassemble the fragmented datagram. The IP identifies each host with a 32-bit IP address. IP addresses are written as four dot-separated decimal numbers between 0 and 255, e.g., 119.65.11.24. The leading 1-3 bytes of the IP identify the network and the remaining bytes identify the host on that network. For large sites, usually sub-netted by major organisations or universities, the first two bytes represent the network portion of the IP, and the third and fourth bytes identify the subnet and host, respectively. Though the IP packets are addressed using IP addresses, hardware addresses must be used to actually transport data from one host to another. The Address Resolution Protocol (ARP) is used to map an IP address to its hardware address. The Network Layer provides logical LAN-to-LAN communications by organising data link frames into datagrams. The data field from the Data Link Layer becomes the entire packet at the Network layer. This packet is called a *datagram*. It consists of a data field with header/footer info. The header/footer fields contain network addresses, routing information and flow control. The fields perform three functions, internetworking, routing and network control.

- Transport Layer: This layer handles error recognition and data recovery. It also repackages long messages when necessary into small packets for transmission and, at the receiving end, rebuilds packets into their original form. The receiving Transport Layer also sends an acknowledgment for receipts. The Transport Layer subdivides the user-buffer into network-buffer-sized datagrams and enforces the desired transmission control. Both transport protocols, Transmission Control Protocol (TCP) and User Datagram Protocol (UDP), could occupy the transport layer. TCP establishes connections between two hosts on the network through 'sockets' which are determined by the IP address and port number. TCP keeps track of the packet delivery order and the packets that must be resent. Maintaining this information for each connection makes TCP a state-full protocol. UDP on the other hand provides a low overhead transmission service, but with less error checking and is stateless. Statelessness simplifies the crash recovery. Reliability and speed are the primary factors differentiating these two protocols. The Transport Layer organises datagrams into segments and reliably delivers them to upper layer services. This layer compensates for delays that occurred in the Network layer. If segments are not delivered to the destination device correctly, the Transport Layer can initiate retransmission or inform the upper layers. The upper layer services may then take the necessary corrective action or provide the user with options. The Transport Layer and the Network Layer work in tandem to provide OSI networking functions.
- Session Layer: This layer allows two applications on different locations to establish and end a session. This layer establishes a dialog for control over the two devices in a session, to regulate each device and how long to transmit.

The session protocol defines the format for the data sent over the network. Remote Procedure Call (RPC) is used as the session protocol. The layer opens a dialog between the sender and receiver to ensure that communications continue. The layer accomplishes this by using three simple steps: connection establishment, data transfer, and connection release.

- Presentation Layer: The Presentation Layer transforms data into a mutually agreed upon format that each application can understand. In addition, it compresses large data and encrypts sensitive information. Translation is the main job of this layer and is involved when two devices speaking different languages open a dialog. External Data Representation (XDR) sits at the presentation level. It converts the local representation of data to its canonical form and vice versa. The canonical form uses a standard byte ordering and structure packing convention, independent of the host.
- Application layer: This layer represents the access level for applications accessing the network services. This layer uses specific networking applications to provide file, print, message, and application database services. It controls for the broadcasting of these services, and ensures their availability. Once the proper service has been identified, the user request is processed. This layer represents the services that directly support applications such as software for file transfers, database access, telnet, DNS, NIS and electronic mail. The Application Layer represents the goal for OSI Networking.

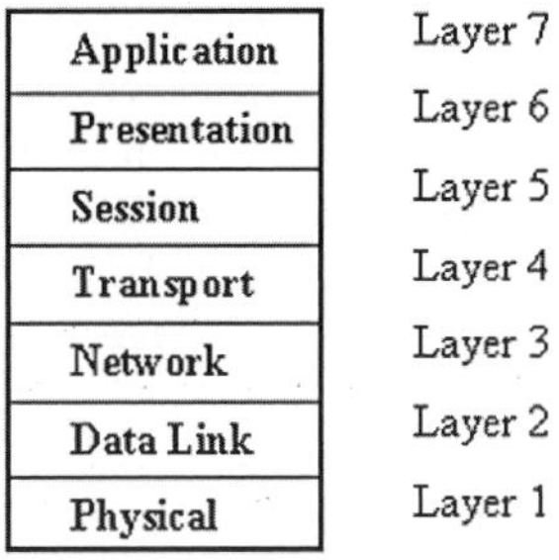

Application	Layer 7
Presentation	Layer 6
Session	Layer 5
Transport	Layer 4
Network	Layer 3
Data Link	Layer 2
Physical	Layer 1

Fig. 18.3. The layered diagram of the OSI Reference Model

18.4 Enterprise Network

Depending on the industrial sector (Refining and Petrochemicals, Iron and Steel, Pulp and Paper, or Pharmaceuticals etc.), the size of the company, diversity of operations and the level of legacy systems in the organisation, deployed Enterprise Networks could vary. Enterprise Networks are a collection of Local Area Networks (LAN), Wide Area Networks (WAN), Metropolitan Area Networks (MAN) or a combination of these with a wide range of communication protocols. Several protocols are frequently and simultaneously used on the same physical platform.

The data carried are diverse and come from Production, Maintenance, Stores and Purchase, Utilities and Chemicals, Marketing and Sales, Finance and Administration, and Transport and Logistics. With web-enabled Internet and Intra-Net operations, the number of network-capable applications continues to grow, rapidly increasing the network traffic manifold. Enterprise Networks are quite expensive. A typical full-scale Enterprise Network model is depicted in Fig. 18.4.

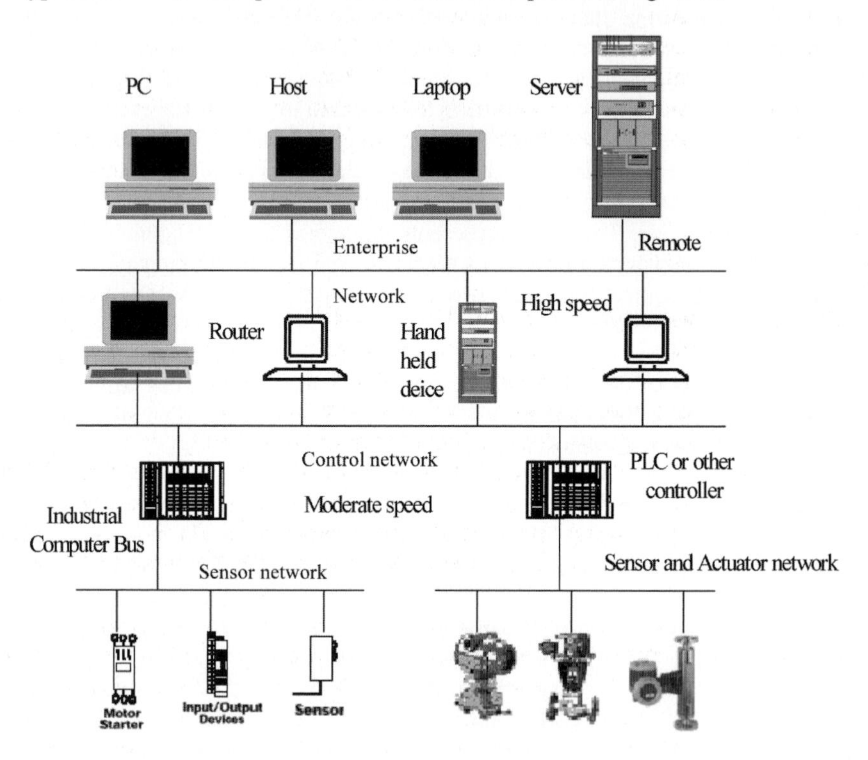

Fig. 18.4. Enterprise wide network

18.4.1 Expanding the DCS and Field Control Systems (FCS) with the Internet (Tan and Tang 2003)

Many manufacturing processes are now widely distributed geometrically, due to economy-related factors in manufacturing and distribution. The layout of an entire plant can now be rather extensive, spreading across continents in certain cases. Therefore, it has become an important challenge to be able to optimise any synergy opportunities in the operations of these distributed systems. In many cases, the same set of processes to manufacture the same product (or to monitor the same process) can be cloned over different plants. This requires close coordination and synchronisation of the distributed operations, as well as an efficient remote monitoring and control facility in place. Thus, an extensive and "borderless" approach towards the effective monitoring of the distributed points is crucial to enhance overall efficiency and operational costs. To expand the scope of DCS/FCS to a higher level, an established and widespread network will be useful to provide the link. The Internet is one viable candidate. It is now widely used as a connectivity tool for educational, commercial, and personal applications. It is an exciting portal that makes it possible for users to access virtually an infinite supply of information. The growth of the Internet is evident in the surge of information and data flow across the Internet, and in the ever-increasing number of network nodes that are set up at everywhere.

Almost anyone with a telephone line is eligible for an Internet connection with many Internet Service Providers (ISP) providing free services. Since its birth in the 1960s, the Internet has evolved from a stage, which merely provided static information downstream (from servers to clients or users) to a stage of tele-presence paradigm; whereby a user can interact and exchange information with the server and among the users themselves. The Internet has indeed pervaded the daily lives of people in many facets. Harnessing the power of the Internet to the networking of automation plants will make it possible to collect more information from the shopfloor and to disseminate it far and wide through every level of the company structure. The fast expanding infrastructure of the Internet, in terms of its high volume of traffic and the large number of network nodes around the globe, makes it highly suited for the networking of plants at different locations. With the growing structure of business functions, the coordination of production and other business functions has become an integral part of the company information technology (IT) structure. Geographically, distributed components with abundance of intelligence are now expected to work together. The utilisation of the Internet for monitoring and control purposes complements perfectly with the development trend of distributed intelligence in the field controllers and devices at the shop floor. As a result, superior control decisions can be made with these readily available resources and information, i.e., historical data, knowledge database, etc. Internet working has become essential for automation and it is poised to change the way plants and factories work. In the subsequent sections, two application cases, where the Internet is used to expand the functionality of fieldbus-based systems, will be elaborated. The first application will illustrate that, with an Internet synergy, the supervisory monitoring of a DCS/FCS can be expanded beyond the vicinity of the local plant to remote sites, which have an Internet access. The second application will carry such remote operations further. It will put forth the concept of Knowledge-based (KB) control to the DCS/FCS and extend the notion of the conventional local knowledge base to a remote one, so that the expert knowledge held can be disseminated from a central remote point to different DCSs/FCSs.

18.5 Synchronous Model of DCS (Raja 2003)

Extensive research has been conducted with the aim of developing appropriate models for real-time systems. From the various models, Raja et al. consider the most commonly used approaches, which take distribution, parallelism and communication into consideration. Keeping the requirements in mind, they classify the various existing approaches as *logical-level* and *system-level.* Approaches falling into the logical-level category are, state-machine based approaches, net-based approaches and synchronous languages. These approaches describe the system at a behavioral level and focus on the specification of the general behavior and verification of certain logical system properties. On the other hand, the system level models are adapted for describing specific system characteristics. These models include scheduling models and distributed systems models. In (Raja 1995) these models are presented and their suitability to modeling requirements are discussed. The approach they have chosen is based on the model used in distributed systems, namely *the synchronous network model.* The adapted model makes care of the real-time aspects and parallelism between computation and communication.

A distributed system in its simplest form can be represented as a set of processors connected over a communication medium. The processors make local computations and exchange messages using the communication medium. Distributed systems can be classified as synchronous or asynchronous. Processors can be synchronous or asynchronous depending on how the local computations are made. The communication medium can be synchronous or asynchronous depending on how the communication between the processors is accomplished. Processors are synchronous if there exists a constant $s \geq 1$ such that for every $s+1$ steps taken by any processor every other processor will have taken at least one step. Otherwise they are asynchronous. If a message sent over a communication medium (at time t) is delivered within a bounded delay D (before $t+D$) then the communication medium is synchronous. If not, it is asynchronous.

Based on the above definitions, four combinations of distributed systems are possible, of which two extensively used models are synchronous (both processors and communication are synchronous) and asynchronous (both processors and communication are asynchronous). We shall adapt the synchronous model, which implicitly contains time-constraints, for modeling the periodic behavior of the fieldbus systems. The synchronous model implicitly assumes that a global clock is available to all processors and that the time takes only integer values. The time instants are referred to as *pulses*. Computations and communication delays are all expressed in terms of pulses. The synchronous model is defined by the following rules:

- Processor Synchrony: The pulses are generated successively. Pulses take the values 0,1,2. They are all available to all processors. At the beginning of a pulse, the processors can send messages to their neighbors.
- Computation: In-between two pulses, a processor can receive messages from neighbors and use them. The computation time of a processor is assumed to be negligible as compared to the time required for transferring the messages.
- Communication Synchrony: Messages sent at the beginning of a pulse i, are delivered between pulses $i + B_l$ and $i + B_u$ (where $0 \leq B_l \leq B_u$). B_l is the lower bound and B_u the upper bound.

In most practical systems, $B_l = B_u = 0$, i.e., the messages sent on pulse i are received before the next pulse $i+1$. The computations are shown in Fig. 18.2.

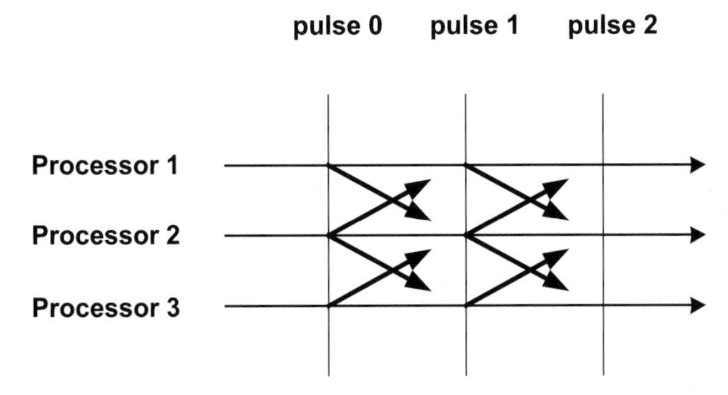

Fig. 18.5. Synchronous computations (Adapted from Raja 2003)

18.6 An Example of Fieldbus: The Local Operating Network (Raji 2003)

Table 18.2: The LonWorks technology offerings (Courtesy: Echelon Inc., USA)

Control Networking Attribute	*Benefit*	*LONWORKS Offering*
An open, robust, and reliable communication protocol	Ensures basic interoperability among devices from different manufacturers since they would all be using the same protocol	LonTalk Protocol
ICs containing the protocol and the devices' applications; Media transceivers (TP, RF, PL, CX, IR, etc.)	Low-cost, off-the-shelf resources. Manufacturers can focus on their application, not the communication obstacles	Neuron Chip (Motorola and Toshiba) Various third-party transceivers available (TP, PL, RF, CX, IR,...)
Network operating system for installing, configuring, maintaining, monitoring, and controlling the network	Uniform network management across all tools and platforms; system interoperability	LNS architecture
Infrastructure devices (routers, repeaters, PC Cards, etc.)	Scalability	Router core module Various third-party routers and interface cards available
Comprehensive interoperability guidelines for manufacturers and system integrators	True open interoperability	LONMARK Interoperability Association LONMARK Interoperability Guidelines

The LonWorks (Local Operating Networks) control networking technology addresses the many technical and business challenges mentioned thus far. It accomplishes this by going beyond simply being a communication protocol, and providing a complete platform on which to build control systems. The LonTalk® protocol, an open and international standard designed specifically for the needs of control, is at the heart of LonWorks networks. It provides a rich palette of features from which control systems manufacturers can pick and choose. The feature set provided by the LonTalk protocol accommodates almost any control application, including building automation, factory and process control, and home automation. The many facets of the protocol provide for security, reliability, and performance while maintaining a small footprint. The Neuron® Chip, available from Motorola and Toshiba, includes an implementation of the. LonTalk protocol along with other built-in features to provide a complete system-on-a-chip solution for control devices. A Neuron Chip and a communication transceiver are usually all that is needed for a LonWorks control device. The LonTalk protocol support many communication media including, twisted pair, power line, fiber optics, coaxial cable, radio frequency, and infrared. Transceivers are readily available from Echelon and many third-party sources throughout the world.

The LonWorks Network Services (LNS) architecture is a powerful network operating system, which provides an object-oriented method for dealing with networked control devices. It permits end-user tool developers to use a unified and powerful API for developing tools for installing, configuring, and maintaining, monitoring and controlling LonWorks control networks. LNS provides for true interoperability among device and tool manufacturers, in the same way that the Microsoft Windows API and Sun Microsystems' Java API provide platforms for interoperability among software applications. LNS clients can run on any platform (PC, MAC, UNIX, embedded, etc.). The LNS Server supports both LonTalk and TCP/IP protocols at the transport layer-1. LNS clients can communicate with the server using either protocol. This means that the same intuitive control network object hierarchy can now be used via *any* TCP/IP connection to the LNS Server. Internet and intranet access to control networks has never been easier.

Infrastructure tools such as routers and network interface cards (NIC) are available from many companies. These products are essential in providing a scalable architecture, which is flexible enough to meet the strict demands of control networks. For example, a LonWorks -based building lighting system might use a twisted pair backbone network which branches off into individual floor subnetworks that communicate over the existing building power line wiring. Each floor might use a twisted pair-to-power line router to not only convert between the two media but also to segment the traffic between floors.

No open networking standard could survive without a set of comprehensive interoperability guidelines, and a neutral, governing body to oversee its development and administration. The LonMark Interoperability Association (http://www.lonmark.org) is a non-profit organisation comprised of members representing the leading control companies in many industries, systems integrators, and end-users. Individual task groups within the association are devoted to establishing a uniform set of guidelines for device, system, and tool manufacturers. A

LonMark certification program ensures interoperability among all certified devices and tools, which dramatically simplifies system integration and maintenance. Table 18.2 shows the key benefits offered by LonWorks Technology. Detail information can be seen at www.echelon.com.

18.7 Acknowledgement

The author acknowledges RS Raji, R Murugesan, KK Tan, KZ Tang, P Raja and G Noubir for their contribution to this chapter. Infact some of the sections in this chapter have been copied from their chapters edited by this author in an earlier publication of the book "Fieldbus Technology" (ISSN 3540401830) by the Springer-Verlag. It was important to include those sections in order to improve the readership of this book. The valuable section contributions by the above authors are greatly recognised.

18.8 References

Echelon Inc. (1995) Product catalogue, Echelon Corporation, 550 Meridian Ave, San Jose CA 95126, USA

Harrison R. and Charles GP (1995) 'Applying LonWorks to the Distributed Control of Manufacturing Machines", Proceedings of the LonUser International Conference, Echelon, California, Frankfurt, Germany, 1995, pp 33-45

http://www.steinhoff.de/fb_comp.htm

http://www.echelon.com

Raji RS (2003) Control network and the Internet. In: Mahalik NP, Fieldbus Technology: Industrial Network Standard for Real-time Distributed Control, Springer Verlag, ISSN 3540401830, Chapter 7, pp 171-182

Raja Prasad and Noubir G (2003) A synchronous model for fieldbus systems. In: Mahalik NP, Fieldbus Technology: Industrial Network Standard for Real-time Distributed Control, Springer Verlag, ISSN 3540401830, pp 271-295

Raja Prasad and Noubir (1993) Static and dynamic polling mechanisms for fieldbus networks, G., ACM SIGOPS, Operating Systems Review, 27:3, July, pp. 34-45

Raja Prasad and Noubir V (1995) Modeling periodic fieldbus systems, Thèse N0 1405 (1995), Ecole Polytechnique Fédérale de Lausanne, Switzerland

Mahalik NP and Lee SK (2002) Flexible distributed control of production line with the Local Operating Network (LON) fieldbus technology: A laboratory study, International Journal of Integrated Manufacturing Systems, Volume 14, No.3, 2003, pp 268-277, MCB Press, UK

Mahalik NP (2003) Application of fieldbus technology in mechatronic systems. In: Mahalik NP, Fieldbus Technology: Industrial Network Standard for Real-time Distributed Control, Springer Verlag, ISSN 3540401830, pp 549-583

Murugesan R (2003) Fieldbus and contemporary standards. In: Mahalik NP, Fieldbus Technology: Industrial Network Standard for Real-time Distributed Control, Springer Verlag, ISSN 3540401830, pp 215-244

Tan KK and Tang KZ (2003) Control of fieldbus-based systems via the Internet, In: Mahalik NP, Fieldbus Technology: Industrial Network Standard for Real-time Distributed Control, Springer Verlag, ISSN 3540401830, pp 215-244

19 Simulation of Distributed Control Systems

E Ould-Ahmed-Vall, BS Heck and GF Riley

School of Electrical and Computer Engineering, Georgia Institute of Technology

19.1 Introduction

A wireless sensor network (WSN) consists of a set of nodes powered by batteries and collaborating to perform sensing tasks in a given environment. It may contain one or more sink nodes (base stations) to collect sensed data and relay it to a central processing and storage system. A sensor node can be divided into three main functional units: a sensing unit, a communication unit and a computing unit.

A distributed control system consists of a set of nodes having sensing, control and actuation capabilities and interacting using an overlapping network (Branicky et al. 2003). In such a system, any of the three main control loop tasks of sensing, control and actuation can be performed in a distributed manner. A wireless implementation of these networks is discussed in this chapter. The use of wireless networks for distributed sensing and control offers several benefits. It allows cost reduction and eliminates the need for wiring. Wiring could become costly and difficult in the case of a large number of control nodes needed for the sensing and control of a large process. However, the use of wireless networks in a control system introduces new challenges. In fact, these networks can suffer from several problems such as unbounded delays and packet losses.

This chapter describes the different architectures that allow the implementation of distributed control systems using wireless sensor networks. These architectures differ according to whether the sensing and/or actuation are performed centrally or in a distributed fashion. In order to analyse the architectures, the Georgia Tech Sensor Network Simulator (*GTSNetS*) (Ould-Ahmed-Vall et al. 2005c) is extended to simulate distributed control using sensor networks. One of the main features of this simulator is its scalability. In fact, *GTSNetS* was shown to simulate networks of up to few hundred thousand nodes. The simulator is implemented in a modular way and the user is allowed to choose from different architectural implementations. If a specific approach or algorithm is not available, the user can easily implement it by extending the simulator. This simulator is demonstrated using an existing Bayesian fault tolerance algorithm.

Fault-tolerance of individual sensing nodes is also considered in this chapter. The Bayesian fault tolerance algorithm is presented and extended to adapt to dynamic failure rates. The enhanced algorithm allows nodes to learn dynamically about the operational conditions of their neighbours. Each node then gives different weight factors (confidence levels) to the information received from each of its neighbours. The weight factors are function of the failure probability of the specific neighbour. The probability of failure is computed from the reliability of the information received from the specific neighbour compared to other neighbours. The simulator facilitates the implementation and evaluation of the algorithm even for a network containing a large number of nodes. The remainder of the chapter is as follows. Section 2 presents distributed control systems and their applications. Section 3 describes the different architectures that can be used to implement distributed control systems with sensor networks. Section 4 discusses the simulation of distributed control systems using sensor networks. Section 5 presents and compares the original and the enhanced fault tolerance algorithms. Section 6 gives an example simulation scenario. Section 7 concludes the chapter.

19.2 Distributed Control Systems

A distributed control system consists of a set of nodes having distributed sensing, control and actuation capabilities and interacting using an overlapping network. This class of networks is also referred to under the names of networked control systems (Branicky et al. 2003), networked sensing and control (Hartman et al. 2005), robotic sensor networks (Sukhatme 2005) and sensor-actuator networks (Sukhatme 2005).

In a distributed control system, any of the three main control loop tasks of sensing, control and actuation can be performed in a distributed manner. Most of the work on distributed control systems implies the use of a wired network such as CAN (controller area network), PROFIBUS, or Ethernet (Chow & Tipsuwa 2001). The use of wireless networks for distributed sensing and control offers several benefits. It allows cost reduction and eliminates the need for wiring. Wiring could become costly and difficult in the case of a large number of control nodes needed for the sensing and control of a large process. However, the use of wireless networks in a control system introduces new challenges such as unbounded delays and packet losses. In addition, the same packet can arrive multiple times through different paths. These problems are less severe when the nodes are less tightly coupled as is the case when the sensor, controller and actuator are within a single node and only occasional and less time-sensitive communication is made between nodes. For a theoretical discussion of the effects of transmission delays and packet losses on the control loop, the user is referred to (Hartman et al. 2005 & Zhang et al. 2001). The applications of distributed control systems include industrial automation, automotive systems, building and environmental systems, aerospace systems and mobile robotics. Two specific applications are described here as examples. The first application consists of an Automated Highway System (*AHS*) (Simsek et al 2001) aiming to improve highway capacity, safety and efficiency without building new roads. In this application, a car can switch between a user-controlled mode and an automated mode. In the automated mode, cars are organised in platoons travelling on the automated lane. Cars in the same platoon travel close to each other, while the distance is higher between two successive platoons. The user can gain back the control of the car at any point, especially when close to the destination. Cars collaborate by exchanging information during the joining and leaving of a specific platoon. An *AHS* has a higher capacity than a regular highway to the decrease in inter-vehicle distance. With a platoon size of 20 vehicles the capacity can be increased by a factor of 4 with a inter-vehicle distance of 1 meter. The lower inter-vehicle distance also helps improving the fuel efficiency by reducing the aerodynamic load. *AHS* also increases the safety since the relative velocity between vehicles in the same platoon is reduced (Simsek et al 2001).

The second application is an underwater robot society (SUBMAR) implemented at the Helsinki University of Technology (Appelqvist 2002). Each node consists of a small robot equipped with various sensors and two tank actuators. The robots use IR sensors to detect the presence of algae and used the tank actuators to administer poison upon detection. Communication between robots is implemented using a point-to-point network.

19.3 Architectures

In this section, the different architectures used to implement distributed control using wireless sensor networks are presented. These architectures differ depending on whether any or all of the sensing, control and actuation functionalities are performed in a distributed way inside the network or centrally at a base station.

19.3.1 Distributed Sensing With Centralised Control and Actuation

This approach consists of using an entire sensor network as the sensing entity in a control system. This allows monitoring a large plant that cannot be covered using a single sensor. It also allows for fault tolerance since sensor nodes can be deployed to have several nodes covering each part of the plant. Information collected by the different sensor nodes is fused and given as an input to the controller, which runs on a supervisor node outside of the sensor network. The controller generates a signal (action) that is applied to the plant by the actuator without involving the network. Fig. 19.1 illustrates this approach. The arrows in the figure indicate the information flow.

19.3.2 Distributed Sensing and Actuation with Centralised Control

This approach is similar to the previous one. The main difference is that any corrective action on the plant is now applied by individual sensor nodes after receiving control commands from the controller. The controller is still run centrally at the supervisor node. Actuation messages are addressed either to all nodes in the network or to specific nodes as a function of their sensor readings. This could be the case, for example, when nodes in a specific region are reporting exceptionally high values. Fig. 19.2 illustrates this approach.

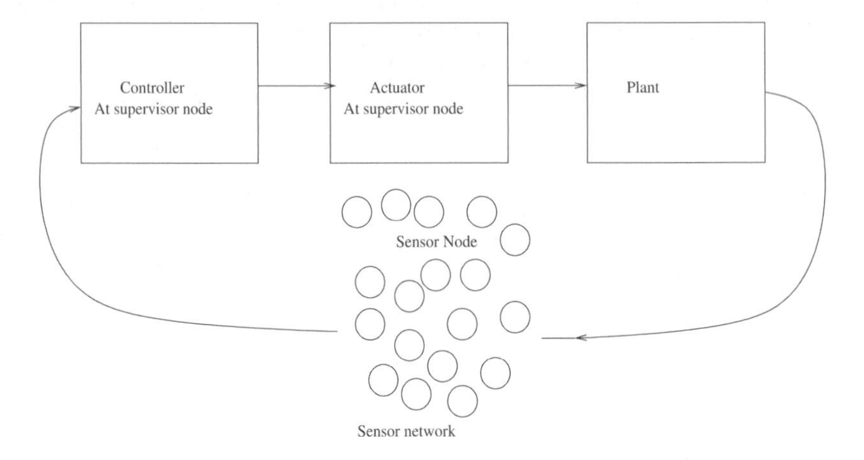

Fig. 19.1. Distributed sensing with centralised control and actuation

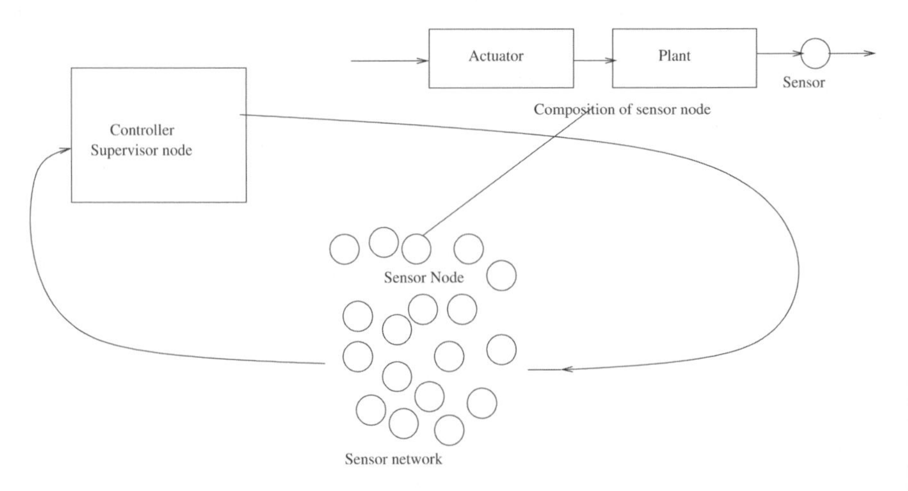

Fig. 19.2. Distributed sensing and actuation with centralised control

19.3.3 Distributed Sensing, Control and Actuation

In this approach, the sensing, the control and the actuation are all performed inside the network. In this case, each control system acts as a sensor node. Each node can collect information about the plant, run control algorithms (act as a controller) and apply any necessary actions (actuator role). These nodes collaborate to control an entire system. However, each node monitors and actuates a specific plant that is a part of a larger system. In this case, each sensor node contains a controller, an actuator and a plant in addition to its normal components. In this approach, the sink node (base station) does not participate in the control process. It plays its traditional roles of collecting information, storing it and relaying it to the outside world when necessary. Fig. 19.3 illustrates this approach.

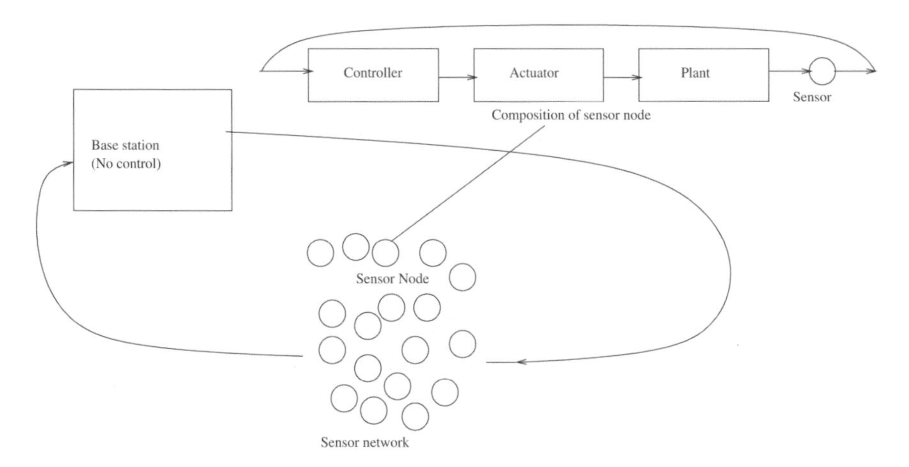

Fig. 19.3. Distributed sensing, control and actuation

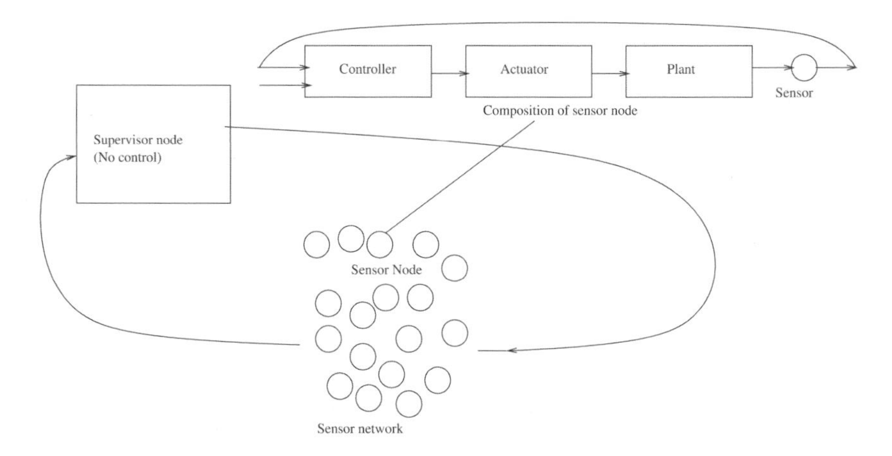

Fig. 19.4. Hierarchical distributed sensing, control and actuation

19.3.4 Hierarchical Distributed Sensing, Control and Actuation

This approach is similar to the previous one, except that a supervisor node can now affect the parameters of the control algorithm executed on individual sensor nodes. This form of hierarchical control is also known as supervisory control. The supervisor node can, for example, change the reference value at individual nodes or load a new control algorithm depending on the current state of the plant as reported by the individual nodes. It can also change the parameters to adjust to changing network conditions. For example, if the energy level becomes low at some of the nodes, these nodes can be asked to sample at a lower rate. This approach is simulated later in this paper. Fig. 19.4 illustrates this approach.

19.4 Simulation

Several sensor network simulators are available and widely used by the research community such as *SensorSim* (Park et al. 2000), *SWAN* (Liu et al. 2000), *TOSSF* (Perrone & Nicol 2002), *TOSSIM* (Levis et al. 2003) and *SENS* (Sundresh et al. 2004). There are also few simulation frameworks for distributed control systems that have been proposed recently. However, none of the existing simulators combine the features of large-scale sensor networks with control applications. The co-simulation of control and network aspects of a distributed control system allows the simultaneous study of the effects of control issues and network characteristics on the system performance. The control design issues include stability, performance, adaptability and fault tolerance. Network characteristics affecting the control system performance include bandwidth, reliability and scalability. This section discusses the two main simulation frameworks for distributed control systems that have been proposed in the recent years (*TrueTime*, a *Matlab*-based simulation framework and an *Agent/Plant* extension to *ns-2*). A new solution extending the Georgia Tech Sensor Network Simulator (*GTSNetS*) is also presented.

TrueTime (Henriksson et al. 2002) is based on *Matlab/Simulink* and allows the simulation of the temporal behaviour of multi-tasking real-time kernels containing controller tasks. It proposes two event-driven *Simulink* blocks: a computer block and a network block. The computer block is used to simulate computer control activities, including task execution and scheduling for user-defined threads and interrupt handlers. The network block is used to simulate the dynamics of a computer network using parameters such as message structure and message prioritizing function. However, this network block is not general enough to simulate various types of networks, especially sensor networks. It also suffers scalability problems. The widely used *ns-2* was extended to allow simulation of the transmissions of plants and controllers in a distributed control system (Branicky et al. 2003). The authors added a plant and an agent classes to *ns-2*. To the best of our knowledge, this solution is not yet interfaced with any of the *ns-2*-based sensor network simulators such as *SensorSim* (Park et al. 2000). In addition, it is not capable of simulating large-scale distributed control systems.

To allow the simulation of large-scale distributed control systems, the Georgia Tech Sensor Network Simulator (*GTSNetS*) (Ould-Ahmed-Vall et al. 2005c) is extended. One of the main features of this solution is its scalability. In fact, *GTSNetS* was shown to simulate networks of up to few hundred thousand nodes. The simulator is implemented in a modular way and the user is allowed to choose from different architectural implementations. If a specific approach or algorithm is not available, the user can easily implement it by extending the simulator. The features of *GTSNetS*, compared to other sensor network simulators include A unified framework of existing energy models for the different components of a sensor node. This allows the user to choose the energy model that best suits his needs.

GTSNetS provides three models of accuracy of sensed data and allows for addition of new models. Modelling the accuracy of sensed data helps understanding the trade-off of quality versus lifetime. This simulator allows the user to choose among different implemented alternatives: different network protocols, different types of applications, different sensors, and different energy and accuracy models. New models, if needed, can be easily added. This makes *GTSNetS* very suitable for simulating sensor networks since such networks are application-dependent and their diversity cannot be represented in a single model. *GTSNetS* is currently, to the best of our knowledge, the most scalable simulator specifically designed for sensor networks. It can simulate networks of up to several hundred thousand nodes. The scalability is the result of active memory management and careful programming.

Finally, *GTSNetS* can be used to collect detailed statistics about a specific sensor network at the functional unit level, the node level as well as at the network level. For a complete description of *GTSNetS* and various simulation examples, the user is referred to (Ould-Ahmed-Vall et al. 2005c & b). To implement the different distributed control architectures presented in the previous section, several classes were added to *GTSNetS*. Due the modularity of *GTSNetS*, the addition of these classes does not have any negative impact on the existing classes. It is facilitated by the possibility of code reuse in *GTSNetS*. The first class (*class Plant*) models a plant. This class is derived from the real sensed object class that is already in *GTSNetS*. The main difference is that the plant can receive a command (signal) from an actuator. A second class implements the actuator (*class Actuator*). An actuator has a plant object associated with it. It can act on the plant in several manners depending on the commands received from the controller. Several methods of actuation are implemented. The user can choose from these methods or implement new ones if necessary. The controller is modelled using an application class (*class ApplicationSNController*) derived from the sensor network application class. Each controller has a specific actuator object attached to it. This actuator object performs actions on the plant depending on the control commands received from the controller application.

The different control architectures are implemented by attaching some or all of these classes to individual nodes. To implement the first architecture (distributed sensing with centralised control and actuation), a plant object is attached to each sensor node. However, since these nodes do not perform the controller and actuator roles, they each have a regular sensor network application. No actuator object is attached to the sensor nodes. A controller application is attached to the supervisor node. An actuator object is attached to this controller. This actuator acts on the plant object, which is sensed by the sensor nodes. In a similar way, the second architecture is implemented by attaching a controller application to the supervisor node. However, this controller application gives commands to actuators, which are attached to individual sensor nodes. Each actuator acts on the plant object attached to its sensor node. In the third architecture, a controller application is attached to each sensor node. This application commands an actuator that acts on a plant, which is attached to the same node. The sink node does not have any role in a fully control application. The fourth architecture is implemented in a similar way to the third one, except that a controller application is now attached to the supervisor node. This application can modify the control algorithms or other parameters at the individual sensor nodes depending on the information they supply.

19.5 Fault Tolerance Algorithms

A key requirement for the development of large-scale sensor networks is the availability of low cost sensor nodes operating on limited energy supplies. Such low cost nodes are expected to have relatively high failure rates. It is, therefore, important to develop fault tolerant mechanisms that can detect faulty nodes and take appropriate actions. A possible solution is to provide high redundancy to replace faulty nodes in a timely manner. However, the cost sensitivity and energy limitation of sensor networks make such an approach unsuitable (Koushanfar et al 2002). Here, the focus is on the problem of event region detection. Nodes are tasked to detect when a specific event is present within their sensing range. Such an event could be detected through the presence of a high concentration of a chemical substance (Krishnamachari and Iyengar 2004). Each node first determines if its sensor reading indicates the presence of an event before sending this information to its neighbours or to a sink node. However, in case of failure the sensor can produce a false positive or a false negative. That is, a high reading indicating an event has occurred when it did not or a low reading indicating the absence of event when one has occurred. A distributed solution proposed by Krishnamachari et al. in (Krishnamachari and Iyengar 2004) is presented in the subsection below. The simulations reported in (Krishnamachari & Iyengar 2004) are duplicated in our simulator, which gives confidence to the accuracy of our simulator. The algorithm in (Krishnamachari and Iyengar 2004) assumes that all nodes in the network have the same failure rate and that this rate is known prior to the deployment. These are unrealistic assumptions, and we demonstrate through simulation that the algorithm introduces too many errors when such assumptions do not hold. We propose an enhanced version of the algorithm where nodes learn their failure rates as they operate by comparing their detection results with the ones of their neighbours. This mechanism is proved to reduce significantly the number of introduced errors.

19.5.1 Distributed Bayesian Algorithm

This solution presented in (Krishnamachari and Iyengar 2004) considers a sensor reading of a high value as an indication of the presence of an event, while a low value is considered normal. It relies on the correlation between the node reading and the readings of its neighbours to detect faults and take them into account. The following binary variables are used to indicate if a node is in an event region (value 1) or in a normal region value (value 0):

- T_i: indicates the real situation at the node (in an event region or not)
- S_i: indicates the situation as obtained from the sensor reading. It could be wrong in the case of failure
- R_i: gives a Bayesian estimate of the real value of T_i using the S_i values of the node and its neighbours

It is assumed that all the nodes have the same uncorrelated and symmetric probability of failure, p:

$$P(S_i = 0 \mid T_i = 1) = P(S_i = 1 \mid T_i = 0) = p \tag{19.1}$$

The binary values S_i are obtained by placing a threshold on the reading of the sensor. The sensor reading when in an event region and when in a normal region have means of m_f and m_n, successively. The error term is modelled as a Gaussian distribution with mean 0 and standard deviation σ. In such a case, p is computed as follows using the tail probability of a Gaussian distribution:

$$p = Q(\frac{m_f - m_n}{2\sigma}) \tag{19.2}$$

Assume each node has N neighbours. Define the evidence $E(a,k)$ as the event that k of the N neighbouring nodes report the same conclusion $S_i = a$. Using the spatial correlation, we have:

$$P(R_i = a \mid E_i(a,k)) = \frac{k}{N} \tag{19.3}$$

Each node can now estimate the value of R_i given the value of $S_i = a$ and $E_i(a,k)$. This is given by:

$$P_{aak} = P(R_i = a \mid S_i = a, E_i(a,k)) = \frac{(1-p)k}{(1-p)k + p(N-k)} \tag{19.4}$$

$$P(R_i \neq a \mid S_i = a, E_i(a,k)) = 1 - P_{aak} = \frac{p(N-k)}{(1-p)k + p(N-k)} \tag{19.5}$$

Three decision schemes are proposed:

- Randomised: determine the values of S_i, k and P_{aak}; generate a random number $u \in (0,1)$; if $u \leq P_{aak}$, then set $R_i = S_i$, else set $R_i \neq S_i$
- Threshold: a threshold θ is fixed in advance; determine the values of S_i, k and P_{aak}; if $\theta \leq P_{aak}$, then set $R_i = S_i$, else set $R_i \neq S_i$
- Optimal threshold: determine the values of S_i and k; if $k \geq N/2$, then set $R_i = S_i$, else set $R_i \neq S_i$

It was shown in (Krishnamachari & Iyengar 2004) that the optimal value of θ in the threshold scheme is $1-p$, which is equivalent to using the optimal threshold scheme. Therefore, only the randomised and the optimal threshold schemes are considered here. Several metrics have been developed to evaluate the performance of this Bayesian solution under different settings. These metrics include:

- Number of errors corrected: number of original sensor errors detected and corrected by the algorithm
- Number of errors uncorrected: number of original sensor errors undetected and uncorrected by the algorithm
- Reduction in errors: overall reduction in number of errors, taking into account the original errors and the ones introduced by the algorithm
- Number of errors introduced by the solution: number of new errors introduced by the algorithm

A full description of these metrics as well as their theoretical values can be found in (Krishnamachari & Iyengar 2004).

19.5.2 Simulation of the Bayesian Solution Using GTSNetS

To simulate this particular solution, only one class (*class ApplicationSNFToler*) needs to be added. The modularity of *GTSNetS* enables the reuse without any change of all the other modules: computing unit, communication unit, sensing unit containing a chemical sensor with the appropriate accuracy model and sensed object. A sensor network of 1024 nodes deployed in a region of 680 meters by 680 meters is considered. Each two neighbouring nodes are separated by 20 meters. The communication range was set to 23 meters. A parameter measuring the fault tolerance range was defined in the *ApplicationSNFTolerance* class. Nodes within this range are considered neighbours and taken into account by the fault tolerance algorithm to produce the prediction value R_i. The definition of this parameter dissociates the neighbourhood region in which neighbours are taken into account by the algorithm from the communication range. This is a more realistic approach than the one used in (Krishnamachari & Iyengar 2004). This parameter is set such that each interior node has 4 neighbours taken into account by the algorithm. One source (sensed object) was placed at the lower left corner of the region of interest. The sensing range was set to 93 meters. These numeric values are only different from the ones in (Krishnamachari & Iyengar 2004) by a scaling factor.

To reduce the size of the exchanged messages, it was decided to run an initial neighbours discovery phase prior to the execution of the fault tolerance algorithm. This avoids having to send node location along with every sensor reading message, which greatly reduces the sensor message size. This helps reducing the energy cost of the algorithm, but requires the existence of an identification mechanism (Ould-Ahmed-Vall et al. 2005b). For the communication, every node communicates only with nodes in its fault tolerance range. In this specific simulation, this range is less than the communication range. Nodes, therefore, broadcast their messages to the neighbouring nodes and there is no need for any routing protocol. When necessary, the user can use one of the sensor network routing protocols implemented in *GTSNetS*, e.g., geographical routing or directed diffusion.

Fig. 19.5 and Fig. 19.6 give some of the performance metrics results for the optimal threshold and randomised schemes for various fault rates. These results were obtained by averaging over 1,000 runs. Both decision schemes correct a high percentage of original errors (about 90% for the optimal threshold and 75% for the randomised scheme at 10% failure rate). These graphs are in accordance with the ones reported in the original paper (Krishnamachari & Iyengar 2004), which increases the confidence in the correctness of *GTSNetS*. The simulations in (Krishnamachari & Iyengar 2004) where conducted using *Matlab* and did not take into account many of the communication aspects and energy constraints present in sensor networks.

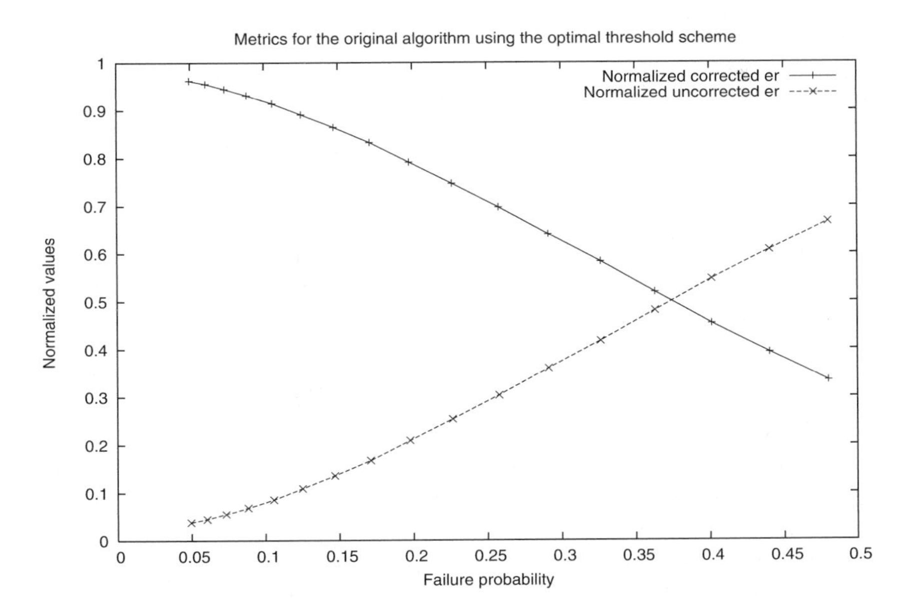

Fig. 19.5. Performance metrics for the optimal threshold scheme

Clearly, the optimal threshold scheme performs better than the randomised scheme. This is expected and is in accordance with the findings in (Krishnamachari & Iyengar 2004). However, the randomised scheme has the advantage of giving a level of confidence in its decision to set $R_i = S_i$ or not. Paak gives this confidence level. This is not possible in the case of the optimal threshold scheme.

19.5.3 Enhancement of the Bayesian Fault Tolerance Algorithm

The Bayesian algorithm assumes that all nodes in the network have the same failure probability *p*. It also assumes that this failure probability is known to the nodes prior to the deployment. These two assumptions are unrealistic. In fact, a node can become faulty with time either because of a lower energy level or because of changing environment conditions. This increases its fault rate. Also, a heterogeneous sensor network can nodes with different operational capabilities and accuracy levels.

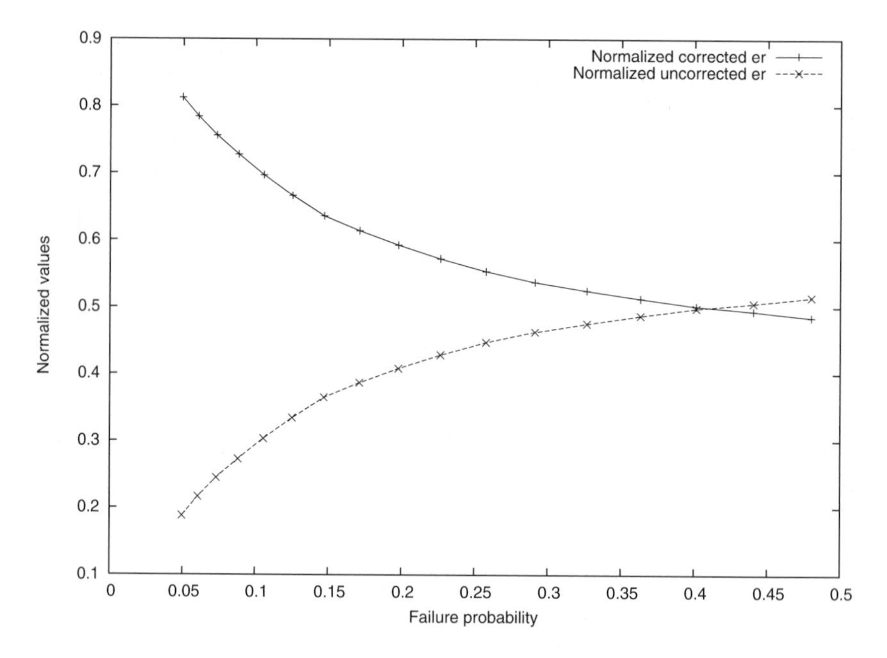

Fig. 19.6. Performance metrics for the randomised scheme

For these reasons, an enhancement to the algorithm under which nodes learn about their failure rates as they operate is proposed. This is done by comparing the sensor reading S_i with the estimated value of R_i, obtained by taking into account the values reported by the neighbours, over time. Different weights are given to information reported by nodes with different fault rates. The reported information

S_j from neighbour j is given a weight of $1- p_j$, where p_j is the current fault rate at node j sent along with its sensor reading.

Several of the previous equations are modified for these dynamic failure rates. Suppose that k of the neighbouring nodes obtained sensor decisions identical with the decision $S_i = a$ at the current node i. For simplification of notation, assume that these nodes are numbered 1 to k and the rest of the neighbours are numbered $k+1$ to N. We define P_{ea} as the probability of the event $R_i = a$ knowing that these first k neighbours have each $S_j = a$. This probability is constant and equal to k / N in the original algorithm. Its value is now given by:

$$P_{ea} = P(R_i = a \mid E(a,k)) = \frac{\sum_1^k (1-p_j)}{\sum_1^N (1-p_j)} \tag{19.6}$$

Here p_j is the current failure probability at the jth neighbour. Each node i keeps track of the current failure probability p_j at each one of its neighbours, by looking at the number of times the reported value S_j by neighbour number j disagreed with the estimated value R_i. This avoids having to send this probability value along with every decision message, which saves the energy. The new value of P_{aak} is given by:

$$P_{aak} = P(R_i = a \mid S_i = a, E(a,k)) = \frac{(1-p^i)P_{ea}}{(1-p^i)P_{ea} + p^i(1-P_{ea})} \tag{19.7}$$

The parameter p^i is the current failure probability of node i. The decision schemes are implemented as in the original algorithm using the new expression of P_{aak}. Since the optimal threshold scheme does not use P_{aak} to decide whether the node is faulty or not, only the randomisation scheme is considered here. For each failure rate level p, as used in the original algorithm, two adjacent probabilities are computed in order to introduce variability as follows:

$$p_1 = p - 0.05 \text{ and } p_2 = p + 0.05 \tag{19.8}$$

This keeps the average failure rate unchanged at p in the overall network. Each sensor is assigned randomly one of these two probability levels. The node does not know in advance the failure probability of its sensor. It starts with a conservative estimate that is equal to the highest of the two probability levels (p_2). A node i learns its real probability level as it operates by looking at the number of times its sensor decision S_i disagreed with the estimate of the real situation R_i. This failure probability, p^i, is update at every round. The performance of the enhanced version of the algorithm is compared to that of the original algorithm as well as with the original algorithm with prior knowledge of exact probability level p that is identical for all nodes. The performance is obtained by averaging over 1,000 runs as in the previous section. The network topology and configuration parameters are the same as those in the previous section.

Fig. 19.7 gives the normalised number of corrected errors under the different experimental settings. Clearly, the number of corrected errors remains relatively constant and does not suffer any degradation from the introduction of the assumptions of unknown and uneven fault rates. The only change is observed for very small probability levels (around 0.05), which is due to the ratio between p and the constant subtracted from (added to) it to produce p_1 and p_2.

Fig. 19.8 gives the normalised number of reduced errors under the different experimental settings. Observe that the performance, as measured by the number of reduced errors, is worst for the original algorithm (for error rates of up to 15%) while the enhanced algorithm gives similar performance to the one for the case where the values of p are identical for all nodes and known prior to the deployment. The degradation in the performance of the original algorithm comes from the increased percentage of errors introduced by the algorithm.

Fig. 19.9 gives the normalised number of errors introduced by the algorithm for different p values. Note that the performance, as observed before, benefits from the introduction of the enhancement to the algorithm. The number of introduced errors is controlled and keeps close to the levels of the algorithm with prior knowledge of error rates that are even throughout the network.

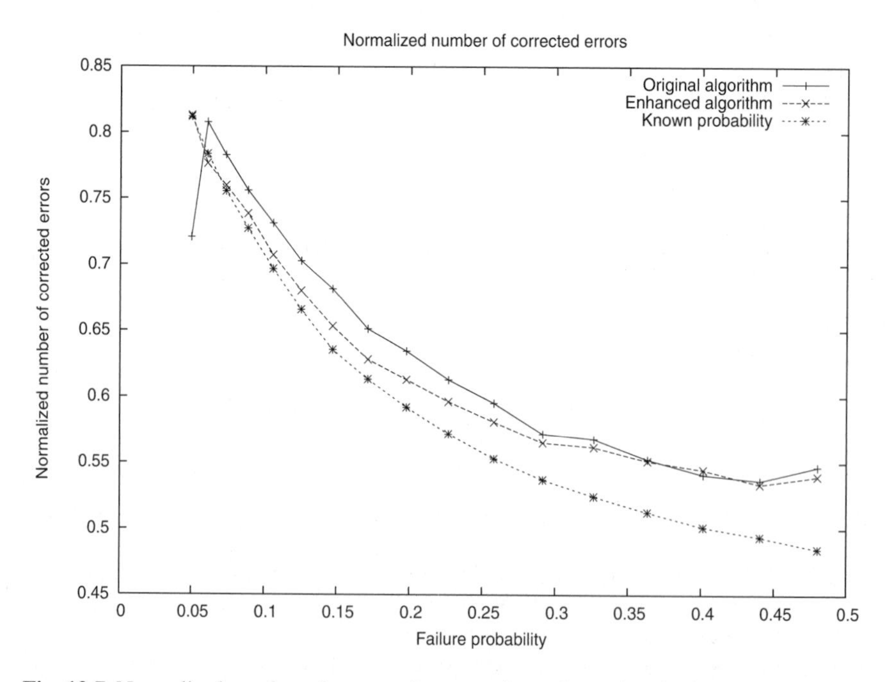

Fig. 19.7. Normalised number of corrected errors using enhanced and original algorithms

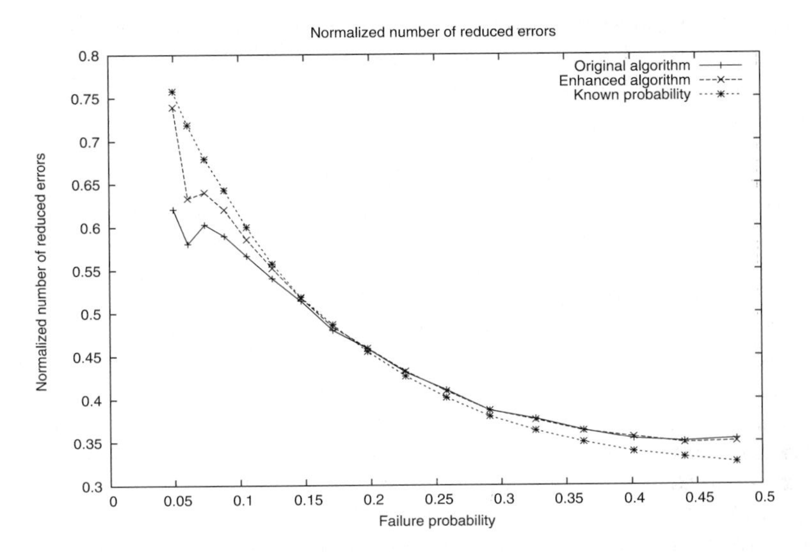

Fig. 19.8. Normalised number of reduced errors using enhanced and original algorithms

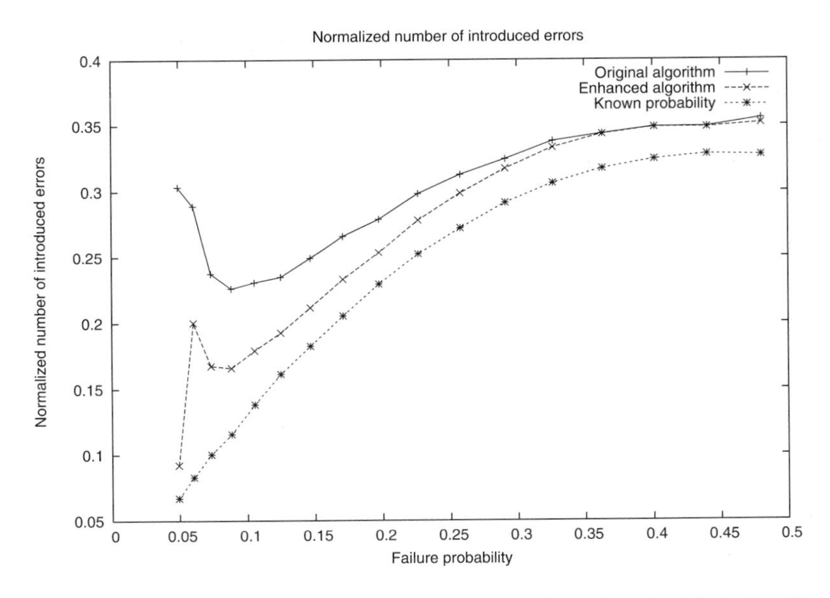

Fig. 19.9. Normalised number of introduced errors using enhanced and original algorithms

In conclusion of this section, it is demonstrated that the prior knowledge of error rates is not necessary for the Bayesian fault tolerance algorithm to perform well and keep the level of errors under control. It is also clear that the assumption of even error rates is not necessary. Relaxing these two assumptions makes the algorithm much more realistic.

19.6 Simulation Example

This section gives an example of how to use *GTSNetS* to study the behaviour of a simple distributed control system. Each node is responsible for sensing, controlling and actuating a plant. A supervisor node is responsible for changing the parameters of the local controllers. For simplicity, a single input single output proportional controller with gain K and measurement y is considered. The plant is chosen as a first order discrete time plant with pole at 1. A hybrid control is added by having the supervisor node modify the gain at a specific node if the local measurement y falls outside of a given Interval, $[I_1, I_2]$. This is a simplistic case of hybrid control where the control is reconfigured depending on the operating region in which it lies. The node informs the supervisor node that it has changed the region by sending a message. The supervisor node sends back a message to update the K value at the specific node.

A geographic routing scheme was used to communicate between the supervisor node and the sensor nodes. The location of the final destination is contained in the message header. Each message is routed through the neighbour node that is closest to the final destination among the neighbours that remain alive. Intermediate nodes route messages in a similar way. The enhanced fault tolerance algorithm, presented in (Ould-Ahmed-Vall et al. 2005a), is incorporated in the control. A node that determines that it is faulty (by comparing its sensor reading with the readings of its neighbours) does not apply any control action and waits until the next reading.

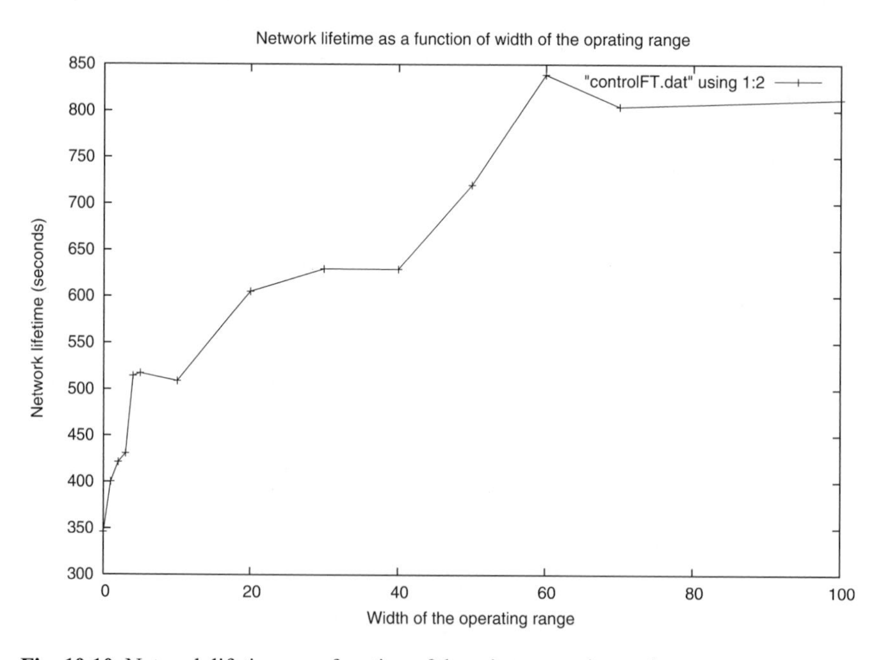

Fig. 19.10. Network lifetime as a function of the min-to-max interval

Each node is powered by a battery containing 2 joules at the beginning. Typical energy consumption parameters given in (Bhardwaj & Chandrakasan 2002) were used for the energy models. Sensors collect data every 30 seconds. Each node broadcasts a heartbeat message every 100 seconds. These messages allow nodes to keep an updated list of neighbours that are still alive. This list is used in the geographical routing protocol. All nodes were, initially, configured with a fault probability $p = 0.05$. However, this probability is dynamically updated using the enhanced fault tolerance algorithm.

The *GTSNetS* simulator can give results on the control performance at each node. In addition, it allows for an exploration of other network related parameters: communication protocols, energy consumption network-wide and at individual nodes, collaboration strategies, etc. Here, for example, we study the lifetime of the network as a function of the width of the operating range $I = I_2 - I_1$ for a network of 1024 nodes. The lifetime, here, is measured as the time from the starting of network operation till the network is considered dead. Here, the base station considers the network dead when more than 0.75 of its neighbours are dead. The death of a node is discovered when it no longer sends heartbeat messages. Fig. 19.10 gives the network lifetime as a function of the tolerance interval for $K = 0.5$ initially. The lifetime increases with the width of the operating range. In fact, for larger operating ranges the sensor reading of individual nodes is less likely to fall outside of the range. This reduces the number of messages sent to the supervisor node, which allows its neighbours to live longer.

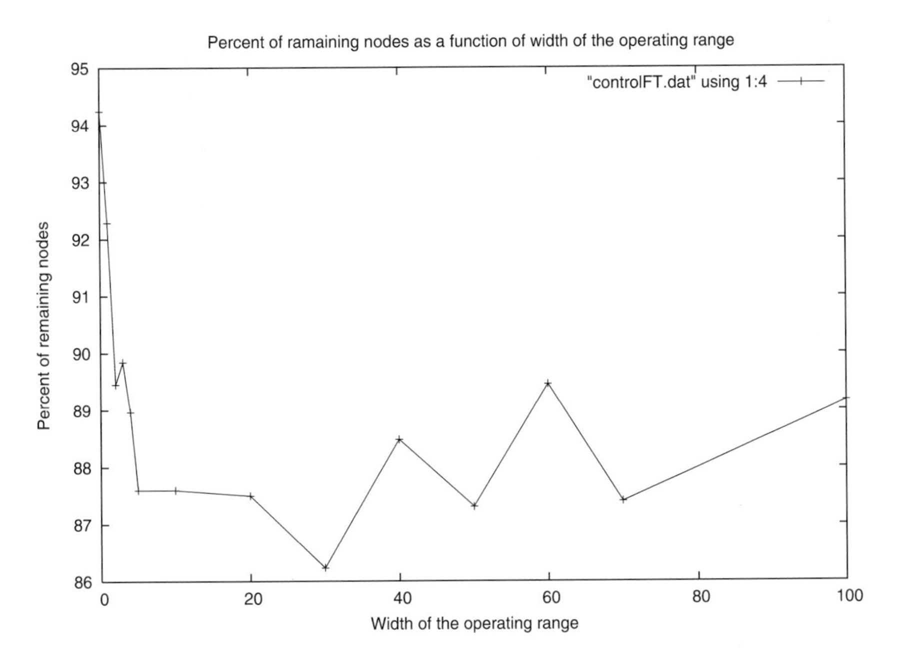

Fig. 19.11. Percentage of remaining nodes as a function of the min-to-max interval

The percentage of nodes remaining alive after the network death is also considered. Since all nodes communicate with the supervisor node, the neighbours of this node are expected to die much sooner than most of other nodes. Fig. 19.11 gives this percentage for an initial value of $K = 0.5$. We observe that this percentage is very high, as expected, and tends to increase with the width of the operating range. However, it remains relatively constant (between 85 and 95% for most values) because the neighbours of the base station still die much sooner than the rest of the network even for large operating ranges. This problem could be overcome by increasing the network density near the supervisor node. This example shows a simple distributed hybrid control system to illustrate how *GTSNetS* can be used to simulate distributed control systems.

19.7 References

Appelqvist P (2002) Mechatronics design of a robot society - a case study of minimalist underwater robots for distributed perception and task execution Ph.D. thesis, Helsinki University of Technology

Bhardwaj M and Chandrakasan AP (2002) Bounding the lifetime of sensor networks via optimal role assignments. In: Proc. of INFOCOM '2002}, pp.~1587--1596

Branicky MS, Liberatore V, Phillips SM (2003) Networked control system co-simulation for co-design. In: Proc. of American Control Conf. (ACC'03)

Chow MY and Tipsuwan Y (2001) Network-based control systems: A tutorial. In: Proc. of IEEE Industrial Electronics Society Conf.

Hartman JR, Branicky MS, and Liberatore V (2005) Time-dependent dynamics in networked sensing and control. In: Proc. of American Control Conference (ACC'05)

Henriksson D, Cervin A, and Arzen KE (2002) TrueTime: Simulation of control loops under shared computer resources. In: Proc. of the 15th IFAC World Congress on Automatic Control

Koushanfar F, Potkonjak M, and Sanjiovanni-Vincentelli A (2002) Fault tolerance techniques for wireless ad hoc sensor networks. In: Proc. of IEEE Sensors

Krishnamachari B and Iyengar S (2004) Distributed Bayesian algorithms for fault-tolerant event region detection in wireless sensor networks. IEEE Trans. on Computers, 53:3

Levis P, Lee N, Welsh M, and Culler D (2003) TOSSIM: Accurate and scalable simulation of entire TinyOS applications. In: Proc. of the 1st ACM Conference on Embedded Networked Sensor Systems (SenSys'03)

Liu J, Perrone LF, Nicol DM, Liljenstam M, Elliott C, and Pearson D (2000) Simulation modeling of large-scale ad-hoc sensor networks. In: Proc. of European Simulation Interoperability Workshop

Ould-Ahmed-Vall E, Heck BS, and Riley GF (2005a) Simulation of large-scale networked control systems using GTSNetS. In: Springer's Lecture Notes in Control and Information Sciences, to appear, Springer

Ould-Ahmed-Vall E, Blough DM, Riley GF, and Heck BS (2005b) Distributed unique global id assignment for sensor networks. In: Proc. of the 2nd IEEE Conference on Mobile Ad-hoc and Sensor Systems (MASS'05)

Ould-Ahmed-Vall E, Riley GF, Heck BS, and Reddy D (2005c) Simulation of large-scale sensor networks using GTSNetS. In: Proc. of 11th Int. Symp. on Modeling, Analysis and Simulation of Computer and Telecommunication Systems (MASCOTS'05)

Park S, Savvides A, and Srivastava MB (2000) SensorSim: A simulation framework for sensor networks. In: Proc. of 3rd ACM International Workshop on Modeling, Analysis and Simulation of Wireless and Mobile Systems (MSWiM 2000)}, pp 104-111

Perrone LF and Nicol DM (2002) A scalable simulator for TinyOS applications. In: Proc. of the 2002 Winter Simulation Conf.

Simsek T, Sousa J, and Varaiya P (2001) Communication and control of distributed hybrid systems. In: Proc. of American Control Conf. (ACC'01)

Sukhatme RS (2005) Sensor-coordinated actuation. In: Bulusu N, Jha S (eds) Wireless Sensor Networks: A Systems Perspective. Artech House

Sundresh S, Kim W, and Agha G (2004) SENS: A sensor, environment, and network simulator. In: Proceedings of the 37th Annual Simulation Symposium (ANSS37)

Zhang W, Branicky M, and Phillips S (2001) Stability of networked control systems. IEEE Control Systems Magazine

20 Wireless Pet Dog Management Systems

Dong-Sung Kim[1] and Soo Young Shin[2]

[1]Networked System Laboratory, School of Electronic Engineering, Kumoh National Institute of Technology, Republic of South Korea
[2]School of Electrical and Computer Engineering, Seoul National University, Republic of South Korea

20.1 Introduction

Recently, wireless sensor networks (WSNs) have attracted much scientific interests as a short-range radio network technique. The traditional sensor system was implemented as one independent device, which only delivers a measured value from the monitoring object without interconnection. The wireless sensor network collects data and delivers to corresponding management control systems for the user. Because of advances in micro electro mechanical systems (MEMS) technology, low energy utilisation and low cost digital signal processors using wireless communication have resulted into a new networked embedded system. It can be applied to measure, process and communicate with each sensor node over wireless communication. A WSN can support an intelligent operation and smart environments using many WSN nodes. A WSN node consists of a sensor and a short-range radio communication module, and it can detect and deliver local conditions such as temperature, sound, light, or the movements of objects. Therefore, there are wide ranges of applications envisioned for WSN. Military interests and applications such as enemy permeation perception and attack perception of the biological or chemical weapon have been studied (Nemeroff 2001). As a management of natural resources, WSNs have been applied for tracking or watching the precious animal (Kumagai 2004; Li 2004; Sikka 2004), the water containment monitoring, forest fire or flood watch (Lorincz 2004). For a health care system, WSNs have been applied for remote medical treatment system (Timmons 2004). In addition, WSNs have been applied to the home automation and industrial control fields such as the intelligent building, the factory and the nuclear plant (Tim 2005; Aakvaag 2005). By the explosive growth of indoor and outdoor wireless networks and network-enabled devices has led to a potential breakthrough in ubiquitous computing possibilities for pet dog management. To the best knowledge of the authors, a wireless management system for pet dog using WSNs has not been reported on yet in the technical literature. Therefore, this research can be a useful reference data for designing and implementing an intelligent pet dog management system using WSNs.

In this article, WSNs are applied to wireless management system for a pet dog. The implemented WSN system is composed of wireless sensing modules, a central control system, and additional devices such as mini guidance robot, automatic feeder and etc. Wireless sensor module is attached on the pet dog neck. It measures data such as the luminance, temperature, and sound, and transmits measured data to the central control system. A graphic user interface (GUI) module is introduced for user monitoring module, which displays measured data from wireless sensor part as graphic form.

20.2 Problem Analysis of Pet Dog Management

By the demands of the efficient management for a pet dog at home, the interest of the intelligent management system for the pet dog has been increased (Tim 2005). However, the intelligence systems for managing the pet animal have been used to the restricted field such as an automatic temperature controller of tropical fish tank and the automatic food feeder using the timer. Recently, the researches of the reaction and the behavior analysis between the person and pet dog have been studied (Tim 2005; Aakvaag 2005). These researches have been focused on the analysis about the conduct of the pet dog based on behaviors, utterance, and physiological signal such as the heart beat count. The results of these researches can be applied to intelligent management system of pet dog. The major problem of pet dog management system lies on a sound occurrence, a food feeding, or disposal of urine/feces in case of the apartment residence. The design requirements of pet dog management system using WSNs are affected by many environmental factors such as temperature, residential types, and so on. Therefore, in this research, an apartment residential type at daytime is assumed mainly. First problem is a noise generated by a pet dog. The pet dog barks to defend the life territory and it guards against unfamiliar persons or objects by instinct. However, it causes inconveniences between apartment residents and dog owner. Table 20.1 shows audible range of a person and a dog, respectively. It shows that the dog's hearing range spreads four times larger than human. Frequency analysis of pet dog sounds can be used to recognise demands and health condition check of pet dog. It can be used for the security purpose also. The wireless intelligence management system using WSNs is composed of multi node connections; it can be extended to the various multi-functional applications for pet dog management.

Table 20.1: Comparison of hearing range

Type	Utterance Range [Hz]	Audible Range [Hz]
Person	85~1100	20~16000
Dog	452~1080	15~60000

Secondly, automatic food feeder based on the timer can perform an unnecessary operation when pet dog goes out. Whenever the person and dog go outside, it needs additional manual setting for the proper operation. If the food feeder can detect the pet dog being automatically, it can provide more efficient way for feeding. Thirdly, environmental change by urine and feces can be a problem. It also caused by the climate change. It will be monitored by temperature sensor, which attaches on the pet dog neck. The temperature data, which is measured from sensor, can be used to control a fan and heater operation. This technique can be applied to the health condition management using the pet dog's body temperature checking. In addition, the operation of the pet dog guidance mini robot for attracting the pet dog interests can be used for decreasing an unnecessary noise at night time and intelligent auto-feeding module. Based on these problem analyses, a design and whole structure of pet dog management system are presented.

20.3 Basic Components

The developed system uses the basic building block for ubiquitous computing environments. These wireless devices include light, temperature, magnetic and acceleration sensors (2 axes). This unit also includes a microphone-based sounder system which can be used for range measurements of pet dog. It adapts the IEEE 802.15.4 communication protocol which is standard for low-speed short-range personal area network (Akingbehin 2005; Lynch 2005). Three modules (Fig. 20.1~Fig. 20.3) is combined for the whole implementation of pet dog management system. The detail specifications of each sensor network platform are shown to Table 20.2.

Table 20.2: Specification of wireless devices

Part Name	Specification
Transmission board	. CPU-ATMEGA 128L . FLASH memory - 128K . EEPROM – 4K . 10bit ADC . Wireless frequency - 868/916 MHz . Data rate – 38.4 K baud . Application range – 500ft . External electricity - 2.7~3.3V
Wireless sensor board	. Acoustic sensor . Temperature sensor . Optical sensor . Buzzer
Programming interface board	. Serial communication . JTAG debugging Tool

Fig. 20.1. Receiver module

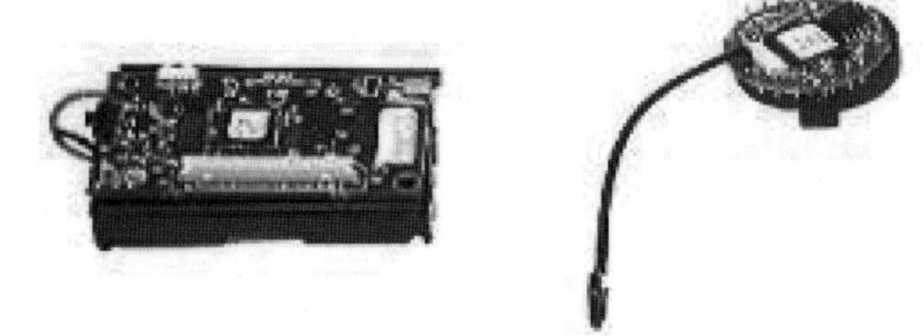

Fig. 20.2. Transmitter module

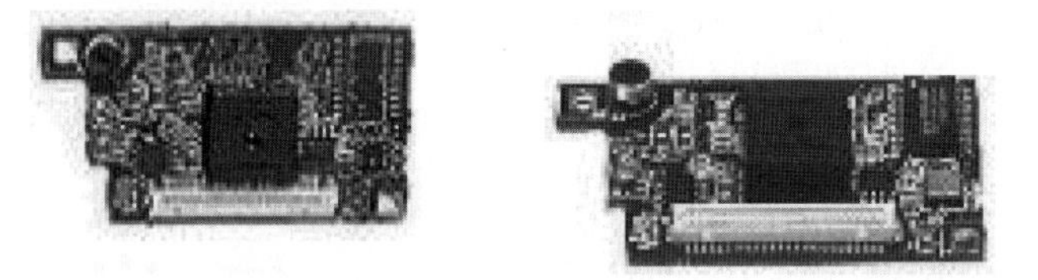

Fig. 20.3. Wireless sensing module

20.4 Design and Structure

The implemented system can gather information such as luminance, temperature or sounds from each wireless sensor node and transmit collected data to a central control system. A control command from the central control system is transmitted to each device such as guidance robot, auto feeder and electric fan or heater. Each controller's operation is intended to attract interests of the pet dog based on mixed sensing information. Table 20.3 shows control target based on sensed data and its user environment. The measured light, temperature, sound information from a pet dog and environment invoke the small-sized pet dog guidance robot and other devices. When the pet dog approaches to the food feeder, sensor node perceives luminous intensity differences and activates the automatic food feeder. When the variation of room temperature is detected, heater or electric fan can be activated to control temperature properly.

Table 20.3: Control target using sensing data

Measurement item	Luminance, temperature, sounds
Control Target	• Luminance: automatic food feeder. • Temperature: electric fan or heater • Sound: mini guidance robot and • Mixed information (temperature and sound): heath or security check
User Environment	• Data measurement and confirmation • Confirmation of Control command • User Interface for Automatic/manual mode selection

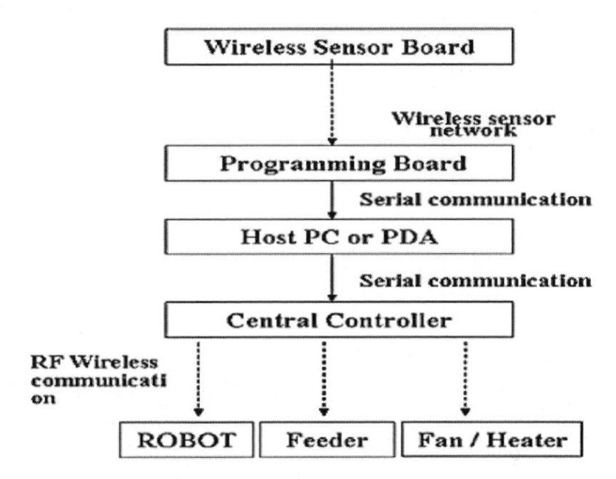

Fig. 20.4. Overall structure of pet management system

A structural block diagram of implemented management system is presented in Fig. 20.4. Whole system includes sensor nodes, a controller, mini guidance robot, automatic feeder, and host PC. It receives the measured data value from the each sensor node and transmits through the radio wireless links. All of the above components are controlled through wireless/wired links. The sensed data such as luminance, temperature, and sound are transferred to the central controller maintaining their sequence. Communications between a control system and small-sized guidance robot or automatic foot feeder are performed via the RF radio communication which uses 433 Hz bandwidth.

20.4.1 Transmitter Module

A bock diagram of data transmission module is presented in Fig. 20.5. It is composed of the measurement part of light, temperature and sound information from sensor module. Each module can support timer and LED operation. In Fig. 20.5, sensor network, which calls subordinate part accomplishes the whole function.

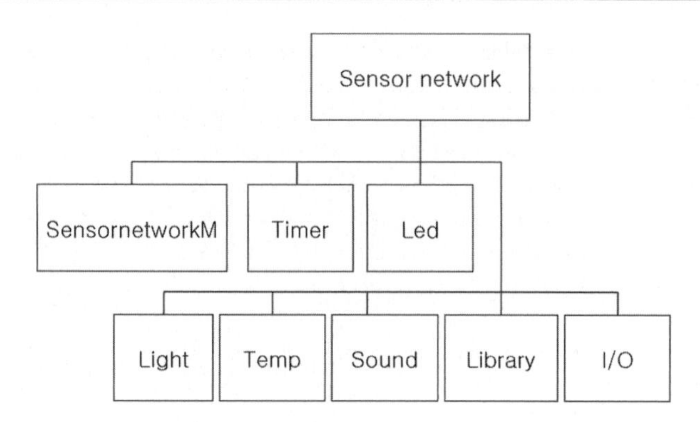

Fig. 20.5. Structure of data transmission module

20.4.2 Receiver Module

The data receiving part is presented in Fig. 20.6. It receives the data values from receiver module and transfer to central controller. And then, central controller send commands to each sub devices such as an auto-feeder, a mini guidance robot, and wireless sensing devices.

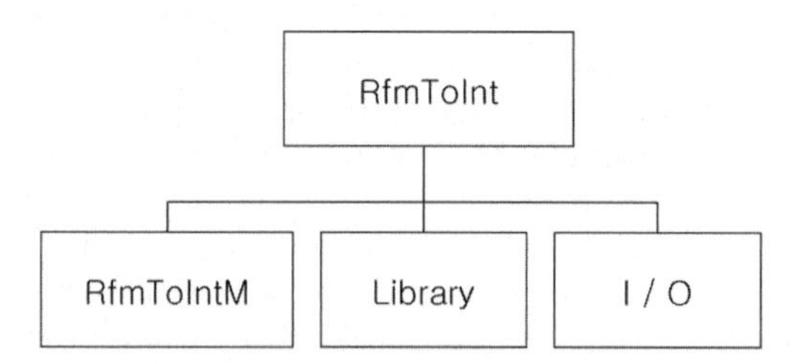

Fig. 20.6. Structure of data receiving module

Fig. 20.6 shows a block diagram of receiver module. It composed of I/O, Library and RfmToIntM module. In case of RfmToIntM, it converts the received I/O data to integer type.

20.4.3 Packet Structure of Wireless Module

The transmission protocol of the WSN system is composed of the 36 bytes. The frame structure and the details of data field are presented in Fig. 20.7. The Add and Type field includes a destination address and data type, respectively. The Group field includes the group ID of Channel. The Length field includes the size of received and transmitted packet length. The Data field includes information of sensed data and its information with 29 bytes long. It allocates information about a light, a sound, and a temperature data. For periodic data transmissions, data packets are transmitted periodically even if there is no sensed information. It includes information about a light, a temperature and a sound in turns.

Add (2byte)	Type (1byte)	Group (1byte)	Length (1byte)	**Data (29byte)**	CRC (2byte)

Light (1byte)	Temperature (1byte)	Sound (1byte)	....	Temperature (1byte)	Sound (1byte)

Fig. 20.7. The details of frame structure and data unit

20.4.4 Graphical User Interface (GUI)

From Fig. 20.8 to 20.10, the implemented GUIs are shown. Fig. 20.8 shows the screenshot of initial setting for wireless communication devices.

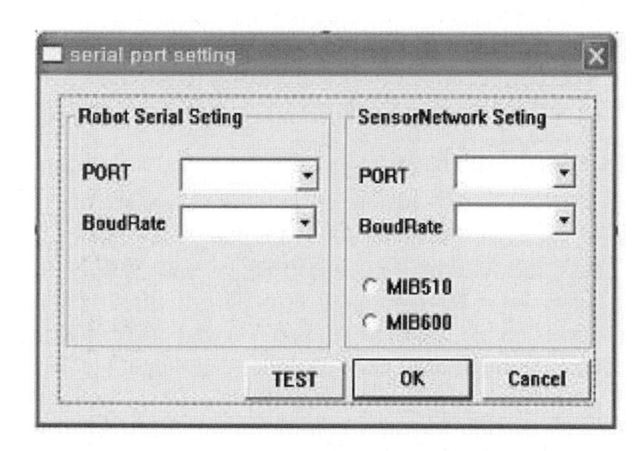

Fig. 20.8. Screen shot of communication setting

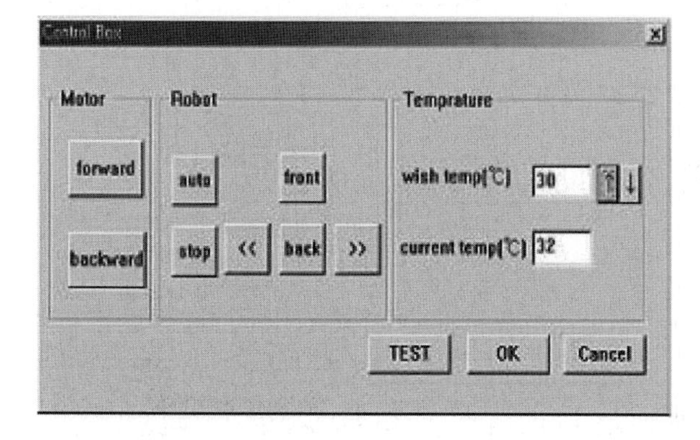

Fig. 20.9. Screen shot of control box

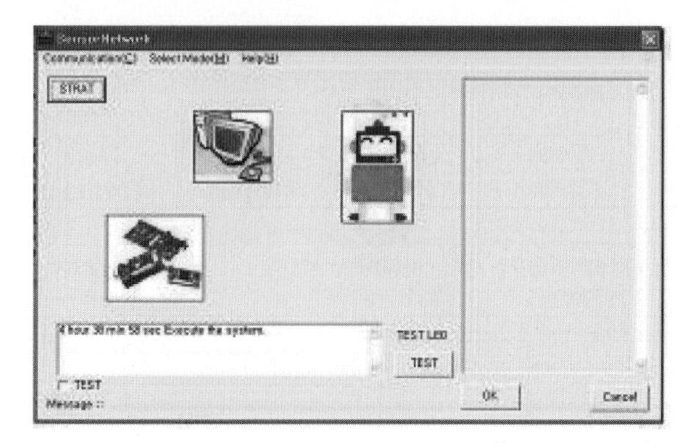

Fig. 20.10. Screen shot of system monitoring program

Fig. 20.9 shows the screenshot of the main control program for the user to monitor the operation status of each control system and sensed data. Fig. 20.10 shows the screenshot of device control module for the user to direct all controllers. It can be preset to automatic or manual mode.

20.5 Sub-control Devices

The entire system includes sensor board, controller, robot, automatic feeder, and host PC. Fig. 20.11 shows overall system structure and its operation. Data communication is performed at RF communication part of wireless platforms and serial communication is connected between receiver and controller. Host PC is also connected to controller via serial link. Controller, automatic feeder, electric fan or heater, and robot are connected via RF wireless links. The wired / wireless system controller creates each control data according to the measured data and delivers those data to each controller. Based on each measured data, guidance robot, automatic feeder and electric fan/heater are operated.

20.5.1 Pet Dog Configuration Radio System

Wireless transceiver installed at a pet dog is combined with a sensor board. The radio transceiver system, located at the pet dog, transmits measured data such as light, heat and sound using wireless communication.

20.5.2 System Controller

Wireless sensor node attached on pet dog is shown to Fig. 20.12. Wired/wireless system controller (Fig. 20.13) received command from host PC to each controller. The CPU supports both serial and wireless communication using 2 UARTs.

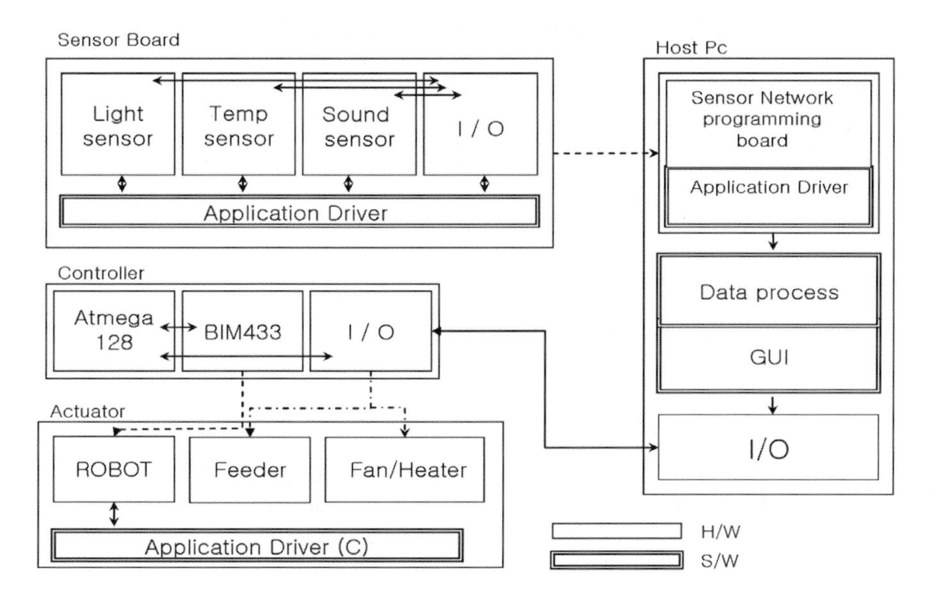

Fig. 20.11. Overall system structure and operation flow

Fig. 20.12. Wireless sensor node attached on pet dog

Fig. 20.13. RF Wireless system controller

The controller of automatic feeder is connected directly to the wired/wireless system controller, and the operation of the pet dog guidance robot uses RF radio communication. Since the manual operation pad for a robot operation is located in controller, the direct control is possible.

20.5.3 Pet Dog Guidance Mini-robot System

If the sensor board detects pet dog's sound, a central controller transmits activating command to the Pet dog guidance robot. Generated command is transmitted to Pet dog guidance robot (Fig. 20.14) by wireless communication through the wired/wireless system controller. At that time, the Pet dog guidance robot is operating according to the predefined self-operation scenarios. It can freely move any direction or be controlled using manual mode by user.

Fig. 20.14. Mini-robot system for pet guidance

The operation of the pet dog guidance robot can be tested under various cases depending on the given conditions. We tested guidance skills using the robot's various movements (circular, forward, backward and complex) and recorded user's voice commands.

20.5.4 Automatic Feeder

An automatic feeder (Fig. 20.15) supplies foods automatically according to typical light condition, which is sensed from developed wireless device. Furthermore, it can feed periodically.

Fig. 20.15. Automatic feeder for experimental test

20.6 Experiments

A pet dog in Fig. 20.16 is used for experimental test. A test environment sets for the developed system are followings. We assume in this case that the test environment is an indoor apartment residential type at daytime.

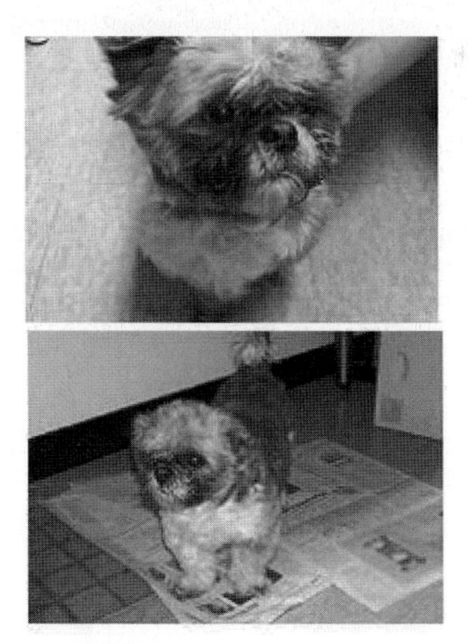

Fig. 20.16. Pet dog for experimental test

- The place where the pet dog eats a food is in a state that the light above 90 % is intercepted.

- When temperature changes ±2°C more than the present, the indoor condition maintenance system is operated.
- If sound measurement comes up 75% of maximum value, central controller judge occurs from pet dog. And then, pet dog guidance robot starts to operate.

In this experimental test, we adjust the automatic feeder starts to operate after dog approaches or other behaviors in the state when the light above 90 % is intercepted, and the pet dog guidance robot starts to operate when sound measurement comes up 75% of maximum value. The robot's various movement (circular, forward, backward and complex) and a recorded voice of person are tested for attract per dog's interest. Also if temperature changes ±2°C more, the electric fan or heater is operated automatically to maintain indoor temperature.

20.6.1 Practical Environment

In the WSN system for pet dog, we developed the user monitoring function which measured data (light, temperature, and sound) from sensor board is displayed in graphic form for easy understanding. General measurement of sample sensor data is shown in Fig. 20.17(a). Each graph displays the information of received light, temperature and sound data. In above-mentioned environment, experimental test result is shown Fig. 20.17(b). The user can identify the measurement data which is displayed on GUI module.

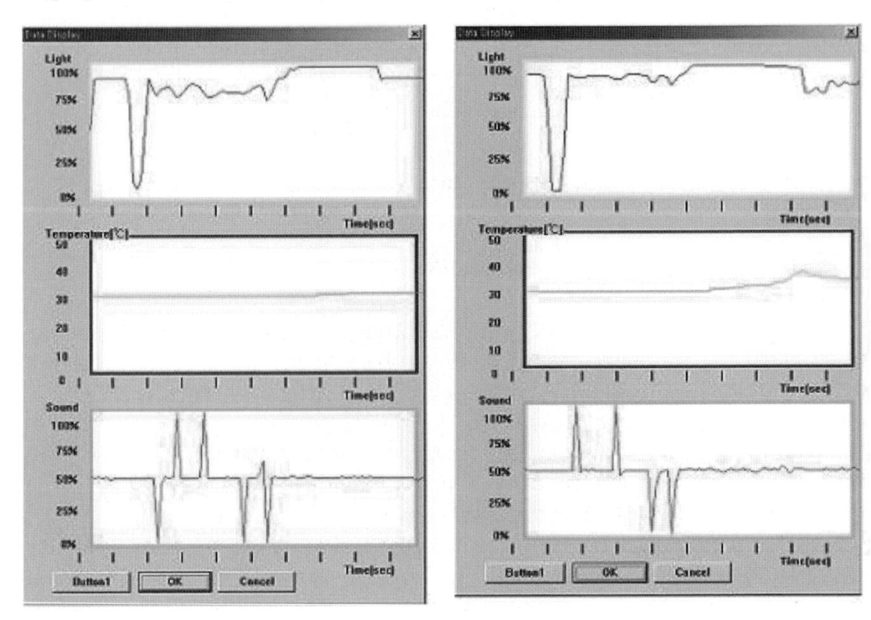

Fig. 20.17. (a) Example of measured sensor data; (b) Screenshots of Experimental measured sensor data

20.6.2 Light Signal Processing

When the developed system detects light changes by pet dog movement, the wireless device sensor board transmits sensing status and information. Received data at central control system is indicated with the graph in the GUI program. System controller generates the automatic feeder activation command. The automatic feeder supplies food to the pet dog automatically. The system structure about the automatic feeder function is given in the Fig. 20.15. In order to find out whether the automatic feeder is worked well or not, it is tested under various angles using various blocks. Finally, the test is conducted by using a small test pet dog shown in Fig. 20.18. In the test result, although there are some differences in sizes, the automatic feeder is worked without an error in most cases.

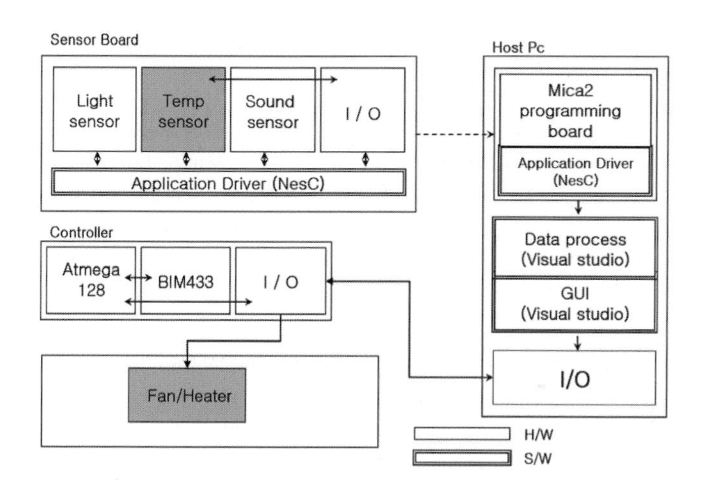

Fig. 20.18. Structure of temperature maintenance

20.6.3 Temperature Signal Processing

In developed system, the host PC processes data by using the measured temperature value. Thus, we developed an expansion part that can control the fan and/or heater according to the measured temperature through wired/wireless system controller. If the temperature value measured from the sensor board is transmitted to the central processor, wired/wireless system controller processes the data. And then, wired/wireless system controller produces the signal to control the operation of each fan and heater. Indoor temperature was 30°C in actual test. We changes surrounding temperature by using heater and fan. If temperature is changed, the order to start or stop the fan and heater is given by the wired/wireless system controller. If the desirable temperature is changed to higher or lower, the commands shown in the Fig. 20.19 are given to the pan and heater according to temperature. As shown in the figure when recent temperature is 30°C, and the user wants to adjust the desirable temperature up to 34°C, fan is stopped and heater is operated. If the desirable temperature is changed to 26°C again, the fan starts to operate. When there is difference about ±2°C between indoor temperature and desired temperature in the controlled item, heater, and fan are operated. Developed part can be expanded to the system that can make alarm and clean the air automatically when pet dog spoils the atmosphere.

```
>> Wish temperature is setting 26
>> Start Aircon
>> Wish temperature is setting 27
>> Wish temperature is setting 28

>> Wish temperature is setting 28
>> Wish temperature is setting 27
>> Wish temperature is setting 26
11 hour 57 min 58 sec Excute the system.

>> Wish temperature is setting 31
>> Wish temperature is setting 30
>> Stop Aircon
>> Wish temperature is setting 29

>> Wish temperature is setting 34
>> Wish temperature is setting 33
>> Start Heater
>> Wish temperature is setting 32
```

Fig. 20.19. Command lines for temperature maintenance

20.6.3 Sound Signal Processing

When the dog sound is sensed in the developed sound-processing module, it is transmitted to programming interface wirelessly. The central controller has the role to transmit a command signal into wireless system controller. Wired/wireless system controller uses the RF wireless communication to operate the pet dog guidance mini-robot. The system structure is shown in the Fig. 20.20.

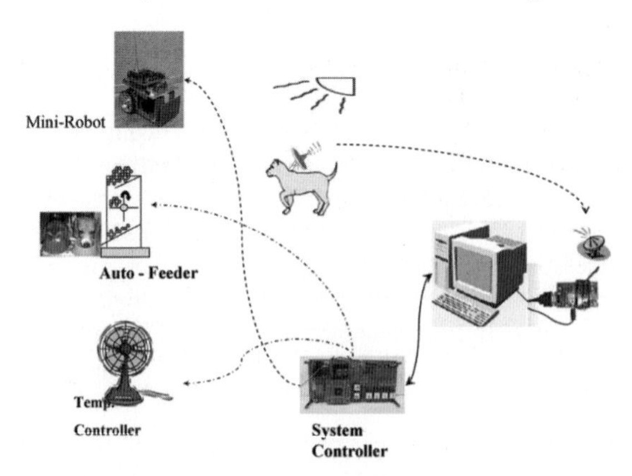

Fig. 20.20. Operation scenario of guidance mini-robot

In developed system, if the sound measurement value exceeds the set value, the central controller determines that the dog is barking, thus, the guidance mini-robot is operated to attract the interest of the pet dog to prevent noise. Specially, in this system, reaction sound signal is set with the value to react only in the frequency domain, which the pet dog barks. If the noise level is higher than the threshold, pet dog guidance mini-robot will be operated by predefined scenario or automatic feeder will be operated.

20.7 Conclusions

In this chapter we tried to find ubiquitous computing possibilities of pet dog management using WSNs. Our goal in designing the pet dog management system was to develop an efficient interface system between a sensed data and each device such as guidance robot or auto food feeder using wireless links. In summary, we have described a wireless management system for a pet dog using wireless sensor network. The developed intelligent wireless management system is composed of a central control system, an automatic feeder, a mini guidance robot, an electric fan, a heater, and some wireless sensing devices. The developed system uses three sensed data types such as light, temperature, and sound from a pet dog and surrounded environment, respectively. The presented design method using those data provides an efficient way to control and monitor the pet dog using wireless links. The implemented system can be used as a design framework of portable devices for the pet-dog management. To enhance the usefulness of implemented system, it is necessary to develop the operation modules using differences of utterance frequency, and physiological signal such as the heart beat count of pet dog.

20.8 References

Aakvaag N, Mathiesen M, and Thonet G (2005) Timing and power issues in wireless sensor networks - an industrial test case. Int. Conf. on Parallel Processing, pp 419 - 426

Akingbehin K and Akingbehin A (2005) Alternatives for short range low power wireless communications. 1st Int. Workshop on self-assembling wireless networks, pp 320 - 321

Clara Palestrini, Emanuela PP, and Marina Verga (2005) Heart rate and behavioral responses of dogs in the Ainsworth's Strange Situation: A pilot study, Applied Animal Behavior Science, 94:1-2, pp. 75-88

Judit Vas, József Topál, Márta Gácsi, and Vilmos Csányi (2005) A friend or an enemy? Dogs' reaction to an unfamiliar person showing behavioral cues of threat and friendliness at different times. Applied Animal Behavior Science, 94:1-2, pp 99-115

Kumagai J (2004) Life of birds - wireless sensor network for bird study. IEEE Spectrum, pp 42 - 49, 41:4, April

Lynch C and O'Reilly F (2005) PIC-based TinyOS implementation. Proceedings of the Second European Workshop on Wireless Sensor Networks, pp 378 - 385

Li Yihan, Panwar SS, and Burugupalli S (2004) A mobile sensor network using autonomously controlled animals. 1st Int. Conf. on Broadband Networks, pp 742 – 744

Lorincz K, Malan DJ, Fulford-Jones TRF, Nawoj A, Clavel A, Shnayder V, Mainland G, Welsh M, and Moulton S (2004) Sensor networks for emergency response: challenges and opportunities IEEE Pervasive Computing, 3:4, pp 16 – 23

Nemeroff G and Hampel D (2001) Application of sensor network communications. Int. Conf. on Military Communications, vol 1, pp 336 - 341

Sikka P, Corke P, and Overs L (2004) Wireless sensor devices for animal tracking and control, IEEE Int. Conf. on Local Computer Networks, pp 446 - 454

Timmons N and Scanlon WG (2004) Analysis of the performance of IEEE 802.15.4 for medical sensor body area networking. IEEE Conf. on Sensor, Ad Hoc Communications and Networks, pp 16-24

Tim Tau Hsieh (2004) Using sensor networks for highway and traffic applications. IEEE Potential, 23:2, pp 13 – 16

21 Agriculture Monitoring

HG Goh, ML Sim and HT Ewe

Mutimedia University, Jalan Multimedia, 63100 Cyberjaya, Malaysia

21.1 Introduction

The world agriculture has experienced several stages of development. It started from the primitive agriculture stage, where human was using stoneware for farming, to traditional agriculture stage, where human started to invent and produce ironwood tools for farming, and now we reach the modern agriculture stage, where information and knowledge are applied in farming activities. Currently, concepts such as precision agriculture (De Baerdemaeker 2001), digital agriculture (Tang 2002), and agroinformatics start to prevail. The application of wireless sensor network (WSN) (Akyildiz 2002) is contributing toward the realisation of those agriculture concepts.

WSN consists of many small and inexpensive nodes. Each of the nodes consists of a simple processor that is equipped with a wireless transceiver and a number of state-of-the-art sensors. Together with its self-organised networking capability it is a suitable candidate for agricultural monitoring. The sensor nodes can sense the environmental changes using a variety of physical, chemical and biological sensors. In agricultural monitoring using wireless sensor nodes, readings from wireless sensor nodes located at different places of a farm will be collected and forwarded to a central server. At the server, the sensor data will be processed and compiled into useful information. This information can then be sent to the farmers through Internet or cellular network. With this technology, farmers can monitor their farm conditions without having to be physically present there. In addition, through the use of appropriate information technology, suitable course of action may be recommended.

In many places of the world, the size of a farm can be as large as several hundred acres. Different portions of the farm may have different microclimates. The traditional way of monitoring a large-scale farm requires a lot of human power. The emergence of WSN applications will enable the farming communities to interact closely with their plants or livestock and gives timely solution to their problems.

A number of WSN applications for agriculture have been implemented worldwide. For example in Canada, Intel Research Laboratory and Agriculture & Agri-Food Canada are using WSN to measure the environmental conditions to help protecting their grapes from frost damage (Burrell 2004). They are also analyzing long-term data of the sensors in order to identify how to harvest the grapes more productively and reduce the use of pesticides and fungicides.

Another example is the recent research carried out by LOFAR-Agro Consortium (Goense 2005) in Netherlands, in which wireless sensor nodes are used to measure the microclimate in potato crops. The detailed information compiled from the sensor readings will be used to improve the knowledge on how to counteract the Phytophtora Infestans fungous disease. Due to the difficulty in obtaining accurate microclimate information under the canopy of the potato crop using remote sensing, WSN is chosen.

Agriculture includes the growing of crops (including grains, fruits, vegetables, flowers, and plants), keeping of livestock, and dairy farming. For different types of agricultural activities, different environmental parameters need to be monitored. Crops are sensitive to weather and environmental changes. Production of crops is thus influenced by those changes. For livestock, the health conditions of the animals are much correlated with the farm environment. In addition, any disease that strikes a particular animal may spread to the whole farm and causes a huge loss. For example, the "Bird Flu" that occurred in many Asian countries in 2005 has caused losses of million of dollars (Reed Business Information Ltd. 2005). Dairy farming needs clean environment, any minor changes will affect the yield of dairy production.

21.2 WSN for Crops Monitoring

The growing of crops can be divided into several stages. In each stage, different parameters are of significance to be monitored, as they will affect the final yield of the crops. Effective monitoring of the farm at different growing stages can help to save cost through the reduction of the usage of excessive resources such as water, pesticide, and fertiliser. In addition, the yield can be maximised due to the use of optimum amount of combination of various resources. For crops monitoring, sensor nodes are usually being placed in a quasi-stationary state. In some of the applications, sensor nodes may be placed into a grid topology as well. The sensor readings are often collected to a central base station or sink node through multihop communication. At the base station, sensor readings are passed to a server for further analysis. Information such as weather data, soil data, images, event detection, and mathematical models can collectively be used as a sophisticated crop management tool for the prediction or recommendation of action to farmers.

Crop such as paddy is very sensitive to the surrounding environmental changes. In a paddy field, it is very important to constantly monitor the on-site field information, which includes the environment conditions, the damage conditions, and the growth states of the crops. Fig. 21.1 shows a prototype of wireless sensor node that is placed in a paddy field for monitoring purpose. The collected information is useful for the preparation of production plan, the prediction of growth and amount of harvest, the prevention of paddy field calamity, and also the improvement of yield efficiency.

Fig. 21.1. A prototype of wireless sensor node for paddy field monitoring

21.3 Livestock and Dairy Farming Monitoring

Livestock and dairy products are another two important sources of food supply chain for human being. There will always be a need for livestock and dairy farmers to monitor their animals and the dairy production as often as possible. Monitoring of animals can achieve two major objectives: animal health monitoring and animal behavior exploration. Animal health monitoring is important for understanding the physical and environmental conditions of the animals, while animal behavoir exploration through observation and assessment of behavior patterns is important in determining health, minimizing stressful or painful situations of animals, and assisting in the improvement of production practices in the farms. The application of livestock and dairy farming monitoring system through the use of WSN can be integrated with the information from databases, mathematical models, and knowledge bases for yield improvement study. A difference between a crop monitoring system and a livestock monitoring system is that animals are mobile while plants are stationary. Although sensors that are used for monitoring the environmental changes such as temperature and humidity can be placed in fixed locations, sensors that are used to monitor the health of the animals have to be attached with the animals. Thus, different design criteria are needed for livestock monitoring as compared to animal behaviour exploration monitoring.

An example of using WSN for animal monitoring is the cattle health monitoring project (Mayer 2004) carried out in Australia. The monitoring system consists of a number of Berkeley MICA2 motes (Crossbow Technology, Inc. 2005) as wireless sensor nodes, a variety of sensors, and an Ultralite GPRS cellular terminal (Call Direct Cellular Solutions Pte. Ltd. 2005) for the transmission of information to the farmers. By creating a small WSN and interfacing it to the existing cellular network, the animal health status and activities can be investigated without interfering them as shown in Fig. 21.2.

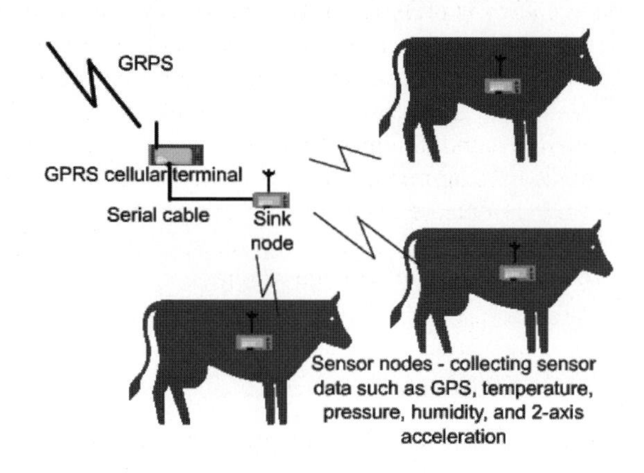

Fig. 21.2. A cattle monitoring system which consists of WSN and cellular network

21.4 Advantages

- Farmers: (i) Can be informed of the environmental changes and make better and earlier decision, and (ii) able to check the farm conditions remotely using mobile devices
- Government: Can predict the yield based on the history of growth record and environmental parameters
- Researchers: (i) Research activities can be carried out more conveniently where sensor readings can be obtained remotely and periodically whenever required without having to be physically present there, and (ii) continuous and long term measurement can be carried out easily

21.5 Issues and Challenges

For agricultural monitoring, sensor nodes are often placed in outdoor environment. This gives different challenges for the outdoor and uncontrolled environment as compared with the indoor and controlled environment. The challenges of using WSN for agricultural monitoring can be divided into the following categories.

21.5.1 Hardware Challenges

- *Limited low cost energy storage:* Currently, most wireless sensor nodes available in the market are using AA alkaline batteries or lithium cells. With proper energy management strategy, these batteries may last for several months. Comparatively, the use of renewable energy source such as solar power or kinetic energy is preferred as it can avoid the trouble of frequent replacement of the battery or even the whole wireless sensor node. However, generally renewable energy converters are much more expensive.
- *Size of the sensor nodes:* The sensor nodes size must be small and suitable for the ease of deployment and minimizing the interruption that might be exerted onto the environment being monitored. The devices must be energy-efficient, with the capability to operate using limited energy source. The sensing devices must produce output that closely represents the parameters being measured under a wide range of input voltage condition.

21.5.2 Protocol Challenges

Sensor data from a far end sensor node can be passed through a number of intermediate nodes using multihop fashion until it reaches the base station or sink node. When all the nodes within the network are using a common radio frequency, interference in one part of the network may corrupt the transmission in another part of the network. The wireless sensor nodes that are at the edge of the network, which have to travel more hops, would have a smaller chance of completing their transmission. In addition, the communication overhead will increase tremendously with the size of the network. These two issues post a question of scalability in a single carrier frequency WSN. Without the expense of higher cost in employing dual- or multi-radio WSN, a more efficient medium access control protocol (Ye 2004), routing protocol (Al-Karaki 2004), and data dissemination or data aggregation protocol (Intanagonwiwat 2003, Sabbineni 2005, Muruganathan 2005) are required. Cross layer protocol design may be required for optimum data aggregation.

21.5.3 Software Challenges

- *Power management:* In every processing task, energy is consumed. A lot of processing is needed for constructing complete information from the sensor readings and thus it uses a lot of energy. However, reducing the processing task may not meet the requirement of applications. Therefore, the microcontrollers in the sensor nodes must be programmed efficiently in order for the application to reduce the energy consumption.
- *Deactivation of sensor nodes:* Sensor nodes in the future are expected to be built in small, cheap, disposable, and discardable nodes for outdoor deployment. Reprogramming and reuse of the sensor nodes would be much difficult

rather than redeploy the new sensor nodes in the interested monitoring fields. It is a challenge to deactivate the old sensor nodes, so that it would not interrupt the functionality of the new sensor nodes.

- *Data sampling and acquisition:* An efficient way in collecting the data samples is needed. For each sampling and acquiring of sensing data, energy will be spent. Frequently acquiring data will cause huge amount of packets being transmitted and this exhausts the energy of the batteries quickly.
- *Dynamic routing for end-to-end data collection:* The destination of the data collection for WSN could be ended anywhere. User may be requesting the sensor readings from a fixed base station or a mobile device that communicate directly to any of the nodes in sensor networks. So, the routing for WSN must not be limited to certain fixed routes. It is a challenging work to come out with a dynamic solution where sensor nodes can freely transmit data to any sensor node using multihop communication with minimum energy consumed.
- *In-network retasking over the wireless:* Since WSN is designed for specific application, the user may need to change certain functionality of individual sensor node from time to time, especially when the sensor nodes are placed at the location where it is not easy to reach. In-network reprogramming will consume a huge amount of energy as compared with other activities in WSN. Thus, an efficient way of in-network reprogramming over the wireless is needed. Retasking can take several different forms of changes to the existing program, such as duty-cycle, sampling rates, making routing decisions, and data processing, dissemination, and aggregation methods.
- *Health and status of the sensor nodes:* For the sensor node to maintain its power budget and local connectivity, a health-monitoring component needs to be running on each sensor device. Health and status of the sensor nodes are checked periodically and usually its data are included in the sensing data packet.

21.5.4 Deployment Challenges

- *Sensor placement:* The challenges in deploying WSN at paddy field are anticipated to be mostly revolving around this aspect. Each sensor must be properly connected to the physical world to ensure close representation of the parameters being measured. This issue requires that each of the sensor nodes be placed in position where it can cater for the need of all sensors attached to it. For example, light sensors must be placed relatively high to avoid being blocked by the plant leaves. Water level sensor, on the other hand, must be close enough to the ground to enable accurate measurement. In tropical countries like Malaysia, frequent raining may cause the sediment to splash and stick onto the surface of various sensors and prevent them from accurately capturing the relevant readings.
- *Nodes positioning and density:* As with any other WSN applications, distribution of sensor nodes throughout the paddy field should provide the end users

with the correct overview of everything that is going on there. Thus, it is important to use the correct number of nodes in a certain size of area, or node density. This will require consideration of radio transmission range, and topography correlation factors.

- *Communication range/environment:* WSN is prone to failure. Limited energy resource and low-rate radio communication have made this kind of communication unreliable in its nature. In order to have reliable communication links, the sensor nodes must be placed close enough to each other to enable reliable multihop communication.
- *Sensor nodes' protective enclosure:* A protective enclosure must be designed. It must be able to protect the sensor nodes from the harsh environment such as varying climate conditions (i.e. extreme heat due to sun radiation and rainfall), muddy soil, and pests' attack.
- *Mobility of the sensor nodes:* WSN for livestock monitoring may require the sensor nodes to be placed on the animal body. This will create two major challenges for the researchers. First is the challenge on how to attach the sensor nodes to the animal body, so that it can stick on the animal body during the monitoring state. Second is the challenge on suitable animal body part to put the sensor nodes, so that it will not interrupt with the animal normal activities.

21.6 Wireless Internet for Agricultural Monitoring

The desire in connectivity has made the Internet grow from wired connection to the wireless connection. For some years, wireless software engineers, communication architects, interaction designers, and mobile content experts have been building a world of wireless Internet applications. By combining the mobile communication networks, Internet, and agricultural monitoring, this will enable the agriculturalists to see their farm directly from their mobile devices. Nowadays, wireless handheld device based on mobile access through cellular network and Internet has emerged as a practical and common mean to collect the real-time sensor data and information at anywhere anytime as compared with the conventional computer-based web access. It is very important to ensure that the user can access to the device as smooth as possible. It will be frustrating if a user needs to wait for long time to receive the requested data and information.

21.6.1 Wireless Internet Applications

A wireless application is software that runs on wireless devices or servers to exchange content and process data over a wireless network. The actual wireless applications are distinguished from one another based on the wireless devices (web phone, handheld, pager, voice portal, web PC, and communicating appliances), networks (WAN, LAN, and PAN), and application families (messaging, web browsing, interacting, and conversing).

Wireless Internet applications can be applied in several fields, such as communication and community (e-mail, calendar, and chat), information (news, weather, traffic, and directories), lifestyle (listings of events, restaurants, movies, and games), transaction (banking, stock trading, purchasing, and auctions), travel (listings of hotel, flights, timetables, and tourist guides), monitoring (home security system, car parking, and health care (Seshadri 2001)) and others (personalised service, location-based services, device-dependent functions, and advertising). Widely applied fields have made this kind of applications benefit to all the society.

21.6.2 Wireless Internet Applications for Agricultural Monitoring

Traditionally, farmers need to spend time manage and monitor their farm. To them, spending whole of their time in taking care of the farms is effort taking with the human power. Even that can be done, a lot of human power needs to be spent. By using the wireless Internet applications in the mobile device for monitoring, farm information can be known anywhere and anytime. Since the cellular network is readily available, WSN just needs to be plugged in and the real time monitoring can be done conveniently. Here is the significance of using the wireless Internet applications for agricultural monitoring:

- *Mobile devices are mobile:* Mobile devices can be brought to anywhere. Application is following the user. This is different from the traditional application, where user needs to reach the fixed phone booth or computer terminal in order to make a communication. The flexibility in mobile devices gives the convenience for the mobile users to access to the Internet at any location as long as the connection is there. This also gives advantage to the farmers to check their farm anytime when they want.
- *24 hours in monitoring farm side:* Once the sensor nodes are deployed, they will be forming self-organised network and operating automatically. Wireless sensor nodes can be programmed to collect periodically sensing readings based on user specification. Traditionally, it is impossible for human to monitor a farm continuously in a whole day. But, by using WSN, sensor nodes can be placed in the farm and running 24 hours non-stop monitoring. Collected data can be sent to the base station for further analysis.
- *Information and critical information reach to the agriculturalists directly:* The mobile device always follows the owner. In most of the case, farmers are directly connected to the mobile devices. Any alert message from the server can reach the farmers directly and instantly.
- *Wide cellular networks coverage:* Cellular networks such as GSM have wide coverage around the world. In many countries, no matter in the city or rural area, the signal from a cellular network tower is able to reach the user's mobile devices. This will enable the farmers to monitor their farm no matter where they are.

- *Small and specific application:* Although the functions of the mobile devices are limited by the processing power, memory, and power resource, small and specific application in mobile devices actually can give a good performance with application for instant information access to the mobile users.

21.7 Design of Agricultural Monitoring Network System

An agricultural monitoring system should not exert any physical or chemical effect on the crops or the livestock. These include shading the plants, intercepting the environment changes, radiating the plants and the animals, or others. Every single part of the system must be well designed and tested before the real implementation. WSN application for paddy field monitoring can be widely used in Asian countries. The key importance of rice is demonstrated by the fact that more than 90 percent of the world's paddy is grown in Asia, where Asian people typically eat two, three or more times of rice daily. Asia produces on average more than 500 million tons of rice per year in the past 5 years (Food and Agriculture Organisation of the United Nations 2006). Table 21.1 shows the average production of rice over various continents. Among all the continents, Asia contributes the highest production. Although in Asia, land utilised for planting the paddy crops is far larger than other continents, in terms of yield efficiency, it is one of the lowest. Thus, any yield efficiency improvement would have significant impact to the income per capita of Asian countries. A well-managed paddy field actually can produce as high as 10 tons of rice per hectare area.

According to IRRI (International Rice Research Institute 2005), five requirements are needed for a good paddy crop management. Table 21.2 lists the requirements and its relevant parameters. The measurement of parameters summarised in Table 21.2 can be manual intensive (International Rice Research Institute 2002, Balasubramanian 2000). With the emergence of WSN technology, the following parameters can first be considered for monitoring: relative humidity, temperature, and water level are still possible to be monitored.

- Relative humidity – Research from IRRI shows that different micro relative humidity of the paddy field will cause different kinds of diseases to grow in the paddy field. High relative humidity and wet leaves will lead to the presence of the spores, such as rice blast (Rice Doctor 2003) and sheath blight (Rice Doctor 2003), in the air. By monitoring this parameter, the disease prevention work can be done more effectively through cultural practices (such as using balanced rates of nutrition) and chemical control.
- Temperature – The suitable temperature for the paddy seeds to germinate is between 10°C to 45°C. Paddy seeds will be destroyed if exposed to the temperature above 45°C for several hours. Necessary action needs to be taken, such as ensuring that the paddy seeds are always properly flooded for high temperature planting area to prevent high failing rate of seeds germination. Besides that, different temperature in the paddy field will create different mi-

croclimates that attract different kinds of insects and diseases. Monitoring this parameter helps to control the usage of suitable pesticide.

- Water Level – For paddy planting, water is the most important element during the germination to emergence, seeding, tillering, stem elongation, panicle initiation to boot, heading, flowering, and milk stages of the paddy. During the vegetative stage, broadcast seeds on the soil without being properly flooded will increase the risk for rats and birds attack. Besides that, well controlled water level is very important for the weed management control at the early stage of paddy planting. Uniform water depth across the field will contribute to a more uniform crop, higher grain yields, and consistent moisture content in the grain.

Table 21.1: Paddy production statistics – average over past 5 years (2000-2004) (Source: Food and Agriculture Organisation of the United Nations 2006)

Continents	Area harvested (hectare)	Production (metric tons)	Yield efficiency (tons/hectare)
Asia	134,927,252.6	537,656,416.0	3.985
Australia	114,260.0	981,740.0	8.648
Africa	8,808,489.6	18,062,869.0	2.072
Europe	579,539.2	3,230,633.6	5.578
America	1,976,920.0	11,949,712.8	6.044
South America	5,355,897.4	20,666.111.2	3.860

Table 21.2: Good paddy crop management components and the associated parameters

0Requirements	**1Control parameters**
Water management	Water depth across the field
Nutrient management	Nitrogen, phosphorus, potassium, zinc, and sulphur levels
Plant population management	Panicles per square meter
Weed management	Water level
Pest and disease management	Temperature, and relative humidity

Agricultural monitoring network system architecture for paddy field monitoring using WSN is shown in Fig. 21.3. This system consists of three major networks, WSN, Internet, and cellular network. WSN is a group of sensor nodes that are deployed in the agricultural farm and forming self-organised network. Internet connects the WSN's gateway with the cellular network through wired connection. Cellular network connects to the Internet and route the information to the users that may be on the move. Another arrangement can also be designed that the gateway is connected wirelessly to the cellular network directly. Once the network of the wireless sensor nodes has been established, sensor nodes will start operating and periodically collecting the sensor readings. Sensor readings can be collected on user request as well. All the sensor readings finally will be sent to a central node called base station or sink node. Data collected are forwarded to the base station through multihop communication among the sensor nodes. In most of the agricultural monitoring applications, such as paddy field monitoring system, sensor nodes are expected to be placed into grid topology. The sensor nodes are usually in fixed or quasi-stationary state.

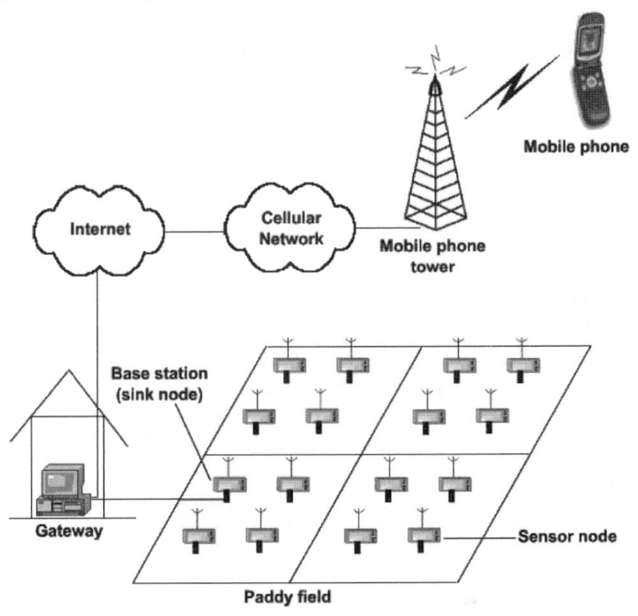

Fig. 21.3. Wireless sensor network architecture for paddy field monitoring

At the base station, data aggregated will be sent to the gateway to be transmitted over the Internet for further data processing. If necessary, knowledge based data processing would be carried out. Database in department of agriculture would be searched for relevant information. In addition, experts from department of agriculture can be consulted. Upon completion of information processing at the server, an alert would be sent at real-time to the farmer together with suitable message and recommendation based on the readings of sensors and information compiled by the server. The farmer can then take immediate actions such as applying fertiliser or pesticide, adjusting water level, and chasing away pests.

The advantages of the system developed are described as follows. Firstly, farmers or workers who are normally lack of technical knowledge can easily deploy it. The wireless sensors can be placed in the paddy field and they will self-organise and be connected to the gateway through sink node and then to Internet and cellular networks. Secondly, the system can provide timely relevant information and intelligent recommendation based on the readings of the sensors. Useful information from department of agricultural database or experts can be utilised to provide real-time consultation to farmers. Lastly, it helps to improve the overall paddy yield while saving cost of unnecessary fertiliser or pesticide.

21.8 Wireless Sensor Network

21.8.1 Hardware and Firmware

Several hardware platforms for WSN, such as i-mote (Intel Cooperation 2006), Berkeley mote (Hill 2000), BWRC picoradio node (Rabaey 2002), and UCLA Medusa MK-2 (Savvides 2002) have been developed for various kinds of application purposes. Monitoring application such as vineyard monitoring (Baard 2003, Burrell 2004) and cattle health monitoring (Mayer 2004) are using the WSN platform from University of California, Berkeley (UCB). This platform was developed by the collaboration from UCB and Intel Berkeley. Currently, many academics and industry players are working together in building their own sensor boards, software, and algorithms for their specific applications. UCB's mote family has evolved over the past few years. It is running in specific-design software for network embedded systems named TinyOS (Hill 2000). TinyOS was created to facilitate the self-organizing of nodes into a sensor network. TinyOS has a programming model tailored for event-driven applications. nesC (Gay 2003) is used as the programming language for TinyOS. nesC supports a programming model that integrates reactivity to the environment, concurrency and communication. For paddy field monitoring, the hardware design must consider the following criteria:

- Easy to be deployed for the farmers
- Sensor nodes must be small and inexpensive
- Robust and protective from the outdoor environment
- Must not have physical and chemical effects for the crops

In order for the application to achieve a better performance in terms of network lifetime for paddy field monitoring, the design of the firmware must consider the following considerations:

- Simple operating system (such as TinyOS and TinyGALS (Cheong 2003)) that consumes very little energy consumption for the operations
- The microcontroller application must be small and specific
- Able to support large network
- For the encryption algorithm, if not necessary, needs not be implemented

21.8.2 Protocol and Algorithm

Networking of wireless sensor nodes involves multiple layers of the protocol stack. Protocol design in WSN is very important to ensure that the sensor data can be aggregated correctly and achieve longer network lifetime. Proper data forwarding will reduce the energy consumption in each of the sensor nodes. The following part presents an overview of multihop routing for end-to-end data collection or dissemination between a base station and fixed sensor nodes. The routing paradigms are categorised based on their functionality and operations.

- *One-To-Many:* One-To-Many multihop for sensor networks is used for data dissemination operation, where the data are needed to send to the entire sensor nodes in an area. This operation is running when there is a new deployment for sensor nodes and the base station is searching for the new sensor nodes; or a new task that is needed to forward to all the sensor nodes. The base station initiates this operation. Whenever a node receives new data, it makes copies of the data and sends the data to all of its neighbours, except the node from which it has just received the data. The algorithm converges when all the nodes in the network have received a copy of the data. Examples of this approach are the classical flooding and Trickle (Levis 2004).
- *One-To-One:* One-To-One multihop transmission or intra-network forwarding is usually used for sending a message from a base station to a specific sensor node; getting a sensor reading from an interested node to the base station; or one-to-one forwarding between two nodes. This operation may integrate with higher layer application such as meta-data or data-centric approach. For data-centric approach, it usually performs discovery, routing and querying procedures. The query is generally in the type of "give me data that satisfy a certain condition". An example of algorithm that uses this approach is the Directed Diffusion (Intanagonwiwat 2003).
- *Many-To-One:* Tree-based is one of the common solutions for Many-To-One multihop forwarding for sensor networks. Tree-based aggregation is the most suitable selection for monitoring application due to the ease in constructing the topology and forwarding mechanism. We categorise the routing algorithms based on different types of networks formed. In Flat-Tree networks, the communication range of a network node can only reach its immediate neighbouring nodes (Goh 2004). In Hierarchical-Tree networks, all nodes can communicate directly and freely with all other nodes in the deployment area (Heinzelman 2000). In Position-Based-Tree networks, network nodes have the locations and energy information (Yu 2001, Beaver 2003). In general, different mutihop routings with different objectives are used for different networks.

For paddy field monitoring, the sensor nodes are placed into a hypothetical 2D grid topology and rooted at a base station. Hierarchical-Tree and Position-Based-Tree networks are not suitable for agricultural monitoring. This is because the sensor nodes' communication range allows them to just communicate with their neighbouring nodes within a certain range. On the other hand, position-based data collection with the help of Global Positioning System (GPS) may cause the device price to rise and becomes too expensive for practical deployment.

We proposed a routing algorithm entitled Information Selection Branch Grow (ISBG) algorithm for energy-efficient data aggregation routing in wireless sensor networks. This algorithm is based on the idea of modified Tree-Based routing algorithm (Goh 2004) to achieve balanced nodes distribution. It is clear that any node that consumes the highest energy will be the first node to run out of energy. By minimizing the energy consumption of the highest energy consumption node, the network lifetime can be prolonged. When this is achieved, all nodes should handle the same amount of network traffic. In tree topology, with balanced nodes distribution among all branches, all child nodes attached to a parent node will handle approximately the same amount of network traffic and thus consume approximately the same amount of energy. This will ensure high network lifetime. The idea of minimizing end-to-end network delay is achieved by developing branches where leaf nodes have minimum number of hops from the base station.

Before the ISBG routing algorithm is discussed, the following terminologies are defined:

- Unmarked nodes: nodes that do not have a parent node.
- Weight: the number of unmarked nodes found at the immediate neighbouring nodes.
- Degree of freedom or "growth space" of a node: the sum of weights of the unmarked neighbouring nodes.
- Unit factor: number of children nodes and itself.
- Total aggregated unit: the sum of the unit factors along the path starting from the base station to the node.

The ISBG routing algorithm requires neighbouring information for constructing a balanced tree topology. Neighbouring information can be obtained by periodically broadcasting the existence of the sensor nodes, for example by sending an IMA (I'm Alive) message in WSNDiag protocol (Chessa 2002). The ISBG algorithm consists of a basic algorithm and a two-stage adjustment. The basic routing algorithm iteratively grows a balanced tree outwards from the base station root. All nodes in the topology will periodically broadcast their existence and its neighbouring nodes' information for each step of the iteration. At each step, the algorithm will first select a potential branch to grow. The potential branch is selected based on descending priority order in a series of Consideration I: i) the lightest weight, ii) the smallest number of child nodes of a base station's branch, iii) minimum degree of freedom, iv) the smallest total aggregated unit, and lastly v) random selection. The next step is to select a potential node to be grown. The potential node is selected based on descending priority order in a series of Consideration II: i) the heaviest weight, ii) the maximum degree of freedom, and iii) random selection. Once the potential branch and potential node are found, a link will be established between them. Finally, the algorithm will update the topology information. The whole process will be repeated until all the unmarked nodes are found. Fig. 21.4 illustrates an example of link establishment in 4×4 grid topology.

The topology that is formed after the ISBG basic algorithm may not be able to achieve balanced distribution for a large topology, such as 10×10 grid topology. Thus, Adjustment I is needed to achieve better balanced distribution and enhance the network lifetime after the construction of network topology from the basic algorithm. Adjustment I is done through pushing the neighbouring branch nodes from the heavier branch to the lighter branch or pulling the neighbouring branch nodes from the heavier branch to the lighter branch. It is also considered as Inter-Branch Adjustment. The network lifetime and the end-to-end network delay are observed to increase after the Adjustment I. Thus, Adjustment II is needed to reduce the end-to-end network delay. Adjustment II is done by moving a node attached from a higher total-aggregated-unit parent node to a lower total-aggregated-unit parent node of the same branch from the base station. It is also called the Intra-Branch Adjustment.

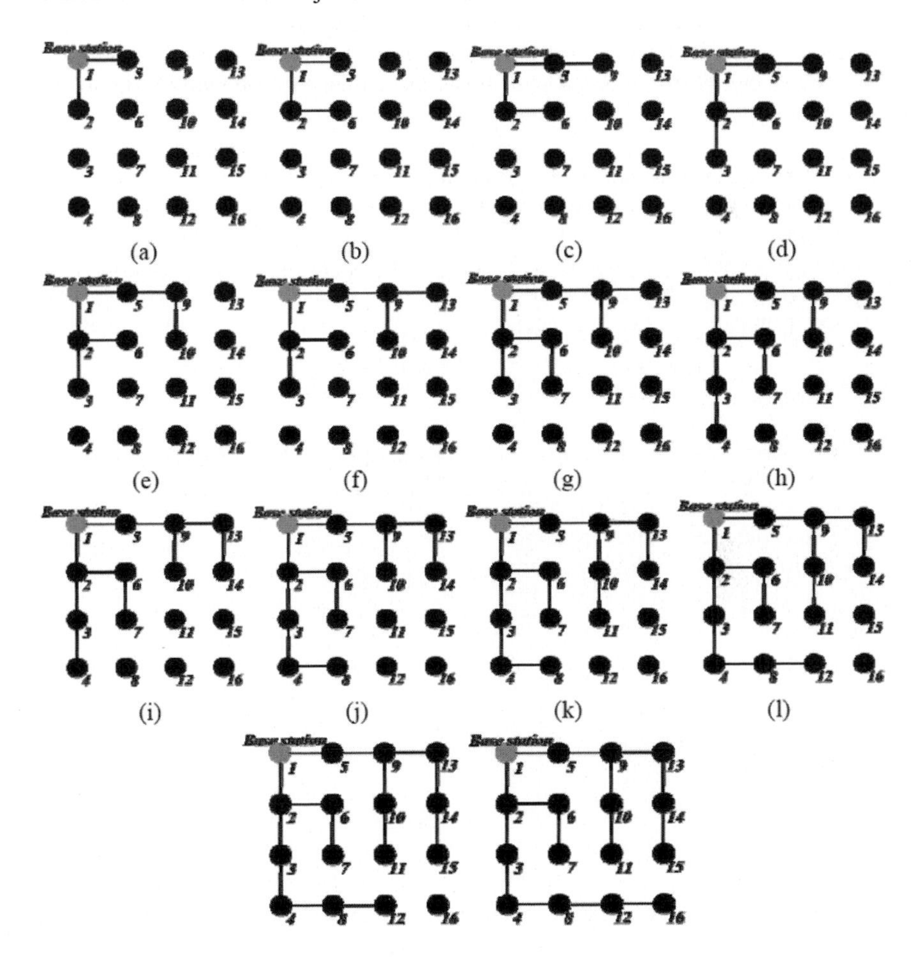

Fig. 21.4. Step by step link establishment for ISBG basic algorithm

For performance analysis, a discrete event based simulator has been developed using C procedural programming language. A 100-node placed in 10×10 grid topology is modeled together with a base station. Distances between the wireless nodes are the same and equal to *d*. First-order radio model (Heinzelman 2000) is used for the calculation of energy consumption. In this simple radio model, it is assumed that the energy dissipated to run the transmitter circuitry, $E_{Tx\text{-}elec}$ is the same as the receiver circuitry, $E_{Rx\text{-}elec}$ where $E_{Tx\text{-}elec} = E_{Rx\text{-}elec} = E_{elec.}$

The equation used to calculate the cost of transmitting a k-bit data packet across a distance d is defined as follows:

$$\begin{aligned} E_{Tx}(k, d) &= E_{Tx\text{-}elec}(k) + E_{Tx\text{-}amp}(k, d) \\ &= E_{elec} \times k + E_{amp} \times k \times d^2 \end{aligned} \tag{21.1}$$

where, E_{amp} is the energy dissipated by transmitter amplifier to achieve an acceptable signal-to-noise. The receiving cost for a k-bit data packet is given by:

$$\begin{aligned} E_{Rx}(k) &= E_{Rx\text{-}elec}(k) \\ &= E_{elec} \times k \end{aligned} \tag{21.2}$$

Each node is assumed to have E_{total} Joule of energy at the beginning. Network lifetime will end if any of the nodes in the topology is considered running out of the energy. The following equation for the time it takes for a packet to be transferred from a transmitter to a receiver is considered. End-to-end packet delay,

$$T = T_{frame} + T_{propagation} \tag{21.3}$$

where T_{frame} = Data packet size / channel transmission rate and $T_{propagation}$ = Link length / signal propagation speed. It should be noted that any delays due to processing, queuing, acknowledgement, or negative acknowledgement are not taken into consideration. Sensor readings will be aggregated by cascading the contents from different sources at the intermediate relay nodes without any analysis or further processing. This process will be repeated at intermediate relay nodes before the data finally reach the base station. Simulation parameters as shown in Table 21.3 are used for the simulation. Several comparisons on average of the network lifetime and the end-to-end network delay have been done for the ISBG routing algorithm with the Stream-Based (Zhang 2003), the Row-To-Column-Based (Zhang 2003), the Cluster-Based (Zhang 2003), the Tree-Based (Goh 2004), and the NCLB (Dai 2003) routing algorithms. The simulation is carried out by selecting the base station at every location of the grid once at a time. For each location where the base station is being selected, the simulation is run for 1000 times.

Fig. 21.5 shows the average network lifetime for six 2D grid routing algorithms. Among all the algorithms, the ISBG routing algorithm performs the best among the others. The ISBG routing algorithm is able to distribute the nodes more evenly over all possible branches and thus able to achieve longer network lifetime. For data aggregation, packet length will be longer when the packet is closer to the base station. So, methods with sequential collection of the sensor readings, such as the Stream-Based routing algorithm will highly suffer from poor network lifetime. For paddy field monitoring, a season of the paddy field from planting stage until harvesting stage is around 120 days or 4 months. Therefore, the deployment of

WSN for paddy field monitoring must at least operate for a season of the paddy planting stage.

Table 21.3: Simulation parameters

Energy dissipated by transmitter circuitry, $E_{Tx\text{-}elec}$	50 nJ/bit
Energy dissipated by receiver circuitry, $E_{Rx\text{-}elec}$	50 nJ/bit
Energy dissipated by transmitter amplifier, E_{amp}	100 pJ/bit/m^2
Initial energy in each node, E_{total}	0.5 J
Link length, d	0.5 m
Channel transmission rate	4096 bps
Signal propagation speed	3.0×10^8 m/s
Data size in each node	4 bytes

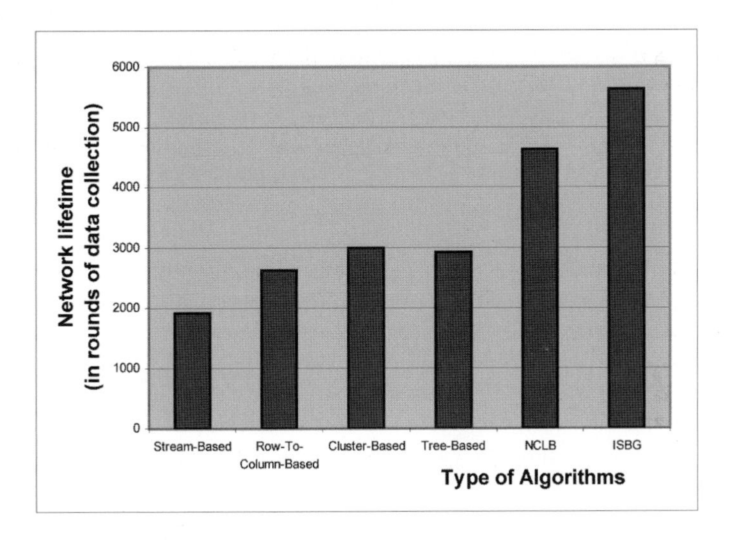

Fig. 21.5. Average network lifetime comparison for stream-based, row-to-column-based, cluster-based, tree-based, NCLB, and ISBG routing algorithms

Fig. 21.6 shows the average of the end-to-end network delay for six 2D grid routing algorithms. The ISBG routing algorithm is able to achieve the shortest end-to-end network delay among all the routing algorithms. The ISBG routing algorithm ensures that the far end of the sensor nodes always use the shortest path to aggregate the sensor data and branches are kept to minimum along the path. The Stream-Based routing algorithm causes a long delay for the data aggregation. This is because the packet needs to be forwarded sequentially and this will cause longer time to collect all the sensor readings back to the base station. The ISBG routing algorithm, which performs reasonably well in terms of network lifetime and end-to-end network delay has been proposed. The results show that overall the ISBG routing algorithm is suitable to be used for data aggregation and it is the best selection for end-to-end data collection. The ISBG routing algorithm is suitable for long-term deployment. It can be used in proactive protocols, where some up-to-date routing information from the neighbouring nodes is maintained. This algorithm can also be extended to other kinds of monitoring applications using WSN.

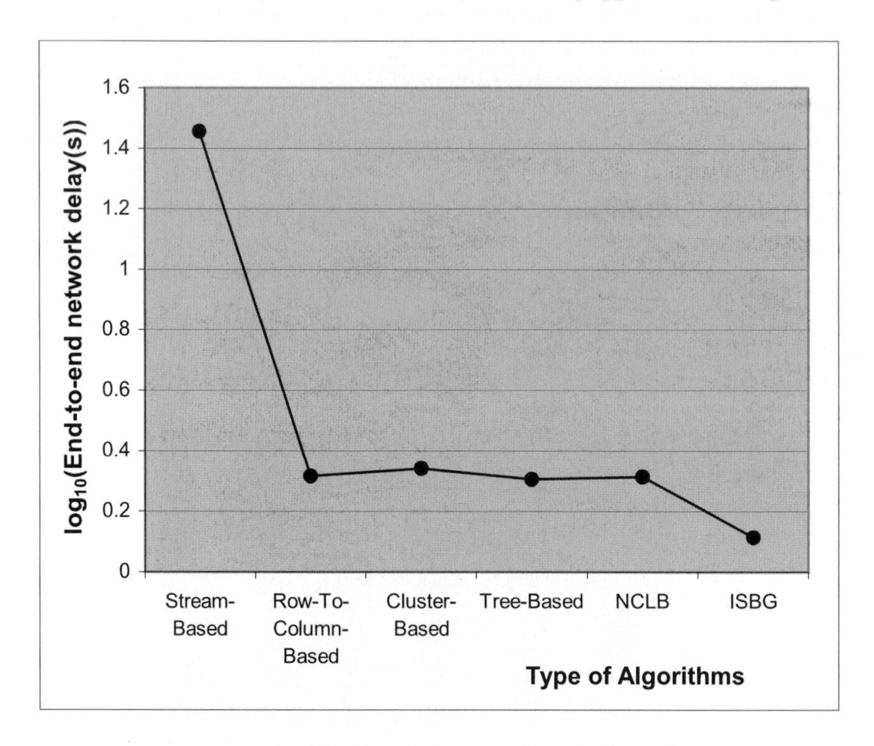

Fig. 21.6. Average of end-to-end network delay comparison for Stream-Based, Row-To-Column-Based, Cluster-Based, Tree-Based, NCLB, and ISBG routing algorithms

21.9 Calibration and Measurement

Calibration is needed to map the raw sensor readings into corrected physical values by identifying and correcting systematic bias. Accuracy and precision in measurement of sensor signal will influence the performance and reliability of a monitoring application. Thus, built-in calibration interface is very important for WSN. Examples of signal calibration can be found in (Whitehouse 2002, Feng 2003). Fig. 21.7 shows an example of calibrated sensor readings for water level measurement. Calibrations can be done at the sensor nodes, the base station or other processing unit. For agricultural monitoring where the calibration is not done at the sensor nodes, the packet payload should consist of source ID, sensor type, and the raw sensor reading. This will ensure that the raw data can be calibrated at a later time.

21.9.1 Internet and Intelligent Processing

Intelligent prediction system: Agroinformatics or sometime referred as agriculture information technology uses the application of computers and information networks to optimally utilise agricultural resources, precisely organise agricultural production, effectively develop sustainable agriculture, and change traditional agriculture into modern agriculture. Various techniques such as geographical information system, optimal simulation, neural networks, generic algorithm, machine learning, knowledge discovery, virtual reality, case-based reasoning (López de Mántaras 1997), or information network can be used for developing intelligent systems for the prediction of the state of the paddy field.

The sensor readings collected by the base station will be sent to the gateway and then through Internet to a remote server. Together with some databases which contain past records of situations of the field, type of seeds planted, season or date of the planting, types of fertiliser and pesticide being used, and the associated amount of production harvested, intelligent prediction of the state of the field, for example whether the field condition is normal or there is a high potential for diseases to develop, can be carried out.

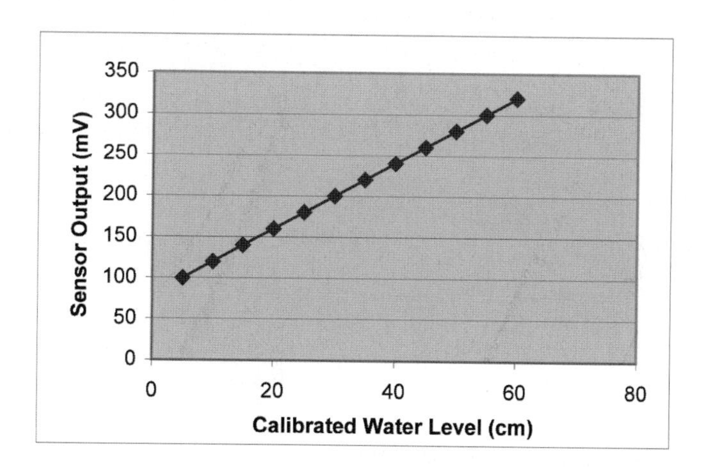

Fig. 21.7. Calibrated sensor readings

21.9.2 Intelligent Consultation System using Mobile Agent

From, Fig. 21.8, after the sensor data are collected and sent to base station/gateway (Fig. 21.8(a)), an intelligent prediction will be made. From the prediction result, mobile agent will be used to search for additional information through cellular network. Databases from agricultural department and weather forecast would be searched and timely consultation will be given to the farmer (Fig.21.8(b)). Finally, the compiled information will reach the farmer through the mobile devices (Fig. 21.8(c)).

21.10 3G Mobile Application

3G networks allow high data rates for packet transmission. It allows transmission of 144kbps in vehicles, 384kbps for pedestrians, and 2Mbps or higher for indoor use. This will allow a large transmission of data such as graphics and audios. In paddy field monitoring, a number of wireless sensors would be installed in the paddy field to collect parameters that would affect the growth and yield of paddy planting. Readings of sensors would be collected and sent to a server over the Internet for data processing. Final result would be sent at real-time to the farmer together with suitable message and recommendation based on the readings of sensors and information compiled by the server.

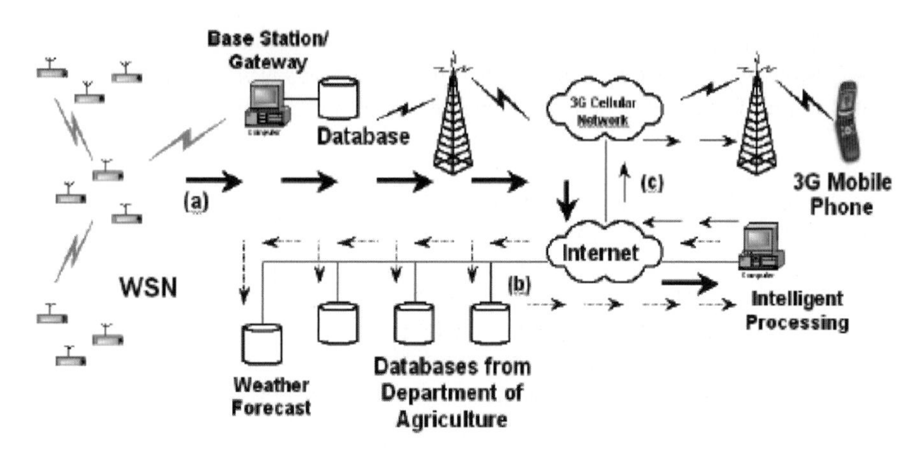

Fig. 21.8. Architecture of intelligent processing system

Fig. 21.9. GUI for 3G paddy monitoring

Since raw data itself are meaningless to a farmer. 3G mobile applications must have a good graphic or animation presentation design. Fig. 21.9 shows two user graphic interfaces (GUI) for 3G mobile applications. The first GUI shows a numerical interface where it presents every detail of the sensor data, while the second GUI shows a graphical interface where an animation will be displayed if an event is found in the field. Both interfaces are built using J2ME MIDP 1.0 (Sun Microsystems, Inc. 2006) and tested using Nokia 7600 3G phone.

Since mobile devices have the constraint on the user display due to the limited screen size, several considerations need to be taken during the development of mobile applications. For paddy field monitoring by government agency and agricultural researchers, numerical interface may be sufficient. The design of the user numerical interface must fulfill the following requirements:

- Time stamp
- Simple numerical data and graph presentations for multiple sensors
- Status of the paddy field after the prediction has been done
- Recommended actions that need to be taken

For the GUI interface, it is aimed to be used by the farmers. The following requirements need to be taken into consideration:

- Numerical data presentation has to be minimised
- Graphical presentation of the paddy field
- Status of the paddy field after the prediction has been done
- Recommended actions after the consultation from the agricultural experts

21.11 Conclusion

The use of wireless sensor network can contribute to modern agriculture. Wireless sensor network for agricultural monitoring system must include the components such as wireless sensor network, intelligent processing, and mobile application. Agricultural monitoring network system offers an alternative way in modern farming. With this technology, we hope it will benefit the farming community in overall yield efficiency.

21.12 References

Akyildiz IF, Su W, Sankarasubramaniam Y, and Cayirci E (2002) Wireless sensor networks: A survey. Computer Networks 38:393-422

Al-Karaki JN and Kamal AE (2004) Routing techniques in wireless sensor networks: A survey. IEEE Wireless Communications 11:6-28

Baard M (2003) Wired news: Making wines finer with wireless. [Online] Available: http://www.wired.com/news/wireless/0,1382,58312,00.html

Balasubramanian V, Morales AC, IRRI; Cruz RT, Philippine Rice Research Institute (PhilRice); Thiyagarajan TM, Tamil Nadu Agricultural University; Nagarajan R, Babu M, Soil and Water Management Research Institute (SWMRI), India; Abdulrachman S, Research Institute for Rice, Indonesia; Hai LH, Ministry of Agriculture and Rural Development (MARD) Vietnam (2000) Adaptation of the chlorophyll meter (SPAD) technology for real-time N management in rice: A review. [Online] Available: http://www.irri.org/publications/irrn/pdfs/vol25no1/IRRN25-1Minireview.pdf

Beaver J, Sharaf M, Labrinidis A, and Chrysanthis PK (2003) Location-aware routing for data aggregation for sensor networks. In: Proceedings of the Geo Sensor Networks Workshop (GSN'03), pp 1-18

Burrell J, Brooke T, and Beckwith R (2004) Vineyard computing: Sensor networks in agricultural production. IEEE Pervasive Computing 3:38-45

Call Direct Cellular Solutions Pte. Ltd. (2005) GSM / GPRS modems for wireless data applications, remote monitoring and control. [Online] Available: http://www.call-direct.com.au/gsm-gprs_modems.htm

Cheong E, Liebman J, Liu J, and Zhao F (2003) TinyGALS: A programming model for event-driven embedded systems. In: Proceedings of the 18th Annual ACM Symp. on Applied Computing (SAC'03), Melbourne, Florida, pp 698-704

Chessa S and Santi P (2002) Crash faults identification in wireless sensor networks. Computer Communications 25:1273-1282

Crossbow Technology, Inc. (2005) Motes, smart dust sensors, wireless sensor networks. [Online] Available: http://www.xbow.com/Products/Wireless_Sensor_Networks.htm

Dai H and Han R (2003) A node-centric load balancing algorithm for wireless sensor networks. In: Proceedings of the IEEE Global Telecommunications Conference (GLOBECOM'03), pp 548-552

De Baerdemaeker J, Munack A, Ramon H, and Speckmann H (2001) Mechatronic systems, communication, and control in precision agriculture. IEEE Control Systems Magazine 21:48-70

Feng J, Megerian S, and Potkonjak M (2003) Model-based calibration for sensor networks. In: Proceedings of the IEEE Sensors, pp 737-742

Food and Agriculture Organisation of the United Nations (2006) FAOSTAT home page. [Online] Available: http://faostat.fao.org/

Gay D, Levis P, Behren R, Welsh M, Brewer E, and Culler D (2003) The nesC language: A holistic approach to networked embedded systems. In: Proceedings of the ACM Conference on Programming Language Design and Implementation, San Diego, California, USA, pp 1-11

Goense D (2005) LOFAR-Agro: Fighting Phytophtora using micro-climate measurements. [Online] Available: http://www.lofar.org/p/Agriculture.htm

Goh HG, Sim ML, and Ewe HT (2004) Performance study of tree-based routing algorithm for 2D grid wireless sensor networks, In: Proceedings of the 12th IEEE Int. Conf. on Networks (ICON'04), pp 530-534

Heinzelman W, Chandrakasan A, Balakrishnan H (2000) Energy-efficient communication protocol for wireless microsensor networks. In: Proceedings of the 33rd Hawaii Int. Conf. System Sciences, Maui, Hawaii, pp 3005-3014

Hill J, Szrwcyk R, Woo A, Culler D, Hollar S, and Pister K (2000) System architecture directions for networked sensors. In: Proceedings of the 8th Internatinal Conference on

Architectural Support for Programming Languages and Operating System (ASPLOS'00), Cambridge, MA, pp 93-104
Intanagonwiwat C, Govindan R, Estrin D, Heidemann J, and Silva F (2003) Directed diffusion for Wireless sensor networking. IEEE/ACM Transactions on Networking 11:2-16
Intel Cooperation (2006) Research - research areas - sensor nets / RFID - Intel® Mote. [Online] Available: http://www.intel.com/research/exploratory/motes.htm
International Rice Research Institute (2005) Rice production, training, management, and extension - the rice knowledge bank at IRRI. [Online] Available: http://www.knowledgebank.irri.org/
International Rice Research Institute (2002) Use of the Leaf Color Chart (LCC). [Online] Available: http://www.knowledgebank.irri.org/knowledgeBytes/lcc/lcc_htm/default.htm
Levis P, Patel N, Culler D, and Shenker S (2004) Trickle: A self-regulating algorithm for code propagation and maintenance in wireless sensor networks. In: Proceedings of the 1st Symp. on Networked System Design and Implementation (NSDI'04), pp 15-28
López de Mántaras R and Plaza E (1997) Case-based reasoning: An overview. AI Communications 10:21-29
Mayer K, Taylor K, and Ellis K (2004) Cattle health monitoring using wireless sensor networks. In: Proceedings of the 2nd IASTED Int. Conf. on Communication and Computer Networks, Cambridge, Massachusetts, USA
Muruganathan SD, Ma DCF, Bhasin RI, and Fapojuwo AO (2005) A centralised energy-efficient routing protocol for wireless sensor networks. IEEE Communications Magazine 43:S8-13
Rabaey J, Ammer J, Da Silva J, Patel D, and Roundy S (2002) Picoradio supports ad-hoc ultra-low power wireless networking. IEEE Computer Magazine 33:42-48
Reed Business Information Ltd. (2005) New scientist report on bird flu. [Online] Available: http://www.newscientist.com/channel/health/bird-flu/
Rice Doctor, Int. Rice Research Institute (2003) Rice blast. Available: http://www.knowledgebank.irri.org/riceDoctor_MX/Fact_Sheets/Diseases/Rice_Blast
Rice Doctor, Int. Rice Research Institute (2003) Sheath blight. [Online] Available: http://www.knowledgebank.irri.org/riceDoctor_MX/Fact_Sheets/Diseases/Sheath_Blight.htm
Sabbineni H and Chakrabarty K (2005) Location-aided flooding: An energy-efficient data dissemination protocol for wireless-sensor networks. IEEE Transactions on Computers 54:36-46
Savvides A and Srivastave MB (2002) A distributed computation platform for wireless embedded sensing. In: Proceedings of the Int. Conf. on Computer Design (ICCD'02), Freiburg, Germany, pp 220-225
Seshadri K, Liotta L, Gopal R, and Liotta T (2001) A wireless Internet application for healthcare. In: Proceedings of the 14th IEEE Symp. on Computer-Based Medical System (CBMS'01), pp 109-114
Sun Microsystems, Inc. (2006) Java 2 Platform, Micro Edition (J2ME). [Online] Available: http://java.sun.com/j2me/
Tang S, Zhu Q, Zhou X, Liu S, and Wu M (2002) A conception of digital agriculture. In: Proceedings of the IEEE Int. Geoscience and Remote Sensing Symp. (IGARSS'02), pp 3026-3028

Whitehouse K and Culler D (2002) Calibration as parameter estimation in sensor networks. In: Proceedings of the 1st ACM Workshop on Wireless Sensor Networks and Applications (WSNA'02), Atlanta, Georgia, USA, pp 59-67

Ye W, Heidemann J, and Estrin D (2004) Medium access control with coordinated adaptive sleeping for wireless sensor networks. IEEE/ACM Transactions on Networking 12:493-506

Yu Y, Govindan R, and Estrin D (2001) Geographical and energy-aware routing: A recursive data dissemination protocol for wireless sensor networks. UCLA Computer Science Department Technical Report UCLA/CSD-TR-01-0023

Zhang J and Shi H (2003) Energy-efficient routing for 2D grid wireless sensor networks. In: Proceedings of the IEEE Int. Conf. on Information Technology: Research and Education (ITRE'03), pp 311-315

22 Intelligent CCTV via Planetary Sensor Network

Ting Shan[1], Brian C. Lovell[1] and Shaokang Chen[2]

[1]Intelligent Real-Time Imaging and Sensing Group, EMI, School of ITEE, The University of Queensland, Australia 4072
[2]National Information and Communications Technology Australia (NICTA)

22.1 Introduction

CCTV (Closed circuit TV) systems cover cities, public transport, and motorways, and the coverage is quite haphazard. It was public demand for security in public places that led to this pervasiveness. Moreover, the adoption of centralised digital video databases, largely to reduce management and monitoring costs, has also resulted in an extraordinary co-ordination of the CCTV resources. It is therefore natural to consider the power and usefulness of a distributed CCTV system, which could be extended not only to cover a city, but also to include virtually all video and still cameras on the planet. Such a system should not only include public CCTV systems in rail stations and city streets, but should also have the potential to include private CCTV systems in shopping malls and office buildings. With the advent of third generation (3G) wireless technology, there is no reason, in principle, that we could not include security cameras feeds from moving public spaces such as taxis, buses, and trains. There should also be the possibility of including the largest and cheapest potential source of image and video feeds which are those available from private mobile phone handsets with cameras. Many newer 3G handsets have both location service (GPS) and video capability, so the location of a phone could be determined and the video and image stream could be integrated into the views provided by the rest of the fixed sensor network.

Another reason to investigate the ad-hoc integration of video and images from the mobile phone network into a planetary sensor network comes from a current project of the authors to use mobile smart phones as a low-cost secure medical triage system in the event of natural disasters. In 2005, a phone-based medical triage system being developed jointly by a commercial partner and the University of Queensland was used by medical officers in major natural disaster areas (ABC News 2005) in the aftermath of 1) the tsunami in Banda Aceh, Indonesia, 2) Hurricane Katrina in the USA, and 3) the earthquake in Kashmir, Pakistan. During these trials the need for the delivery of person location services based on robust face recognition through the mobile phone network became apparent. For example such a service could have proved invaluable to quickly reunite families and help determine the identities of missing persons. In major natural disasters, millions of people may be displaced and housed in temporary shelters, as was indeed the case after hurricane Katrina devastated New Orleans. In such extreme disasters is extremely difficult to rapidly determine who has survived and where they are physically located.

22.2 Review

In addition to our mobile phone based medical triage system, a possible testbed for intelligent CCTV is the emerging experimental planetary scale sensor web, IrisNet (Gibbons et al. 2003). IrisNet uses internet connected desktop PCs and inexpensive, off-the-shelf sensors such as webcams, microphones, temperature, and motion sensors deployed globally to provide a wide-area sensor network. IrisNet is deployed as a service on PlanetLab (www.planet-lab.org), a worldwide collaborative network environment for prototyping next generation Internet services initiated by Intel Research and Princeton University. Gibbons et al. (Gibbons et al. 2003) envisage a worldwide sensor web in which many users can query, as a single unit, vast quantities of data from thousands or even millions of planetary sensors. IrisNet stores its sensor-derived data in a distributed XML schema, which is well suited to describing such hierarchical data as it employs self-describing tags. Indeed the robust distributed nature of the database can be most readily compared to the structure of the Internet DNS naming service.

Wide area person recognition and location services are a valuable application that could be deployed on IrisNet. Apart from the obvious use of the technology for public security by law enforcement officers, a case can be made for access by the general public as well. For example, a mother who has lost her child in, say, a shopping mall could simply upload a photograph of her child from the image store in her mobile phone and the system would efficiently look for the child in an ever-widening geographic search space until contact was made. Clearly in the case of IrisNet, there is no possibility of humans being employed to identify all the faces captured by the planetary sensor web to support the search, so the task must be fully automated. Such a service raises inevitable privacy concerns, which must be addressed, but the service also has the potential for public good as in this example of reuniting a worried mother with her lost child. Now we will focus on some of the crucial technologies underpinning such intelligent CCTV services - automatically detecting and recognizing faces in image and video databases.

22.3 Robust Face Recognition

In order to build a robust face recognition system suitable for deployment on multiple unmatched camera sensors with uncooperative subjects, we need to fulfill the four key requirements of accuracy, robustness, scalability, and speed. Our system has three major components comprising: 1) a Viola-Jones (Viola and Jones 2001) face detection module based on cascaded simple binary features to rapidly detect and locate multiple faces from the input still image or video sequences, 2) view-based Active Appearance Models (AAMs) (Cootes and Taylor 1996) to estimate facial pose and compensate for extreme pose angles, and 3) Adaptive Principal Component Analysis (Chen and Lovell 2004) to recognise faces since the method is robust to poor lighting and extreme facial expressions (Fig. 22.1).

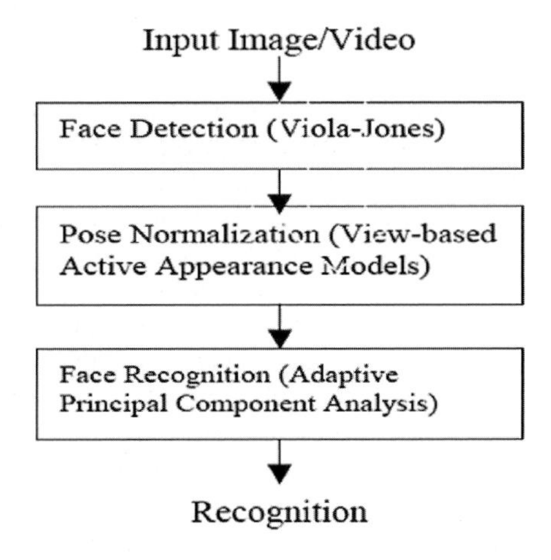

Fig. 22.1. The framework of robust face recognition system

22.3.1 Face Detection

Face detection is a challenging task that has attracted much attention in recent years. It is a necessary first-step in a face recognition system to locate a face or faces from cluttered backgrounds. It also can be used in diverse areas such as human-computer interaction, content-based image retrieval, and intelligent surveillance. Techniques for face detection can be divided into three categories: feature-based approaches, template matching, and image-based approaches (Hjelmas and Low 2001). Feature-based approaches such as using edges (Govindaraju 1996; and Huang et al. 1996), skin color (Lee et al. 1996), motion (McKenna et al. 1995) etc, are suitable for real-time systems due to their fast feature extraction, but they suffer from low detection rates. Edge detection is the first step in edge representation. After detection, these detected edges need to be labeled and matched to a face model. Facial features such as pupils, lips, and eyebrows are normally darker than the regions around them. This property can be exploited to detect facial parts (Hoogenboom et al. 1995) and also the face itself. A number of colour models have been used, including RGB (Satoh et al. 1999), normalised RGB (Sun et al. 1998), HSI (Lee et al. 1996), YIQ (Dai and Nakano 1996), YES (Saber and Tekalp 1998), YUV (Abdel-M and Elgammal 1999). Naturally such colour models are ineffective if the light is not white or the camera is monochrome. Two main template-matching approaches are used. The first approach is "feature searching"

(Jeng et al. 1998) based on relative positions of facial features. This technique first detects prominent facial features, and then uses knowledge of face geometry to verify the existence of a face by searching for the less prominent facial features. Eyes are most commonly used due to their unique appearance. The second major approach is using various deformable face models, such as snakes (Gunn and Nixon 1994; Nikolaidis and Pitas 2000; Yokoyama et al. 1998), deformable templates (Yuille et al. 1992) and point distributed models (Cootes et al. 1996). Image-based approaches treat face detection as a pattern recognition problem and avoid using face knowledge directly. The central idea is to use supervised learning to train a face/non-face classifier. Various statistical methods have been used including Eigenfaces (Turk and Pentland 1991; Sung and Poggio 1998; Yang et al. 2000), neural networks (Rowley et al. 1998; Feraud et al. 1997; Roth et al. 2000), and support vector machines (SVMs) (Osuna et al. 2000). These techniques can generally achieve good performance, but most of them are computationally expensive and thus are not suitable for real-time applications.

In 2001, Viola and Jones (Viola and Jones 2001) proposed an image-based face detection system which can achieve remarkably good performance in real-time. The main idea of their method is to combine weak classifiers based on simple binary features, which can be computed extremely quickly. Simple rectangular Haar-like features are extracted; face and non-face classification is done using a cascade of successively more complex classifiers, which are trained by the AdaBoost learning algorithm. Our face detection module is based on the Viola-Jones approach using our own training sets.

22.4 Cascade Face Detector

The cascade face detector uses a sequence of binary classifiers, which discard non-face regions and only send likely face candidates to the next level of classifier. Thus it employs a "coarse-to-fine" strategy (Fleuret and Geman 2001). Simple classifiers can be constructed which reject the majority of non-face regions at the very early stage of detection, before the use of more complex classifiers with higher discriminative capability. Mre discriminating and complex classifiers concentrate their processing time on face-like regions as illustrated by Fig. 22.2.

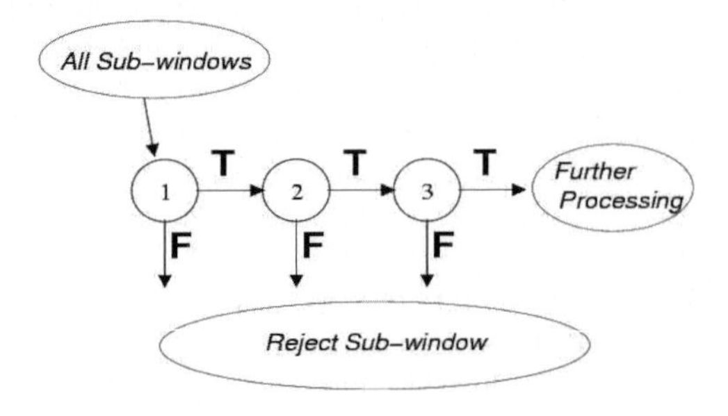

Fig. 22.2. The cascade detection process

22.4.1 Adaboost Classifier

The Haar-like wavelet features (Mallat 1989) extracted from the image subwindows are an image representation method, characterising the texture similarities between the regions by computing the sum of pixel values in different regions.

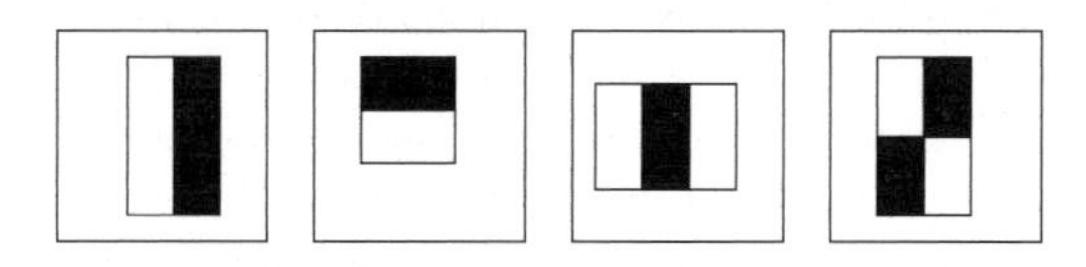

Fig. 22.3. Four types of rectangle features defined in a sub-window. The value of the feature is the difference between the sums of pixels within the white and black rectangles

The value of a two-rectangle feature is the difference between the sums of the pixel values within the two rectangular subregions. The value of a three-rectangle feature is the difference between the sums of pixel values within the two outside rectangles and the sum of pixels of the centre rectangle. A four-rectangle feature computes the difference between sums of pixel values in diagonal pairs of rectangles (Fig. 22.3). Given a subwindow whose size is 24*24 pixels, the exhaustive set of rectangular features is 116,300 (86,400 for the two-rectangle features, 27,6000 for thethree-rectangle features, and 2,300 for the four-rectangle features), which is over-complete. The integral image, also known as the "summed-area table" (Crow F 1984) in the domain of computer graphics, can compute the Haar-like rectanglr features very quickly. Placing the origin at the top left corner of the image, the value of the integral image at location (x, y), denoted ii(x, y), is calculated as the sum of the pixel values contained in the rectangular region bounded by the origin and (x,y). The calculation of the integral image can be calculated efficiently by the recursion:

$$s(x,y) = s(x,y-1) + i(x,y) \text{ and } ii(x,y) = ii(x-1,y) + s(x,y) \quad (22.1)$$

where s(x, y) is the sum of the column pixel values, with initialisation values s(x, -1) = 0, and ii(-1, y) = 0. Thus we can calculate the integral image representation of the image in a single pass (Fig. 22.4). The value of the sum of pixel values in any arbitrary rectangle region can be easily recovered from the integral image.

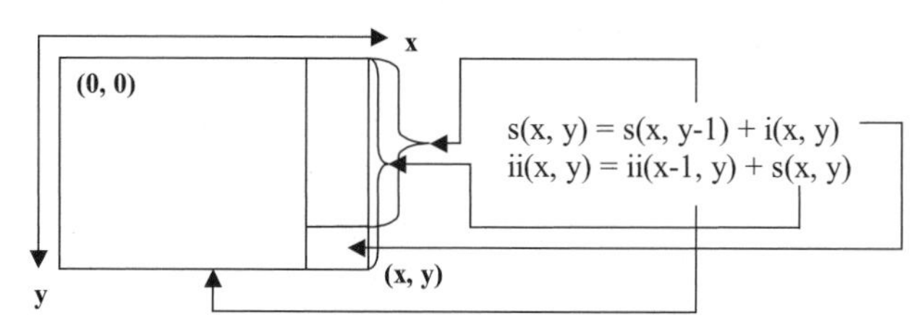

Fig. 22.4. Representation of s(x, y) and ii(x, y)

In Fig. 22.5 the sum of pixel values in rectangle D can be computed by:

$$ii(4) + ii(1) - ii(2) - ii(3) \tag{22.2}$$

where ii(1) is the sum of pixel values in rectangle A, ii(2) is sum of pixel values in A and B, ii(3) is sum of pixel values in A and C, and ii(4) is the sum of pixel values in A, B, C and D.

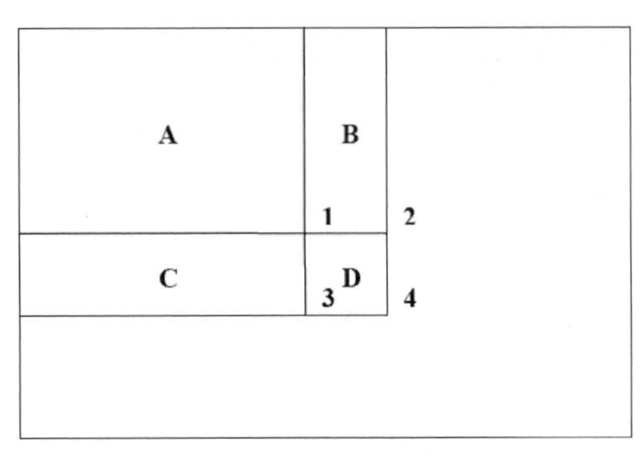

Fig. 22.5. The sum of pixels in rectangle D can be computed by: $ii(4) + ii(1) - ii(2) - ii(3)$

Similarly, to compute two, three and four rectangle features, we need only 6, 8 and 9 integral image values respectively.

22.4.2 Adaboost Learning Algorithm

The Adaboost learning algorithm is used to select the best rectangle features and linearly combine these features into a classifier. Adaboost is a boosting learning algorithm, which can fuse many weak classifiers into a single more precise classifier. The main idea of Adaboost is as follows. At the beginning of training, all training examples are assigned equal weight. During the process of boosting, the weak classifier with the lowest classification error is selected and the weights of the samples, which are wrongly classified by the weak classifier, increase. The final classifier is a linear combination of the weak classifiers of all rounds, where classifiers with lower classification error have a higher weight. Details of the learning algorithm can be seen in Table 22.1. A weak classifier h_j contains a feature f_i, a threshold, θ_i and a direction ρ_i

$$h_j = \begin{cases} 1 & if \quad \rho_i f_i(x) < \rho_i \theta_i \\ 0 & otherwise \end{cases} \tag{22.3}$$

Here x is a 24*24 pixel sub-window of an image.

Table 22.1: The Adaboost learning algorithm. T hypotheses are constructed each using a single feature. The final hypothesis is a weighted linear combination of the T hypotheses where the weights are inversely proportional to the training errors.

- Given example images $(x_1, y_1), \ldots, (x_n, y_n)$ where the labels $y_i = 0, 1$ for negative and positive examples respectively.
- Initialize weights to $w_{1,i} = \frac{1}{2m}, \frac{1}{2l}$ for training example $y_i = 0, 1$ respectively, where m and l are the number of negatives and positives respectively.
- For $t = 1 \ldots T$
  1) Normalize weights, so that w_t can be treated as a probability distribution
  $$w_{t,i} \leftarrow \frac{w_{t,i}}{\sum_{j=1}^{n} w_{t,j}}$$
  2) For each feature j train a classifier h_j which is restricted to using a single feature. The error is evaluated with respect to w_t, $\varepsilon_j = \sum_i w_t \,|\, h_j(x_i) - y_i \,|$.
  3) Chose the classifier h_j with lowest error ε_t.
  4) Update weights according to:
  $$w_{t+1,i} = w_{t,i} \beta_t^{1-e_i}$$
  where $e_i = 0$ if example x_i is classified correctly, 1 otherwise, and $\beta_t = \frac{\varepsilon_t}{1-\varepsilon_t}$
- The final strong classifier is:
$$h(x) = \begin{cases} 1 & \sum_{t=1}^{T} \alpha_t h_t(x) \geq \frac{1}{2} \sum_{t=1}^{T} \alpha_t \\ 0 & otherwise \end{cases}$$
where $\alpha_T = \log \frac{1}{\beta_t}$

22.5 Training and Detection Results

22.5.1 Training Database

The face training database includes 4916 hand labelled faces downloaded from Peter Carbonetto's website (Carbonetto 2005). The negative training data were randomly collected from the internet and do not contain human faces. Some example face images are shown on Fig. 22.6.

Fig. 22.6. Example of some face images used for training (Carbonetto 2005)

22.5.2 Structure of the Detector Cascade

After training, our final detector has 24 layers with a total of 2913 features. Some detection results by our implementation are shown in Fig. 22.7.

22.5.3 Pose Normalisation

Pose normalisation refers to compensating for the pitch, roll, and yaw motions of the head to allow for non-frontal viewing conditions. It acts as a bridge between the face detection and face recognition modules. Our face recognition system is exceptionally insensitive to illumination and facial expression changes and can attain very high recognition rate with frontal view face images.

Fig. 22.7. Detection results on several photographs

22.5.4 In-Plane Rotation

Many facial features can be used to detect in-plane face rotation, including the eyes, mouth, and pupils. In our system, we train an eye localiser and rotate the face image to the eyes horizontal position by computing the angle between two eyes and horizontal baseline.

22.5.5 Eye Localiser

Our eye localiser is trained using the same method with face detection module but different feature set due to the property of eye region. Feature set we used as shown in Fig. 22.8. We used the "BioID" (BioID 2005) face database which contains 1,521 face images (some of them wearing glasses). We manually cut two eye regions from every face image and rescaled them to 32*16 to get 3,042 positive samples. Because in our system, the eye localisation is always carried out after the face detection module, only on the face-like regions, we use the remaining part of the 1,521 face images as the negative training samples. Some eye training samples can be seen in Fig. 22.9. The final eye localiser contains 36 stages and can process images very fast. It doesn't consume significant computation time as it only operates on the face-like regions whose size is relatively small compared to the whole image and faces are also a rare event in the video stream. Also, if the eye localiser can't locate eyes in the face candidate region, the candidate would be discarded as it is likely to have been wrongly detected. In other words, eye localiser acts as a verifier for the face detection module.

22.5.6 Rotation

After we locate the eyes in the face regions, we use the coordinates (x_{left}, y_{left}), (x_{right}, x_{right}) of the eyes to calculate the rotation angle θ (Fig. 22.10) by:

$$\theta = \arctan\left(\frac{y_{right} - y_{left}}{x_{right} - x_{right}}\right) \tag{22.4}$$

Then we rotate the image to become a vertical frontal face image. A face-like region detected by the face detector and the rotated by the in-plane normaliser can be seen in Fig. 22.11.

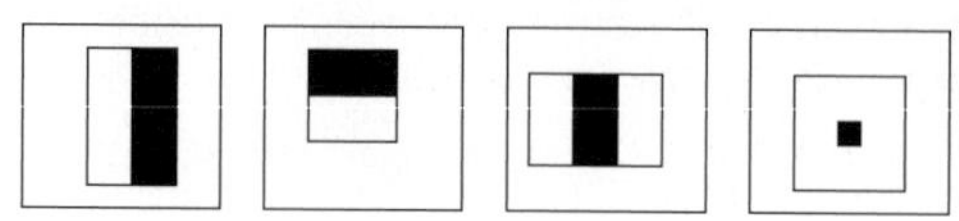

Fig. 22.8. Features set used by eye localiser

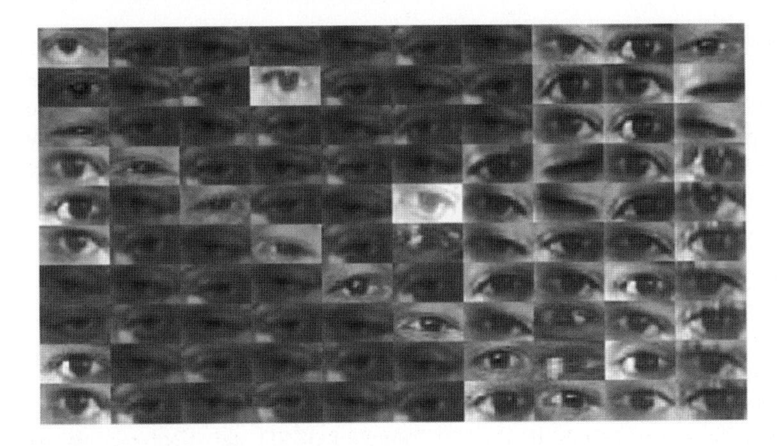

Fig. 22.9. Eye samples for training eye detector

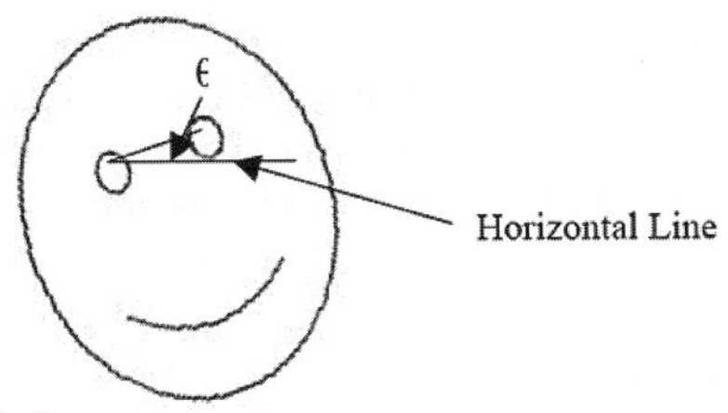

Fig. 22.10. Rotation angle θ

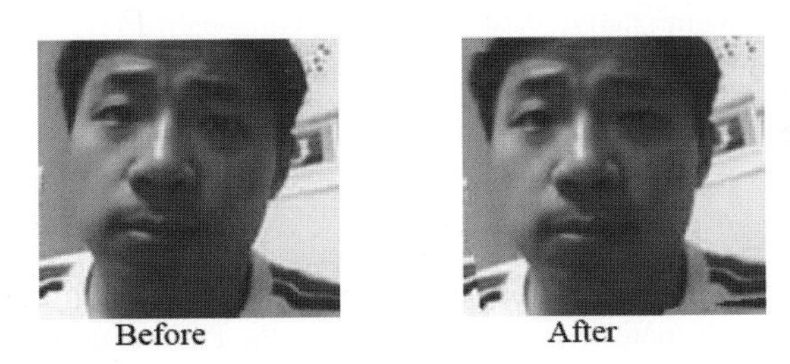

Fig. 22.11. A face sample before and after in-plane image rotation

22.5.7 Out-of-plane Rotation

The out-of-plane rotation problem is much more complicated than in-plane rotation as it requires the construction of a 3D face model. Recently Cootes *et al.* have shown that human faces from different view angles (from left profile to right profile) can be modeled by 3 distinct Active Appearance Models (AAMs) (Cootes et al. 2000), these models can be used to estimate head pose and track faces by switching between the different models and synthesizing any new view of a face from a single view. Motivated by Cootes, we propose a similar approach to solve our out-of-plane rotation problem. In our approach, we apply our face detector to a given input image to locate the initialisation position before the AAM search is carried out. We only use one face model which can represent face pose variation from -45 degree to 45 degree horizontally and -30 degree to 30 degree vertically to compensate for the pose change problem. We argue that it is meaningless to attempt to recognise a person from a face image with too large a pose angle when we only know the person's frontal image — this is even a difficult task for human beings.

22.5.8 Building the Active Appearance Models

The Active Appearance Model is a powerful tool to describe deformable object images, it was originally introduced by Cootes and Taylor (Cootes and Taylor 2001). They demonstrate that a small number of 2D statistical models are sufficient to capture the shape and appearance of a face from any viewpoint (Cootes and Taylor 2001). The Active Appearance Model uses principal component analysis (PCA) on the linear subspaces to model both the shape and texture changes of a certain object class. Given a collection of training images for a certain object class where the feature points have been manually marked, a shape and texture can be represented by applying PCA on the sample shape and texture distributions as:

$$x = \bar{x} + P_s c \text{ and } g = \bar{g} + P_g c \tag{22.5}$$

where $\bar{x}$ is the mean shape, $\bar{g}$ is the mean texture and P_s, P_g are matrices describing the respective shape and texture variations learned from the training sets, and the parameters, **c** are used to control the shape and texture change.

Cootes *et al* demonstrate that an Active Appearance Model trained on near frontal face images can handle pose change of up to 45 degree each side (Cootes and Taylor 2001). In our trials, we collected 20 frontal face images from Feret face database (Feret 2005). We first applied our face detector on these images, and then labeled each of them with 58 points around the main features including eyes, mouth, nose, eyebrows and chin. Some sample training face images can be seen in Fig. 22.12.

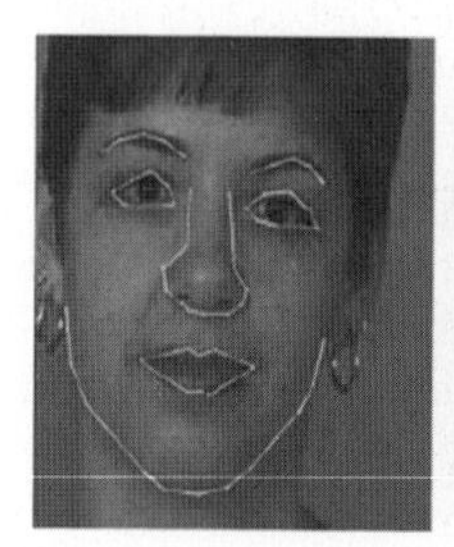
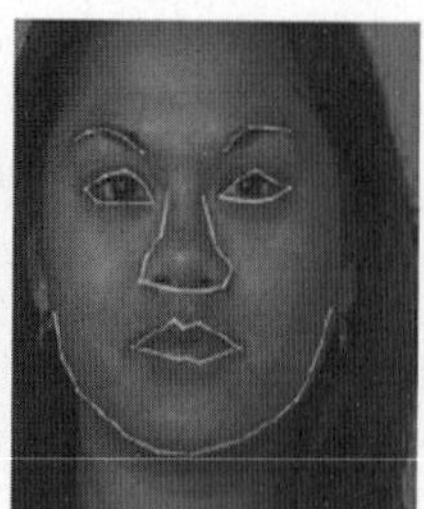
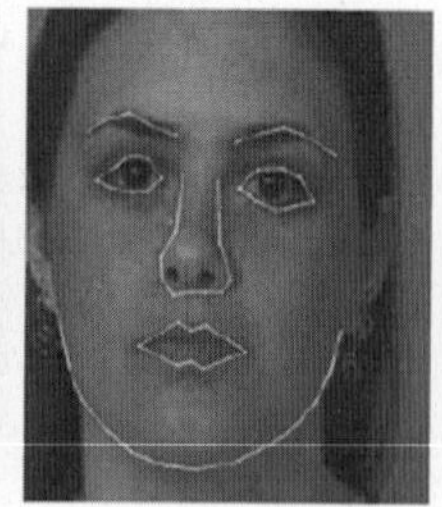
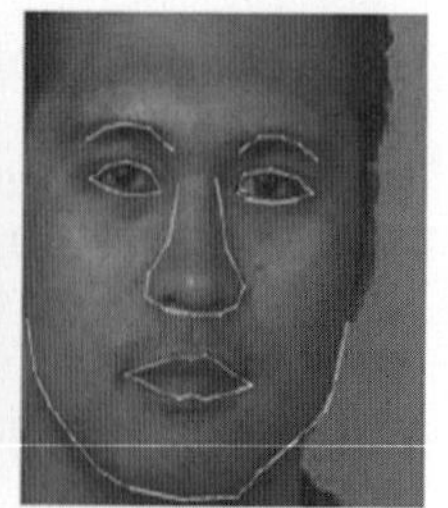

Fig. 22.12. Sample training face images with labeled 58 points on main facial features

22.5.9 Combination of Face Detector with Active Appearance Model Search

The initialisation of the Active Appearance Model Search is a problem since the original AAM search is a local optimisation. Some failed AAM searches due to the poor initialisation can be seen on Fig. 22.13. We solve the initialisation position problem by using our face detector to provide initialisation. The face detector finds the location of a human face in an input image and provides a good starting point for the subsequent AAM search which precisely marks the major facial features, (mouth, eyes, nose etc.). Some results from our combined face, eye detector and AAM search can be seen in Fig. 22.14.

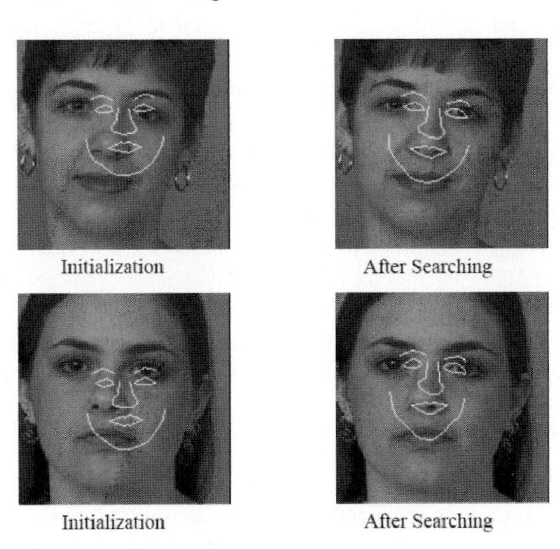

Fig. 22.13. Failed AAM searches due to poor initialisation

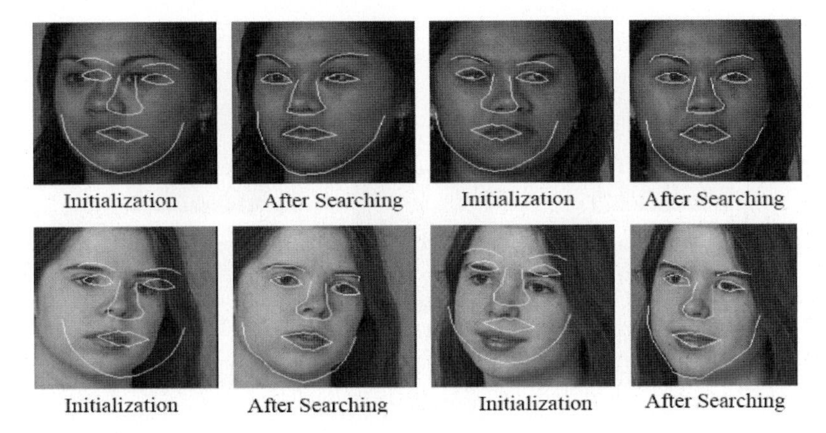

Fig. 22.14. Some AAM search results on Feret Face Database

22.5.10 Predicting Pose

Here we follow the method of Cootes *et al* (Cootes et al. 2000). They assume that the model parameters are related to the viewing angle, θ, approximately by:

$$c = c_0 + c_c \cos(\theta) + c_s \sin(\theta) \tag{22.6}$$

where c_0, c_c and c_s are vectors which are learned from the training data. (Here we consider only head turning. Head nodding can be dealt with in a similar way).

Given a new face image with parameters **c**, we can estimate orientation as follows. Let R_c^{-1} be the left pseudo-inverse of the matrix $(c_c \mid c_s)$, let $(x_\alpha, y_\alpha)' = R_c^{-1}(c - c_0)$, then the best estimate of the orientation is $\tan^{-1}(y_\alpha / x_\alpha)$

22.5.11 Predicting Fontal View

After we estimate the angle θ, we can use the model to synthesise new views, here we will synthesise a frontal view face image, which will be used for face recognition. Let c_{res} be the residual vector not explained by the rotation model, $c_{res} = c - (c_0 + c_s \cos(\theta) + c_s \sin(\theta))$. To reconstruct at a new angle, α, we simply use the parameters: $c(\alpha) = c_0 + c_c \cos(\alpha) + c_s \sin(\alpha) + c_{res}$. Here α is 0, so the equation will be: $c(0) = c_0 + c_c + c_{res}$. By changing parameter **c**, we can reconstruct the new frontal face image. Some synthesised results can be seen in Fig. 22.15.

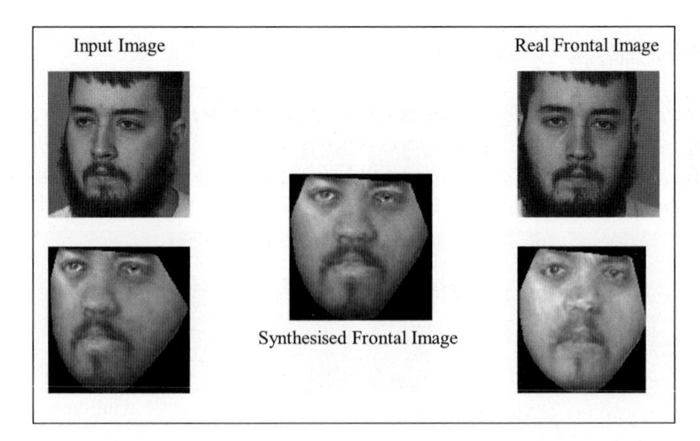

Fig. 22.15. Some synthesised frontal face images from the Feret Face Database

22.6 Face Recognition

Robust face recognition is a challenging goal because of the gross similarity of all human faces compared to large differences between face images of the same person due to variations in lighting conditions, view point, pose, age, health, and facial expression. Most systems work well only with images taken under constrained or laboratory conditions where lighting, pose, and camera parameters are strictly controlled. Recent research has been focused on diminishing the impact of nuisance factors on face recognition. Many approaches have been proposed for illumination invariant recognition (Yilmaz and Gokmen 2000; Gao and Leung 2002) and expression invariant recognition (Beymer and Poggio 1995; Black et al. 2000). But these methods suffer from the need to have large numbers of example images for training, which is often impossible in many situations when only few sample images are available such as in recognizing people from surveillance videos from a planetary sensor web or searching historic film archives. In the last several years, research on face recognition has been focused on diminishing the impact of changes in lighting conditions, facial expression, and pose. Chen and Lovell developed APCA (Chen and Lovell 2004) and Rotated APCA to compensate for illumination and facial expression variations.

22.6.1 Adaptive Principal Component Analysis

We first apply Principal Component Analysis (PCA) (Turk and Pentland 1991) for feature abstraction because of its good generalisation capacity. Every face image can be projected into a subspace with reduced dimensionality to form an m-dimensional feature vector $s_{j,k}$ with $k = 1,2,\cdots K_j$ denoting the kth sample of the class S_j.

22.6.2 Bayes Decision Rule

After constructing the face subspace for image representation, we need to warp this face space to enhance class separability. The Bayes classifier is the best classifier which achieves minimum error rate for pattern recognition if prior probabilities are known. The conditional density function is:

$$p(s \mid S_j) = \frac{\exp[-\frac{1}{2}(s - u_j)^T \operatorname{cov}_j^{-1}(s - u_j)]}{(2\pi)^{\frac{m}{2}} \mid \operatorname{cov}_j \mid^{\frac{1}{2}}} \tag{22.7}$$

where, u_j is the mean of class S_j and cov_j is the covariance matrix of S_j.

22.6.3 Whitening and Eigenface Filtering

In order to compensate for the influence of between-class covariance on the estimation of pdf, we introduce a whitening power p to control the distribution, that is

$$\mathrm{cov} = diag\{\lambda_1^{-2p}, \lambda_2^{-2p}, \ldots \lambda_m^{-2p}\} \tag{22.8}$$

where, $\lambda_i \, (i = [1 \ldots m])$ are the eigenvalues extracted by PCA. Consequently, the whitening matrix Z is: $Z = diag\{\lambda_1^p, \lambda_2^p, \ldots \lambda_m^p\}$, where the exponent p is determined empirically.

The aim of filtering is to enhance features that capture the main differences between classes (faces) while diminishing the contribution of those that are largely due to nuisance variations (within class differences) such as lighting. We thus define a filtering parameter γ, which is related to identity-to-variation (ITV) ratio. The ITV is a ratio measuring the correlation of a change in person versus a change in variation for each of the eigenfaces. For an M class problem, assume that for each of the M classes (persons) we have examples under K standardised different lighting conditions. Let us denote the ith element of the face vector of the kth lighting sample for class (person) S_j by $s_{i,j,k}$. Then,

$$ITV_i = \frac{BetweenClassCovariance}{WithinClassCovariance} = \frac{\frac{1}{M}\sum_{j=1}^{M}\frac{1}{K}\sum_{k=1}^{K}| s_{i,j,k} - \varpi_{i,k} |}{\frac{1}{M}\sum_{j=1}^{M}\frac{1}{K}\sum_{k=1}^{K}| s_{i,j,k} - \mu_{i,j} |} \tag{22.9}$$

$$\varpi_{i,k} = \frac{1}{M}\sum_{j=1}^{M} s_{i,j,k} \quad \mu_{i,j} = \frac{1}{K}\sum_{k=1}^{K} s_{i,j,k}$$

Here $\varpi_{i,k}$ represents the ith element of the mean face vector for lighting condition k for all persons and $\mu_{i,j}$ represents the ith element of the mean face vector for person j under all lighting conditions. We then define the filtering matrix γ by, $\gamma = diag\{ITV_1^q, ITV_2^q, \ldots ITV_m^q\}$, where q is an exponential scaling factor determined empirically. After transformation, the conditional pdf is given by:

$$p(s \mid S_j) = \frac{\exp[-\frac{1}{2}\sum_{i=1}^{m}\frac{(s_i - \mu_{i,j})^2}{\lambda_i^{-2p} ITV_i^{-2q}}]}{(2\pi)^{\frac{m}{2}}\prod_{i=1}^{m}\lambda_i^{-p} ITV_i^{-q}} \tag{22.10}$$

and the distance d between two face vectors $s_{j,k}$ and $s_{j',k'}$ is defined by the Euclidean distance of their transformed vectors:

$$d_{jj',kk'} = \| Z\gamma(s_{j,k} - s_{j',k'}) \|_2 \tag{22.11}$$

The final transformation matrix is, $U' = Z\gamma V$, where, V is the set of eigenvectors extracted by PCA.

22.6.4 Cost Function

The whitening matrix Z controls the overall scatter of all samples and tends to make the subspace isotropic, while the filtering parameter γ is designed to enhance the separability of classes and may stretch the space. There should be a trade off between these two effects. We use the following cost function which is a combination of error rate and the ratio of within-class distance to between-class distance and optimise empirically using an objective function defined by:

$$OPT = \sum_{j=1}^{M} \sum_{k=1}^{K} \sum_{m} \frac{d_{jj,k_0}}{d_{jm,k_0}}, \forall m \in d_{jm,k_0} < d_{jj,k_0}, m \in [1 \cdots m] \tag{22.12}$$

where d_{jj,k_0} is the distance between the sample $s_{j,k}$ and $s_{j,0}$ which is the standard image reference for class S_j. Fig. 22.16 shows the large improvement in robustness to lighting angle.

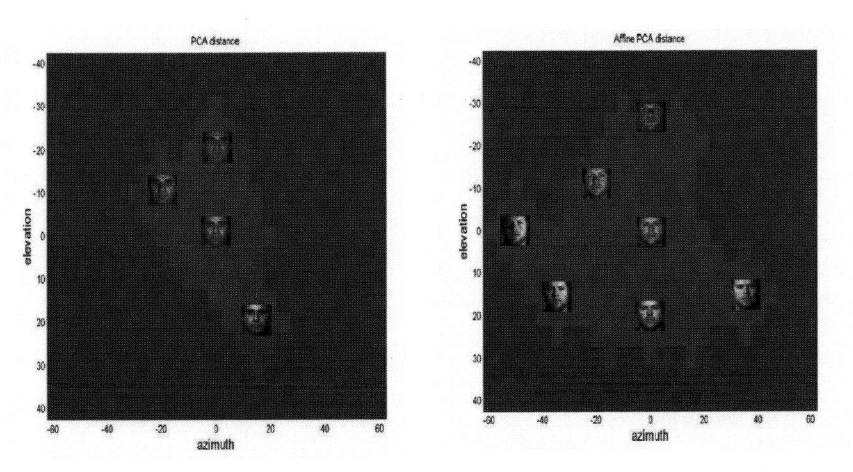

Fig. 22.16. Contours of 95% recognition performance for the original PCA and the proposed APCA method against lighting elevation and azimuth

22.6.5 Rotated APCA

We applied similar techniques to face images with variations in expression, but could not attain the levels of performance comparable to those obtained on illumination variant faces. This is because Eigenfeatures extracted by PCA from face images with illumination variation naturally cluster into two groups: 1) features strongly related to within-class covariance, and 2) features strongly related to between-class covariance. Usually the first three eigenfaces are strongly related to illumination (within-class) variation. Therefore, it is easy to find the eigenfeatures that represent within-class variation and suppress these with eigenfiltering. However, for expression change, since different people display the same expression in different ways, PCA does not successfully separate between-class and within-class features.

We therefore rotate the feature space according to within-class covariance to enhance representativeness of the features and to improve estimation of the conditional pdfs. After rotation, some features represent predominantly within-class variation and by selecting these via eigenfiltering the influence of between-class variation on estimation are diminished. Moreover, after rotation, features are highly distinguished in terms of their ITV and compression of within-class features will affect within-class covariance more than between-class covariance and hence improves separability. The rotation matrix R is a set of eigenvectors obtained by applying singular value decomposition to the within-class covariance matrix. R transforms every face vector s into the new space:

$$r = R^T s \tag{22.13}$$

Fig. 22.17 shows significant recognition performance gains over standard PCA when both changes in lighting and expression are present.

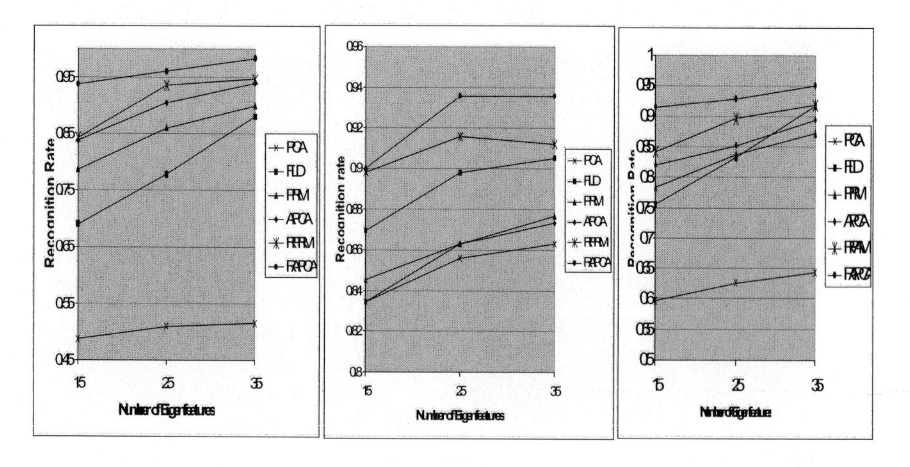

Fig. 22.17. Recognition rates for RAPCA, APCA, and PCA versus number of eigenfaces with variations in lighting and expression from Chen

22.6.6 Tracking

Firstly, we define a recognition confidence output, which indicates the confidence of the recognition result from the face recogniser. The confidence output is calculated from the angle θ, which measures how closely the input face image subspace matches the recognised face subspace. Two thresholds are selected to provide three levels of recognition confidence (low, medium and high). Depending on the level of confidence there are three possible outputs. If recognition confidence is low, the face will be surrounded with a red rectangle with the text "Unknown" appearing below. If medium, the rectangle will be yellow with a likely identity appearing below followed by a question mark. If high, the rectangle is green and the identity appears below without the question mark. Some selected frames are shown in Fig. 22.18.

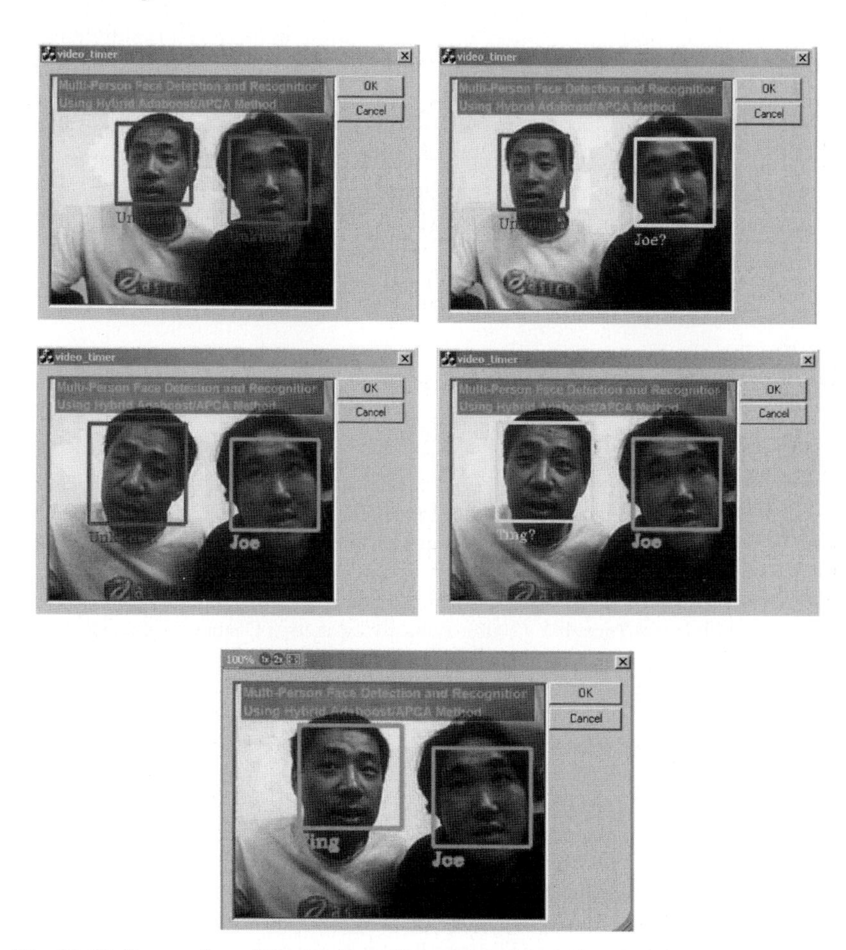

Fig. 22.18. Some selected frames (from left to right, up to down) showing different recognition confidence levels using different colour rectangles

Only the face sub-window with high confidence will be registered for tracking. Here we adopt a simple strategy for face tracking that if the position of the corresponding face sub-windows from successive frames is smaller than a certain threshold, we will accept that these face images are from the same person. The confidence output decreases over time if the recognition result falls to the low or medium level.

22.7 Summary

In this chapter we describe some technologies underpinning the pattern recognition engine of a system for locating persons on a planetary sensor network. A fully automated system must be highly robust to nuisance problems such as lighting, expression change, pose, and camera variation. Although there is a rapidly emerging need for reliable pattern recognition technology due to the explosion of digital multimedia data and storage capacity, the technologies are mostly unreliable and slow. Yet, the emergence of handheld computers with built-in speech and handwriting recognition ability, however primitive, is a sign of the changing times. The challenge for researchers is to produce pattern recognition algorithms, such as face detection and recognition, reliable and fast enough for deployment on data gathering networks of a planetary scale.

22.8 References

ABC News (2005): http://www.abc.net.au/pm/content/2005/s1283572.htm [last visited 24-Nov-2005]

Abdel-Mottaleb M and Elgammal A (1999): Face detection in complex environments from colour images. Proc: Int. Conf. on Image Processing, vol. 3, pp: 622-626, 24-28 Oct

Beymer D and Poggio T (1995): Face Recognition from One Example View. Proc. Int'l Conf. of Comp. Vision, 500-507

BioID (2005): http://www.humanscan.de/company/index.php [last visited 14-Dec-2005]

Black MJ, Fleet DJ, and Yacoob Y (2000): Robustly estimating Changes in Image Appearance. In: Computer Vision and Image Understanding, 78(1), 8-31

Carbonetto P (2005): http://www.cs.ubc.ca/~pcarbo/ [last visited 14-Dec-2005]

Chen SK and Lovell B (2004): Illumination and Expression Invariant Face Recognition with One Sample Image per Class. In: 17th Int. Conf. on Pattern Recognition (ICPR' 04) – vol. 3 pp. 300-303

Cootes TF and Taylor CJ (1992): Active shape models—'smart snakes'. In: Proc. British Machine Vision Conference. Springer-Verlag, pp 266-275

Cootes TF and Taylor CJ (1996): Locating faces using statistical feature detectors. Proc of the 2nd Int. Conf. on Automatic Face and Gesture Recognition, pp: 204

Cootes TF, Walker L, and Taylor CJ (2000): View-Based Active Appearance Models. 4th Int. Conf. on Automatic Face and Gesture Recognition, pp: 227-232, March

Cootes TF and Taylor CJ (2001): Active Appearance Models. In: IEEE PAMI, 23:6, pp. 681-685

Crow F (1984): Summed-are tables for texture mapping. Proc of SIGGRAPH, Volume 18(3), pages 207-212

Dai Y and Nakano Y (1996): Face-texture model based on sgld and its application. In: Pattern Recognition 29, pp. 1007-1017, June

Feraud R, Bernier O, and Collobert D (1997): A constrained generative model applied to face detection. In: Neural Processing Letters 5(2): 11-19

Feret (2005): http://www.itl.nist.gov/iad/humanid/feret/ [last visited 23-Nov-2005]

Fleuret F and Geman D (2001): Coarse-to-Fine Face Detection. In: Int. J. of Computer Vision, 41:85-107

Gao YS and Leung MKH (2002): Face Recognition Using Line Edge Map. In IEEE PAMI. 24(6), June, 764-779

Gibbons PB, Karp B, Ke Y, Nath S, and Sehan S (2003), IrisNet: An Architecture for a Worldwide Sensor Web. In: Pervasive Computing, 2:4, 22-23, Oct – Dec

Govindaraju V (1996) Locating human faces in photographs. In: Int. J. of Computer Vision, 19:2, August, pp: 129-146

Gunn SR and Nixon MS (1994): A dual active contour for head boundary extraction. In: IEE Colloquium on Image Processing for Biometric Measurement, pp: 6/1 – 6/4, 20 Apr.

Hjelmas E and Low BK (2001) Face Detection: A Survey. In: Computer Vision and Image Understanding, 83:3, Sept., pp.236-274 (39)

Hoogenboom R and Lew M (1996): Face detection using local maxima.In: 2nd Int. Conf. on Automatic Face and Gesture Recognition, Oct 14-16, Killington, Vermont, USA

Horward J (2005): http://smh.com.au/news/national/howard-backs-more-security-cameras/2005/07/24/1122143730105.html [last visited 23-Nov-2005]

Huang J, Gutta S, and Wechsler H (1996) Detection of human faces using decision trees. In IEEE Proc. of 2nd Int. Conf. on Automatic Face and Gesture Recognition, Vermont

Jeng SH, Liao HYM, Liu YT, and Chen MY (1998): An efficient approach for facial feature detection using geometrical face model. Proc: 13th Int. Conf. on Pattern Recognition, vol. 3, pp: 426-430, 25-29 Aug.

Lee CH, Kim JS, and Park KH (1996): Automatic human face location in a complex background. 2nd Int. Conf. on Automatic Face and Gesture Recognition, Oct 14-16, Killington, Vermont, USA

Lv XG, Zhou J, and Zhang C S (2000): A novel algorithm for rotated human face detection. In: Computer Vision and Pattern Recognition, vol. 1, pp: 760 – 765

Mallat SG (1989): A theory for multi-resolution signal decomposition: The wavelet representation. In: IEEE Transactions on Pattern Analysis and Machine Intelligence, 11(7): 674-693

McKenna S, Gong S, and Liddell H (1995) Real-time tracking for an integrated face recognition system. In: 2nd Workshop on Parallel Modelling of Neural Operators, Faro, Portugal, Nov.

Nikolaidis A and Pitas I (2000): Facial feature extraction and pose determination. In Pattern Recognition 33, 1783-1791

Osuna E, Freund R, and Girosi F (1997): Training support vector machines: An application to face detection. Proc: Computer Vision and Pattern Recognition, pp: 130-136, June

Roth D, Yang MH, and Ahuja N (2000): A SNoW-based face detector. In: Advances in Neural Information Processing Systems 12, pp: 855-861, MIT Press

Rowley HA, Baluja S, and Kanade T (1998): Neural network-based face detection. Proc. IEEE Computer Society Conference on Computer Vision and Pattern Recognition, pp: 203-208, 18-20 June

Saber E and Tekalp AM (1998): Frontal-view face detection and facial feature extraction using colour, shape and symmetry based cost functions. In: Pattern Recognition Letters, 9:8, pp: 669 - 680

Satoh S, Nakamura Y, and Kanade T (1999): Name-it: Naming and Detecting Faces in News Videos. IEEE MultiMedia,6:1, pp. 22-35, Jan-Mar

Sun QB, Huang W M, and Wu J K (1998): Face detection based on colour and local symmetry information. Proc. of 3rd Int. Conf. on Face & Gesture Recognition, pp: 130

Sung KK and Poggio T (1998): Example-based learning for view-based human face detection. ITEE Trans. on Pattern Analysis and Machine Intelligence

Terrillon J, Shirazi M, Fukamachi H, and Akamatsu S (2000): Invariant face detection with support vector machines. In: 15th Int. Conf. on Pattern Recognition, vol. 4, pp: 210-217, 3-7 Sept.

Turk M and Pentland A (1991): Face recognition using eigenfaces. Proc. Computer Vision and Pattern Recognition, pp: 586-591

Wong C, Kortenkamp D, and Speich M (1995): A mobile robot that recognises people. Proc: 7th Int. Conf. on Tools with Artificial Intelligence, pp: 346-353

Yang MH, Ahuja N, and Kriegman D (2000): Face detection using mixtures of linear subspaces. Proc. 4th IEEE Int. Conf. on Automatic Face and Gesture Recognition, pp: 70.

Yilmaz A and Gokmen M (2000): Eigenhill vs. eigenface and eigenedge. In Procs of Int. Conf. Pattern Recognition, Barcelona, Spain, 827-830

Yokoyama, Yagi Y, and Yachida M (1998): Facial contour extraction model. Proc of 3rd Int. Conf. on Face & Gesture Recognition, pp: 254

Yuille AL, Hallinan PW, and Cohen DS (1992): Feature extraction from faces using deformable templates. In: Int. Journal of Computer Vision, 8:2, pp: 99-111

23 Modulation Techniques and Topology: Review

N. P. Mahalik[1], Changwen Xie[2] and Kiseon Kim[1]

[1]Department of Information and Communications, Gwangju Institute of Science and technology, Republic of South Korea
[2]Changwen Xie, Wicks and Wilson Limited, Morse Road, England

23.1 Introduction and Background

Wireless communication has existed for a long time. However, new wireless communication systems are being developed more rapidly than ever. The wireless sensor networks (WSNs) are used for such varied application areas as health, military, home, agriculture, and others. WSN is usually a radio communication scheme, which uses spectral resources. For different application areas, there are different technical issues that researchers and developers are currently resolving (Akyildiz 2002). The network is composed of a large number of sensor *nodes*, which are densely deployed either inside a system or very close to it. A sensor node represents a significant improvement over the traditional sensor. The potential growth in communications and microelectronics have enabled the development of low-cost, low-power, multifunctional tiny sensor nodes that can be small in size, and can communicate with each other in a co-operative way when placed at short distances. A typical node performs sensing, data processing, and communication functions by using a sensory element, a physical processor, (a microprocessor or a microcontroller) and a communication module (transceiver, network port, etc.). Traditional sensors send raw data, although in many situations the data are conditioned to some extent, while a sensor node takes advantage of current processing abilities to carry out locally, and transmit only the required and partially processed data. Over time, many algorithms and protocols have been proposed. The algorithms and protocols may not be fully suited for the sensor network. This is due to the fact that sensor networks entail different sets of attributes in order to meet the unique features and application requirements of target applications. Importantly, WSNs are intrinsically different from traditional distributed systems due to the strict resource constraints on the sensor nodes. Resources are primarily constrained by energy consumption, hardware size and cost. The lifetime of the system is about a couple of weeks, requiring power-aware hardware and software solutions. The cumulative hardware cost of the system needs to stay low, even though the number of nodes employed in a particular real-world application is relatively large (Sallai 2003). Apparently, sensor network applications require traditional wireless ad-hoc networking techniques. Akyildiz et. al. (Akyildiz 2002) enlists the following attributes as far as differences between sensor networks and ad-hoc networks are concerned. Many researchers and developers are developing schemes that fulfill these requirements.

- The number of sensor nodes in a network can be several orders of magnitude higher than the nodes in an ad-hoc network.
- Nodes are densely deployed.
- Nodes are prone to failure.
- The topology of a sensor network changes very frequently.
- Sensor nodes mainly use a broadcast communication paradigm, whereas most ad-hoc networks are based on point-to-point communications.
- Sensor nodes are limited in power, computational capacities, and memory.
- May not have global identification (ID) because of large amounts of overhead and large number of sensors.

23.2 RF and Communication Modules

Wireless technology utilises RF (Radio Frequency) signals, an electromagnetic (EM) signal that operates in the range of 3 KHz (mostly 9KHz) to 300 GHz. Typical communication systems consist of three main building blocks, including transmitter, receiver, and communication media. The transmitter transmits the baseband signal by adopting appropriate modulation techniques. The modulated signal is transmitted to the channel through an *impedance-matching* unit. The impedance is defined as the opposition that a circuit offers to the modulated signal, or any other varying current, at a particular frequency. The receiver receives the modulated signal from the channel and demodulates it to obtain the original signal. The types of channels are optical fibers, conducting wires, cables, and air. When air is used as the channel, antennas are required at both ends; one at the transmitting end to transmit the modulated RF signal and another at the receiving end to receive it. The antenna can be considered as the impedance matching unit. Such communication is defined as a wireless communication system, and the frequency of operation of the system falls under the RF range (Table 23.1).

Table 23.1: RF frequency of operation

Frequency range	Designation	Wavelengths
9 kHz - 30 kHz	Very Low Frequency (VLF)	33 km - 10 km
30 kHz - 300 kHz	Low Frequency (LF)	10 km - 1 km
300 kHz - 3 MHz	Medium Frequency (MF)	1 km - 100 m
3 MHz - 30 MHz	High Frequency (HF)	100 m - 10 m
30 MHz - 300 MHz	Very High Frequency (VHF)	10 m - 1 m
300 MHz - 3 GHz	Ultra High Frequency (UHF)	1 m - 100 mm
3 GHz - 30 GHz	Super High Frequency (SHF)	100 mm - 10 mm
30 GHz - 300 GHz	Extremely High Frequency (EHF)	10 mm - 1 mm

Traditionally, the basis of communication deals with the transmission of the spectral power of the desired frequency component or band (group of consecutive frequency components) from one point to another in an effective way. The transportation of power is accomplished by the use of a set of communication modules, whose roles are very explicit in terms of achieving the final goal, i.e., communication. The block diagram of a typical communication system accommodating various modules is illustrated in Fig. 23.1.

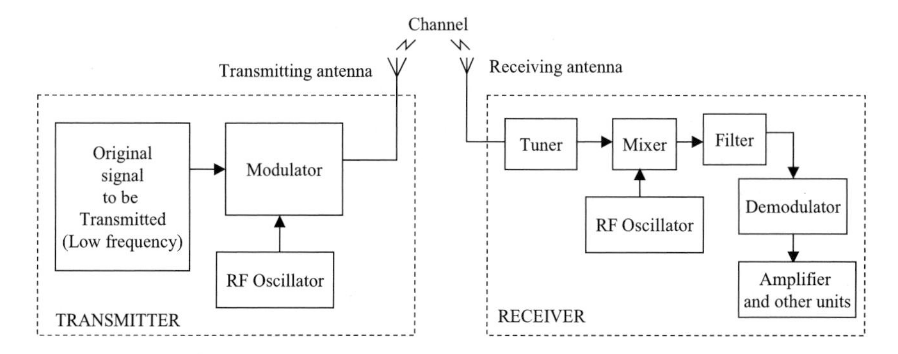

Fig. 23.1. A typical communication system

23.3 Review on Modulation

Modulation is a process of changing the parameters of a high frequency signal called a carrier *signal*, with respect to the intensity of a given weak signal called an *original baseband* signal or *modulating* signal. The high frequency signal is usually a sinusoidal signal. The parameters are simply the amplitude, frequency, and phase. Modulation is essential in communication systems, where a weak signal is transmitted by the use of a carrier signal. The carrier signals are usually sinusoidal in nature. There are many forms of modulation such as Amplitude Modulation (AM), Frequency Modulation (FM), and Phase Modulation (PM). When a high frequency signal has amplitude varied in response to the intensity of a low-frequency weak signal, the modulation is called AM. When the frequency is varied in accordance with the intensity of the weak modulating signal, the modulation is referred to as FM. A similar definition can also be given for PM. The process of recovery of the original baseband signal from the modulated signal is called *demodulation*. There exists another modulation scheme, called pulse modulation; pulse modulation is categorised under two categories, namely, Pulse Amplitude Modulation (PAM) and Pulse Width Modulation (PWM). Here a constant-amplitude and constant-width pulsed signal is used instead of high-frequency sinusoidal signal. When the amplitude of the constant-amplitude constant-width pulsed signal is varied in response to the intensity of the given modulating signal, keeping the width of the pulse constant, then it is referred to as PAM modulation.

Similarly, if the width of the pulse is varied in response to the intensity of the given signal, keeping the amplitude of the pulse constant, then it is referred to as PWM modulation. All the above schemes are called analog modulation schemes.

23.4 Digital Modulation

23.4.1 ASK

Amplitude-shift keying (ASK) is a digital modulation method in which the carrier wave is multiplied by the digital signal d(t) (Fig. 23.2(a)). Mathematically, the modulated carrier signal s(t) is given by:

$$c(t) = d(t)Sin(2\pi f_c t + \varphi) \tag{23.1}$$

The refined version of the above equation can be expressed as follows.

$$c(t) = \sqrt{\frac{2E}{T}} \cos\left(2\pi f_c t + \frac{d_n m(t - nT)}{T}\right) + \theta;\ nT \le t \le (n+1)T \tag{23.2}$$

E is the energy of the signal, T is the symbol duration, f_c is the carrier frequency, m is the modulation index, d_n is the nth data bit, and θ is a constant phase shift. Fig. 23.2(a) illustrates how modulation is obtained. At the receiving end ASK operates as a switch. The presence of a carrier wave indicates a binary "1" and its absence indicates a binary "0". This type of modulation is therefore called on-off keying. Amplitude modulation has the property of translating the spectrum of the modulation to the carrier frequency. Fig. 23.2(b-c) is the frequency domain representation of the modulating and the modulated signal. Note that the bandwidth of the signal remains unchanged. The modulation process can impart to a sinusoid signal on more than two discrete amplitude levels. These are related to the number of levels adopted by the digital message. The disadvantages of ASK are that the scheme is susceptible to sudden gain changes, and inefficient modulation techniques regarding data. There are sharp discontinuities at the transition points. As a result of which the modulated signal introduces an unnecessarily wide bandwidth. Band limiting is therefore adopted before transmission occurs. ASK can be used to transmit digital data over optical fibers.

23.4.2 FSK

Frequency-shift keying (FSK) is considered as a form of frequency modulation in which the modulating signal shifts the output frequency between predetermined values. In the FSK method, the two binary states, logic 0 (low) and 1 (high), are each represented by an analog waveform at different specific frequencies, as shown in Fig. 23.3. This is referred to as BFSK (Binary FSK). Thus, data are transmitted by shifting the frequency of a continuous carrier in a binary manner to two discrete frequencies, f_1 and f_2. $f1$ and $f2$ are offset from carrier frequency fc by equal but opposite amounts. The two frequency levels are designated as the "mark" and "space" frequency. By convention, mark corresponds to the higher RF (Watson 2005). We can encounter non-coherent and coherent forms of FSK. When the instantaneous frequency is shifted between two discrete values, as stated above, it is known as non-coherent FSK. On the other hand the coherent form of FSK exists in which there is no phase discontinuity in the signal. Fig. 23.3(b) shows an example, and Fig. 23.3(c) illustrates the frequency domain representation of a typical modulated signal employing the principle of FSK.

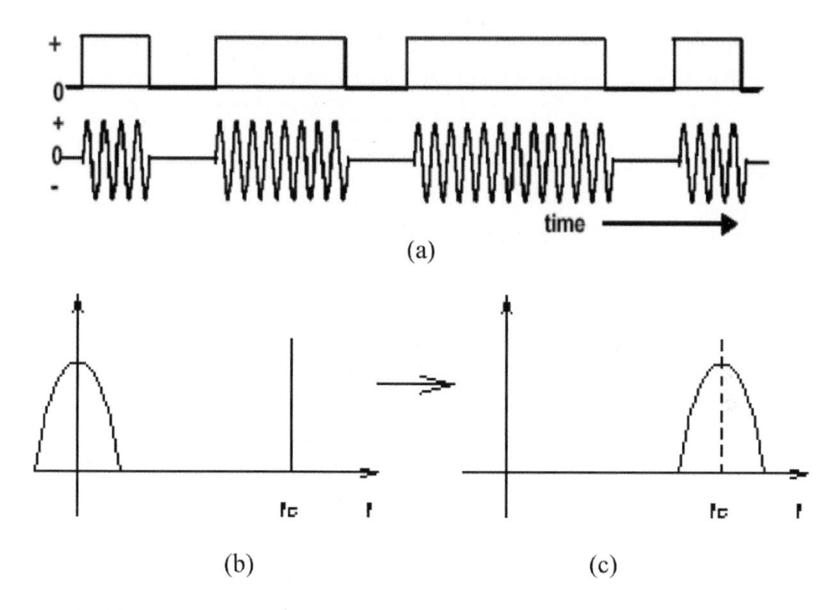

Fig. 23.2. (a) Amplitude-shift keying; Amplitude-shift keying frequency domain, (b) Original signal, (c) Modulated signal (Source: http://www.cs.ucl.ac.uk)

Multiple frequency-shift keying (MFSK) (Fig. 23.3(d)) is a variation of FSK that uses more than two frequencies. The FSK in which two original signals are multiplexed and transmitted simultaneously among four frequencies are called double frequency-shift keying (DFSK). The Bell Telephone System first used this technique in their Model 103 modem, but FSK can also be used in sensor networking. FSK is robust, simple, multi-path transmission system with constant transmit power. It can operate at more than 1 million bps over on a wireless system. Although, FSK has good characteristics, it is not generally suitable for applications where bandwidth efficiency is seen to be central.

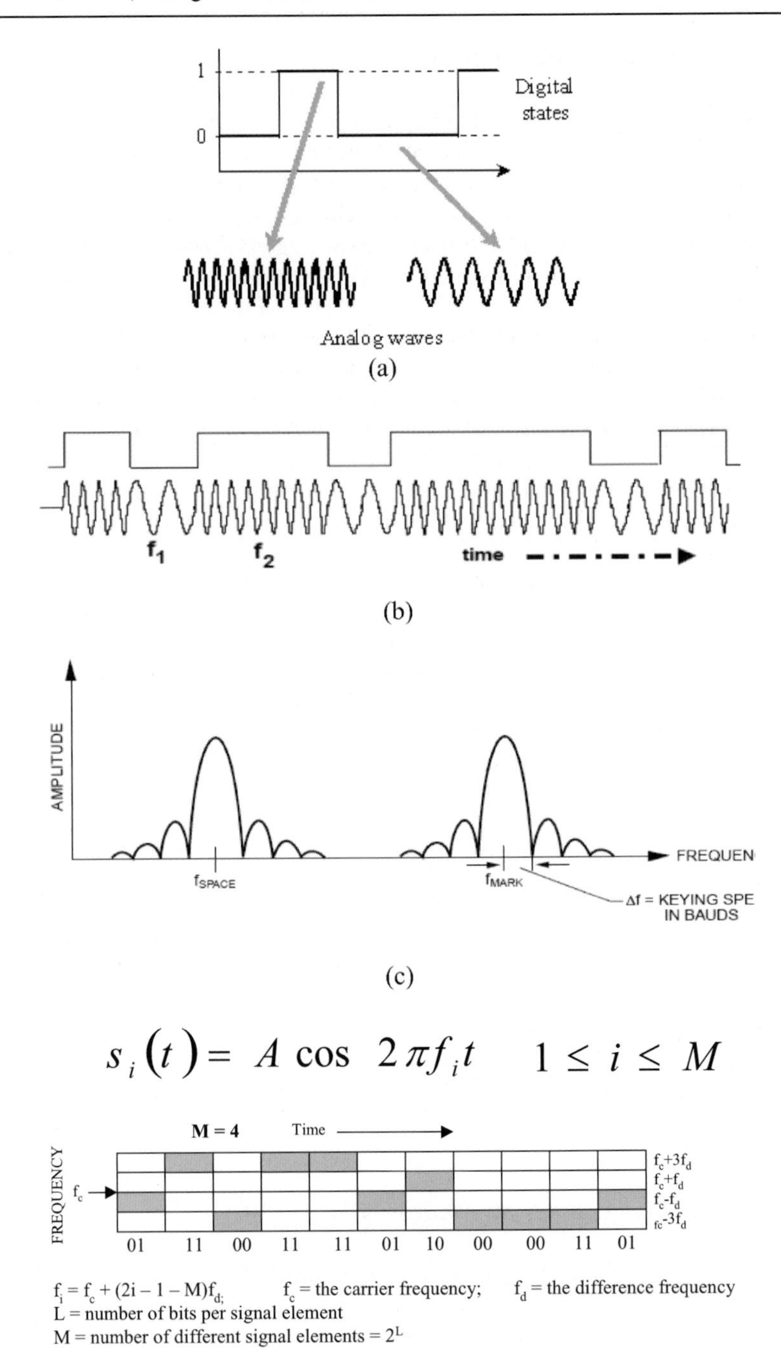

$$s_i(t) = A \cos 2\pi f_i t \quad 1 \le i \le M$$

$f_i = f_c + (2i - 1 - M)f_d$; f_c = the carrier frequency; f_d = the difference frequency
L = number of bits per signal element
M = number of different signal elements = 2^L

Fig. 23.3. Frequency-Shift keying, (a) Analog waves representing binary digits, (b) An example of a bit pattern, (c) The frequency spectrum of a typical FSK signal

23.4.3 Phase Shift Keying (PSK)

Phase-shift keying (PSK) is a technique, which shifts the period of a wave. The representing characters such as bits are transmitted by a shift in the phase of an electromagnetic carrier wave with respect to a reference. The amount of shift corresponds to the symbol to be transmitted. The PSK system is designed such that the carrier can only assume two different phase angles. Each change of phase carries one bit of information. There are several methods that can be used to accomplish this.

The easiest one is the binary phase-shift keying (BPSK). This scheme uses two opposite signal phases such as 0 and 180 degrees. The phase shift could be -90° for "0" and +90° for a "1". The digital signal is broken up time-wise into individual bits. More sophisticated forms of PSK employ M-ary or multiple phase-shift keying (MPSK). In such cases there are more than two phases, there are usually four or eight phases as follows.

- Four = 0, +90, -90, and 180 degrees
- Eight = 0, +45, -45, +90, -90, +135, -135, and 180 degrees

When m = 4 MPSK mode is called quadrature phase-shift keying or quaternary phase-shift keying (QPSK), and each phase shift represents two signal elements. Eight phases-based MPSK is known as octal phase-shift keying (OPSK), and each phase shift represents three signal elements. As is obvious, by employing MPSK, data can be transmitted at a faster rate. Mathematically, the PSK signal can be represented by

$$c(t) = \sin\left(2\pi f_c t + \varphi(t)\right) \tag{23.3}$$

The IEEE 802.15.4 standard spells out a direct sequence spread spectrum (DSSS) transmission scheme using BPSK for 868/915 MHz and offset-quadrature phase shift keying (O-QPSK) for 2.4 GHz (Malan 2004). The ZigBee protocol, a low-power wireless network of monitoring and control devices, works with the IEEE 802.15.4 standard that focuses on low-rate personal area networking (PAN), and defines the lower physical (PHY) and the medium access control (MAC) layer protocol.

23.4.4 Gaussian Frequency-Shift Keying (GFSK)

Using bandwidth-efficient modulation, communication satellites can transmit signals through a smaller frequency band. The Gaussian frequency-shift keying (GFSK) method in this respect is good. GFSK is simply frequency-shift keying but the input is first passed through a Gaussian filter. Binary 1 and 0 are positive frequency shift and negative frequency shift from base frequency, respectively. The impulse response of the Gaussian filter is given by:

$$h(t) = \frac{e^{\left(\frac{-t^2}{2\delta^2}\right)}}{\sqrt{2\pi\delta}}, \text{ where, } \quad \delta = \frac{\sqrt{\ln(2)}}{2\pi BT} \tag{23.4a}$$

where, B is the 3 dB bandwidth of the filter.

Gaussian filtering is one standard way for reducing the spectral width. BT in the above equation controls the bandwidth of the filter. Bluetooth radio modules use Gaussian Frequency-Shift Keying (GFSK) for modulation (Meheta 2003). The data can be transmitted at a symbol rate of 1 Mb/sec. Gaussian minimum-shift keying (GMSK) employs a form of FSK. It has high spectral efficiency. GMSK modulation focuses on phase as the modulation variable, while GFSK focuses on frequency. GMSK seems better in terms of its spectral efficiency, but GFSK is easier to implement. GMSK is typically used for cell phones while GFSK is often employed in cordless phones (Perrott 2003; Lee 2004).

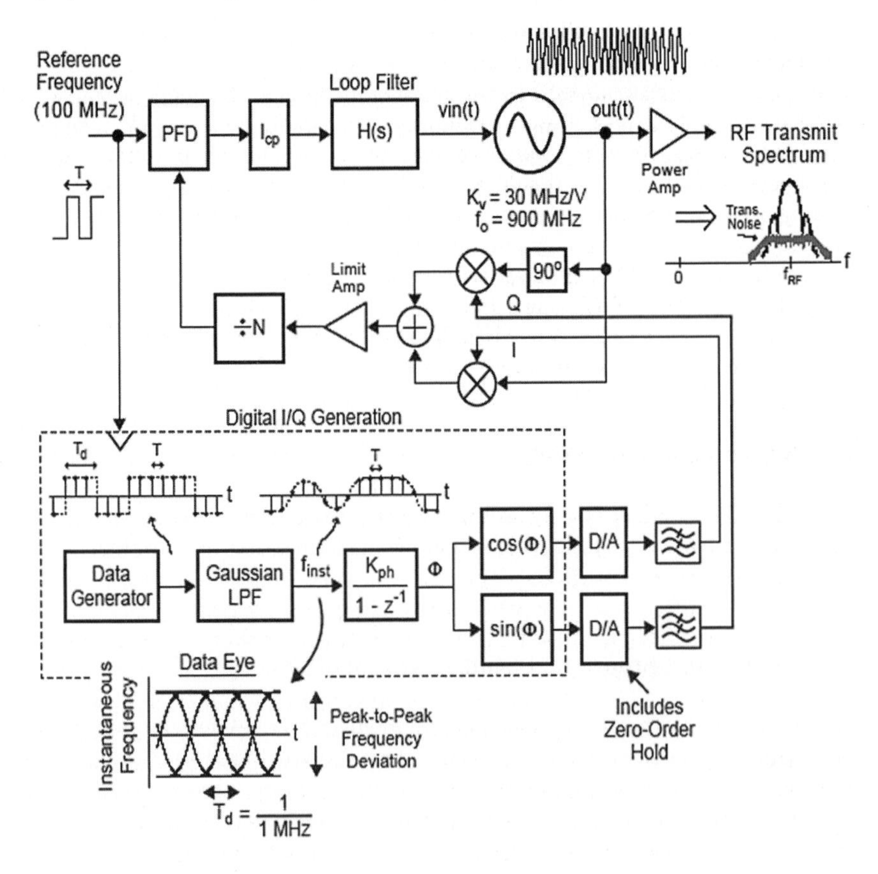

Fig. 23.4. (a) A typical GMSK modulator

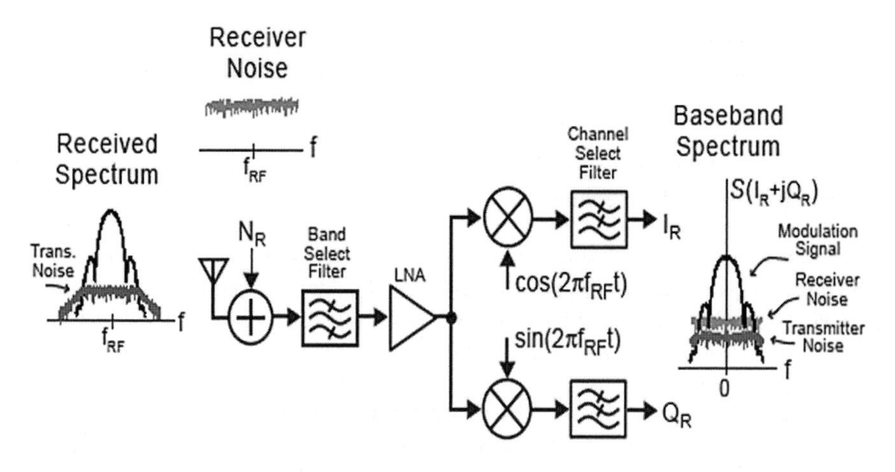

Fig. 23.4. (b) Demodulator (Adapted from Perrott 2003; ©: MH Perrott 2003 - MIT))

Fig. 23.4 is a typical GMSK modulator and demodulator as designed by Perrott. The factors used to compare the encoding scheme are signal spectrum, clocking, and signal interference, as well as noise immunity, cost, and complexity. Since the original signal usually has low frequency components, less bandwidth is required. Clocking is important, which can make ease of determining the beginning and end of bit position. The scheme should perform well in the presence of noise. It is also equally important to consider cost and complexity factors since the higher the signal rate to achieve a given data rate, the greater the cost. The attributes, which determine how successful a receiver will be in interpreting an incoming signal, are signal-to-noise ratio (SNR), data rate, and bandwidth. One can note that an increase in SNR, data rate and bandwidth decreases bit error rate (BER), increases bit error rate, and increases data rate, respectively.

23.4.5 Discussion on dFSK

The distributed frequency-shift keying (dFSK) modulation technique is good in dense sensor networks (Fig. 23.5). Typically, a sensor broadcast protocol allows all nodes to agree on a common stream of bits to send. The nodes can synchronise their transmissions so as to generate a strong signal at the farther end. Alternatively, the nodes listen to a few transmissions by other nodes to resolve the temporal estimation of when to schedule their respective transmissions. The aggregate waveform that results because of the coherent superimposition of a large number of these weak transmissions, generated by each of the sensors, appears as a prespecified set of zero-crossings. The information is essentially conveyed to the far receiver through the location of these zeros. This mode of operation is analogous to that of standard FSK. It is through spectral properties of the transmitted symbol that information is conveyed to the far receiver. The signal $R_N(t)$ at the output of the channel is modelled as (Hu 2004),

$$R_N(t) = bA_N(t) + n(t) = b\left(c_N \sum_{i=1}^{N} p(t - T_i) \right) + n(t) \tag{23.5}$$

where, $A_N(t)$ is the basic network pulse, and $b \in \{-1, 1\}$ is one bit of information shared among all transmitters obtained as a result of an execution of a sensor broadcast protocol, N is the number of nodes used to maintain transmissions appropriately so that the total amount of power to be radiated by the entire network is bounded, c_N is a constant that depends on the number of nodes, $[T_1 \ldots T_N]$ is a vector of random relative offsets, and n(t) is noise, primarily Gaussian.

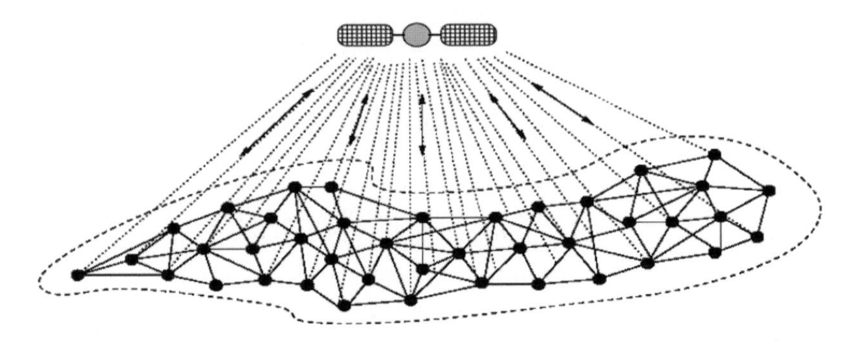

Fig. 23.5. Typical illustration of a dense sensor network (Servetto 2004)

Synchronisation: The main challenge is to achieve network-wide synchronisation of all the nodes in the network. In this context the goal is to define an aggregate signal $A_N(t)$ that is received by all network nodes simultaneously. It has to contain the information needed to synchronise all nodes. This is accomplished by setting up a distributed estimation problem. A detailed description is reported by Hu (Hu 2003). There the behavior of $A_N(t)$ has been studied in the asymptotic regime as N tends to infinity. The knowledge of the zero-crossing locations allows solving the synchronisation problem. $A_\infty(t)$, a deterministic limit aggregate waveform, is continuous everywhere and takes positive, negative, and zero values. One can note that all the nodes can detect the equispaced zero-crossings of $A_\infty(t)$ at the same time. Hence, the synchronisation problem can be solved using this information. The generation of data observed at i^{th} node is governed by the following state equations,

$$s_{n+1,1}^{c_1} = s_{n,1}^{c_1} + 1$$

$$t_{n,i}^{c_1} = \alpha_i (s_{n,1}^{c_1} - \bar{\Delta}_i) + \Psi_i (s_{n,1}^{c_1}) \tag{23.6a-b}$$

LHS of the first equation is the time of the nth transition operational counter $s_i(t)$. c_k is the clock at the kth node. LHS of Eq. (23.6b) above denotes the time at which the ith node looks at the jth transmission. α_i is the constant frequency offset between the two clocks. Since the two clocks have been started at different times, there should be a term to model that behaviour, and Δ does this. $\psi_i(t)$ is white Gaussian noise. The noise has zero mean and explicit variance. In order to maintain synchronisation, the ith node observes m consecutive zero-crossings and then makes a minimum variance unbiased estimate of the next pulse transmission time. For some integer n the m observations, $Y = [y_1......y_m]$, takes on the form (Hu 2004; Servetto 2003),

$$Y = \begin{bmatrix} \alpha_i(s_{n,1}^{c_1} - \bar{\Delta}_i) + \Psi_i(s_{n,1}^{c_1}) \\ \alpha_i(s_{n,1}^{c_1} - \bar{\Delta}_i) + \alpha_i + \Psi_i(s_{n,1}^{c_1} + 1) \\ \alpha_i(s_{n,1}^{c_1} - \bar{\Delta}_i) + 2\alpha_i + \Psi_i(s_{n,1}^{c_1} + 2) \\ \cdots\cdots\cdots\cdots\cdots\cdots \\ \alpha_i(s_{n,1}^{c_1} - \bar{\Delta}_i) + (m-1)\alpha_i + \Psi_i(s_{n,1}^{c_1} + (m-1)) \end{bmatrix} \tag{23.8}$$

In order to know how the dFSK modulation technique works, consider a case where the communication system deploys M-ary signaling. Let us concentrate on one interval i.e., $t = \tau_o$ and $t = \tau_o+1$. In M-ary signaling, $q = \{0, 1,\ldots, M-1\}$ and $R = 2q+1 \in \{1, 3,\ldots, 2M-1\}$, where R is the number of zero crossings, and q is a positive integer. Each node looks for a zero-crossing in a small interval around its estimate based on a primary zero-crossing when other zero-crossings are placed between $t = \tau_o+k$ and $t = \tau_o+k+1$. The node i still observes the zero-crossings at the same interval, i.e., $t = \tau_o+k$ and $t = \tau_o+k+1$. This in turn implies that the synchronisation properties are maintained using the zero-crossings at $t = \tau_o+k$ for $k \in Z$ (Hu 2004).

23.5 Spread Spectrum Communication (SSC)

The conventional modulation system employs a carrier frequency, which does not change with time except for small, swift fluctuations that occur as a result of modulation due to the known bandwidth of the original baseband signal frequency components. A broadcasting channel transmitted at 99.1 MHz on an FM system stays at 99.1 MHz with swift but small fluctuations. The frequency of such a system is kept as constant as possible within certain tolerances, so that the modulated signal can easily be demodulated by synchronizing or mixing the carrier frequency again at the receiver. There are two important problems with this type of conventional communications. The communication is subject to,

- Interference
- Intercept

Interference can occur deliberately or accidentally. As a consequence, there was a need to establish a secure communication system in which information must be kept confidential between the source and destination. This was the origin of the spread spectrum idea (Fig. 23.6). To overcome the problems that arise from traditional communications systems, the frequency of the transmitted signal is purposely varied over a comparatively large range of the RF wave. This variation is actually done according to a specific pattern. This pattern is referred to as the frequency (carrier) versus time pattern. In order to intercept the signal, a receiver must be tuned to frequencies that vary precisely according to this variation adopted at the transmitting end. This in turn implies that the receiver has to understand the frequency-versus-time patterns of the carrier function employed by the transmitter, and also know the starting-time at which the pattern began. Although some methods were designed to be resistant to noise, interference, jamming, and unauthorised detection, a reliable wireless performance is a combination of range, noise, transmission power, antenna design, and nevertheless, security. The other major advantages of SSC can be summarised as follows (Glas 1996).

- Low power spectral density
- Interference limited operation
- Privacy due to unknown random codes
- Applying spread spectrum implies the reduction of multi-path effects
- Random access possibilities
- Good anti-jam performance

Hedy Lamarr reported a Spread Spectrum (SS) communication method in 1941. However, most of the work done in this field was during the 50s to 70s, and was heavily backed by the military. The method was first used for commercial purposes in the 1980s. Later, the US FCC (Federal Communications Commission) opened up the ISM (Industrial, scientific, and medical) frequency bands for unlicensed SS. ISM equipments are appliances, which are designed to generate RF energy for industrial, scientific, medical, or other similar purposes, excluding the applications in the field of telecommunication. Typical ISM applications are industrial heating equipment, ultrasonic equipment, diathermy, or magnetic resonance equipment, and other electrical appliances. Over time the ISM equipments have been operated in the following frequencies and bands.

Operating frequency	Tolerance
13,560 KHz	+7 KHz
27,120 KHz	+163.0 KHz
40,680 KHz	+20.0 KHz
2,450 MHz	+50.0 MHz
5,800 MHz	+75.0 MHz
24,125 MHz	+125.0 MHz

The FCC allows the use of SS technology in three radio bands, namely 902-928, 2400-2483.5, and 5752.5-5850 MHz, for a transmission power less than 1 Watt. Over the last decade a new commercial marketplace based on SS technology has been emerging. This field covers the art of secure digital communications that is also being exploited for industrial purposes. Typical communication applications are short-range data transceivers, which also include global-positioning systems (GPS), 3G mobile telecommunications, W-LAN (IEEE802.11a, IEEE802.11b, IEEE802.11g), and Bluetooth. Fundamentally, SS is an RF communications system in which the original baseband signal bandwidth is purposefully spread over a wider bandwidth by mixing a high-frequency carrier signal. Typical to the method is that the term "spread spectrum" refers to the expansion of the signal bandwidth, by several orders of magnitude (Maxim Inc. 2003).

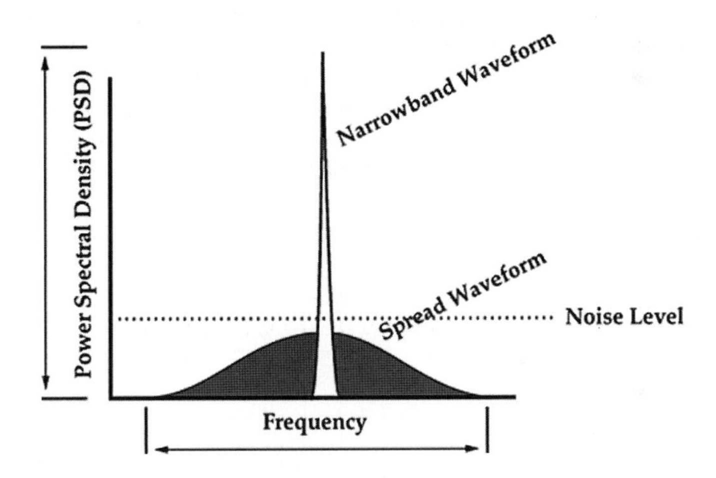

Fig. 23.6. Principle of spread spectrum (Adapted from Price 1995)

Most commercial SS systems transmit an RF signal bandwidth as wide as 20 to 254 times the bandwidth of the information to be sent. Some SS systems can employ bandwidths even 1000 times the information bandwidth. Typically there are two types of SS techniques: Direct Sequence Spread Spectrum (DSSP) (Fig. 23.7) and Frequency Hopping Spread Spectrum (FHSS). Some combination of these two types, called a hybrid spread spectrum is also possible (SSS online 2004). These communication methods are called CDMA (Code Division Multiple Access) systems.

DSSP is one of the most widely used techniques as the modulation module is relatively simple to design. A very narrow band carrier signal is modulated by a predefined code sequence. The phase of such carrier changes abruptly in accordance with this code pattern. The code pattern is generated based on a pseudorandom process. A pseudorandom process is a process that appears random but in reality is not. The length of the sequence is fixed. After a predefined number of bits the code repeats itself. The frequency spectrum of a data-signal is spread using a code uncorrelated with that signal. The energy used in transmitting the signal is spread over a wider bandwidth, and appears as noise. The ratio between the spread baseband and the original signal is called processing gain, and the typical values ranges from 10 dB to 60 dB. In practice, the SS technique injects the corresponding code before the antenna. The process of mixing or injection is called the spreading operation. Eventually, the effect of the process diffuses the information in a larger bandwidth. At the receiving end, the spread spectrum code is despreaded in order to obtain the original modulating information. The process of despreading operation is to reconstitute the information in its original bandwidth. The data signal is multiplied by a pseudorandom noise code, called PNcode or chipping code. In other words, the data signal at the sending station is combined with a higher data rate bit sequence (PNcode) that divides the user data according to a spreading ratio. Note that the PNcode is a redundant bit pattern for each bit that is to be transmitted. The speed of the code sequence is called the chipping rate, measured in chips per second (cps). The amount of spreading depends upon the ratio of chips per bit of information. In summary, the information is recovered by multiplying the signal with a locally-generated replica of the code sequence.

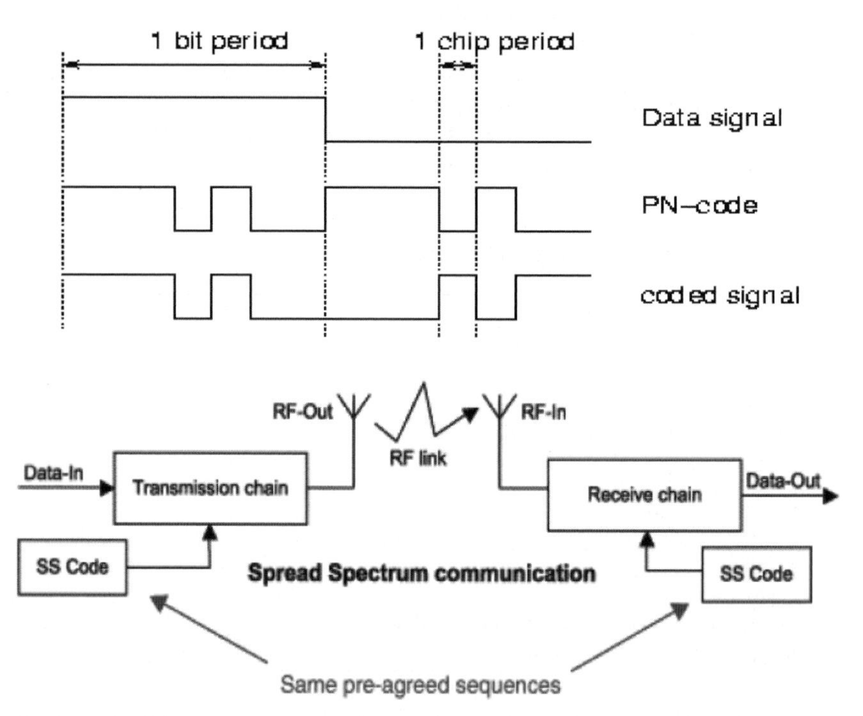

Fig. 23.7. (a) Direct-sequence spreading (Adapted from Glas 1996); (b) Schematic of a typical spread spectrum communication system (Adapted from Maxim/Dallas Feb. 2003)

DSSS: DSSS has been designed to increase the available speed on a wireless network. It may divide the available 2.4 GHz band 83.5 MHz spectrum, in most countries, into 3 wide-band (usually 22-24 MHz) channels. The method uses an

11-bit spreading code in order to reduce the interference on signals in the wideband channel. DSSS is typically chipped using BPSK. Note that SS communications is not an efficient means of utilizing bandwidth. Spread spectrum was initially developed to be used by the military because it uses wideband signals that are difficult to detect and also resist attempts at jamming.

FHSS: FHSS is also resistant to noise, interference, jamming, and unauthorised detection. In this case the range of available frequencies in the ISM band of 2.4 GHz is divided into 79 separate and distinct channels of 1 MHz. Transmissions are sent over each of these channels in a pseudorandom sequence. For instance a typical transmission can be made over Channel 1, Channel 7, Channel 52, Channel 21, Channel 60, Channel 2, etc. The system switches frequencies many times a second, transmitting on each channel for a fixed amount of time. Once it transmits over a particular channel the transmission proceeds on to the next channel, based on a predefined pseudorandom order covering all of the channels before repeating the sequence again and again. It becomes difficult for any non-participating agent to decode the transmission without any clear knowledge on dwell time and hopping pattern. The dwell time is the duration the transmission stays over a channel, and the hopping pattern establishes the way the channel sequence is followed. There could be up to 20 hop patterns a typical communication system can implement and the user can chose from. A minimum hop rate of 2.5 hops per second is typical. The use of different hopping patterns, dwell times, and/or number of channels allows two disjoint wireless systems to exist in a physical domain with less chance of causing interference and fear of data from one communication being seen by the other.

Table 23.2: Comparison of DSSS and FHSS (Source: RAYLINK AND RAYTHEON ELECTRONICS)

FHSS	DSSS
• FHSS systems have 79 channels with the 2.4 Ghz ISM band • Operates at 1 MHz of bandwidth • Choice of 1 MHz versus 22 MHz, a 1 Mhz design is easier to implement • FHSS typically uses GFSK • Non-linear, high efficiency Class C amplifier is adequate for its use • FHSS do not encounter the effects of delay • FHSS is allowed to hop in 400 msec • Ideal for high user density areas	• DSSS systems can allow 3 channels within the 2.4 Ghz ISM band • 2 Mbps DSSS occupy 22 MHz of bandwidth • The circuit design requires good passband characteristics • BPSK modulation common • Use of low efficiency ClassA or Class AB amplifiers. This implies that DSSS is more power hungry • 100 nsec delay is fairly common. Close to the DSSS chipping rate

23.6 Adaptive Modulation Scaling Scheme

Modulation scaling is an efficient wireless communication technique to adapt to dynamic traffic loads by adjusting the modulation levels, i.e., bits per symbol in an

M-ary system (Yang 2005). It can be utilised to reduce energy consumption at the cost of increased delivery delay. The concern is to determine the optimal modulation level for each node in WSN such that the overall energy consumption is minimised without violating the QoS requirements in terms of mean end-to-end latency and packet loss ratio. Based on a classical queuing theory, a model has been proposed by Yang et al. in order to analyse the relationship of modulation levels used by the nodes in the delivery path, energy consumption, packet delivery latency and loss.

23.7 Key Points on Digital Communication

Transmission rate is determined by how fast the voltage (or symbol type) can be varied on the channel. The design criteria of a communication system involves the knowledge of frequency content of the signal such that the filtering behaviour of the channel should not attenuate and distort too much of the signal. Multiple cables increase throughput or allow the same throughput at a lower bandwidth, which is probably cheaper. Data transmission should not be limited to binary format over a channel. We see two types of communication methods: binary (2 levels e.g. on/off) and multi–level (4 or more levels). These two methods can further be classified as follows.

- Binary signalling using a single cable
- Binary signalling using many parallel cables
- Multi-level signalling using a single cable
- Multi-level signalling using many parallel cables

Using 4 voltage levels, for instance, we can uniquely encode 2 bits into each of the levels, such as 00 = level P, 01 = level Q, 10 = level R, 11 = level T. Each time the symbol state is changed, two bits of information are conveyed, as compared with only one for a binary system.

Thus, one can send information twice as fast for a given bandwidth. Multi-level signalling over many cables with an increase in throughput can be effectively used. In principle, it is possible to use any number of symbols states for conveying digital information. For instance 1024 different voltage levels will convey $\log_2 1024 = 10$ bits. The important aspect of this method is its ability to resolve the individual states (voltages, frequencies, light intensities, etc.) accurately at the receiver, which eventually depends on the level of noise and distortion introduced in the communication channel. The concluding remarks are:

- A higher information rate is possible for a given symbol rate and corresponding channel bandwidth, and
- A lower symbol rate can be obtained leading to reduce bandwidth requirements.

23.8 Network Topologies

Network topology plays important role in sensor network. Topology of a network is defined as the manner in which the nodes are logically connected. Typical wired topologies are star, token passing, peer-to-peer (Fig. 23.8) and tree. Wireless topologies are similar, but the difference is that there are radio links instead of physical connections. In star topology, each node has a connection leading back to a central hub. The hub (also called master) failure can knock out the entire network. Stars are relatively expensive because of the cost of the hub. This topology allows tight traffic control because no node is allowed to communicate unless requested by the hub. No communication is allowed between slaves except through the master. In a token ring, access to the ring is controlled by a special bit pattern called the *token* circulating in an idle ring. To transmit a packet, a node grabs the token from the ring and releases it after transmission. While this scheme has considerable flexibility in terms of packet priorities, token passing introduces an overhead that can increase delay and hence decrease throughput. Peer-to-peer (P2P) type topology embodies Carrier Sense Multiple Access (CSMA) protocols. This kind of protocol may occur in one of three types, namely non-persistent, 1-persistent, and p-persistent (LonWorks Bulletin 1995). A node wishing to transmit listens to the transmission channel to detect if any other node is transmitting. The node's transceiver receives the signal from the transmission medium while transmitting and compares the received data with that just transmitted. If a collision occurs, the received data will be in error and the transmission aborted (LonWorks Bulletin 1995). Measurements have shown that all nodes equally share channel capacity, but the probabilistic nature of the operation means that it is not possible to guarantee a finite delay on a particular message. Also, the access for a particular message is probabilistic and does not include any priority mechanisms. CSMA/CD (Carrier Sense Multiple Access/Collision Detection) networks are designed to expect collisions and to handle them by re-transmitting frames when necessary. The algorithm handles these re-transmissions automatically, which should be placed locally and transparent to the user. All transmitting nodes causing those nodes to halt transmission for a short random time period before attempting to transmit again will detect any collision (LonWorks Bulletin 1995). Advantages cited for CSMA/CD are lower delays for large propulsions of burst traffic sources, and more robust management. If there is a priority assignment to the nodes then the highest priority message must be transmitted first. A tree topology combines characteristics of P2P and star topologies. It consists of groups of star-configured networks connected to a linear bus (P2P) network. A node failure or cable failure in midstream will down the entire network. Also, if it is required to add a node in the middle of the chain, the network will go down while the node is added. We commonly encounter three types of protocols for wireless networks. Wireless sensor systems can use any of these following protocols (Manges 2000), namely; (i) Master-slave protocols (MSP), (ii) Broadcast type, and (iii) P2P (Web architecture). In MSP one can think that the node gives the commands, and another node executes them. The host is considered the master, and the sensors behave as slaves. Broadcast networks are close to master-slave relationships. In this case the master sends commands to more than one slave at a time. On the other hand, in P2P protocol all nodes are viewed equally. A node can become a master at a particular moment and then be reconfigured at another time. Although P2P networks offer reasonably good flexibility, they are sometimes difficult to control as any node can communicate directly with any other node. The IEEE-802.11 standard was the first wireless standard that promised to bring the interoperability of Ethernet connectivity to wireless networks.

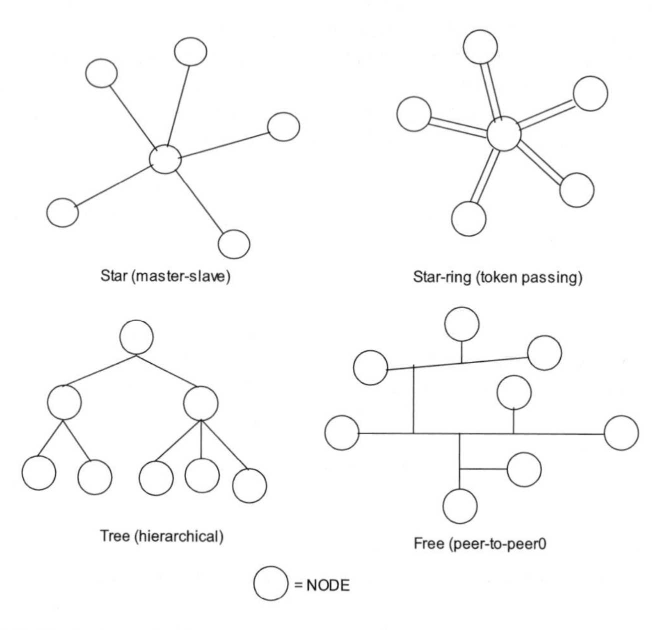

Fig. 23.8. Typical topologies

As observed, network topologies perform well when they look conformant to the topology of the application at hand. If the application entails a hierarchical strategy, then a star topology could be most suitable. Conversely, if the target demands a collection of cooperative peers, then web architecture will work best. Many issues affect network performance besides the network topology. Some useful network-based questions are (Manges 2005):

- Full vs. half duplex vs. simplex: Can the nodes transreceive simultaneously?
- Analog vs. digital: In what form does the signal enter the medium?
- Baseband vs. broadband: Whether or not a carrier is to be used to increase the number of channels that can be put on the network medium?
- Master-slave vs. broadcast vs. P2P: How do nodes interact with each other?
- Circuit switched vs. packet switched: How long a node can own a communications channel?

23.9 References

Akyildiz IF, Weilian S, Sankarasubramaniam Y and Cayirci E (2002) A Survey on Sensor Networks, IEEE Communications Magazine, Aug

Echelon Inc., LonWorks Bulletin 1995, www.echelon.com

Glas J (1996) Non-cellular wireless communication systems, Ph.D. Thesis, Delft University of Technology, USA

http://www.sss-mag.com/primer.html

http://en.wikipedia.org/wiki/Phase-shift_keying

http://www.cs.ucl.ac.uk/staff/S.Bhatti/D51-notes/node12.html

Hu AS, Servetto SD (2004) dFSK: Distributed frequency shift keying modulation in dense sensor networks. In: Proc. of the IEEE Int. Conf. on Communications (ICC), pp 222-7

Hu AS, Servetto SD (2003) Asymptotically optimal time synchronisation in dense sensor networks. In: Proc. 2nd ACM Workshop on Sensor Networks and Applications (WSNA), San Diego, held in conjunction with ACM MobiCom

János S, György B, Miklós M, Ákos L, Branislav K (2003) Acoustic ranging in resource-constrained sensor networks, DARPA/IXO NEST program (F33615-01-C-1903), MobiCom

Lee TH (2004) Planar microwave engineering: A practical guide to theory, measurement, and circuits, Cambridge University Press, Aug.

Manges WW (2005) Oak Ridge National Laboratory, Wireless Sensor Network Topologies, Sensor Magazine, http://www.sensorsmag.com/articles/0500/72/main.shtml

Manges WW, Allgood GO (2000) Wireless sensors: Buyer beware, Oak Ridge National Laboratory, Sensor Magazine, Available online: http://www.sensorsmag.com/articles/0401/18/

Mehta P, Clint OCH, Alan S (2003) Personal area connectivity with Bluetooth wireless technology, Dell's technology white papers, Jan. 14

Michael PH (2003) GMSK transceivers, Project No. 2, Department of Electrical Engineering and Computer Science, Massachusetts Institute of Technology, April 25

Malan R (2004) Here comes wireless sensors, Barrington Partners, Management Inc., Available online: http://www.machinedesign.com/ASP/strArticleID/56796/strSite/MDSite/viewSelectedArticle.asp

Maxim/Dallas, An introduction to direct-sequence spread-spectrum communications, Application note 1890, Maxim Integrated Products, Inc. Feb 18

Price HE (1995) Digital communications

Servetto SD (2004) Time synchronisation and distributed modulation in large-scale sensor networks, RPI Workshop, 29 April

Servetto SD (2003) Distributed signal processing algorithms for the sensor broadcast problems. In: Proc. 37th Annual Conf. Inform. Sciences Syst. (CISS), Baltimore, USA

Schilling DL, Raymond PL, Milstein LB (1990) Spread spectrum goes commercial, The IEEE Spectrum of Aug.

SSS Online Inc, (2004) http://www.sss-mag.com/ss.html#tutorial

Watson B (2004) FSK: Signals and demodulation, Tech-note: The Communications Edge™, WJ Communications, Inc., www.wj.com

Yoshikawa S and Mita A (2005) Digital sensor network using delta-signal modulation for health monitoring of large structures, II Eccomas thematic Conf. on smart structures and materials, Lisbon, July 18-21

Yang Z, Yuan Y, Jianhua H and Wenqing C (2005) Adaptive modulation scaling scheme for WSN, Special Section on Ubiquitous Networks, IEICE Trans. On Commn., E88:3, March

Index

Printing: Krips bv, Meppel
Binding: Stürtz, Würzburg